Harrison Learning Centre
City Campus
University of Wolverhampton
St Peter's Square
Wolverhampton
WV1 1RH
Telephone 0845 408 1631
Online renewals:
www.wlv.ac.uk/lib/myaccount

Telephone Renewals: 01902 321333 or 0845 408 1631
Online Renewals: www.wlv.ac.uk/lib/myaccount
Please return this item on or before the last date shown above.
Fines will be charged if items are returned late.
See tariff of fines displayed at the Counter.

Strategic Airport Planning

Cover photo © 1995 SoftKey International Inc.
Plan of Austin-Bergstrom International Airport reproduced with permission.

Strategic Airport Planning

Robert E. Caves
Department of Aeronautical and
Automotive Engineering and Transport Studies
Loughborough University of Technology
Loughborough, Leicester, U.K.

Geoffrey D. Gosling
Institute of Transportation Studies
University of California
Berkeley, CA, U.S.A.

1999
Pergamon
An imprint of Elsevier Science
Amsterdam – Lausanne – New York – Oxford – Shannon – Singapore – Tokyo

ELSEVIER SCIENCE Ltd
The Boulevard, Langford Lane
Kidlington, Oxford OX5 1GB, UK

© 1999 Elsevier Science Ltd. All rights reserved.

This work and the individual contributions contained in it are protected under copyright by Elsevier Science Ltd, and the following terms and conditions apply to its use:

Photocopying
Single photocopies of single chapters may be made for personal use as allowed by national copyright laws. Permission of the publisher and payment of a fee is required for all other photocopying, including multiple or systematic copying, copying for advertising or promotional purposes, resale, and all forms of document delivery. Special rates are available for educational institutions that wish to make photocopies for non-profit educational classroom use.

Permissions may be sought directly from Elsevier Science Rights & Permissions Department, PO Box 800, Oxford OX5 1DX, UK; phone: (+44) 1865 843830, fax: (+44) 1865 853333, e-mail: permissions@elsevier.co.uk. You may also contact Rights & Permissions directly through Elsevier's home page (http://www.elsevier.nl), selecting first 'Customer Support', then 'General Information', then 'Permissions Query Form'.

In the USA, users may clear permissions and make payments through the Copyright Clearance Center, Inc., 222 Rosewood Drive, Danvers, MA 01923, USA; phone: (978) 7508400, fax: (978) 7504744, and in the UK through the Copyright Licensing Agency Rapid Clearance Service (CLARCS), 90 Tottenham Court Road, London W1P 0LP, UK; phone: (+44) 171 436 5931; fax: (+44) 171 436 3986. Other countries may have a local reprographic rights agency for payments.

Derivative Works
Tables of contents may be reproduced for internal circulation, but permission of Elsevier Science is required for external resale or distribution of such material.
Permission of the publisher is required for all other derivative works, including compilations and translations.

Electronic Storage or Usage
Permission of the publisher is required to store or use electronically any material contained in this work, including any chapter or part of a chapter. Contact the publisher at the address indicated.

Except as outlined above, no part of this work may be reproduced, stored in a retrieval system or transmitted in any form or by any means, electronic, mechanical, photocopying, recording or otherwise, without prior written permission of the publisher.
Address permissions requests to: Elsevier Science Rights & Permissions Department, at the mail, fax and e-mail addresses noted above.

Notice
No responsibility is assumed by the Publisher for any injury and/or damage to persons or property as a matter of products liability, negligence or otherwise, or from any use or operation of any methods, products, instructions or ideas contained in the material herein. Because of rapid advances in the medical sciences, in particular, independent verification of diagnoses and drug dosages should be made.

First edition 1999

British Library Cataloguing in Publication Data
A catalogue record from the British Library has been applied for.

Library of Congress Cataloging-in-Publication Data

Caves, Robert E.
    Strategic airport planning / by Robert E. Caves, Geoffrey D. Gosling.
        p.   cm.
    Includes bibliographical references and index.
    ISBN 0-08-042764-2 (HC)
    1. Airports--Planning.   I. Gosling, Geoffrey David.   II. Title.
TL725.3.P5C35   1999
387.7'36--dc21   99-10050
                                                                  CIP
ISBN: 0 08 042764 2

♾The paper used in this publication meets the requirements of ANSI/NISO Z39.48-1992 (Permanence of Paper).
Printed in The Netherlands.

CONTENTS

List of Tables xi
List of Figures xi
Preface xiii

1	**Introduction** 1	
1.1	Context of Airport System Planning 1	
1.2	Scope of the Book 4	
1.2.1	Aviation system planning 5	
1.2.2	Motivation for airport system planning 6	
1.3	Development of the Airport Planning Process 6	
1.3.1	Master planning 7	
1.3.2	Strategic system planning 8	
1.4	Challenges to the Traditional Planning Approach 9	
1.4.1	The changing context of airport planning 9	
1.4.2	New aspects of old challenges 10	
1.4.3	Uncertainty 13	
1.4.4	Planning methods 14	
1.5	Format of the Book 15	
2	**The Evolving Context of Airport Planning** 17	
2.1	Traditional Institutional Roles 17	
2.1.1	Airline regulation 17	
2.1.2	Airports as a public facility 18	
2.2	Market Imperfections 19	
2.2.1	Lack of competition 19	
2.2.2	Direct and indirect subsidies 19	
2.2.3	Shortage of investment funds 20	
2.2.4	Political decision distortions 21	
2.3	Environmental Concerns 21	
2.3.1	Global concerns 21	
2.3.2	Aircraft noise 22	
2.3.3	Air and water quality 23	
2.3.4	Energy use 24	
2.3.5	Ground traffic 25	
2.3.6	Safety 25	
2.4	Economic Considerations 27	
2.4.1	Efficiency 27	
2.4.2	Economic benefits of air transportation 28	
2.4.3	Externalities 29	
2.5	Institutional Changes in the Aviation Sector 30	
2.5.1	Airline deregulation and ownership 30	
2.5.2	Airport corporatisation and privatisation 32	
2.5.3	Sustainability 33	
2.5.4	Intermodalism 35	
3	**Towards a Strategic View of Planning** 39	
3.1	The Nature of Strategic Planning 39	

3.1.1		Defining a strategic view of airport planning	39
3.1.2		The strategic planning process	46
3.2		Stakeholders, Goals and Objectives	48
3.3		Measuring System Performance	52
3.4		Risk and Uncertainty	54
3.5		Competition	55
3.5.1		The airline industry	55
3.5.2		Airports	57
3.5.3		Other modes	58
3.5.4		Access to resources	59
3.6		Constraints	60
3.6.1		Operational capacity	61
3.6.2		Environmental concerns	62
3.6.3		Financial	67
4	**Understanding Aviation System Behaviour**		**69**
4.1		The Ultimate Consumers	69
4.1.1		Air travel demand	69
4.1.2		Passenger choices	71
4.1.3		Air cargo	71
4.2		Airline Decisions	73
4.2.1		Competition	73
4.2.2		Route structure	76
4.2.3		Strategic alliances	79
4.2.4		Non-scheduled operations	81
4.2.5		Future fleets	82
4.3		Other Aircraft Operators	84
4.3.1		General Aviation	84
4.3.2		Military aviation	87
4.4		National Government Roles	88
4.4.1		Airline regulation	88
4.4.2		Air Service Agreements	89
4.4.3		Airport development	90
4.4.4		Air traffic management	91
4.4.5		Aircraft certification and personnel licensing	93
4.4.6		Airport certification and regulation	93
4.5		Implications for Airport Planning	94
5	**Community Response to Airport Development**		**101**
5.1		Airport Benefits	101
5.1.1		Economic activity	101
5.1.2		Stimulation of the economy	103
5.2		Airport Disbenefits	109
5.3		Jurisdictional Issues	113
5.4		Political Response to Conflicting Interests	115
5.4.1		Difficulties in achieving a balance of interests	115
5.5		The Increasing Costs of Environmental Mitigation	119
6	**Economics of Airport Development**		**129**
6.1		The Airport as a Business	129
6.2		Airport Revenues and Costs	133

6.2.1	Costs	133
6.2.2	Revenues	137
6.3	Implications for Investment Strategies	141
6.3.1	Performance objectives	141
6.3.2	Competitive position	143
6.4	Comparative Performance Measures	145
6.5	Public Investment Priorities	147
6.5.1	Justification for public sector role	147
6.5.2	Programmatic considerations	148
6.6	Managing Capacity Constraints	152
6.6.1	Slot allocation	152
6.6.2	The role of yield management	155
6.6.3	Technological solutions	155
7	**Regional Airport System Planning**	**163**
7.1	Metropolitan Airport Systems	165
7.1.1	Evolution of multi-airport systems	165
7.1.2	Opportunities for new airports	167
7.2	System Planning Process	169
7.2.1	Policy objectives and study design	170
7.2.2	Inventory of facilities	170
7.2.3	Forecasts of future demand	171
7.2.4	Definition of alternative plans	172
7.2.5	Evaluation of alternative plans	175
7.2.6	Selection of the preferred alternative	177
7.2.7	Implementation of the plan	178
7.2.8	Monitoring	179
7.3	Activity Allocation Models	180
7.3.1	Air passenger airport choice	180
7.3.2	Airline service choice	183
7.3.3	Based aircraft airport choice	185
7.4	Forecasting in a System Planning Context	186
7.4.1	Effect of supply on demand	186
7.4.2	Technology	187
7.4.3	Timescale	188
7.4.4	Judgement	189
7.4.5	Scenario writing	189
8	**National Airport System Planning**	**195**
8.1	National Planning Concerns	196
8.1.1	Role of air transport	197
8.1.2	Investment strategies	201
8.1.3	Development of gateways	202
8.1.4	Modal priorities	203
8.1.5	System capacity	204
8.1.6	National standards	204
8.2	Implementation Strategies	205
8.2.1	System operation and management	205
8.2.2	Financial support	206
8.2.3	Regulation	206

8.3	Institutional Considerations	207
8.3.1	Departmental responsibilities	207
8.3.2	Regional and local governments	208
8.3.3	Supra-national issues	208
8.4	Role of Analysis in Policy Formulation	208
8.4.1	Scale and complexity	208
8.4.2	Stakeholder acceptance	209
8.4.3	National models	210
8.5	Information Dissemination	211
8.5.1	Closed versus open processes	212
8.5.2	Availability of study results	213
8.5.3	Availability of data and analysis tools	214
8.5.4	Opportunity to comment	214
8.5.5	Role of information technology	215
8.5.6	Summary	216
9	**United States Experience**	**219**
9.1	The US Airport System Planning Process	220
9.1.1	National airport system planning	222
9.1.2	State aviation system planning	223
9.1.3	Regional airport system planning	224
9.2	The National Plan of Integrated Airport Systems	225
9.2.1	Airport development funding requirements	225
9.2.2	Airport system performance	227
9.2.3	Airport Capital Improvement Plan	229
9.3	The California Aviation System Plan	230
9.3.1	Technical support	232
9.3.2	The Central California Aviation System Plan	233
9.3.3	Multimodal transportation planning	234
9.3.4	The Capital Improvement Program	234
9.3.5	Evolution of the state role	236
9.4	The Minneapolis/St Paul Dual-Track Process	237
9.4.1	The dual-track strategy	240
9.4.2	Addressing uncertainty in the planning process	241
9.4.3	Implementing the dual-track strategy	243
9.4.4	Abandoning the new airport option: the legislature intervenes	244
9.4.5	Traffic growth overtakes the process	247
9.4.6	Lessons from the dual-track experience	252
9.5	The Seattle "Flight Plan" Project	254
9.5.1	Forecast demand	257
9.5.2	Development of alternatives	261
9.5.3	The Sea-Tac Airport Capacity Enhancement Plan	263
9.5.4	Evaluation of alternatives	266
9.5.5	Recommended system evolution	268
9.5.6	Implementation	268
9.6	Conclusions	269
10	**The United Kingdom Case**	**273**
10.1	The London System	273
10.1.1	The airports and their historic roles	273

10.1.2	Heathrow developments	281
10.1.3	The search for further runway capacity	290
10.1.4	House of Commons Transport Committee hearings	294
10.1.5	Policies, options and difficulties	296
10.2	National Planning	305
10.2.1	The airports and their roles	305
10.2.2	Attempts at national airport planning	306
10.2.3	The 1986 Airport Privatisation Act	310
10.2.4	London and the regions	313
10.3	The Planning Process	314
10.3.1	Planning frameworks	314
10.3.2	Implicit goals for the UK airport system	318
10.3.3	The performance of the UK airport system	320
10.3.4	Future prospects	322
10.4	Conclusions	324
10.4.1	Heathrow	324
10.4.2	London	325
10.4.3	The UK	325
10.4.4	UK planning methodology	326
11	**The European Union Case**	**329**
11.1	The Airports, their Utilisation and their Settings	329
11.2	Policy Considerations	334
11.2.1	Competition and fiscal policies	334
11.2.2	Sustainability	340
11.2.3	Subsidiarity in third country relations	341
11.2.4	Transport policy and Community cohesion	342
11.3	Airline Strategies	343
11.3.1	Airline responses to liberalisation	343
11.3.2	Hinterland hubs	344
11.3.3	Gateway hubs	345
11.3.4	Multi-airport metropolitan areas	346
11.3.5	Low cost new entrants	347
11.4	Power Relationships	349
11.5	Towards a Realistic European Air Transport Network	350
12	**National Airport Planning Cases**	**353**
12.1	Brazil	353
12.2	Canada	356
12.3	Germany	357
12.4	Greece	360
12.5	Japan	363
12.6	Norway	367
12.7	Spain	368
13	**Competing Roles for Airports**	**371**
13.1	Introduction	371
13.2	Glasgow, Scotland	372
13.3	Belfast, Northern Ireland	374
13.4	Port Authority of New York and New Jersey	375
13.5	Edmonton, Canada	377

13.6	Saõ Paulo, Brazil	377
13.7	Paris, France	379
13.8	Montreal, Canada	380
13.9	Northwest England	381
13.10	Oslo, Norway	385
14.	**Towards Improved Strategic Planning**	**389**
14.1	Planning Principles and Processes	390
14.1.1	The application of planning principles	390
14.1.2	Planning processes and frameworks	393
14.1.3	Participatory planning	393
14.1.4	Power relationships	395
14.1.5	The role of planners	398
14.2	Planning Techniques and Practices	398
14.2.1	Formulating goals and objectives	398
14.2.2	Dealing with the future	399
14.2.3	Multisector influences and evaluation techniques	400
14.3	Airport System Design	402
14.3.1	Expansion of existing airports	402
14.3.2	Development of new airports	403
14.3.3	Role of airports in regional development	404
14.3.4	New technological options	404
14.3.5	Airport classification and appropriate roles	405
15	**Conclusions**	**407**
15.1	Lessons from Hindsight	407
15.1.1	Neglect of system planning principles	407
15.1.2	Implications of 'market place' planning	410
15.2	Adapting Creatively for the Future	413
15.3	Recommendations	417
References		**421**
Index		**455**

LIST OF TABLES

Table 3.1	Styles of planning	46
Table 3.2	The strategic planning process	47
Table 3.3	UK airport annual passengers and growth by size class	54
Table 4.1	Percentage of short haul flights in western Europe and the USA by market type	75
Table 4.2	Comparative European hub performance	78
Table 4.3	General Aviation activity in the United States	86
Table 5.1	Economic impacts of two major US airports	105
Table 5.2	Monthly trips per employee in high technology industries	106
Table 6.1	FAA project appraisal weightings	150
Table 6.2	Criteria for evaluating projects	151
Table 6.3	Stand productivity per turnround	159
Table 7.1	Travel choice model parameter comparison	181
Table 8.1	Passenger growth (%) at European airports, 1994/1993	203
Table 9.1	Flight Plan Phase III alternatives	265
Table 10.1	The Roskill Commission cost benefit analysis	276
Table 10.2	Comparative growth factors, 1994/1990 a) Scheduled b) Charter	284
Table 10.3	Traffic projections with and without expansion at Heathrow a) The base case b) The Heathrow option	293
Table 10.4	Seat factors and yields at Gatwick relative to Heathrow	302
Table 11.1	European airport runway capacity utilisation	332
Table 11.2	UK international scheduled traffic	337

LIST OF FIGURES

Figure 1.1	Historic trend in real fares	2
Figure 1.2	Worldwide traffic growth	2
Figure 2.1	Change in noise impact at Heathrow	23
Figure 3.1	Dimensions of airport system planning	41
Figure 3.2	The continuous planning process	44
Figure 3.3	Planning relationships	45
Figure 3.4	Risk criteria for communities around airports	66
Figure 4.1	The effect of hubbing on narrowbody turnround times	96
Figure 5.1	Options for Manchester's new runway	123
Figure 6.1	Operating costs of UK airports, 1994/95 a) Operating expenditure b) Operating expenditure per work load unit	135
Figure 6.2	Operating surplus per work load unit	142
Figure 7.1	Conversion of Bergstrom Air Force Base, Austin, Texas	168
Figure 7.2	Typical ratios of annual to peak hour traffic	174
Figure 9.1	MSP 2005 noise contours - No action alternative	238
Figure 9.2	MSP airport configuration - No action alternative	239
Figure 9.3	Proposed MSP development alternative	245
Figure 9.4	Location of new airport alternative	246
Figure 9.5	MSP 2010 Long-Term Comprehensive Plan	248
Figure 9.6	MSP 2020 Concept Plan	249

xii *Strategic Airport Planning*

Figure 9.7   Projected average aircraft delay at Seattle-Tacoma International Airport 255
Figure 9.8   Flight Plan Project Schedule...256
Figure 9.9   Forecast of air travel demand - Puget Sound Region..259
Figure 9.10 Locations of alternative airport sites - Puget Sound Region 264
Figure 10.1 The London area airports..274
Figure 10.2 Third parallel runway options for Heathrow 286
Figure 10.3 Cumulative percentage departures per week at Heathrow 287
Figure 10.4 Shares of London Airport passengers 297
Figure 10.5 Stansted shares of origin zone passengers
             a) International scheduled trips through Stansted from surrounding zones 298
             b) International charter trips through Stansted from surrounding zones.............299
Figure 10.6 Gatwick share of London domestic routes 300
Figure 10.7 Gatwick share of London international routes..301
Figure 10.8 Effect of frequency on Gatwick share of London short haul market.................301
Figure 10.9 Regional shares of UK international markets
             a) Scheduled passengers b) Charter passengers 309
Figure 13.1 Traffic at southern Scottish airports 373
Figure 13.2 Traffic at Belfast airports..374
Figure 13.3 Traffic at New York airports 376
Figure 13.4 Traffic at São Paulo and Rio...378
Figure 13.5 Traffic at Paris airports...379
Figure 13.6 Traffic at Montreal and Toronto 381
Figure 13.7 Shares of regional passengers 382
Figure 13.8 Liverpool shares of northwest passengers 383
Figure 13.9 Leeds/Bradford shares of northwest passengers 383

PREFACE

The aim of the book is to identify the strengths and weaknesses of past strategic planning of airport systems, and to attempt to provide guidance on how the concept of strategic system planning can be used to advantage in the future. It is an attempt to return to the ground which was so well covered some 20 years ago by Richard de Neufville in his book: "Airport System Planning". The need to update and extend this work, as well as the challenge in doing so, arise from the subsequent changes in the structure of the air transport industry and the contexts within which it must work. The industry is increasingly becoming liberalised, privatised and globalised. However, the intended competition is sometimes seriously constrained by lack of physical and environmental capacity or by the economic forces that shape the behaviour of the operators. At the same time, the planning context is becoming more sensitive to sustainability issues and to calls for integrated transport solutions to address increasing levels of congestion.

The need for a strategic systems view has never been greater, as entrepreneurial stakeholders attempt to create and take advantage of their own comparative advantages. Those in government need to understand the system behaviour and the extent to which it may be necessary to intervene in the provision of facilities, and how air transport may best be fitted into other transport policies. Equally, the entrepreneurs need to understand what their natural roles may be and what they would need to do to move away from those roles.

The content of the book stems from studying past attempts to prepare national and regional strategic system plans in a variety of contexts, as well as from noting the lack of a systems context in many individual airport master planning studies. The ideas presented in this book have been honed by discussions with colleagues and students on postgraduate and short courses in Berkeley, Loughborough and ITA in Brazil, as well as numerous professional colleagues throughout the aviation industry, though all responsibility for these ideas rests with the authors. Among those at Loughborough who have influenced the work are Norman Ashford, David Gillingwater, Lloyd Jenkinson and David Pitfield. Colleagues at Berkeley whose work and ideas have shaped the thinking in this book include David Gillen, Mark Hansen and Adib Kanafani. Thanks go to them, to Henrique Gennari whose PhD helped to structure some of the strategic planning ideas, to Darren Rhodes for his PhD work on an integrated aircraft design model, and to all others whose work has been relevant, whether it is attributed in the text or has trickled into our consciousness in a less formal way. Thanks also to Mary Ashworth who compiled and formatted the text. Finally, thanks to those whose lives have been so disrupted by the prolonged production of the text, namely Anna Caves and Katie Korzun, without whose tolerance and good humour the task would never have been completed.

1

INTRODUCTION

Few things better characterise the changes in transportation and communication that have taken place during the twentieth century than the growth and changes in air transportation. In 1903 two bicycle mechanics and self-taught aeronautical engineers were able to achieve sustained powered flight for 40 yards at Kitty Hawk, North Carolina. At the time, the fastest train journey from Washington DC to San Francisco took over four days, while the transatlantic journey from New York to Hamburg took seven days. Today, aircraft have replaced both trains and ocean liners for all long distance travel, with everyone from diplomats to students routinely crossing oceans and continents in a matter of hours. Modern long haul aircraft can fly a third of the way around the globe without stopping. Nor are these changes restricted to passenger travel. A business package can be picked up in Los Angeles one afternoon and delivered in Paris two days later.

1.1 CONTEXT OF AIRPORT SYSTEM PLANNING

The advances in the technology of air transportation have been matched by a progressive reduction in its cost, as shown in Figure 1.1. Not surprisingly, these changes have led to a dramatic increase in air travel, which are projected to continue into the foreseeable future, as shown in Figure 1.2. Together with this growth in air travel has come the need for ever increasing airport capacity to handle this traffic. At first, airports were constructed in a somewhat ad hoc manner. As the technical requirements of successive generations of aircraft became clearer and the volumes of passengers and cargo being handled increased, the need for common and accepted procedures for airport planning become apparent, and a well-defined body of technical literature emerged, supported by the standardising influence of such agencies as the International Civil Aviation Organisation and United States Federal Aviation Administration.

Figure 1.1: Historic trend in real fares

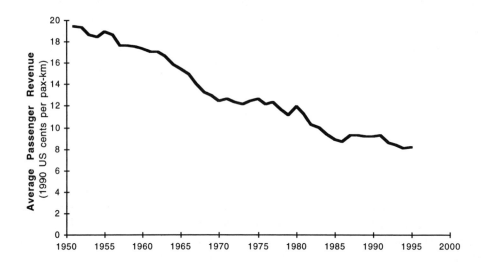

Source: International Civil Aviation Organisation, ICAO Bulletin; Civil Aviation Statistics of the World, various annual issues.

Figure 1.2: Worldwide traffic growth

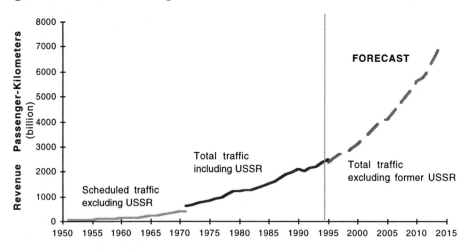

Sources: International Civil Aviation Organisation, ICAO Bulletin; Civil Aviation Statistics of the World, various annual issues.
Douglas Aircraft Company, 1996.

However, these procedures and guidance material largely focus on the planning of individual airports. The interaction between airports, and the planning of airports at a regional or national level, has been less well defined and given much less attention. Even so, over the past three decades the need to plan individual airports within the context of the airport system as a whole has become more widely recognised. This has been driven largely by two concerns. The first is the emergence of national or statewide funding programmes to support airport development, and the consequent need for a systematic approach to allocating that funding among the many eligible airports. The second concern is the emergence of systems of multiple airports serving large metropolitan regions. As air traffic has increased beyond the point where it can be handled by a single airport, or simply due to the geographical extent of the region, the traffic has become distributed between several airports.

Travel and tourism, much of it using air travel, now employ more than 10% of the world's workforce and generates the same percentage of the world's Gross Domestic Product. Similarly, a large proportion of the world's freight value is moved by air. The air transport industry estimates that the total impact on the world economy in 1992 was US $1,000 billion and accounted for 22 million jobs (ATAG, 1993). A recent survey by the Airports Council International showed that 85 of their members operated 246 airports which handled 40% of world passengers, employed 71,000 people and generated some US $11.4 billion in 1995, half of which came from non-aeronautical sources (ACI, 1997). This activity is generated by the travel habits of a very small proportion of the world's population. Hardly any of the two billion Chinese or Indians have flown, and even in the United States, where air transportation is widely used for domestic travel, only 15% of US citizens hold a passport (Lovett, 1995). Increasing wealth and lower prices will cause continuing increases in demand. The International Civil Aviation Organisation (ICAO) reports that passengers carried have been increasing at 5.0% and air transport movements (atm) at 3.5% per annum worldwide for the last decade.

The world's most mature air travel market is the US, but its traffic is still expected to increase at 4.0% annually. This will call for very large absolute increases in capacity, even at the busiest airports. While the average passenger growth of all the member airports of the Airports Council International (ACI) was 5.0% in 1997, the growth at the four airports with the greatest number of passengers, namely Chicago O'Hare, Atlanta, Dallas Fort Worth and Los Angeles was 1.8%, 7.7%, 4.2% and 3.7% respectively (ACI, 1998). Their average traffic growth of some 3 million passengers per annum (mppa) was therefore greater than the total 1997 traffic of such airports as Adelaide, Christchurch, Madras, Ankara, Faro, Larnaca, Milan Malpensa, Prague, Brasilia, Guadalajara, or Oklahoma City. Until the recent economic setbacks, the highest growth rates have been occurring in the developing world, particularly in Asia and the Pacific Rim, as these countries climb towards a level of income of US $10,000 per capita. Traffic in China has been growing at over 20% per year for two decades. These countries are beginning to feel the environmental constraints which have been a prime concern in the western world for the last two decades, as cities expand into the area around their airports. Though only Seoul is in the world's busiest 10 airports, and only Bangkok, Taipei, Jakarta, Beijing and Kuala Lumpur are in the top 50, many cities are considering expensive and remote sites for new replacement airports to support the further expansion of traffic, in air freight as well as passengers.

The air transport network also continues to expand and change. Worldwide there were 1,470 new routes started during the year ending September 1993, due chiefly to liberalisation or to the opportunities presented by low labour and aircraft costs during the recession, while 1,250 routes were eliminated. Several of the new routes were long haul direct services. There are some 1,300 airlines in the world, 150 of which started up in 1993 while 80 ceased operating. Of the total start-ups, 40% were pure freight services The deregulated US accounted for 20% of the start-ups, the bulk of the others being in the CIS, China, India, Canada, UK, Pakistan and Nigeria. Between 1985 and 1993, the number of nonstop flights between Europe and the Asia/Pacific region grew from 2,860 to 17,650.

The ACI survey indicated that the existing airports are having to spend at the rate of twice their operating surplus, equivalent to 40% of their annual revenue, in response to the growth of demand. Estimates of the requirement for expansion of the US system alone in the five years to 2002 range from the Air Transport Association's figure of US $25 billion to the American Association of Airport Executives figure of US $50 billion. These estimates do not include the many new large airports which are planned or under construction which would lift the annual capital spend per passenger from the range of US $3 to US $10 up towards US $40. Planning the airport infrastructure required to keep this huge and dynamic industry moving forward clearly presents a major challenge.

1.2 SCOPE OF THE BOOK

The book addresses the role of airport system planning in developing a strategic approach to meeting this challenge. However, effective strategic planning requires careful integration into the entire airport development process. Strategic plans frequently are not implemented; at least, not in the way they were originally intended. Often, this is because 'the strategy is not a strategy at all, the strategy itself is not implementable, or the strategy is not owned by the implementors' (Giles, 1991). The book addresses the need to improve the front end of the process of providing appropriate airport facilities, recognising that an airport's strategic plan needs to be set in an understanding of the airport's role within a system and that system planning is a vital part of the process. It is concerned with putting in place effective methodologies and frameworks to improve the prospects of achieving an appropriate balance which can allow growth in civil aviation while respecting the needs of other interest groups and, indeed, allow win/win situations to be created. Particular attention is paid to the economic viability of projects, to the increasing need to cover the full social costs of aviation operations, to ensuring that planning frameworks allow a 'level playing field' for assessing aviation projects and that jurisdictional implications of the planning process do not reduce the efficiency of implementation. Planning today must be done in a much more turbulent setting than when the seminal book on airport system planning by de Neufville (1976) was written. The present book offers an approach designed to take account of the trends in deregulation, privatisation, environmental awareness and the need to plan for a sustainable future.

While this book addresses airport system planning from the perspective of planning activities for regional, state and national systems of airports, it should be recognised that these terms tend to overlap. While the definition of a nation is clear, the term state may mean different things in different counties. Similarly, regions may be areas within a state, or groups of states, or even groups of nations. These distinctions are semantic, and not important beyond the need to be clear in each specific case how the terms are being used. To avoid confusion in this book, we will use the term state to mean a jurisdictional entity within a nation with its own level of government one tier below the national government (termed a province in some countries) and the term region to mean any defined geographical area, whether it is entirely within a state, or includes parts or all of several states, or even several nations. We will not use the term state as a synonym for nation (although this is often done in other contexts), but refer to national system planning or national governments. When we speak of local governments, we mean jurisdictional entities below the level of a state, including counties, cities, and so forth. It should be noted that some nations, the United Kingdom being a case in point (at least at present), are not divided into states in the foregoing sense, and thus only have national and local governments.

It should be clear from the foregoing discussion that a dominant aspect of airport system planning is that it almost invariably involves many different government jurisdictions, and that the intergovernmental relations inherent in this are a crucial aspect of system planning. In the extreme case of a multinational region, such as parts of Europe, airport system planning may well involve delicate issues of international relations. Yet the very speed and ubiquity of air travel make the need for such planning imperative. Airport developments in one location have impacts on the traffic using airports hundreds, or even thousands, of kilometres distant.

It is not intended that this book should be a 'how to do it' manual of strategic system planning. Rather, it is meant to be an exploration of issues and practices, based in critical commentary of the available theory and of the attempts that have been made to plan individual airports or systems of airports. Equally, it is not intended that it should deal with detailed planning and design. These areas are covered fully in existing texts (Ashford and Wright, 1992; Horonjeff and McKelvey, 1994).

The geographic scope is large. The main case studies are based in the developed world, but the lessons are interpreted in a wider context. The implications of the different planning settings are fully acknowledged. All scales of airport systems are examined, and attention is given to situations requiring incremental change as well as those requiring large scale expansion.

1.2.1 Aviation system planning

Some system planning studies use the term aviation system planning. Often this is simply a synonym for airport system planning, but in some cases it implies a broader perspective that encompasses the development of the airspace and air traffic control system, and even the way in which air service is provided. Of course, these issues are also germane to planning airport development and an airport system plan should consider the way in which air service is provided and air traffic management is performed. However, in this book we will restrict the use of the term aviation system planning to studies that encompass a broader set of decisions than just questions of airport development. Thus an aviation system plan might also address investment in navigation aids and air traffic control infrastructure, or aviation weather monitoring and information systems.

These topics go beyond the scope of what we can hope to cover in this book without short-changing the treatment given to airport planning issues. They will not be ignored entirely, but their treatment will be restricted to those aspects that impact airport system planning.

1.2.2 Motivation for airport system planning

Why undertake airport system planning? Why not let each airport evolve as its management sees fit, in response to market forces and the actions of competing airports and indeed other transport modes? The answer to these questions rests in the belief that planning the development of a system of airports in an integrated way will result in a more efficient use of resources and a better provision of air transportation services. This answer however in turn raises the questions of how we measure efficiency and what do we mean by "better" air service. It is also necessary to examine the role and purpose of the systems analyst in those nations where control has been largely devolved to the market. These are fundamental questions that we shall address in this book.

1.3 DEVELOPMENT OF THE AIRPORT PLANNING PROCESS

In the early days of flying, airport development tended to be driven by the military and the airlines in a rather ad hoc manner. As air travel grew after the second world war, the implications for land use and the need for consistency in standards required the aviation community to adopt a more formal approach to infrastructure provision. The need to interact with existing land use planning processes and to justify investment led to the adoption of master planning for individual airports. Meanwhile those states or nations with a tradition of central planning developed processes for the strategic planning of their whole airport system.

In the United States, the first efforts at planning the airport system, as distinct from individual airports, began to emerge following the Civil Aeronautics Act of 1938, which established the Civil Aeronautics Authority (later renamed the Civil Aeronautics Administration) and directed its Administrator to conduct a detailed survey of the airport needs of the nation (Horonjeff and McKelvey, 1994). An advisory committee was appointed composed of federal and state aviation officials, airline representatives, airport managers, and other interested parties, and a report was submitted to Congress in March 1939. Further studies were conducted during the 1940s and in 1946 the Federal Airport Act established a federal grant program to fund airport development from general revenues. This program was extended several times between 1946 and the late 1960s, but no formal planning requirements were established. In 1958 the Civil Aeronautics Administration was combined with the recently established Airways Modernization Board to form the Federal Aviation Administration (FAA).

At the Annual Meeting of the National Association of State Aviation Officials (NASAO) in September 1967 several states announced their intention to initiate preparation of state airport system plans, and following the meeting NASAO formed a joint working group with representatives of the FAA to develop guidance on preparing these plans. The resulting report (FAA, 1968) established a well defined airport system planning process that remains largely unchanged to this day.

By this point the need for a strong federal program to support airport development to handle the rapidly increasing traffic levels was widely recognised. The passage of the 1970 Airport and Airway Development Act established both an Airport Development Aid Program funded from aviation taxes and a Planning Grant Program, which in addition to supporting the preparation of airport master plans provided funds to states and localities to undertake the preparation of airport system plans. The Act also required the FAA to prepare a National Airport System Plan to identify the nation's future airport development needs. In May 1970, shortly after the passage of the Act, the FAA supplemented the existing guidance on planning the state airport system by issuing an advisory circular on Planning the Metropolitan Airport System (FAA, 1970). Between 1970 and 1975, 109 planning grants were approved for the preparation of airport system plans (FAA, 1976). The first edition of the National Airport System Plan was published in September 1973. Over the years since then, the system planning process has continued to evolve, as described in chapter 9, and all the states and most of the larger metropolitan regions have prepared airport system plans, which have been updated from time to time.

1.3.1 Master planning

Development of an airport master plan is recommended by ICAO (1985a; 1987) and the US Federal Aviation Administration (FAA, 1985) as the basis for the comprehensive planning of individual airports. The objectives of master planning are to allow orderly development compatible with the framework of local, regional and national economic and transport plans and with national and international aviation policies, while protecting and enhancing the environment. The process should also inform public and private interests of aviation requirements, providing a planning framework which enables affected political entities to participate in the planning, and result in optimal use of land. It demonstrates an airport's commitment to a sound business plan and so to the airlines, to the users and to its contribution to the local and regional plans, helping to remove uncertainties in the community.

The master plan includes the phased physical planning of airspace, airfield, passenger and cargo terminals, circulation, support and service facilities, ground access, and the suitable disposition of these facilities. It also includes economic planning of capital requirements and the associated cost and revenue streams, together with an assessment of environmental impacts and potential mitigation measures. All these aspects of the planning follow from forecasts of aviation activity. Several options are usually taken forward to an evaluation stage, where a decision is taken to select one particular solution to present for public examination. It should not be expected to incorporate detailed facilities planning, since its purpose is to obtain strategic authorisation (Smith, 1998). Indeed, it must be expected that it will need to be subject to continual revision as matters of operational and commercial efficiency are faced and as priorities and conditions change.

A prime objective of master planning is to determine the ultimate site capacity and then to protect it from the consequences of ill-considered development of facilities on the airport and from encroachment of incompatible land uses around the airport which might restrict either its physical expansion or result in traffic limits due to environmental impacts.

1.3.2 Strategic system planning

Wherever an authority has responsibility for the funding, ownership or management of more than one airport, it needs some structured way of determining investment priorities, promoting cost-effective development of air transport and its infrastructure, and establishing standards and roles for the airports. This often results in categorisation of airports by size and function, with the implication that airports in the same category would be treated equally with respect to the criteria used for making decisions about their facilities and operational capability. This is often particularly true in meeting the ICAO requirements for airports categorised as international gateways, and for deciding the extent to which domestic airports should also meet those standards. Some countries carry this top-down process through to the production of individual airport master plans which fit the system-derived roles and budgetary capability. Others incorporate locally determined master plans into national plans to the extent necessary for the specific administrative responsibilities taken by the central authority. Either way, a system plan has to be compatible with viable master planning of each facility. The primary problems faced in national system planning have usually been the division between local and national funding coupled with differences in the perceived role of airports across the various regions of a country.

As the number of airports has increased within individual regions of a country, it is common for a mix of large and small airports to evolve serving a variety of functions. This occurs particularly in large and growing metropolitan areas, as land becomes increasingly scarce and environmental impacts from aviation operations become particularly noticeable. In this context it becomes necessary to ensure that the aviation activity uses the facilities in the way which is best for the whole community and that any necessary expansion of capacity be provided where it does the least harm and the most good. Objectives such as these can only be met by getting all the affected parties to agree to cooperate in a system planning exercise which accepts inputs from a wide range of interests and strives to reach a consensus decision. It helps if there is a defined framework for this process and the studies which should be incorporated in it. In the US, the FAA advisory circulars on statewide system planning and metropolitan system planning, discussed in chapters 7 and 9, provide this guidance.

The appropriate level of capacity at each airport cannot be decided without a system-wide analysis, whether the desire to invest in new capacity arises from individual airports (bottom-up) or from national or statewide concerns (top-down). The interactions, in both supply and demand, between the system elements are normally too strong to be ignored. It is becoming more difficult to add capacity to existing airports due to a combination of factors, including shortage of land, environmental impacts from aircraft and ground traffic, opposition from those questioning the need for further air travel, inappropriate planning processes, uneven power relationships which attempt to manipulate the process, funding problems, and investment risks in the face of uncertainty about future traffic. Also, airport capacity has been shown to be somewhat elastic in the face of increased demand, and further opportunities are available to relieve congestion at existing airports. In some cases, it may be appropriate to increase capacity by developing alternative sites, even though there are usually access penalties which make it hard to convince airlines that they should provide service there. Also it is more difficult to convince people not accustomed to aircraft noise that they should tolerate it than it is to continue to afflict those who already are accustomed to it. All these matters need to be addressed through system planning.

1.4 CHALLENGES TO THE TRADITIONAL PLANNING APPROACH

1.4.1 The changing context of airport planning

In the era of strict economic regulation, the air transport industry tended to experience orderly growth of traffic, which facilitated the development and use of formal master and system plans. Airport roles and their levels of traffic were largely in the hands of predictable route licensing regulators. The airport management function was seen as a demand-responsive provision of service and facilities, thus minimising the risks of over provision. The public purse was expected to cover the remaining risk in the interests of continuity and the wider value of an adequate air transport system. The consumer had to accept the regulators' judgement of the necessary costs of providing a safe and regular service that met social needs, together with the limited choices which were deemed suitable. Communities around airports, and the wider community of environmentalists, tended to have to defer to near universal decisions in favour of providing transport capacity on demand so as to support the air transport industry and economic expansion. The setting in which air transport now has to function, at least in the western world and increasingly elsewhere as well, is very different.

Capacity must be provided in an adequate and appropriate manner if the industry is to continue to grow in a way that is useful and acceptable to society. Pressure on resources has meant that it is no longer possible to provide capacity on demand. As the requirements of the various system stakeholders are examined later in the book, it will become clear that the test of adequate capacity is one which will result in a much more globally sustainable air transport system. One interpretation of this might be that the system should not be constrained to operate in such a way that the internal and external social costs without additional capacity exceed the costs of additional capacity and the corresponding internal and external social costs. Strategic planning should be clarifying these costs and the planning process should allow this judgement to inform decision making.

Appropriate capacity implies a distribution of capacity in which the level of its utilisation is justifiable by local social cost/benefit analysis. Strategic system planning should provide a basis for decisions affecting the geographical distribution of facilities, and in the balancing of capacity across the system elements at each location, establishing best practice and assessing each location's potential to meet the demands imposed on it by the different development options.

Constraints to growth have, in the past, been mostly physical or financial. They are now increasingly environmental as well. Some 400 US airports have some form of noise restriction (*Commuter World, April-May 1990, pp 5-10*), despite the considerable benefits flowing from the retirement of the noisier aircraft which were certificated under the so-called Stage 2 rules of the Federal Aviation Regulations covering aircraft noise (FAA, 1992). In Europe, 77% of the airports responding to a survey by ACI reported some kind of operational restriction due to noise (ACI Europe, 1995), while Zürich and Stockholm's Arlanda airports are examples of those where activity is limited due to air pollution. The limit or cap depends on the local and national regulations on allowable pollution, and on whether the method of control is regulatory or based on price. The safety of the surrounding communities is beginning to be seen as a serious consideration which may limit future expansion. The communities are now able to gain support from the rising tide of environmental concern and those questioning the appropriate use of resources. They are also suspicious of airport motives and the strength of planning agreements. Even if a proposed development appears relatively innocuous, communities have learnt to suspect the 'thin end of the wedge'. Consequently they ask simple but difficult questions such as: "would it matter if the airport did not get another runway?" or "if they get a second runway, will they not then later need a third and a fourth?".

The economic liberalisation of airlines, allowing the industry to set its own fares and frequencies, and to enter and leave routes at will, has allowed a major restructuring of the air transport network, resulting in large perturbations in the traffic at many airports. The new opportunities for airlines to compete on price and quality of service imply considerable changes in the quantity and distribution of revealed demand, now being more driven by the desires of the consumers themselves than by the industry's and consumers' needs as perceived by regulators. Global alliances between major carriers cause the smaller nations to worry about losing control over their own air transport system's destiny. These changes make it more difficult for airports to predict the demands for which they should prepare.

Global economic competition at the level of individual regions within a country, together with a realisation of the role which an airport can play in furthering regional ambitions, has caused the public authority owners of the airports to increase their marketing and expansion efforts. They have often been frustrated in this by lack of finance. More recent tendencies towards the privatisation of the airports, partly to alleviate the finance constraints, have created an even greater desire to compete and expand.

1.4.2 New aspects of old challenges

Financing development is still a major problem. Historically, the air transport system's economics have been dominated by the operating cost of the airlines and particularly, by the direct operating cost of the aircraft. However, the balance of assets between the infrastructure and its users has always been much more equal, one estimate being that the world's airlines hold approximately 60% of the total assets in the industry value chain, which includes aircraft manufacturers (Borgo and Bull-Larsen, 1998). Even in the UK, with a large proportion of its flying being outside UK airspace, the combined assets of the airports and the Civil Aviation Authority (CAA) have equalled those of the airlines, as reported in the CAA annual reports and airline and airport statistics. A 'heroic' estimate of US $100 billion for total world investment in airports between 1990 and 2000 (Wheatcroft, 1990) was almost certainly too low; as it implied that only 20% of new investment would go into infrastructure. ICAO (1992) estimated that the global investment in airport and route facility infrastructure over the period 1991-2010 will be between US $250 and US $350 billion in 1991 dollars, while the comparable figure for the 11,000 aircraft required will be some US $800 billion. If true, even the ICAO estimate would substantially reduce the likely present 50% contribution of infrastructure to total system assets, so putting additional pressure on capacity and increasing congestion.

In many countries, the problem of finding the funds to finance necessary new capacity has, in the past, been partly hidden by the use of general funds, justified by the belief that there are induced benefits which flow to a country from the provision of an adequate airport infrastructure. Increasingly in the developed world, air transport is being required to pay its own way, even when the airports remain in government hands. Yet it has been said that although investment in air traffic control (atc) is a good 'bankable' proposition, the same could not be said of airports (Wheatcroft, 1990). This conclusion was based on the presumption that 3% of airline revenues would go to atc and 5% to airports, the airline contribution to airports being matched by concession revenue. The rate of return on the US $100 billion needed between 1990 and 2000 would then be only 6.5%, compared with approximately 12.5% from atc. While the most successful airport companies already generate more revenue from concessions than assumed in the calculation, the actual investment required is certainly greater than Wheatcroft's estimates and will increase further in the future, since the cost of each new unit of capacity increases as the most productive options are used up. It has been estimated that the EU airports alone will need to invest US $66 billion between 1996 and 2005 (Feldman, 1997a). China plans to invest US $8.4 billion between 1996 and 2001 (*Jane's Airport Review, May 1998, pp 9-11*). Russia has closed over 500 of its 1,300 airports since 1992 and traffic has dropped to a quarter of former levels, but the Airport Association which represents 134 airports in the former USSR plans to spend US $5 billion before 2010 in bringing them up to international standards (*Jane's Airport Review, May 1998, p 5*).

Other commentators increasingly believe that airports are a good investment, but that it may be less wise to invest in airlines. Oslo's Fornebu airport has the best revenue/operating expenditure ratio of 25 European airports at 2.72, while the eighth best managed a ratio of 2.45 (Lobbenberg and Graham, 1995). The best airlines in 1995 were Southwest and British Airways with ratios of 1.15 and 1.09 respectively. As explained in chapter 5, a large part of airports' annual costs are incurred in capital charges, so it must be expected that their operating ratios are higher than those of the airlines. However, their Returns on Invested Capital (ROIC) are estimated to be some three points higher than their cost of capital, while the airlines' ROIC is three points below their cost of capital Incidentally, the best returns appear to be obtained by the owners of computer reservation systems, followed by the aircraft manufacturers and lessors (Borgo and Bull-Larsen, 1998). It is little wonder that airlines feel that, with all the pressure on them to cut costs, the airports should also play their part in improving efficiency. Paradoxically, it is mainly those major airports which are most vital for a nation's international links and which need the majority of the investment that are also the only ones capable of being fully self-supporting. The maintenance of a complete system of airports requires firm financial discipline from the smaller airports, consideration of the degree of justifiable subsidy and whether the subsidy should come from general funds or from cross-subsidy by larger airports. If the choice is cross-subsidy, does that make it harder to privatise the system, and, if so, how else can government funding restrictions be circumvented?

Capacity. Planners have been concerned with the need for more capacity for a long time, as shown by the London Area case in chapter 10. The 1986 US National Airport System Plan contained the requirement for four new primary and 11 other new airports for commercial services (FAA, 1987). European studies (AEA, 1987; SRI, 1990) confirmed serious capacity problems at London, Frankfurt, Düsseldorf and Madrid, and many other airports will reach their runway capacity before the year 2000 unless urgent action is taken to expand them or build new airports. Despite the capacity shortages, the industry appears to be reasonably satisfied that the expected future growth will be sustainable. Airbus certainly feels that the situation will be manageable except in the cases of European runway capacity, and of airspace in the short term. Boeing anticipates that air transport demand will continue to grow at over 4% per annum in Europe (Boeing, 1998) and that congestion will not affect traffic growth any more than in the past, but will affect operational solutions to it, with alternate routes and airports and larger aircraft. However, a TRB panel of aerospace manufacturers felt that congestion is both the most important and most difficult issue to forecast in determining the number and size of new aircraft deliveries between 1998 and 2007 (TRB, 1993).

By 1998 only one major new airport has been built in the US (Denver) and one in Europe (Munich II) in the prior 20 years. The Gardermoen replacement for Oslo's Fornebu and Bergstrom airport at Austin, Texas were the only new major commercial airports under construction in Europe and North America, both being on existing military sites. Instead of more runway capacity being provided at the existing busy airports where market entry by new airlines would do most to bring the benefits of competition, planners are tending to fall back on enhanced use of more remote and under-utilised facilities. In the past, planners predicted that airlines and their customers would shift to more remote facilities at many other locations (e.g. Mirabel at Montreal, Palmdale at Los Angeles, and Prestwick at Glasgow) but in most cases have been proved wrong despite the ability to regulate routes. The developed world's airport systems are thus unbalanced with, on the one hand, crowded airports with few plans for expansion and, on the other hand, under-utilised airports which the airlines decline to serve.

The problem of under-utilised capacity can be put down in part to an inheritance of military airfields which were poorly located for civilian use when released for other uses, together with a lack of appreciation by airport planners of the factors shaping airline network strategy. However, the decision to use them has been largely forced upon planners as the only environmentally acceptable solution to expansion. The dense land use patterns which surround busy airports and the ecological arguments against the construction of new airports in existing open spaces, together with the consequent exposure to aircraft noise of communities not previously subject to it, conspire to produce overwhelming environmental arguments against expansion of existing facilities or completely new airports. It required the hearing of some 30,000 individual complaints and quite extreme environmental mitigation measures before Munich II could be built, even though a prime reason for its construction was the environmental unacceptability of the then existing airport of Munich Riems.

On balance, there is a real possibility that rising congestion will constrain the future growth of traffic as the ratio of infrastructure to airline assets falls. The resulting traffic would then be distributed through the system in a way that will be determined by the availability of capacity rather than by the unconstrained preferences of users or society. The evolution of the air transport system would increasingly be dictated by local decisions on expansion of capacity. Clearly there are already serious barriers to market entry in the present system, aggravated by a shortage of infrastructure. If the relative infrastructure assets were to reduce further, this will tend to raise the barriers higher, unless the application of new technology or management innovations can be made to play a significant role in creating more capacity or better managing the use of the existing capacity.

1.4.3 Uncertainty

Most carriers surveyed by the International Air Transport Association (IATA) in 1993 were more concerned with the volatility and unpredictability of the operating environment and lack of data (TSUG, 1995) than with capacity constraints, which did not feature significantly in their traffic modelling methods. Airport forecasters share the same lack of data on actual fares and on future socio-economics, but are much more interested in how lack of capacity at their airport would affect demand. The experience of attempts to generate new airport capacity, at least in the developed world, suggests that airport entrepreneurs will face increased risk from three main uncertainties:

- user demand and its satisfaction by the airlines in a competitive setting
- capacity and cost implications of environmental protection
- refusal of planning permission after lengthy preparation, or duplicative permission leading to excess capacity.

Can these uncertainties be managed so that there is an efficient future scale and distribution of capacity, or will the risks drive the system towards maximising the utilisation of the existing infrastructure at the expense of the best interests of the users?

The appropriate plan for an airport system should depend on the way the system is to be used, unless society is prepared to take the attitude that the consumers will have to plan their requirements around the system rather than vice versa. Airport planning almost always assumes that the system will continue to be used in the same way in the future, though this almost never turns out to be the case. Likely exogenous influences on the system in the future will include changes in regulation, modal competition, telecommunications, differential economic growth, globalisation of industrial production, oil availability and price, demographics, tourist behaviour, and political and social norms. Important changes within air transport will be seen in aircraft technology, sea-air transhipment, airline and airport mergers, network development and increasingly scarce capacity.

The uncertainty in predicting the future stems from likely but unforeseeable structural changes in these factors. Not only are these difficult to predict, but, if they were predictable, much of their benefit in promoting and complementing change would be diluted. Yet the 'myth of predictability' (Gifford, 1993), where forecasters take the option of presuming that the most likely future is the 'business as usual' one, almost always results in technological obsolescence and under-utilised facilities. Most importantly, it distorts the development of the market, perpetuating the present situation by investing in it, so that beneficial change is thwarted by having to overcome the drag of sunk investment. This raises questions about the possibility of developing planning methods which recognise the probability of structural change, which promote solutions that retain maximum flexibility without causing undue planning blight, and which allow decision-makers to adopt them without being accused of not doing their job.

The future is bound to contain surprises. If it is to cope with change, infrastructure planning must be a continuous, adaptive process rather than a one-off attempt to generate a blueprint to formulate the future. Are the managers of the system and of individual airports ready to adopt these ideas, or will the requirements of an established planning process, short term political concerns, and the need to satisfy financial investors continue to sustain the myth of a predictable future?

1.4.4 Planning methods

One of the central questions to be addressed is whether individual airports can plan adequately without the guidance of national-level strategic planning, i.e. can roles, and the competition for them, be resolved through numerous and independent planning studies of a market which may only become known 10 years later? Traditionally, it has been presumed that bottom-up and local planning requires a clearly defined setting. Formal master planning leads to the need to expose all the airport's long term strategy to public examination. This may make some airports loath to adopt the formal master planning path for fear of alarming the public unnecessarily about the follow-on consequences of allowing a relatively small initial development. Further, now that airports are more concerned with profits and competition, the long term plans may be regarded as commercially confidential, or, alternatively, they may be regarded as a bold play to attempt to overwhelm potential competition. Either way, the classic master planning approach is likely to be compromised by the new setting in which the airports find themselves, yet bottom-up system planning can only function from the base of competent individual master plans. The likely compromise will be to commit to a limited time horizon for master planning, which rather defeats the purpose of the exercise. Equally, it must be asked whether national-level strategic studies could give sufficiently accurate guidance on roles while not unduly constraining local initiatives. Indeed, can sensible strategic planning be attempted at all in a liberalised environment whose very ethos is innovation at the level of the individual firm?

Another issue closely coupled with entrepreneurial planning is whether the planning process is able to deal adequately with large infrastructure projects that have national and international implications. This is a highly interactive and multilayered problem, posing questions of local versus national interest and of how to balance these interests across the groups of stakeholders. It has been said that "neither the invisible hand of urge for profit, nor the presumed wisdom and goodwill of god-like civil servants, can be trusted to steer us clear of follies and fallacies. Humble, disinterested seekers after truth are our only hope" (Sharman, 1986). Like Pilate, the

Introduction 15

difficulty remains in defining "what is truth?". The answer might be explored in the sort of open discussion that a UK Minister of Transport had in mind when he said: "If you are going to have some sort of strategic sustainable framework into the next century, then it has got to be built on an element of common ground and you are not going to get that unless you have this sort of debate" (Mawhinney, 1995). Openness and participation in planning were identified as necessary some long time ago (e.g. Hoover and Altshuler, 1977), but how is that debate to be organised in a way that satisfies all the stakeholders?

A successful debate implies, among other things, a judgement on the extent of the full social costs of air transport. It is a move in the right direction for airlines to begin to regard these costs as a responsibility rather than as a penalty, but is there a consensus on the scale of the costs and how they will be recovered? Can the industry invest in large scale infrastructure without an assurance that policies on these and other planning issues will not change radically?

The aim of the book is to throw some light on these questions, to investigate how far the proper application of existing best practice and systems analysis tools can provide a better understanding of the issues, and to indicate where there needs to be further development of techniques and processes.

1.5 FORMAT OF THE BOOK

The book follows a logical sequence of discussion of issues involved in the strategic planning of airports. It starts by describing the changing context in which the planning must take place. It then goes on to define a strategic view of planning in the following chapters. Chapter 3 introduces the overall concepts of strategic and system planning, with a complete view of the process, the interested parties in the process and the difficulties that must be faced in applying the process. It is emphasised in chapter 4 that strategic planning requires a full understanding of the air transport system, and the behaviour of each of its stakeholders is therefore examined so that the implications of their behaviour on the airport system can be understood. The wider interests of the community are then considered in chapter 5, where the value of airports and of air transport are discussed, together with the evolving relationship of the community in airport planning. Then the airport's viability is reviewed in chapter 6, both economically and from the point of view of efficient management of capacity enhancement. With this understanding of the strategic planning process and the behaviour of those groups involved in it as background, the state of the art of planning for two levels of airport systems is then described, namely regional systems in chapter 7 and national systems in chapter 8. In each case, emphasis is given to the important stakeholders, the type of analysis that is appropriate and the pathways to decision and implementation.

The second half of the book presents a series of case studies. The purpose of these cases is partly to inform readers of the efforts being made to plan airports strategically in a variety of settings, so that they may infer lessons that can be transferred to their own setting. It is then possible to refer to this case study material in the preceding chapters, rather than having to present it several times. However, the case studies also serve to illustrate the generic system planning processes, to point up the way that each context requires individual tailoring and to explore the imperfections of the process as it has been applied.

The first setting is the US, where the main emphasis is on formal system planning studies at national, state and regional levels. The case studies examine the relationship between state and national system plans and detailed master planning of individual facilities. The US has a long history of formal airport system planning theory and application, so there is much to learn from its experience in implementing and improving the process. A very different perspective is given in chapter 10, where the UK experience of airport strategic planning is described, both for the London area and for the UK as a whole. There has been little attempt to plan the UK airports as a system, and now the fragmented private ownership makes it difficult for any cohesion to be imposed on the system, yet somehow the airports have to come to an accommodation on the roles they should play. Chapter 11 explores the emerging attempts to define a European airport system in the light of the generic theories and the lessons which can be learnt from the US and UK experience, particularly in terms of understanding what factors shape the airports' roles and the policy levers which are available to realise any preferred system.

Chapter 12 presents a range of shorter case studies of national system planning, each drawing out the defining features and influences in the planning processes and the implications for the resulting system. Brazil's major airports are planned and operated by a single entity, but the airports' roles are likely to change with the beginning of route liberalisation. Canada has taken a new approach to a system which had been very reliant on subsidy but which is now dependent on the success of largely autonomous airport authorities. Germany's airport expansion is seriously cramped by environmental lobbies and a lengthy planning process. The Greek airport system has to serve tourist and social purposes, but the local authorities have considerable political influence which makes it difficult for the Civil Aviation Authority to take a real system view of the set of airports for which they are responsible. Japan has strong central planning and national integration policies to give it direction, but suffers from equally strong environmental objections to airport development which forces it into ever more expensive solutions. Norway similarly has a strong policy to use aviation for social integration but, with a very low population density and equally difficult terrain, has developed a system of short takeoff and landing airports. Spain also plans and operates its airports under a single entity, but has tried to match the roles to the type of traffic and to explore the relative benefits of different scenarios for provision of the airport system. Similarly, chapter 13 explores several regional situations where there has been the potential for airports to compete or complement one another. The forces which shape these systems are identified, the difficulties associated with the demise of route regulation is highlighted, together with the difficulty of achieving a balanced system, given the market power of the main airport in a system.

The book concludes by drawing out the lessons from the case studies in chapter 14 and relating them back to the generic theories of strategic system planning. Suggestions are made in the concluding chapter for how the analytical content and the process might be improved, and of how the air transport system as a whole may have to reorient its objectives in the light of the general pressure for a sustainable future.

2

THE EVOLVING CONTEXT OF AIRPORT PLANNING

The development of aviation and its planning have gone hand in hand. The political, economic and social setting within which aviation functions has also changed over time. This chapter examines the changing context of airport planning. It traces the traditional roles of government and the market imperfections which have accompanied the regulatory context they imposed. It then describes the increasing concerns over environmental impacts and the economic and other inefficiencies which flowed from the interference with the market place. These concerns have led to many important institutional changes which have dramatically altered the setting for airport management and planning. The changes considered in detail are those affecting airline behaviour, airport ownership, and attitudes toward sustainability and intermodalism.

2.1 TRADITIONAL INSTITUTIONAL ROLES

2.1.1 Airline regulation

Regulation of US domestic air service was prompted in the 1920s by concerns for safety, lack of funds for infrastructure, poor financial viability of carriers and a perceived need to protect and subsidise an emerging industry (Wolfe and NewMyer, 1985). Most other countries adopted similar policies, protecting their investments and ensuring socially necessary services by controlling route entry, route exit, fares, capacity and frequency. Distinctions were made between nationally important domestic services, like those providing guaranteed service under the public convenience and necessity licences granted to the trunk and local carriers in the US, and the smaller interregional or feeder services, the latter being subject to less regulation.

International routes were regulated economically for much the same reasons, with most countries subsidising their flag carriers very heavily, fearing that stronger countries' airlines would capture larger market shares. The international regulations were based on the International Conference for Air Navigation in 1919 which was reinterpreted in 1944 by the Chicago Convention on Civil Aviation which brought the International Civil Aviation Organisation into being in 1947. Among other things, members agreed to Freedoms of the Air, allowing the carriage of traffic between countries by the airlines of those countries (third and fourth freedoms). Additional agreements, often of a bilateral nature, laid down more specific conditions. These bilateral Air Service Agreements (ASA) often contained pooling arrangements between the licensed carriers which effectively stopped all competition except for the rare event when fifth freedom rights were granted to a carrier from a third country to carry traffic in the market. In general, each country took an equal share of the capacity on a route and operated some pooling of revenue. There were usually restrictions on the number of gateways which could be used. In return, as well as showing the flag abroad, the airlines were often expected to provide domestic services, often with uneconomic fare levels to meet social goals. They were also used as instruments of government policy, including their use for defence reserves, as a source of employment and as a showcase of a country's technological capability.

2.1.2 Airports as a public facility

The aviation infrastructure has long been regarded as a public facility, similar to roads in providing access for public administration and commerce. It was not always so. The early landing strips and navigation aids in the US were set up by the airlines flying the mail. Nor does the philosophy of public ownership of transportation infrastructure extend to all modes. The US rail network has always been in private hands, as was the UK rail network for a long time. However, the importance of unconstrained access to airports of adequate quality for the support of interstate commerce was soon appreciated by the US federal authorities. At the same time, it was considered that a federal contribution to the infrastructure could only be made if it was in public ownership and therefore guaranteed to remain in use for the specified purpose. The states and city authorities also realised the importance of having an airport in order to be 'on the map'. A system of divided responsibilities grew up, with funding shared between the federal, state and local authorities and the airlines, but with purely local ownership and management.

In Europe, many airfields were started out of civic pride, but they were often requisitioned by government during wartime and then offered back to the communities as the military need diminished. The cities therefore obtained a 'peace dividend' of military investment, although developments in aircraft technology often meant that the runway layouts were frequently not ideal for civilian use. Over time, the local or national public purse has put a lot of investment into these airports, as well as often not fully recovering the operating costs even when all the airport revenues are taken into the general purse. The perceived justification for this expense has been that the community and its representatives have overall control over the airport's operations. Any attempt to place the airports in private hands leads to fears that this control would be compromised. On the other hand, the communities could benefit from a cash windfall, which may not be justifiable, depending on who made the initial investments.

Meanwhile, national and civic pride often caused airports to be overbuilt relative to the revealed demand.

Neither in Europe or elsewhere is it possible to generalise about ownership and management, except that until very recently it was almost entirely in public hands. Some countries centralise ownership and management, others have central ownership and local management, others have shared ownership, and yet others have local ownership and management. The institutional role of the national government in these circumstances ranges from planning and managing at a system level to the provision of funding.

2.2 MARKET IMPERFECTIONS

While few would argue with the need for safety regulation, economic regulation is often accused of condoning inefficiencies in exchange for a dubious effectiveness in meeting social goals. The deregulation movement had its origins in the feeling that domestic air transport had developed to the point where its protection from the forces of market competition was no longer justified. Any competition allowed under a regulatory regime tends to be perfunctory, not producing the efficiency and choice enjoyed by consumers of products in more genuinely competitive markets. The market tends to be further distorted by overt and hidden subsidies, while public ownership often distorts the flow of investment funds. Imperfections generate an airport system whose planning is then constrained to perpetuate it. This section considers these factors in a little more detail.

2.2.1 Lack of competition

The obvious result of regulation is the lack of competition, the only differentiation of product between airlines normally being in cabin service and punctuality, even if there is more than one carrier per route. There is then no incentive to improve service or efficiency except for the ultimate sanction of withdrawing the licence, which is impractical if there is no other substantial carrier of the same nationality. There is not even an incentive to grow market share unless, like the US airlines, there are private shareholders. While the technical quality of the protected and regulated airlines was often high, due to the safety regulation and the enthusiasm of the employees, there was little control over cost because control was mostly exerted over excess profits. Thus there was always the temptation for the airline to fulfil the ordained role and to take any surplus as salary.

2.2.2 Direct and indirect subsidies

The social service obligations imposed on the carriers, and the need for airports to accommodate the services, required direct subsidy to be provided to the carriers and to the airports, particularly in the early days when traffic levels were low. The Essential Air Service programme in the US causes airport facilities to be made available for very low levels of traffic.

Indirect subsidies like the tax-exempt bonds used for airport investment in the US distort normal investment criteria, while fuel tax exemption distorts mode choice. The interest rate on taxable bonds in the US is 2.0 to 2.5 percentage points higher than tax-exempt debt with equivalent risk (Dillingham, 1996). Cross-subsidy of domestic or thin routes from international or dense ones, often necessary to fulfil the social obligations of a certificate of public convenience and necessity, also distort the system. Similarly, so do heavy discounts on airport charges for domestic services, since they are often much greater than the additional costs of providing international facilities.

2.2.3 Shortage of investment funds

As the traffic grew in the stronger economies, governments began to find it difficult to release funds at a sufficient rate to keep up with the very high growth rates in the industry. The funds were also needed for other more pressing social purposes, often being diverted for reasons of short term political expediency. While political judgement may have been correct in deciding to prioritise investment of public money away from aviation to, say, hospitals or to check inflationary pressures, the consequence is that economically viable investments have not been made. In this sense, the funding constraints have been artificial, particularly since some countries do allow government entities to borrow on the open market. In most countries there is a constraint on the extent to which private funds can be used for government owned facility expansion, though there has always been a healthy mix of public and airline funding of US airports.

In fact, the distinction between public and private spending for airport projects has much more to do with who pays the cost of airport investment than whether airports are more important than other public objectives. Whether funds flow into the national treasury from aviation user taxes and then get distributed in airport capital investment grants or the investments are made with privately raised capital which is then repaid from user charges, the users of the aviation system end up paying either way. Ultimately, investments in airport facilities withdraw resources from the national economy, whatever the funding mechanism. This is not to say that the mechanism by which these investments are made is irrelevant or of no concern to national policy makers. Use of tax-exempt revenue bonds, as is common in the US, deprives the national and state treasuries of the taxes that would have been paid on the interest or capital gains if taxable funding instruments, such as private bonds or stock, had been used. Likewise, use of general fund revenues spreads the cost across the entire taxpaying population, while the use of user taxes distributes the cost on the aviation system users according to the structure of the taxes. The use of an *ad valorem* ticket tax, as has been the practice in the US for many years, results in travellers on more expensive tickets contributing more to airport infrastructure than those on less expensive tickets. Whether this is fairer than charging everyone the same, as occurs with the fixed Passenger Facility Charge used at many US airports, is a political not a financial question.

To the extent that private sources of capital generally expect a higher rate of return than is required by public borrowing (and of course tax revenues are simply transfer payments that require no rate of return at all), the use of private funding represents a greater transfer of wealth from the users of the aviation system to the owners of the capital. Whether this is appropriate involves such questions as the relative efficiency of privately funded airport investments compared to publicly funded ones, and the relative contribution to social welfare of investments in airport infrastructure compared to other investments that compete for private capital. While these questions may seem far removed from a decision of whether to build a new runway or not at a particular airport, they are central to how any national programme of airport development should be structured and funded.

2.2.4 Political decision distortions

The airport system which would result from a classical planning process is often distorted by ad hoc politically motivated decisions which overturn the logical solutions. In some cases the interference is warranted, as when the process has wrongly identified public feeling or underweighted it. However, the decision often suffers from unwarranted distortion because of undue influence being exerted in a largely covert way. Apart from apocryphal stories of airport locations being chosen to coincide with the constituency of a minister of transport, an obvious example is the influence of the military, who sometimes invoke the cloak of classified national defence requirements to oppose the joint use of facilities or other developments that would require them to modify their operations. National air traffic control authorities are often in a similar position of unchallengeable power, even when not under the direct control of the military. However open the overt planning process, it is seldom immune to this sort of interference. While these political distortions cannot be entirely avoided, the planning process will be much less vulnerable to their influence if the underlying issues are formally recognised and addressed, rather than leaving the affected stakeholders no recourse for their concerns other than direct appeal to a political decision.

2.3 ENVIRONMENTAL CONCERNS

Aviation only began to attract serious criticism for its effect on the environment when the first generation jets brought high levels of noise and smoke to airports. Many of the impacts have since been reduced substantially, but the antagonism but has not reduced commensurately, partly because the increased activity has diluted the benefits. Meanwhile, new environmental issues have come to the fore. This section describes the main current concerns at airports after first considering the global impacts.

2.3.1 Global concerns

Society at large is increasingly concerned about the deterioration of the global environment and about aviation's use of resources, though the main concerns in aviation have been associated with the movement of aircraft on and near airfields, together with the ground activity required to support the aircraft.

Many countries have regulations for environmental control of the unwanted products of transport. The regulations support Agenda 21, the action plan adopted at the 'Earth Summit' in Rio in 1992. There, the UN Framework Convention on Climate Change, signed by 150 states, showed a concerted effort to control the emissions of greenhouse gases. The recognised threat to the biosphere has resulted in a plethora of regulations to control the impact on the global environment of airborne toxins, the depletion of the ozone layer and the increase of greenhouse gases. For a long time there have been regulations in individual countries to limit noise and air pollution from industrial and transport sources.

The main concerns of the UN conference were the ozone (O_3) layer and the 'greenhouse effect'. The release of chlorofluorocarbons (CFC) and other gases deplete the ozone layer, so letting harmful ultraviolet radiation penetrate to the surface of the globe. This concern is also expressed by the EC's Environment Directorate. Subsonic jets, flying in the troposphere below the O_3 layer, could be an important source of nitrous oxides (NO_X). The original ICAO environmental regulations did not consider the full flight cycle, but its Committee on Aviation Environment Protection (CAEP) is working to develop high altitude emission limits. A National Aeronautics and Space Administration (NASA) study showed that, while the main problem for the ozone layer is CFC production, a fleet of supersonic transports (SSTs) might reduce the O_3 layer by 15% (Paylor, 1991). However, there is considerable uncertainty over the chemical reactions involving nitrous oxides (NO_X) and carbon dioxide (CO_2) at high altitudes. The NO_X is being closely studied following the Montreal Protocol on Substances which Deplete the Ozone Layer, while the Intergovernmental Panel on Climate Change is monitoring CO_2. Certainly a reduction in the ozone in the stratosphere leads to an increase in harmful ultra violet radiation at ground level, and nitrous oxides do break down ozone, but NO_X can also be a source for O_3 (Williams, 1988).

2.3.2 Aircraft noise

As air transport activity increased and more long haul jet operations were operated, the communities around airports became increasingly annoyed. The industry responded by the noise certification of all new types of aircraft and later banning those which could not meet the earliest (Stage 1) regulations. Rules also exist for the rate at which 'Stage 2' types should be phased out. In most countries this should be complete by 2002, but this is likely to take much longer in less developed countries because of the costs of premature obsolescence. This will leave only aircraft certificated to the current Stage 3 regulations in the developed world. The reduction of the proportion of Stage 2 and non-certified aircraft in the fleets at airports, as shown in Figure 2.1 for Heathrow, has caused very substantial reductions in the noise footprints, but already there is pressure for at least a Stage 'three and a half'. The footprints are calculated on integrated sound energy levels over the average working day (Leq). There are many detailed variations on these noise metrics. Despite the universal use of these integrated metrics which take account of the number of events as well as the intensity and duration of each event to guide land use decisions, it is usually found that noise complaints are much more associated with individual noisy movements.

Figure 2.1: Change in noise impact at Heathrow

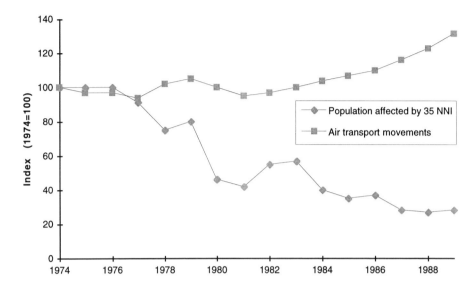

Source: British Airways, 1996

Aircraft noise results mostly in annoyance rather than damage to health, but its effect is difficult to measure since its character varies with the type of sources and people vary enormously in their reaction to it. Communities around airfields vary greatly in population and their tolerance to aircraft noise. Their variable reaction causes uneven control of impacts from place to place.

A particular problem exists with helicopters. Their noise has a distinctive quality due partly to blade slap, and their noise is associated with poor images of safety and being 'spies in the sky'. On the other hand, communities can be very tolerant of helicopter noise when the use is beneficial to society e.g. air/sea rescue. It is very difficult to quieten helicopters, and the noise duration is long due to the slow flying speed.

2.3.3 Air and water quality

Air pollution causes local concentrations of toxins e.g. lead, hydrocarbons (HC), carbon monoxide (CO), volatile organic compounds (VOC), sulphur dioxide (SO_2), and particulates. In addition, nitrogen oxides and SO_2 are carried a long way by the winds to cause acid rain damage far from the point of production. Aviation has already put a great deal of effort into minimising its impact on air and water quality in the airport environment.

The contribution of aircraft to air pollution is highest on the airport because of the concentration of activity and the effects of low engine power. Hydrocarbons and carbon monoxide outputs become very low at high power, but it is difficult then to control smoke and nitrous oxides. There are certification requirements for maximum pollutants per defined Landing and TakeOff (LTO) cycle for large aircraft. The air pollution from small aircraft is quite negligible and is very unlikely to result in concentrations of pollutants which would reach the recommended maximum levels set by the European Commission, the World Health Authority or individual states.

Large emissions of some volatile organic compounds occur in aircraft maintenance, e.g. paint strippers and cleaning fluids. The ground support of aircraft operations also burns a lot of hydrocarbon fuel.

Impressive advances have been made in the control of aircraft exhaust emissions, despite the facts that the majority of the pollution around airports comes from ground vehicles, that aircraft contribute less than 1% of air pollution in the cities they serve (Longhurst and Raper, 1990) and that the concentration of pollutants near airports is generally lower than the averages for the cities they serve (Williams, 1988). In two decades, the weight of emissions per passenger during a standard LTO cycle for aircraft certification as defined by ICAO in 1981 (ICAO, 1993) has been reduced by 90% for hydrocarbons and by 65% for carbon monoxide as engine technology has improved. Nitrous oxide emissions have only been held constant due to the increase in engine pressure ratios (Snape, 1990), but smoke has been almost eliminated.

Impressive though these advances are, they are unlikely to satisfy the environmentally concerned policy makers. The recent measurements of air pollution do not yet fully reflect these reductions, because the average age of the aircraft fleet at most airports is at least 10 years. Research shows that the LTO cycle defined by ICAO is no longer representative. Taxiing aircraft now tend to use lower power settings, and no account is taken of the engine start. On the other hand, air traffic flow control tends to reduce the taxi time and operators often shut some engines down after landing (Caves, Jenkinson and Brooke, 1996). Odour is now seen as an important impact. The ICAO standards do not address carbon dioxide, sulphur dioxide, water vapour or trace elements (Dobbie, 1995). Amendments to the Clean Air Act in the USA are, on balance, likely to result in pollution capping of airports, and restricting chemicals used in the production and maintenance of aircraft (Rheingold, 1991).

Ground water contamination is also important. Individual firms within the industry are concerned to meet new regulations on waste. Again, there are regulations which any industrial plant must obey and which limit concentrations of contaminants. The problems at airports usually come from glycol being used to de-ice runways and aircraft and also from fuel spills and leaks. Efforts are being made to minimise these problems by using materials which are more environmentally friendly (Karrman, 1991; Gould, 1992), collecting and recycling the waste products (e.g. sealed trenches around deicing areas) or by calling a halt to operations when the concentration levels are found by monitoring to be too high.

2.3.4 Energy use

In the period of rapid aviation growth during the 1950s and 1960s, energy was cheap and plentiful. After the 1973 oil crisis, many agencies created policies to minimise energy use at airports, typified by the FAA. Now most airports have some form of energy policy, often driven by pollution regulations or cost rather than by energy consumption per se. More than half of a sample of 104 airports responding to an ACI Europe survey (ACI Europe, 1995) now generate energy on site and use the surplus for heating, though only six use renewable energy sources.

The EC (1992), in both a Green Paper and a subsequent White Paper, is clear that the best way of reducing air pollution is to burn less hydrocarbon fuel. This will only happen with less use of petrol and diesel powerplants together with greater efficiency of those powerplants which remain powered by hydrocarbon fuels. Road vehicles and aircraft are seen as the primary targets for reduction by switching mode use to rail and switching road vehicles to alternative power sources, e.g. battery power. Any solutions of this sort might well compromise level of service.

2.3.5 Ground traffic

The impact of airports as traffic generators on the local and regional road system was not great when traffic levels were only a few million passengers per year. As traffic levels grew and airports became prime employment sites, traffic levels increased to the point where they contributed significantly to the traffic volumes on the roads. Whereas aircraft engine emissions have been reduced to the level where they make only a minority contribution to air pollution around airports (Raper, 1996), particularly in relation to the general urban industrial environment, the road traffic bringing passengers, cargo, workers and visitors to the airport gives a much larger contribution to emissions. This is increased by the emissions from service vehicles on the airport. The groundside pollution can be reduced by both more efficient use of vehicles and conversion to less polluting power sources, particularly in the case of airport service vehicles.

2.3.6 Safety

It is beginning to be evident, as the noise of individual aircraft movements falls, that there is an underlying reason for complaints about aircraft movements which is harder to articulate, namely concern about the risk of an aircraft crashing onto populated areas outside the airfield boundary. This fear may be triggered by any evidence of the presence of an aircraft by sight or sound, but more so if the sound is very loud, intermittent or startling.

A few communities have long expressed these concerns explicitly, due to direct experience of such accidents. It is one of the reasons for the relocation of Munich airport, and the crash of a Viscount into the middle of Stockport in the 1960s has caused the close examination of safety issues at the recent public enquiry into a new second runway at Manchester. Some work had already been done to assess the risk associated with expansion of Amsterdam's Schiphol airport before the Boeing 747 freighter that crashed into a block of flats in the early 1990s brought the issue into stark focus. Media concern is exemplified by a large article which translated all recent UK air crashes into a Heathrow setting, pointing out where they would have impacted (*Mail on Sunday, 27 August, 1995*). On the other hand, a housing estate has been built on the site of a 1972 Trident crash near Heathrow: perhaps in the belief that 'lightning will not strike in the same place twice'!

The system which has evolved to control risk in aviation has had the prime objective of protecting the occupants of aircraft from being involved in accidents and in minimising their risk of death in an accident. It has developed by historic precedent and been subject to change through lessons learnt in accidents or by the need to accommodate new types of aircraft. There appears to have been an implicit presumption that the risks to third parties would then be sufficiently low to avoid controversy, though there have even been suggestions in the past for covering cities with nets to catch falling aircraft. There is, however, a considerable history of studies into the probabilities and consequences of aircraft impacting sensitive targets such as nuclear power stations. In the last decade, third party risk has been raised as an issue more frequently during public enquiries into the advisability of allowing increases in aircraft operations. Prior to this, few countries appear to have seen fit to draw up guidelines for limiting the increased risk of the public around airports relative to the background risk from aircraft en-route operations.

There has been increasing pressure for authorities to limit incompatible land use around airports, but this has been focused on the desire to allow expansion or increased activity at airports without physical or environmental constraint. ICAO's Annex 16 contains recommendations for compatible land use and many countries have developed their own guidelines. These are based entirely around noise nuisance criteria.

Only the UK and a few other countries have developed and applied a policy of Public Safety Zones (PSZ) off the ends of busy runways. The intention is that developments in land use should not result in significant congregations of people within the PSZ. Considerable judgement has been employed in determining just what constitutes a 'significant congregation', particularly with respect to sporadic activities such as race meetings and the passage of passenger trains.

The UK CAA recently revised the methodology for estimating third party risk at ground sites (Slater, 1993). This found an airport-related impact rate of 0.7 per million atm, based on accidents in the UK between 1981 and 1992 where there was substantial damage to or destruction of the aircraft. Distributions of accident sites in terms of distances along and normal to the extended centreline of the runway were fitted to the accident data, so allowing contours of risk of impact to be developed for each type of movement for each threshold. Among other results, the study found the following risks of impact per thousand years on a site of 0.1 sq km:

2,000 metres from touchdown at Heathrow	0.97
Rural site in southeast England	0.012
Rural site in Scottish Highlands	0.0051
Central London	0.0038

the variations being due to traffic density and type of traffic.

The above CAA work used a UK data base rather than a world-wide one. The latter increases the size of the data base, but risks of all sorts are higher in the developing world. Certainly the most notable accidents with third party fatalities since the Amsterdam 747 have been in developing countries. During 1996, a Moscow Airways Antonov An-32 crashed into a market in Kinshasa, Zaire, killing 220 people on the ground after trying to abort a takeoff (*Aerospace, March 1996, p.6*) and a DC-8 freighter killed 20 third parties when its pilot lost control while carrying out an unauthorised engine-out training takeoff in Asuncion, Paraguay (*Flight, 13 March, 1996, p.5*).

The CAA work has been extended by the UK Department of the Environment, Transport and the Regions to include assessment of the risk of ground fatalities and to develop criteria for assessing the acceptability of the risk (DETR, 1997), following studies of this sort particularly in relation to Amsterdam Schiphol (Smith, 1991; Hillestad et al, 1993; Piers et al, 1993) and Manchester, UK (Purdy, 1994). These studies have at least some elements of each of four sub-models which combine to assess the risk and its acceptability, except that Piers et al do not address the acceptability of risk. The sub-models address the following issues:

- the probability of an accident
- the likely distribution of accidents around an airport
- the consequential risk of fatalities
- the acceptability of the risk.

The adequacy of each of these submodels is reviewed elsewhere (Caves, 1996). The validity of each of the sub-models is, in fact, open to question.

2.4 ECONOMIC CONSIDERATIONS

As aviation matured, institutions and the public as represented by the consumer organisations began to take a more considered view of aviation. They questioned the efficiency of the operators relative to other businesses. They also questioned the mode's value to society relative to the external costs it imposed on society. It is therefore necessary to consider the efficiency of regulated aviation and the arguments for and against aviation's impact on society, these being primarily economic and environmental respectively.

2.4.1 Efficiency

The main problem with a fully regulated air transport system is that there is no easy test of its efficiency, since the regulator has no benchmark against which to compare the airlines' costs, particularly when they are being asked to behave differently from a profit-maximising operator. Any comparative analysis is subject to the difficulty that most of the players operate in near monopoly conditions. Indeed, it was only the availability of comparisons with the less regulated intrastate and commuter carriers in the US that provided the evidence of the inefficiency of the regulated carriers and the lack of compensating comparative benefits in service standards or safety. Similarly, the European carriers have been compared with the deregulated US trunk and European charter carriers in the run up to the liberalisation measures instituted by the European Commission, both comparisons being hard to justify, particularly with regard to the quality/price relationship. It should be noted that a majority of European passengers, namely those on charter airlines, have always enjoyed most of the benefits of de facto deregulation.

Under economic regulation, there was no incentive for airlines to be particularly efficient, even when in private hands, since the regulations limited the profit that airlines could make. Similarly, public ownership of airports used to be no guarantee of efficiency, though the residual cost method of operating practised at many airports in the US means that the airlines keep a watch on the costs, since they meet any deficit. There may be no intrinsic reason for public utilities being less efficient, but there is much anecdotal evidence that this is so. The public has become more aware of this possibility, and is now keen to control outgoings from the public purse. Capital and operating costs are now scrutinised by users, backed by IATA policies of only accepting fee structures which relate to justifiable costs. It is felt in some quarters that private entrepreneurs can find solutions to the problem of cost control which are beyond the capabilities of a government entity, while quality can be assured by competition. It is by no means certain that this latter conclusion can be drawn from the experience of deregulation of air transport in the US or that the UK experience of privatisation described in chapter 10 fully supports the former conclusion.

One way which competition is likely to improve the system's efficiency is in allowing the market to set the bond ratings for airport projects. This rests on the expectation that the market is likely to assess the worth of an investment more efficiently than formal project evaluation procedures applied by a bureaucracy. However, it must be recognised that this is no panacea. The desire to obtain a favourable bond rating can discourage entrepreneurial risk, which is one of the key factors that contribute to the supposed greater efficiency of private sector management. Also bond ratings depend entirely on cash flow projections, and thus necessarily ignore wider social benefits. Perhaps the most certain way to improve system efficiency, though possibly to the detriment of other goals, is the elimination of subsidy in all its forms. If the other goals are deemed to be important, it will normally be more efficient to apply direct subsidy to them, rather than using aviation as an instrument. If this is impractical, then at least the degree of subsidy needs to be made transparent.

2.4.2 Economic benefits of air transportation

With the trend to scrutinise the need for further investment in aviation, it becomes more important to recognise the economic benefits that flow from increases in aviation activity, so that a balanced appraisal can be made. Benefits accrue to users and to society as a whole. Theoretically, since any transport is a derived demand and most users would prefer not to travel in order to undertake a desired activity, transport is only used when the benefits of the activity outweigh the costs of travel. Air transport is used either when it is the only way to derive a net benefit or when it offers the greatest net benefit compared to other modes, this rationale holding for both passengers and freight.

Thus, the user benefits are made self-evident by the revealed demand. However, the benefits should not be overstated. They extend only to the difference in utility derived from using air transport rather than the next most convenient mode, and can be simply calculated (Butler and Kiernan, 1988). Equally, the benefits should not be understated. Sometimes the whole utility of the activity could not be experienced if air transport did not offer a sufficiently low deterrence to travel. A common example of this is a professional meeting that draws participants from an international audience. Garrison, Gillen and Williges (1997) discuss how the availability of affordable air travel has stimulated a broad spectrum of economic activity in the recreational, hotel and restaurant industries.

Over and above the benefits to the users of air transport, society as a whole may benefit from the activities which become available when the deterrence to travel is sufficiently low. Once again, however, the benefits should not be overstated. These benefits already counted as user benefit cannot justifiably be double counted on society's behalf.

The benefits to users and to society may be economic or social. Most attention has been devoted to the economic benefits, but it is by no means certain that these are more valuable than the social benefits (Caves, 1993). The economic benefits are often overstated by being in gross terms rather than net, are often double counted and are, in any case, much more easily quantified.

The most quantifiable benefits in the field of airport development are those derived from increased local employment, both direct and indirect, together with any stimulation of the local and regional economy. These benefits have traditionally justified local airport subsidy, the communities feeling that they should not fall behind in the provision of infrastructure and air service if they wish to ensure their share of investment. As the competition between regions and cities for trade and tourism increases (Ohmae, 1995), these arguments for investment in airports become more persuasive. The arguments can be used just as well by a private developer to answer accusations of negative environmental impacts as by local government owners. It is therefore important that the arguments should be properly understood. They are examined further in chapter 5.

2.4.3 Externalities

When energy was cheap and aviation activity was low, little attention was paid to the costs imposed on other users from congestion or on society in general from environmental impacts. Other modes had reached levels of activity which raised these debates a long time ago, indeed, complaints about the noise of chariot wheels were made in ancient Rome. Aviation activity increased substantially just at the time when it began to be realised that the environment, even at a global scale, had only a finite carrying capacity.

The direct social costs cause the majority of complaints against the growth in aviation. Social costs of aviation may accrue from considerations of safety, severance of communities, visual intrusion, disruption of the ecology, resource depletion, air and water quality, and noise, as well as the indirect effects brought on by development. All of these factors may be important in specific cases. It has been realised that the market does not provide an efficient valuation of the costs of these externalities, though much work is now being done to quantify noise and air pollution impacts. There is a growing recognition that there should be a "level playing field" in evaluating investments in different modes of transport, with all forms of transport internalising the cost of externalities.

The industry has recognised the need to strike a realistic balance between the costs and benefits of airport development. The Director of the International Air Transport Association's Infrastructure Action Group said: "In many cases it will be necessary for additional capacity and genuine environmental concerns to achieve an acceptable and cost effective trade-off responding to local requirements. Solutions can only be found if we respect environmental needs, in particular the genuine concerns caused by aircraft noise and emissions. Equally, airport authorities will need support in explaining the economic benefits that the air transport industry can bring to their areas." (Meredith, 1991) This view was endorsed by the Director General of the European Region of the Airport Council International's predecessor (AACI) at their Airport Capacity Conference (Yates, 1992).

The balance that must be struck between the benefits and the costs of expanding the aviation industry will strongly influence the nature and extent of the expansion that will be possible. The ACI is aware that "what is required is a measured assessment of all transport modes in order to determine sensible and achievable methods of further reducing the inevitable impact of transport on the environment" (AACI, 1992a). The view of the European Commission and some of the European Governments is that the balance is currently in favour of aviation and against the environment (Koppert, 1992), as described in the European case study.

2.5 INSTITUTIONAL CHANGES IN THE AVIATION SECTOR

2.5.1 Airline deregulation and ownership

The dismantling of regulation seems to be as inevitable as its rise, both trends being logical in their own era. The cost of retaining the protection of carriers' routes and capacity as described in section 2.1.1 began to seem too high relative to the benefits to commerce and those associated with social integration. Not all commentators agreed that large net benefits would flow from deregulation, though support actually came from some airlines, who could see that they would be able to adopt more efficient route structures. Fears were expressed over the ability to transfer the efficiency of intrastate carriers to the interstate market, the development of an oligopoly situation, the lack of security for the traditional revenue bond method of airport financing, the loss of service to small communities and hence of feed traffic to the trunk carriers (Taneja, 1976).

Despite these fears, and those of most of the carriers that their level of service would fall as they responded to competition, control was progressively relaxed in the latter half of the 1970s, starting with cargo services. In the US, the Civil Aeronautics Board (CAB), as the regulating authority, had to phase itself out, hand over its residual responsibilities, and declare its own 'sunset' in 1984. Meanwhile, the CAB was crusading for a similar philosophy to be applied in international air transport: signing a liberal bilateral air service agreement (ASA) with the Netherlands in March 1978, followed rapidly by 22 others (Dresner and Trethaway, 1992).

However, many drawbacks have been identified. Airline oligopolies have developed, giving serious barriers to entry. The adoption of the hub concept by US passenger and cargo carriers allowed them to retain feed and protect themselves from local competition. Even so, airline finances have become more volatile, strain has been thrown on system capacity at those hubbing points which have constraints on development, airport traffic levels have become less predictable, and the evidence of benefit to the users is sufficiently conflicting that the only conclusion which can be drawn is that there have been winners and losers. Some communities have gained service, others have lost it. Some passengers pay higher prices, others pay less. Some airlines have gone into bankruptcy, either permanently or into reorganisation under 'Chapter 11' of the US bankruptcy laws. Other airlines have prospered, some relatively, others in an absolute sense.

Thus, in a deregulated regime, the institutional role needs to shift from ensuring that devolved rights are not abused to ensuring that monopoly powers are not abused. Protection of the industry is presumed to be no longer necessary. Efficiency is presumed to be ensured by the positive encouragement of competition, relieving the regulators from the difficult task of judging the justification of prevailing industry cost levels. There is a consequent onus on the institutions to ensure that conditions are conducive to encourage competition.

Many other countries have since taken the same path with their domestic transport, including Australia and Chile. Internationally there has been less movement. However, the EU has been encouraging the breakdown of the economic regulation in international intra-EU air transport since their 1979 report, "Air Transport: A Community Approach", culminating in April 1997 with cabotage rights allowing EU airlines to operate virtually any sector in the EU. There is uncertainty, as there was in the US, over how the residual regulation is to be effected, i.e. questions of monopoly abuse, anti-competitive practices, market access, and non-EC relations. There is no longer a distinction between scheduled and charter airlines within Europe.

Liberalisation is occurring in various degrees in many parts of the world. In Brazil, the airline TAM has bought into a Paraguayan airline in order to be able to offer competitive service into the US, even though it means a transit stop in Asunción. It similarly bypasses the domestic regulation barring direct regional airline competition with the trunk carriers on major routes by serving São Paulo's downtown Congonhas airport via Campinas, as explained in chapter 13. In Nigeria, domestic competition has, as in Canada, been offered for some time by nominally charter carriers, and now two of them have formed foreign-owned subsidiaries in Sierra Leone and Liberia to compete for the international traffic for which Nigerian Airways holds all the bilateral rights (*Airline Business, May 1996, p 18*).

Although several countries other than the US have had privately owned 'second force' airlines for a long time, British Airways was one of the first major carriers to be privatised. The efficiency benefits of privatisation were shown by its almost unique resilience in the face of the recession in the early 1990s. Privatisation of airlines is spreading as quickly as official and unofficial liberalisation, with equity participation or less intrusive collaboration by foreign airlines, as governments offload their financial responsibilities. The Baltic states exemplify the trend. Air Latvia has 49% of its equity in foreign hands, 66% of Estonian Air is being sold, and Lithuanian Airlines is codesharing with Finnair (*Airline Business, July 1996, pp 48-51*). The Hungarian government is offering further shares in Malev, having already sold 30% to Alitalia and another 5% to an Italian investment agency (*Flight, 14 February 1996, p 29*). It is mainly the unattractiveness of the investment opportunity which limits the further spread of privatisation of the smaller airlines.

The introduction of the commercial ethic into airlines has, in fact, caused fundamental changes in their nature. "In the old days, the flag carrier ....was a real company which you could see. You could count its aircraft.....Today,...its principal assets are all invisible. They are its route licence network, its reservation system, and its brand name. Intellectual property is real, and...land and buildings....(are) now regarded as ephemeral and eminently disposable." (Webb, 1998). The route licences in the quotation can be extended to include runway slots. Webb goes on to point out that the brand may extend to code sharing and franchising partners so that the passenger is no longer clear which airline he or she is actually flying with. As he puts it, "airlines have gone from the Cheshire Cat to the smile on the face of the Cheshire Cat, and even that smile, you will recall, eventually disappeared."

It can be seen from the above that, in a privatised setting, the institutional role becomes one of seeking a balance between continuing to promote the national interest via the major airlines, including setting limits on the degree of foreign ownership, and the need to ensure the expression of sufficient competition for the consumer to share in any gain in efficiency.

2.5.2 Airport corporatisation and privatisation

The funding constraints and the distortions due to subsidy can be eased by privatisation. In some contexts, corporatisation can overcome funding limits, but it is becoming more common to consider full privatisation as a way of bringing genuine risk capital into airports. The UK case study illustrates the process's risks and rewards.

The trend to airport privatisation is already strong in the UK and is under way in many other countries. It is a trend which is proceeding in parallel with liberalisation, both in airlines and in airports: indeed, there are possibilities for privatising atc as well. It is seen as a way of improving internal efficiency and of making funding for expansion and renewal more available than when under government borrowing limits. When an airport's ownership and management is uncoupled from the local or national community by privatisation, its own preferences also become important, whether the change is simply corporatisation, commercialisation or a partial or full equity sale.

The privatisation trend was initiated in the UK in 1986 with the British Airports Authority (BAA), an entity serving the London area and Scotland, following a government White Paper (HMSO, 1985). Since then, Vienna (with nearly 49% owned by the public) and Copenhagen have both put some of their shares on the stock exchange, and several UK regional airports have completed the privatisation process by selling all or part of the shares which went to their original owners when they were corporatised in 1986. New Zealand corporatised its airports in 1988 (*Aviation Week, July 11, 1994, p 47*). Five different forms of airport privatisation have taken place: the sale of existing airports, long term leases of airports to private firms, contract operations, creation of new terminal facilities by build-operate-transfer consortia, and the creation of new airports as private business ventures (Poole, 1992). Prime examples of each of these are BAA, Atlantic City, Lockheed Air Terminal's contract at Burbank, Toronto Terminal 3 (and Birmingham Euro-Hub) and London City respectively. Recently there has been a spate of share sales in UK airports, as documented in chapter 10.

The trend to privatisation shows no signs of slackening. Russia, for example, intends its airports to pass to local authorities as joint stock companies, with eventual private participation through the sale of some of the government's remaining interest (*Flight, 19 June 1996, p 16*). It is selling 49% of Moscow's Domodedovo airport, with more to follow, as a way of unwinding the traditional joint airline/airport enterprises (*Jane's Airport Review, December 1996, pp 57,59*). Romania is corporatising 17 airports (*Airport Business Communiqué, July 1998, p 11*). China is also planning to allow up to 49% foreign ownership of its airports. Several Latin American countries have similar ambitions, following the sale of the Bolivian system to the US-based Airport Group International (Gill, 1998) and the complete Argentinian system to a multinational consortium. The Brazilian airport company INFRAERO is privatising all its non core aeronautical activities. Mexico is organising the privatisation of its airport system in four groups, in each case with a maximum of 15% of the shares being held by joint foreign/Mexican corporations to inject expertise through a 50 year operating concession, the remaining shares being sold through the market (*Airport World, June 1998, pp 16-17*). Australia has sold a 50 year lease on the first three of its large airports and is in the process of doing the same for another seven airports. The New Zealand government is selling its stakes in Auckland and Wellington airports (*Jane's Airport Review, July/August 1998, p 3*). There is a voluntary option for the privatisation of up to five US airports, following the BAA's involvement in managing Indianapolis and the international terminal at Pittsburgh (*Passenger Terminal World, Jan-March, 1997, pp 44-47*) and stakes in terminal developments by British Airways at New York Kennedy and Northwest Airlines at Detroit (*Air Transport World, March 1998, pp 59-63*). The German federal government now plans to sell its shares in several major airports (Enders, 1996), and some cities are also likely to sell their shares in those airports.

The general trend to supporting the ideology of privatisation, together with the belief that governments should concentrate on their core strengths of legislating rather than managing airlines and airports, is causing a similar change in the role of institutions towards airports. There is an even greater possibility of airports having and abusing monopoly positions than with airlines, so the institutional role has to change from planning and facilitating to regulating profits.

2.5.3 Sustainability

It has been suggested that the most important goal of freedom of choice relates to access to opportunities (Wilson, 1967), but there is a growing opinion that all modes of transport are consuming more resources than could be justified if the long term resource costs were truly reflected in the price. Thus, some take the view that if the main objectives of transport policy are prosperity, accessibility, cohesion, safety and fair competition, then it must reflect a proper balance between freedom and excessive consumption (Group Transport 2000 Plus, 1990). Planners are coming to the conclusion that ever greater mobility is no longer an acceptable social goal (Nijkamp and Reichman, 1987). Similarly, freight transport stands accused of generating unnecessary trips, because pricing does not reflect true costs (Hägerstrand, 1987). The rearrangement of total logistic chains around the Just-in-Time concept is criticised on the same grounds (Hillsman and Southworth, 1990).

It is not necessary to hold the ultimate introspective view that 'the final delusion is movement, change, and variety for their own sakes alone' (Merton, 1987), to see that there is a self-fuelling process at work in the availability of mobility which is of no more benefit to society than to the individual. The so-called 'paradox of universality' (Nijkamp and Reichman, 1987) refers to the phenomenon that those who managed to improve their well-being following a transportation improvement are bound to lose it sooner or later to others who follow their example. A new attempt to maintain their advantage by buying further mobility sustains the process. Ultimately, chaos would reign if everyone had a magic carpet (Hägerstrand, 1987). A good example of this situation would be the future success of the Bede supersonic single seat aircraft. In other words, the existence of an unsatisfied demand is not a sufficient condition to justify further transport investment.

Sustainability is the yardstick by which the appropriate level of mobility is coming to be judged. This concept was formulated in a report by the World Commission on Environment and Development (WCED, 1987), which stated that any future social or economic activity should be capable of meeting the needs of the present without compromising the ability of future generations to meet their own needs. In terms of mobility, this has become interpreted as working to reduce the harmful effects of providing transport whilst sustaining all necessary social and economic interactions. Since pollution and resource depletion are increased by the use of energy, policies have been directed to the promotion of low energy modes of travel, as well as to the increased efficiency of, and reduced demand for high energy modes (Gillingwater, 1996). In the political arena, aviation is often seen as benefiting the rich, while offering little to the poor, or as being an elitist mode which increases inequality (Graham, 1995). This ignores the representational nature of much business travel, i.e. the travel is not so much improving the utility of the traveller as it is facilitating economic production. It also ignores the fact that 70% of air travel is for leisure, much of it undertaken by the lower and middle social groups. As for promoting inequality, this seems to be a charge which could be laid against any mechanised mode of transport. Aviation is also seen as being resource-thirsty in comparison with other modes. On this score, it will have to demonstrate a willingness to improve as well as to maintain pressure for more realistic assessments of relative energy efficiency. It has certainly suffered from misrepresentation in the past, notably by the European Commission as noted in the European case study, but also in the advice given to emerging economies which so often ignores the resource costs associated with constructing fixed guideways (highways or track).

The most worrying development in the rising influence of the philosophy of sustainability is the intrusion of value judgements based in this philosophy into the role of those who should be taking an impartial view. The inspector at a recent inquiry into a proposed development of a business park, which would have hampered the development of East Midlands airport in the UK, included the following statement in his report to the Minister:

> " The growth in domestic and international air travel will depend in part on economic, political, technical and environmental factors that are probably undreamt of today. In particular, I have reservations as to the continued political and moral acceptability in the long term of what will still be a small minority of the earth's population consuming a disproportionate amount of the earth's resources on air transportation. 'This Common Inheritance' (1990) explained the importance of ensuring that development is sustainable and of remembering the moral duty of stewardship of the planet. In such a context, although I accept that there will be a continued growth in air travel, I do not accept that the rate of growth of the past years could or should be allowed to continue."

The Minister accepted the inspector's recommendation to allow the business park, and this was also confirmed on appeal.

Even without such judgemental inputs, society will require policies on full social cost recovery and limits on the most harmful activities. If the worst fears of ozone depletion and global warming are well founded, they could result in controls on altitude or number of flights. The former would put pressure on en-route capacity while the latter would lead to larger aircraft to the detriment of the network and the smaller operator. Both may result in punitive taxes on the burning of hydrocarbons. The lead has been taken by the Clean Air Act in the US in setting targets for reduction of pollutants and for taxes to penalise high emission generation, by Sweden with carbon taxes and by the State of California. Much is being done in turbine engine design to reduce emissions at source. So far, few countries have followed these examples in imposing taxes on domestic aviation, and they are being withheld from international services on the basis of keeping a level playing field. However, global interest bodies could force worldwide taxation of air transport. The EC is also considering a carbon tax among other policy measures in its search for 'Sustainable Mobility' (EC, 1992a and 1992b). The UK House of Commons (1995) believes that aviation should pay a greater proportion of the environmental costs, and is backing the EC's desire to tax aviation kerosene. The EC has agreed to develop a cost benefit approach to assessing such taxes.

2.5.4 Intermodalism

Historically, the planning for each transportation mode has proceeded in a fairly uncoordinated way, with separate agencies responsible for different parts of the transportation system. This has been particularly true in the case of airport planning and aviation generally, where very little attempt has been made to view air transportation as part of a larger system. Of course, airport planning has addressed the need to link the airport to the ground transportation system, although even this is usually primarily concerned with meeting the needs of private vehicles. As larger societal concerns over highway congestion and the environmental impacts of highway traffic have begun to shape transportation policies, many of the larger airports around the world have begun to give increased attention to improving service by public modes, particularly by providing rail links between the airport and the regional or national rail system. However, national or regional airport planning that views air service as one element of an integrated transportation system, to be developed in a coordinated way, has been the exception rather than the rule. This is hardly surprising when national civil aviation agencies are often independent of the department responsible for transportation. Even in the United States, where the Federal Aviation Administration is an agency of the Department of Transportation, aviation funds are allocated by separate Congressional committees and aviation tax revenues flow into a separate trust fund that cannot be used for surface transportation modes.

However, in spite of these constraints there is a growing recognition of the need to view the development of the aviation system in a broader context. In the last analysis, all transportation costs, for aviation and surface modes alike, must be borne by the economy, and since very few air trips begin or end at the airport, even air travellers have to use surface modes for some part of their trip. Funds invested in airport development are not available to improve the surface transportation system and vice versa. As air traffic continues to increase and it becomes more and more difficult to find sites for new airports near large metropolitan areas, or to expand those that presently exist, the question of whether some air trips would be better handled by surface modes must be faced. Even if new airport sites can be found, they are almost always much further from the metropolitan area than the existing airports, and in consequence the issue of how to provide effective surface transportation access becomes more pressing.

One would imagine that the major European countries, facing the need for massive investments in both increased airport capacity and the development of high-speed intercity rail systems, would have made these issues a central focus of transportation policy. Certainly, steps in this direction are being and have been taken. Many of the larger airports have existing or planned rail access links (Coogan, 1995), the Swissair Fly-Luggage programme allows baggage to be checked through to train stations throughout Switzerland (Jud, 1994), and the construction of a station for the Train à Grande Vitesse (TGV) at Paris Charles de Gaulle airport will facilitate rail access to that airport from a wide area of western Europe. However, efforts to improve the coordination between air and rail services are still in their infancy. For many years Lufthansa has run trains between cities in the Rhine Valley and Frankfurt Airport, which has a rail station for both intercity and suburban trains under the main terminal building. These trains appear in the airline schedules and are operated as an integral part of the route network, allowing through ticketing and baggage handling. Yet such services are the exception rather than the rule. Indeed, their existence appears to depend on a dominant national flag carrier that can establish what is effectively a monopoly relationship with the rail service. It remains to be seen whether the privatisation of the railways in the UK will stimulate the emergence of airlines operating feeder train services to their major hubs.

In the United States, passage of the 1991 Intermodal Surface Transportation Efficiency Act (ISTEA) has caused transportation agencies at all levels to give increased attention to the integration of different modes into a coherent transportation system, including the interface between the air and surface components of the transportation system. In many major metropolitan areas, increasing concern over the impacts of airport generated traffic on the surrounding street and highway system, as well as the emissions generated by those trips, is forcing airports to pay more attention to strategies to reduce or mitigate ground access traffic. However, while the responsibility for planning and funding airport facilities clearly rests with the airport authority, responsibility for developing and operating the airport ground access system typically is divided between a large number of public and private agencies, which may not view the airport as their most important concern. This increases the complexity of the planning task. Yet the importance of the efficient functioning of the ground access system to the successful development of large airports means that airport authorities must play a leading role in developing a coordinated approach to solving ground access problems, that will require them to address a much broader and more complex range of issues than they have typically had to deal with in the past (Gosling, 1997).

The emphasis on improving intermodal facilities resulting from the ISTEA legislation, and continued in the subsequent 1998 Transportation Equity Act for the 21st Century, has created new opportunities to fund improvements in airport ground access links (Coogan, 1994). In addition to greater flexibility in the way federal transportation funds are used, ISTEA also strengthened the cooperative planning process between states, metropolitan planning organisations, and airport authorities (Lacombe, 1994). However, airport access projects must still compete with other priorities, and developing projects that can generate the necessary institutional support remains a challenge. Elsewhere in the world, and particularly in Europe, the emphasis being given to developing or improving urban and intercity rail services has tended to focus attention on ways to better link airports with these systems, without giving much consideration to alternative strategies.

Perhaps the most progress in intermodal coordination has occurred in the air cargo industry with the emergence of the integrated package express carriers, such as Federal Express, DHL and United Parcel Service. These companies operate their own fleets of vans and trucks to handle the pick-up and distribution function, and even truck cargo over shorter distances when time and traffic loads permit. The movement of these companies into heavier freight and provision of total logistics services allows them to better balance their loads and increase the utilisation of their aircraft and trucks by moving lower priority cargo during the day or when space permits on the overnight express movements. This sector of the industry has a particular importance for airport system planning because unlike passenger airlines, which tend to move much of their cargo traffic on scheduled passenger flights and therefore the cargo needs to be handled at the major airports, the integrated package express carriers do not need to be located at major regional airports, and indeed may prefer to operate at smaller, less congested airports.

In summary, effective intermodal planning for intercity travel and freight movement will be driven by one of two forces. The first is a shift in government policies to prioritise transportation infrastructure investment across modes and to seek ways to make each mode internalise the costs of its social and environmental externalities. The second is market opportunities. The overnight express industry developed because Federal Express and its competitors discovered that shippers were willing to pay a high premium for the convenience of pick-up and guaranteed delivery. The first passenger airline that discovers that air travellers will pay a corresponding premium for the convenience of being picked up at an agreed time and delivered to their final destination, and can figure out how to do this for the mass market at a profit as Virgin Atlantic does for first class passengers, will transform the air transportation industry, and not incidentally the type of facilities required at airports. Likewise if airlines discover that they can increase their market share and profits by operating trains or integrating them into their network through agreements with other train operators, this will impact not only the facilities required at airports, but the composition of the air traffic that must be handled.

3

TOWARDS A STRATEGIC VIEW OF PLANNING

The planning contexts described in the previous chapter will apply regardless of whether the planning is operational, tactical or strategic. The thesis of this book is that the other forms of planning can be performed more effectively if they are themselves set in the context of strategic planning, which implies taking a broad multidimensional view of the future. In the spatial scale, this means considering the relationship between airports within some grouping which is normally referred to as a system of airports. This chapter considers the nature of these systems and the multidimensional futures in which they might have to function. It also considers the process of strategic planning and the challenges in producing plans for an efficient system of airports.

3.1 THE NATURE OF STRATEGIC PLANNING

3.1.1 Defining a strategic view of airport planning

Most of the advantages of taking a strategic view of planning airports flow from defining the role that each airport should play within a group of related airports. An understanding of the behaviour of the system which is to be developed or improved is therefore essential to its successful planning. Systems analysts usually probe deeply into the workings of a system in order to clarify its scope and function, and it is instructive to take this approach with airport systems. The questions that are addressed by systems analysts (Open University, 1983) include:

- Can the system and its components be described?
- What decides whether something is part of it?

- What interactions are there with wider systems:
  - financial?
  - input, output?
  - constraints?
- Are there functional sub systems?
- How are conditions controlled at each stage?
- What measures the efficiency of the process?
- What people, groups are involved, what are their powers and how do they interact?
- How have chance events affected it?
- What relevant external changes might there be?
- What are the main objectives, and can the achievement of these be measured?
- What are the main conflicts?
- How can changes be implemented?
- Whose values determine what is desirable?
- On what criterion can one decide what change is desirable?

The number of ways in which the scope of a system involving airports can be defined is almost infinitely large. The situation can be formalised, as illustrated in Figure 3.1, in the three non-orthogonal dimensions of spatial scale, transport system elements and sectors of the economy.

Systems which are very large and multidimensional are usually incapable of meaningful analysis because the nature of many of the interactions is inadequately understood, e.g. the effect of the provision of more flights on a regional economy. On the other hand, too narrow a system definition can omit aspects which are ultimately more important than those being studied, e.g. the effect of political intervention on demand, or the effect of changing aircraft technology on airport roles. The dichotomy is illustrated by the divergent views within the air transport industry on the value of taking a holistic approach to planning. On his retirement as IATA's Director General, Gunter Eser said: "We are components of an industry, a world travel and transport system and not a collection of individuals". When American Airlines' Chief Executive, Bob Crandall, took over, he said: "the mistake people make is to talk about an industry. You can't ask a sensible question by talking about an industry". This debate is, in fact, rather hollow and misses the point of systems analysis, which is not so much that the whole is more than the sum of the parts, but rather that the behaviour of the system cannot be derived from that of the parts at all: 'the system is an independent framework within which the parts are placed' (Angyal, 1941).

Even when it is possible to achieve agreement on goals among the users of airports, there is often the problem that airports in separate administrative systems affect one another. This creates challenges for the weaker airport, which can be turned into opportunities if the administrative and ownership barriers can be dismantled. Bratislava airport has considerable spare capacity which can be used to supplement Vienna airport, now that Slovakia's borders with Europe are open. Problems can be caused by mismatch between planning jurisdiction boundaries and the functional influence of airports, as instanced by the boundary of two FAA regions that divide the geographical area in Missouri and Illinois served by St Louis airport (Keller, 1998). In fact, the way an airport system is used bears little resemblance to the

Figure 3.1: Dimensions of airport system planning

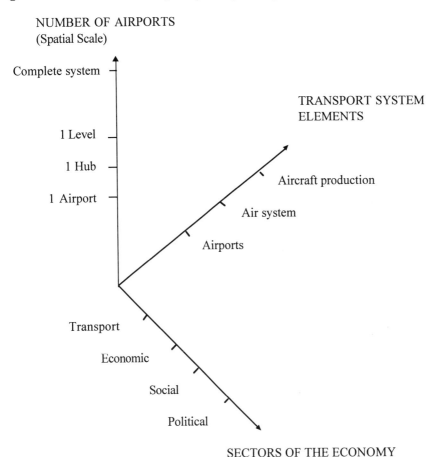

Source: compiled by the authors

boundaries of the authorities responsible for planning and operating the airports. The low cost passenger carriers and night freight operators are making use of secondary airports in many major cities, but there is no obvious shared responsibility for ensuring the provision of capacity for this growing system, or the management of the wider planning issues which these activities bring with them (Marchi, 1998).

Airports should be seen as integral parts of the total air transport system. The system consists of physical components, their owners and operators, the controlling authorities and the rules under which it operates. It is strongly influenced by the needs of its ultimate consumers, the social and economic characteristics of the national setting within which it operates, the impact it makes on the local and global environment and its acceptance by the communities it serves.

The ultimate consumers are the airline passengers, the employers who pay the business passengers' fares, the freight shippers and the business-related aspects of general aviation. The entities which speak for these interests are all stakeholders in the system, though their influence varies with the setting.

If the boundary of the system were to be drawn around only air transport and its users, which is the sort of system some would like to believe to be free-standing, the preferences of the users would be the primary determinant of the system's shape and performance. Planning such a system not only requires full knowledge of the future preferences of the users, which may well not be expressed in revealed demand, but it ignores the rightful, and probably powerful, voices of the other stakeholders in the wider system. Unless their interests can in some way be represented by shadow utilities, methods other than quantitative systems analysis need to be used in order to develop an airport system which will be the best compromise for society as a whole.

Systems are dynamic. Static equilibrium is usually difficult to achieve, because of unsynchronised changes in the variables influencing the system and also because of the tensions which exist between the various interested parties, or 'actors'. In the case of airports, examples of the changing variables are population distribution, technology, other modes of transport, passenger travel behaviour and political regimes. Some of the typical areas of tension are between consumer needs and the supply offered; national and local interests; consumers and non-users. One could argue that a primary reason for airport system planning is formally to recognise and resolve as many of these conflicts as possible prior to implementation, even more than to harmonise standards (e.g. of facility provision, safety and reliability) and to produce an economically efficient system by ensuring the use of best practice.

Systems normally achieve dynamic stability by closing the disturbance/response loops with feedback. Performance indicators are necessary to provide the comparison between goals and outcomes as well as to monitor the efficiency of an airport system, its quality of service and its positive and negative impacts on society. If the correct criteria are set as tolerable limits to these indicators, they should act as triggers for change.

The discipline of planning arose from a need to foresee and prevent future problems which might arise from an uncoordinated set of developments. This becomes more important as projects get bigger and take longer to come to fruition, if indeed they ever do so. Their justification requires the identification and quantification of benefits as well as costs, and also a clearer idea of the objectives. Thus the systemic approach becomes applicable. The aims and scope of system planning are stated succinctly in the proceedings of a recent symposium (TRB, 1992):

- Aviation system planning is a continuous and iterative process that requires coordination and cooperation among federal, state, regional and local aviation planning agencies.
- The planning process involves both top-down guidance and bottom-up identification of needs, options, and proposed developments.

- Aviation system planning should cover the needs of all sectors of civil aviation and reflect a balance of their individual interests and the national need.

- System plans should look beyond aviation demand and infrastructure needs and take into consideration economic and social objectives to be advanced by commercial and private air transport.

- At each level from local to national, aviation system plans should describe current conditions, present a vision of the future, state the goals to be met, enumerate the criteria of success, and lay out a path of evolution from where we are to where we want to be.

- Strategic planning can be used as a thinking tool, evaluating options via 'what if' scenarios. It should be useful in establishing and defending priorities, and should be a corollary of business and marketing plans.

The strategic planning process is well described in the FAA advice on state airport system planning (FAA, 1989), illustrated in Figures 3.2 and 3.3.

System planning should be concerned with the big questions, without ignoring the local operational consequences and constraints. A good example of the questions which should be asked was given in 1965 by the then Fleet Planning Manager for BEA (later the Chief Executive of BA) (Watts, 1965):

- One would question the need to operate a local network into London from nearby towns if this used up air space at the expense of the main domestic and international services.

- With present aircraft size the London-Paris frequency by all carriers will be 50 per day in 1970. Is this reasonable if it has the effect of precipitating more and more airports?

- In the long term look at air transport it is also important to take a realistic look at the integration of road and rail with air. Air feeder services are notoriously expensive.

- It seems reasonably clear that a large vertical takeoff and landing aircraft, given the right cost level, is a requirement for the future if a multiplicity of airports, each one farther away from London than the last one, is to be avoided.

If more attention had been given to these larger questions, and to the planning framework which would be necessary in order to address them, many of the mistakes identified in the London case study could have been avoided. A comprehensive system planning exercise would have addressed these issues.

All forecasts will inevitably be wrong. That itself is not a problem. Unjustified belief in them, probably because it is difficult to complete a traditional master planning exercise without a notionally firm forecast to hang the calculations on, is a problem. "This fact is the fatal flaw of

44 *Strategic Airport Planning*

Figure 3.2: The continuous planning process

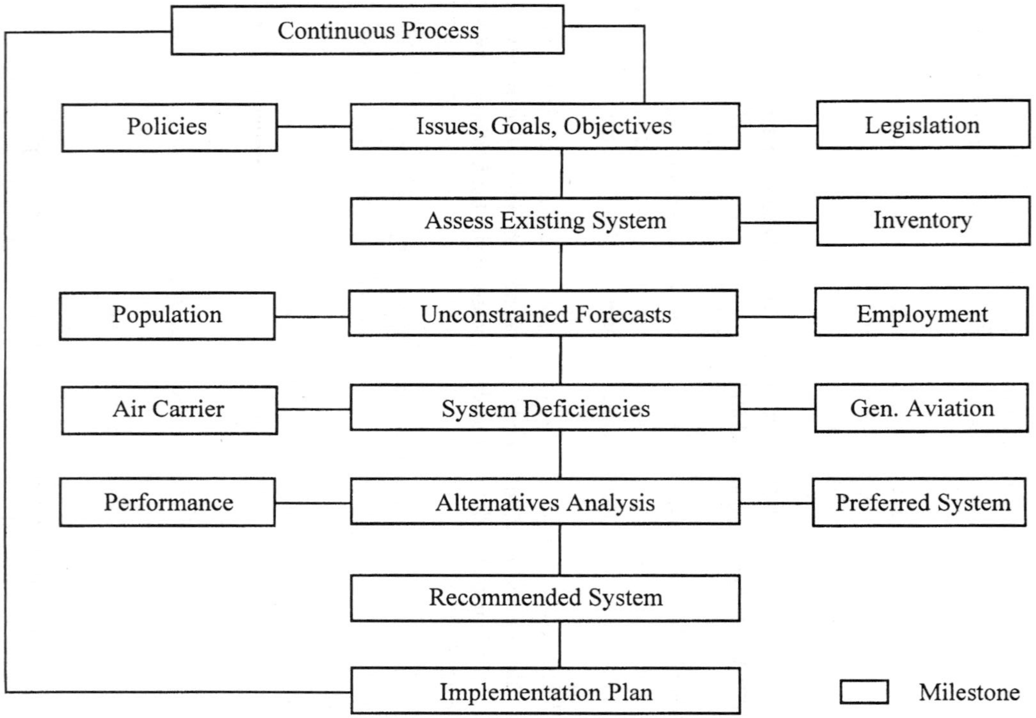

Source: FAA (1989)

Figure 3.3: Planning relationships

Source: FAA

the master planning process" (de Neufville, 1990). A better approach is to see strategic planning as an opportunity to ask 'what if' questions by developing a range of potential scenarios, each justified by a rigorously argued feasible path to it from the present day. The secret of a useful set of scenarios is that they should not be constrained by present conceptions but should include potential changes in technology, management, regulation, politics, personal tastes and attitudes, as well as the more obvious economic factors. Scenarios are considered further in chapter 7.

Airports may be influenced by interactions with other airports which are outside the control of the administrative organisation with responsibility for their investments and quality. Thus national system planners should recognise not only the competition between gateway airports within the nation but also the competition with other nations' gateway airports for through traffic. Similarly, regional or state planners must contend with internal competition between the primary and secondary airports and also with established national airports (Caves, 1993). Municipal airport planners have to recognise regional and national competition if they are to construct realistic future scenarios. Planners analysing metropolitan multi-airport systems have to confront the additional problem of competition within that area as well as the other interactions. The outcome of the local competition will itself affect the external interactions as the attractiveness of the total local system to airlines and their passengers changes. These critical relationships between airports are investigated in detail in the chapter 13 case studies.

The context within which airport strategic planning must function varies enormously among different countries, that may be classified along various dimensions, such as: importers/exporters; low/high income; regulated/liberal; autocracy/democracy (Gennari, 1989). The mix of characteristics will affect the type of transport provision, the perceived role of air transport and the way it is regulated, the source and extent of infrastructure finance, the system's ownership, the importance given to the roles of the various stakeholders and the type of planning framework. At one extreme, the framework may be based in entrepreneurial, bottom-up, demand-responsive processes adopted by individual stakeholders. At the other extreme, it may rely on centralised, top-down, goal-oriented and multisectoral processes. Many other mixes are possible between these extremes. Planning methods must reflect the context, as summarised in Table 3.1 (Hill, 1986).

Table 3.1: Styles of planning

Power distribution	Central		Fragmented	Dispersed
	Strong	**Weak**		
Style of planning	Command	Policies	Corporate	Participatory
Control	Budgetary, Statutory	Criteria, Incentives	Normative compliance	Voluntary compliance
Orientation	Plans	Policies	Processes	Processes
Role of planner	Specialist	Advisor	Negotiator	Organiser

Source: Hill, 1986

The style will affect the number of clearance points in the decision process, the conceptual distance between decision and implementation, the accountability, the principal beneficiaries, the degree of assumed consensus and the opportunity for participation, as well as the role of the planner. In the developed world, the planning styles evolved over time in pace with changes in society, from rigid master planning through the reliance on a systems analysis to an emphasis on an evolving future with continuous participation by the stakeholders (Teitz, 1974). The underlying objective of planning has been seen to be the appropriate allocation of limited resources among a number of competing users, relying on quantitative analyses and functional rationality (Friedmann, 1973). Airport system planning in the US still largely reflects the characteristics of strong, central power distribution shown in Table 3.1, although in practice the power distribution is somewhat more fragmented. The UK, as explained in the chapter 10 case study, more closely follows a dispersed power distribution, where the planner's role is to set up a globally acceptable framework within which entrepreneurs can practice their skills, and then to act as a facilitator.

3.1.2 The strategic planning process

Traditionally, airport strategic planning starts with setting goals. Then an inventory is taken of the quantity and quality of existing infrastructure and the way it is being used. This will include the use of performance indicators tailored to determine those aspects where the system is falling short of its goals. The future traffic generation and its distribution through the system must be predicted, so that any shortfall of capacity can be determined and the additional required facilities can be projected and the associated costs estimated. The distribution of traffic and to some extent its generation will depend on the supply of flights, which in turn will depend on airline strategy. Decisions on route starts, and on frequency and aircraft size will depend on the strength of demand and the probability of competition, so all of these factors also need to be predicted. Alternative solutions should be generated, designed and evaluated.

The chosen solution has to be implemented and the system has to be monitored against the objectives, the objectives themselves being subject to change over time.

Figure 3.2 illustrates the system planning process, including the feedback loops which are necessary to ensure the system's internal coherence and to keep it aligned with its changing external environment. There are difficult issues to be addressed at each stage of this planning process, some of the more difficult ones being identified in Table 3.2. Perhaps the most important area is policy formation. The traditional process is basically deterministic and unidirectional, even if it is iterative, so if the first step is not correct it is unlikely that the other stages can be completely successful.

Table 3.2: The strategic planning process

Elements	Issues
Goals	Who sets them?
Objectives	Are they feasible?
Criteria	Can they be implemented?
Inventory	Quality and utilisation?
Demand	Predicting independent variables
Alternative solutions	Omission of inconvenient alternatives
Evaluation	Avoidance of bias Measures of effectiveness
Implementation	Financial resources Responding to Change Strategic /operational conflicts National /local conflicts Political/cultural issues
Monitoring	Responsibility

These issues are considered further in chapters 7 and 8, as well as being exemplified in the case studies in chapters 9-13.

3.2 STAKEHOLDERS, GOALS AND OBJECTIVES

The physical components of the aviation system are the aircraft, the airports and the available airspace. The mix of aircraft is largely determined by the airlines or private operators, sometimes also influenced by government policies on trade. The airports are largely inherited assets which are upgraded on the basis of level-of-service policies or on demand, to the extent that resources are available and that the necessary permissions can be obtained. The available airspace is determined by government, priority often being given to military requirements. The capacity of the airspace is dependent on the level of technology and staffing, limited by the state of the art of atc.

The stakeholders of the system comprise:

- the owners of the system's components
- the operators of the system's components
- the suppliers of the system's components
- the users
- the regulators
- those affected by the system.

Ownership of the system's components is usually mixed. Airspace is a national asset, and atc ownership and operation normally reflects this, though there are several instances of corporatisation (e.g. in New Zealand) and competitive operation of local airport traffic control (e.g. in the UK). The large majority of flying in uncontrolled airspace and a significant fraction of that in controlled airspace, at least in the US, is by privately owned aircraft, and ownership of the airlines is increasingly in private hands. Airport ownership is also moving out of the public sector as local and national governments decide that there are other priorities for spending and that they can have a competent airport system without needing to own it.

Operation of the airports is often in local or private hands, even when they are publicly owned. Control of the airport operations is, in any case, achieved by a complex mix of national government, local government and airport owner policies and the regulations which support them. Typically, national governments influence operations by controlling route rights, by airport and airline licensing, by the designation of international gateway status and the staffing to support it, by granting of permission to expand, and by their influence over local government policies and spending. Local government controls the land use, reviews the costs and benefits which the operations provide for the local community, and often has a direct input to the management of an airport.

The airport owners will have policies aimed at the satisfaction of their own goals. Often the ownership mix of national government, state government, local government and private enterprise results in the policies of the various owners being in conflict with each other and with the expectations of the professional airport management.

It might be presumed that government goals would include national, regional and local economic development, social integration, protection of national airline interests, protection of the environment and the minimisation of governmental spending. Private owners would be much more interested in obtaining a good rate of return on investment, with consequences for the preferred mix of traffic and use of assets. They would also wish to be proactive in defining their airport's roles, rather than government owners who would tend to regard their role as passive service providers.

Government environmental policies are influenced by the global and local lobbies as well as by the internationally agreed policies on sustainability. Each airport community will strike its own balance between environmental, economic and social benefits and costs, which may well be in conflict with national policies. In many cases, the communities impose additional operating constraints to those recommended nationally and internationally.

Airlines all over the world are increasingly motivated by profit as well as the maximisation of market share, while being prime customers of airports which may or may not share these goals. The airlines wish to minimise infrastructure costs but, at the same time, they wish to minimise delays, to organise their airport presence so as to retain control of their operations, and to promote their own brand image. To this end, British Airways argued for a relatively lenient control on BAA's charges by the UK CAA, for fear that the airport group would not consider further investment in capacity at Heathrow if the profit potential were too low. It is not uncommon for 3% of an airline's passengers to buy 40% of their tickets and generate 65% of their revenue, so they wish to have the facilities to cater particularly well for at least this portion of their traffic. If the airline is hubbing at an airport, it will wish to have particularly close control over the aircraft turnrounds and the transfer of passengers and baggage. Indeed, wherever airlines have exclusive use of facilities, as is common in the US, the consequent desegregation of traffic by airline means that the design hour flows have to be derived more in relation to each of the individual airline's traffic than to the total traffic. This means that economies of density in terms of spatial provision are less available, except in the centralised processing areas.

Airlines investing in an airport's terminal facilities, as is common in the US and is beginning to occur in Europe, require a secure environment, a say in the planning of new facilities and in the method of cost recovery. In particular, they will not be keen to contribute to a development fund which will allow new capacity to be provided for competitors. Change of ownership of the airport will also cause them concern, so that the possibility of privatisation may influence their investment strategies. British Airways is selling its stake in the hubbing terminal which was specially built for it at Birmingham, now that many of the shares are passing from the local authorities to Aer Rianta, the Irish Airport Authority.

Airlines are becoming increasingly aware of the need to mitigate their impacts on the environment, both in their internal auditing of pollutants and 'cradle to grave' management of resources, and in their desire to contribute to being a good airport neighbour as the implications of environmental capacity capping begin to bite. They argue strongly that the main mitigation policy should be to minimise congestion. British Airways burns an extra 60,000 tonnes of fuel per year due to air traffic delays in the London terminal area, of which 50% is burnt in the delay itself and the other 50% is burnt in carrying the extra fuel in case of delay (British Airways, 1996), despite the existence of flow control which aims to keep aircraft on the ground until there is a free slot along the route. They burn an additional 6,600 tonnes due to inbound taxiing delays and a further 15,000 tonnes outbound.

Passenger desires and cost minimisation are important determining factors in an airline's network planning, and one would expect them to become dominant in the era of privatisation and competition. Legislation, fleet capability and management strategies for survival and growth also influence the decision. The major airlines are increasingly opting to concentrate their services and to use frequency and hubbing as prime competitive tools. In addition, they are stimulating traffic with low fares, using sophisticated yield management systems. One societal concern is that the present competitive system's emphasis on frequency to the detriment of aircraft size, and the consequent additional congestion, is wasteful of energy, though the passenger may assess the considerable benefits of frequency as being worth more than the costs of congestion.

General Aviation (GA) and other tenants are also customers of an airport. GA tends not to generate much revenue for an airport. Other tenants' contributions will depend on the nature of the lease agreements and on the view taken by management on the synergy that the tenant's presence brings to the airport's business. This leads to a tendency to price GA and marginal tenants out of busy airports, often after having encouraged them in the earlier stages of an airport's development.

The ultimate goal of consumers is to minimise their overall disutility of travel. This implies minimising time, cost and discomfort, and maximising safety, security, punctuality, choice and convenience. In general, this requires the availability of competitive airlines and airports close to the desired trip origin, with frequent direct services to their preferred destination. Since the optimum supply of transport will often not be available, the users are forced to make tradeoff decisions. Individual users then attempt to minimise their disutility, based on the information they have on the options available.

Policy formulation requires goals, objectives and criteria for the whole system to be established. This is often a very difficult task because of the disparities in viewpoint of the various stakeholders, particularly when they do not appreciate the need for help to resolve their differences. A common source of disagreement is the tension between national and local government policies, so that national objectives can be thwarted by, for example, local opposition to noise. Local politicians will adopt or ignore national policy as it suits their ambitions, usually wishing to retain their individuality and regarding other local authorities as competitors for scarce national resources. Another difficulty is the coordination between strategic, tactical and operational planners, in terms of understanding what is possible at the 'coal-face'. There will be occasions when the implicit goals of the stakeholders are more important than the explicit ones aired for public consumption. It will often be necessary to forsake uniformity for pragmatism, for example in the definition of adequate level of service in the face of ability to pay the consequent ticket price.

It is not the recommended role of a systems analyst or planner to generate the goals and objectives, but this may be necessary if they cannot be elicited from the policy makers. It is often difficult to get a clear statement of goals, but one which is frequently implied is the satisfaction of consumer desires within the constraints of available resources and respect for the goal of sustainability. It is not an easy task to find out what the consumers' real desires are, or to differentiate between desires and essential need, or to convert these into the inevitably compromised policy goal of how much desire to satisfy, given budgetary constraints. Revealed demand studies ignore the potential user, and tend to imply that the users are satisfied with the present system, which has evolved usually through a demand-responsive approach to capacity planning. Stated preference methods, on the other hand, presume that real desires can be elicited in a hypothetical situation and that the present population can speak for future passengers. Yet if the desires and needs are not understood, it is unlikely that the system will have the appropriate distribution of capacity, and that it will be used in the way the planners have envisaged. A simple analogy is the worn grass where pedestrians have chosen a more direct path than the paved ones laid down by some planner with more interest in preserving symmetry.

An objective which might flow from the above goal is adequate accessibility, particularly when the system's ownership and operation are within the public sector. The corresponding criterion might then be a certain level of equal opportunity to access scheduled air transport. These objectives and criteria are frequently used in urban transport planning. The most common problems with objectives and criteria are that they tend to be optimistic relative to the ability to afford them and that the indicators are not sufficiently discriminating to identify weaknesses.

Transport planners in other modes and scales have worked not only with access to a network (as in the UK with motorways and in the USA with airports), but with the much more relevant concept from a behavioural standpoint of access to necessary destinations. This is reflected in the historic growth of a tree-shaped road network based on the need to administer regions and districts from the centre of power in a country or the development of trade routes or drovers' trails. Even more relevant to aviation studies is the concept of a space-time geography within which the individual traveller is constrained by a time budget. One indicator of this is the ability to make day-return trips (Hägerstrand, 1987), as shown in the Norwegian case study. The Governor of Wyoming state in the US takes this access question so seriously that he is 'open to the idea of a state-run air service' to combat the loss of small city air service (*Air Transport World, May 1998, p 84*).

Another public goal which might be adopted would be to minimise the disparity in wealth between regions by minimising the disparity in accessibility between regions, as instanced in the rationale behind the EC's Transport Networks (TEN), thus attempting to reverse the spiral of spatial disadvantage which tends to build up when transport investment merely follows revealed demand. Thus more investment would be channelled to regional airports according to a criterion which might be stated in terms of allowing equal opportunity to access specific destinations.

Unfortunately, there are many examples of policies which achieve aims opposite to those intended (Dunsire, 1980). A prime example in airport planning is the development of airports remote from city centres in order to increase the scope for traffic growth by relieving noise nuisance, only for the traffic to bypass the city altogether as in the case of Montreal, or for the environmental capping to be almost as severe due to the greater sensitivity of newly affected people or the encroachment of incompatible land uses. These and other effects are exemplified further in the case studies.

3.3 MEASURING SYSTEM PERFORMANCE

Performance indicators are necessary to monitor the economic efficiency of an airport system, its quality of service and its positive and negative impacts on society. If the correct criteria are set as tolerable limits to these indicators, they should act as triggers for change.

According to Gillen (1997), a properly defined and implemented Performance Measuring System (PMS) should provide feedback on process performance, should facilitate learning and should drive continuous improvement. It should indicate efficiency, effectiveness and changeability, measuring respectively how economically the resources are being used, the extent to which customers' needs are being met and how prepared and flexible the airport is with respect to future change. Rosenhead (1989) takes the view that the monitoring of performance means answering the following questions:

- is it the right thing to do? (effectiveness)
- does it work? (efficacy)
- are the minimum resources being used? (efficiency)

The relevant indicators will cover a multiplicity of result-based measures of the success of an initiative and determinant-based measures of competitiveness. The UK Treasury (*Economic Progress Report, No 188, UK Treasury, Jan-Feb 1987, pp 6-7*) uses the following definitions:

- Efficiency may be defined as output divided by resources consumed.
- Effectiveness is the extent to which the criteria are met.
- Economy is the comparison of actual input costs with planned or expected costs.

In strategic airport planning, it has been common for the trigger for new capacity to be a prediction that the level of service is about to fall below some criterion of acceptability. The level of service is normally stated in terms of acceptable delay of aircraft and passengers in some arbitrarily defined 'peak hour' (Ashford and Wright, 1992). The additional capacity required to continue to avoid exceeding the acceptable delay has then been calculated and planned to come on stream in discrete steps, the size of the steps striking a compromise between the benefits from economies of scale and the difficulty of getting an early return on the investment. A commonly used design criterion for aircraft delay has been an average of four minutes, based on economic analysis of the costs of expansion relative to the costs incurred by delay, but this has risen to 10 minutes in the peak at Heathrow as the airlines are prepared to compromise standards in order to allow more runway slots to be made available. It is necessary for there to be consultation between an airport and its operators on the preferred compromise between level of service and declared capacity. Internationally recommended standards exist for the definition of the congestion levels of service (ACI/IATA, 1996). A prime difficulty is in the consistent definition of delay and the distinction between that caused by lack of capacity with that caused by operational factors. The use of flow control clouds the issue further, as does adjusting schedules to obtain better punctuality ratings at the expense of lower utilisation of resources, particularly since the airlines' profits are affected not only by the increased costs but also by the loss of revenue resulting from lower aircraft utilisation. The European Civil Aviation Conference (ECAC) has issued a useful guide to consistent delay analysis (ECAC APATSI, 1996), but it does not address all these concerns.

In a commercial enterprise, economic efficiency is measured by normal financial indicators such as return on investment, operating ratio and benchmarks of acceptable levels of unit costs and revenues for the various streams of business. The performance of the separate subsidiaries of the company would be assessed individually, and internal profit centres would often be set up within each subsidiary. Quality of service, or effectiveness, is monitored in terms of speed of response to orders, percentage of order dates missed, numbers of complaints, number of accidents, days lost by illness, etc., all contributing to internal quality audits. An equivalent PMS can be defined for airport terminals, so that time and space standards are set, according to the desired level of service, and then monitored for compliance. Concessions revenue or lease income per passenger and per metre of selling space should also be monitored (Doganis, 1992). The negative impacts on society would also be monitored by auditing environmental fallout, and the positive economic impacts by, for example, jobs created. Much of the information will, in any case, be required in order to satisfy safety, environmental or trading regulations.

In contrast, a government entity is often only subjected to such monitoring as the government itself or its consumer 'watchdog' organisations see fit to apply. Airport systems which are centrally owned and operated will normally have standards of safety set by another arm of government, though sometimes within the same agency. Sometimes there is quality and efficiency oversight by government committees. Alternatively, statutory bodies may be charged with auditing government departments, exemplified by the General Accounting Office (GAO) in the US, or monopoly situations in general. If performance is to be maintained in centrally planned, owned and operated systems at the levels to be found in commercial enterprises in the more developed nations, the organisation of the mechanisms for monitoring those systems needs to be carefully constructed. It will often not be easy to institute an equivalent set of indicators to those used by a commercial enterprise, partly because the necessary information cannot be separated from the other functions undertaken by the same government departments and employees responsible for airports.

Most airport systems fall somewhere between these two extremes. This makes it difficult to develop a consistent set of indicators for benchmarking across different systems of airports, particularly when there is a mix of centralised air traffic management, partly corporatised airports, fully corporatised airlines and fully private aviation users. This is discussed further in chapter 7.

54 *Strategic Airport Planning*

The opportunity exists in a centralised airport system not only to measure the systemwide societal implications of options for the distribution of capacity, and to balance these against the implications for the users, but also to influence that distribution. It is, however, difficult to capture the differing sensitivities of the stakeholders throughout the system with any convenient indicators.

3.4 RISK AND UNCERTAINTY

Even in the era of fully regulated services there were considerable perturbations in the traffic at smaller airports, even though the linkages were predominantly to larger cities. The US experience shows that deregulation produced an even greater turbulence and instability in airport market shares (de Neufville and Barber, 1991), inevitably resulting in greater risk for any given investment. This is also the case in the UK, and is particularly so among smaller airports. Table 3.3 analyses the recent history of the traffic by class of airport in the UK, showing both the greater growth rate of those airports below one million passengers per annum (mppa) in 1984 and the greater variability of the smallest classes of airports. The traffic through this smallest class of airports in the UK is dominated by charters on the one hand and by scheduled domestic routes on the other hand: there is relatively little use of these airports for scheduled international services. The history of individual lower density routes shows an even greater variability over time, often making their use for revealed demand studies rather dubious, even where the data are readily available. The faster growth of the middle classes of airport has been due largely to the strong trend towards regionalising the holiday charter services and the consolidation of scheduled services at regional hubs rather than local airports. These data are further elaborated in the UK case study.

Table 3.3: UK airport annual passengers and growth by size class

Airport class[1]	1984[2]	1992[2]	1995[2]	1995/1984	1995/1992
1.0m-5.0m	1,650	2,549	3,082	1.87	1.21
0.5m-1.0m[3]	528	2,332	3,890	7.37	1.67
0.2m-0.5m	328	507	736	2.25	1.45
0.1m-0.2m	147	220	349	2.37	1.59
5,000-0.1m	40	19	26	0.64	1.38
<5,000	4	5	3	0.82	0.60

Notes: 1. Traffic classes based on 1984 passengers
        2. Passengers in thousands
        3. This class consists only of Stansted

Source: CAA Annual Airport Statistics, modified by the authors

UK government forecasts (DTp, 1997) correctly reflect considerable uncertainty in likely UK aggregate traffic levels in 2015, but do not reflect any of the indicated additional volatility at individual smaller airports. Also, they do not attempt to account for individual airline behaviour, though presuming certain route start behaviour associated with predicted route demand.

In view of this volatility, the individual airport entrepreneur certainly cannot presume some consensus level of traffic growth in return for investing in expansion. The utilisation of new capacity will depend crucially on the decisions of the airlines and their users, i.e. the very factors deliberately avoided in the DTp forecasting. These same factors cannot be ignored even by the airports in the London area, either in terms of the distribution of traffic within the London system or in terms of competition between London and other major continental hubs.

The uncertainty of traffic levels and the factors controlling them requires a planning approach which accepts it as a fact rather than pretending it were not so. Not only must this be recognised in forecasting, as discussed in chapter 7, but the whole planning process should reflect this reality. This requires a continuous planning process and also a willingness to debate the value of potential normative futures.

3.5 COMPETITION

Increased competition is accepted in most countries as a desirable goal, but it is not so easy to achieve. The adjustments made by a competitive industry are among the most important factors generating turbulence in the distribution of activity across the airport system. The competitive behaviour of airlines, airports and other modes therefore needs to be understood if planning is to be able to deal with them.

3.5.1 The airline industry

The US experience of the increased competition allowed by deregulation in a mature domestic marketplace has been assessed by many analysts (Bailey, 1989; Borenstein, 1989; Dempsey, 1990; de Neufville and Barber, 1991; GAO, 1991; Levine, 1990; Morrison and Winston, 1989; TRB, 1991; Windle, 1991). Despite considerable disagreement between commentators, some conclusions can be drawn. The major airlines have adopted a mixture of offensive and defensive behaviour. The most noticeable defensive trends in a liberalised airline industry has been the formation of a hub and spoke network based on one or more 'fortress hubs'. The fortress allows the carrier a local monopoly and the opportunity to raise local fares in order to compensate for lower yield per mile for transfers across the hub. This has been supported by strategic alliances which have resulted in an oligopoly. It is difficult to enforce antitrust laws in these situations. On the other hand, the majors have competed hard for indirect traffic over their hubs and have also, on occasions, created new hubs to serve markets dominated by other major carriers. There remains some competitive movement in the system as some hubs become too expensive to expand despite their marketing advantages, and as low cost point-to-point carriers increase their traffic share. The hubbing strategy has not ensured large profits for the majors, despite consumer claims that concentrated hubbing dilutes the contestability of local markets. The more promising results of liberalisation of fares and route entry have come from the rise of the new low cost carriers, who have tended to create new markets as much as they have diverted passengers from the majors. The leading low cost carrier in the US, Southwest Airlines, has also had a better profit record than the other US major carriers.

It is still to be seen how, in the long term, the market is divided between the hub carriers and the low-cost point-to-point carriers. This perhaps depends on how well the major carriers succeed in strengthening their hubs with international long haul traffic and on how much the regulators intervene in local monopoly situations. The residual controls still available to limit market excesses after deregulation have only been applied very lightly, particularly in the matter of mergers which have allowed even greater concentration at hubs. The European experience, detailed in chapter 11, shows that major airlines have tended to follow the US response to deregulation, and that the regulators have been similarly lenient.

The monopolistic nature of international hubs has always been apparent, though some airlines have never maximised their online transfer opportunities. These hubs will continue to be monopolies in countries which do not liberalise their international agreements, as well as in many liberalised situations, because access to the hub will tend to depend on 'grandfather' privileges. This can be threatened in two ways. One is the provision of new capacity at the hub, which is one of the reasons why United and Continental were not initially in favour of the new airport at Denver. It is a delicate balancing act for a hub airline to have enough slots for its own expansion without encouraging competition. The other way that hub dominance can be threatened is with competing hubs in the same geographical area. Amsterdam's Schiphol airport, with its relatively low local demand base, is very aware that KLM's over-the-hub traffic could be siphoned off by Heathrow, Frankfurt, Charles de Gaulle and others if both KLM and the airport do not ensure that every aspect of service to the through traffic is of the right quality.

Passengers have been affected by hubbing in a number of ways. It is difficult to say if hubbing has reduced average costs, but fares on local services at hubs are high relative to indirect services over hubs. Some proportion of these additional fares are justified by the additional fixed costs incurred at large hubs (Reiss and Spiller, 1989), and by the welfare benefits of greater flight frequency, easier connections and non stop flights (Borenstein, 1989). Also, many more points have direct service from quite distant hubs, due to the importance of small traffic points in adding to a hub's synergy. Passengers at those small traffic points chosen as spokes have obtained benefits of greater system access, often across several competing hubs. Other points in the US have lost all service due to the low priced competition at well-connected airports unless protected by the Essential Air Service subsidy. In general, service has become more variable geographically, operationally and in terms of price. The increase in the number of losers as well as winners is a reflection of the market at work. Passengers' perceptions of hubbing have been investigated using choice models to obtain trade-offs between direct and transfer flights in US domestic markets and also to examine passenger choice between gateway airports for international markets, as described in chapter 7.

3.5.2 Airports

The opportunity for competition between airlines raises many possible consequences for airports, depending on the extent to which the airlines take advantage of the opportunity. Except when they are constrained by government intervention, airlines plan their network development very differently from the way a government agency, or private initiatives, would plan a system of airports. They are likely to take advantage of the more transferable nature of their assets to retain flexibility in scale and route structure.

Some of the increased competition between airlines has been transferred across to airports through association of based carriers with their airports, but also through the search for airport capacity by those carriers intent on expansion. An opposing trend has been for the rationalisation of route structures to result in loss of service at smaller airports, causing them to realise that they must offer a competitive advantage if they wish to regain the service. Competition has also come to airports in their own right, as restrictions on route rights are relaxed and opportunities open up for new gateways. As the airport industry becomes more competitive, it no longer is advisable for an airport just to develop reactively in response to any arbitrary inherited role. To do so will almost certainly mean having to accept those roles which other, more proactive, airports have discarded. This will tend to result in traffic with the least desirable characteristics: peaky, low yield, noisy, financially fragile and volatile. Thus, even those cities prepared to continue with some form of subsidy in return for broader social and economic benefits may lose patience as the subsidy rises while the traffic fluctuates and the positive impacts fail to materialise.

Airports will therefore increasingly wish to influence the role they want to play in the national and international airport system, bearing in mind that the role needs to be sustainable if it is not to result in wasteful investment. Schiphol sees itself as able to create and to sustain it's 'mainport' role in Europe despite the environmental cap at approximately 40 million passengers per annum (mppa) in 2015 by progressively developing a multimode hub to transfer short haul traffic from air to rail (Amsterdam Airport, 1995). Aeroporti di Roma also intends to change its marketing (aided by Alitalia selling its stake in 1995), not only to be more client orientated but by launching the Leonardo da Vinci airport as 'the hub of the Mediterranean', with terminal expansions which will give very high levels of service at 38 m^2 per peak hour passenger (*The hub of the Mediterranean, Airport Business, April/May 1996, pp 19-23*). Manchester decided in 1990 to change the role as defined in the Development Strategy document of 1985, to become a hub even without a strong based airline. This required major revisions of its master and business plans, calling for a second runway at a throughput of 15 mppa rather than being able to expect that its single runway would support over 20 mppa. Denver's new airport, on the other hand, assumed its role to be a straight replacement of Stapleton's role, but with no operational or capacity constraints. In the short term, its success has been somewhat muted due to the withdrawal of Continental Airlines and the imposition of considerably higher user fees, though delays appear to have reduced very substantially and there has been no noticeable reduction in traffic (*Open for business, Flight, 26 April 1995, pp 34-35*).

Even the major airports often find themselves unable to dictate their role within the overall air transport market. The smaller airports are bound to find it much harder to create the conditions which will allow them to develop their preferred role, even if they have the expertise to use the most effective marketing strategies. It will be crucial to their success that they should be able to identify sustainable roles. They will need to understand the natural roles of other airports in their system and be able to judge when it would be better to complement them rather than to compete. It may for instance, be better for an airport and for its community to act as a spoke to a few major hubs, or to remain a predominantly charter or freight airport, than to attempt to develop a range of thin direct scheduled services with fragile airlines. Within the airport's regulatory freedoms, it's role will depend on the strength of local demand, how it prefers to express itself and on how attractive the airport makes itself to airlines.

3.5.3 Other modes

The Head of the European Commission's Airport Policy Unit in DG VII wants to see the environmental capacity problem alleviated by a shift from air to rail (AACI, 1992). High speed rail will certainly reduce air traffic on dense short haul routes, if the TGV experience (Bonnafous, 1991) and US analysis (Gomez-Ibanez and Pickrell, 1987) are good general indicators, assuming approximately equal factor input prices and return on investment. Airbus Industrie suggests that a one hour advantage increases market share by 20% (Reed, 1990). The Japanese experience, with quite similar fares for air and rail, is that air takes the greater share of traffic than rail at about 750 km (Feldman, 1995). However, the increasing size of the fast rail network in Europe is not necessarily a more severe competitor for air transport than the car in the US, except that it competes over longer distances. Certainly there is a switch of high yield traffic from air when door-to-door time by rail becomes less than three hours, but interlining and natural growth in traffic can still leave a significant air market. Orly to Nice, Orly to Marseille and Orly to Brussels remain three of the densest routes in Europe (de Wit, 1995) despite the high speed rail opportunities, and BA's shuttle from Heathrow to Manchester has more than a million annual passengers despite trains averaging 160 km per hour over only 300 km, so that British Midland has at last decided to compete on the route. The resilience of air is due partly to the diffused nature of the demand within metropolitan areas (at least 50% is home-based rather than office-based) and partly because for many trip ends access to downtown rail stations is often more difficult than to airports.

Lufthansa has considered dropping all air routes with less than two hours ground travel time (Chuter, 1995), though the rail operators are being slow to provide the necessary level of service for transfer traffic. However, many important air city pairs have too little traffic, or are over terrain which is too difficult to ever justify direct high speed train services at the necessary fares. Less than 10% of the European scheduled airline capacity is threatened. If it all switched, that would be equivalent to only two years' traffic growth (Veldhuis et al, 1995). On the routes where rail does compete well, it can serve to reduce the strain on air capacity.

There are many instances of improved rail access to airports, including high speed intercity services at Amsterdam, Lyons, Charles de Gaulle and the mature examples of Frankfurt, Gatwick, Birmingham and the Swissair/Swiss Rail integrated operations. In the future, improved interlining between air and rail may be enhanced with through ticketing, as with Prestwick's £5 rail feeder offer from anywhere in Scotland (Whitaker, 1995). This might become easier to achieve after privatisation of rail companies, but many questions will be raised over barriers to entry and airlines may continue to compete for interline traffic with any such services. This might even create more pressure for slots at airports as the airlines remaining on the routes would have to compete with rail on frequency, though in general it is more likely that advantage will be taken of the opportunity to release slots for more lucrative services.

Telecommunications may be able to replace some physical travel, but it is generally accepted that, over short haul distances, any substitution effects are compensated by stimulation of travel (Salomon, 1986), though there is some evidence that teleconferences may obtain as much as 11% substitution and only 1% stimulation (Khan, 1987). The greatest impact in the US is expected to be in intra-company connections, where some 30% of air trips may be substituted by 2010 (*Aviation Week, 29 November, 1993, p 40*). Any fear that the wider use of advanced telecommunications will prove to be a more effective way of equalising the spatial distribution of settlements than aviation appears to be groundless (Goddard and Gillespie, 1986; Nijkamp and Salomon, 1989; Robins and Hepworth, 1988).

The potential substitution from telecommunications is more likely to affect long haul air travel. Leaving aside any thoughts that virtual reality may compete with the actual experience of tourism and be more environmentally friendly, the differential advantages of teleconferencing must be greater at long distances, particularly with falling prices as the large capacity of fibre optic cables becomes available. Already, with only conventional telephones, the traffic between the US and Europe and between the US and the Far East grew by 23% and 27% respectively on an annual basis between 1970 and 1990 (Owen, 1991). This is one area where technology may improve more quickly than in aviation, while the lower energy requirement of telecommunications will also exacerbate price differentials as well as reducing high altitude pollution. Telecommunication cost and other disutilities are almost constant with distance, which cannot be said for long haul air travel (Salomon et al, 1991). The concern for aviation would be that a relatively small impact on the high yield business market might exert a pressure for increased fares in the highly elastic leisure market.

3.5.4 Access to resources

Another form of competition that must be considered in a more strategic view of airport system planning is the broader competition for resources within the national economy. Funds invested in airport development are not available to support other societal needs, whether schools, hospitals, national defence, and so forth. At the level of public expenditures, this becomes an issue of national budgets and taxation policy. With the increasing trend toward the use of private funds for airport development, as discussed in chapter 2, airports will also find themselves in the financial marketplace competing for investment funds with the full spectrum of the private sector.

Thus airport planners need to consider how well justified their proposed projects are, not from the narrow perspective of their eligibility for a dedicated source of funds, but in the light of the comparative worth of other uses of those funds. In the case of the private financial market, these issues translate into the dual criteria of the return on investment and the associated risk. While an airport may be able to take advantage of a monopoly position to exact excessive user charges to provide an acceptable return on an otherwise poor investment, the risk to the investor is that a competitor may emerge that weakens the monopoly position and forces the airport to choose between either seeing its traffic base erode or reducing its charges, either of which will reduce the return on the investment. This risk will in turn drive up the cost of borrowing.

In the case of public funds, political pressures will ultimately determine how governments allocate funds across programmes. If airports are widely perceived to be contributing valuable benefits to society and using resources frugally, it will be much easier to justify increased levels of investment. As those who have to be convinced become more sophisticated in their appreciation of these issues, it will become increasingly difficult to make this case on the basis of simplistic claims of the inherent worth of aviation and the flawed logic of most economic impact studies. This is not to say that airports do not contribute significantly to the economy, but that contribution is more subtle than simply adding up all the payroll and spending by airport related organisations (which anyway is clearly a cost to society not a benefit).

Another critical area where the aviation system competes for resources is in land near metropolitan areas for new airport sites. Airports need large areas of land, preferably well served by surface transportation links, as close as possible to the developed urban area and yet not already surrounded by residential or other noise sensitive uses. These land areas are of course highly attractive for other potential uses, and in many cases may already be used for purposes that society deems both important and scarce enough to give special protection, such as public parks, wildlife refuges or prime agricultural land. No matter how much money is available to purchase the land, it may not matter if those with control over the land refuse to sell or allow it to be used for airport development. Therefore it is important that the trade-offs involved in developing alternative sites, or indeed taking no action, are fully and fairly assessed. In some cases, extraordinarily expensive compensation and mitigation measures may be justified by the even larger benefits to be gained from acquiring a particular site. However, for these analyses to be credible to all the parties involved, it is imperative that they be performed by analysts that are widely accepted as independent of the various vested interests. Only in this way can the political process be properly informed as to the full consequences of each course of action.

3.6 CONSTRAINTS

An important aspect of strategic planning is understanding the constraints under which the system operates and the assessment of the extent to which these and other constraints will influence the system in the future. The most important implication of the constraints is the effect they have on the provision of capacity. The constraints arise from the technical characteristics of the operations, from environmental concerns and from the ability to generate funds for expansion of capacity.

3.6.1 Operational capacity

The operational capacity must be adequate all the way from the airspace to the end of the ground portion of the users' trips, in order to avoid bottlenecks. The normal way to assess the adequacy of capacity is to measure delay, calculate the consequent costs and determine the economic justification for the necessary spending to reduce the delay. The US Federal Aviation Administration (FAA) has preferred to use annual delay as the primary indicator of the need to enhance capacity, rather than relying on the estimates of hourly capacity which are subject to many sources of variability. Congestion expresses itself in delay and in other more hidden inefficiencies such as non-optimum flight tracks and lower productive utilisation of aircraft. Also, airlines often open their schedules in order to improve reported on-time performance. It is therefore important that delays are all measured relative to best practice in free-flow conditions. Weather was the primary cause of 66% of delays over 15 minutes in 1991 and terminal air traffic volumes accounted for 27%. The weather delays were largely the result of instrument approach procedures that are much more restrictive than the visual procedures in effect during better weather conditions. The average measured flight delay was 4.1 minutes for a total cost to the airlines of over US $700 million. On average each flight was also subject to a nine minute delay in taxiing and a one minute hold at the gate.

Each airport subsystem can provide the critical bottleneck in specific cases. Some are more amenable than others to solution by smart management or extra investment. Perhaps the most intractable problem is augmentation of runway capacity beyond the capabilities of the existing runways. According to a 1990 Stanford Research Institute (SRI) report for IATA, at least 16 European airports will be constrained by insufficient capacity by 2000. The Outline Plan of the Trans European Airport Network for the European Union countries plus Switzerland, Austria, the former Yugoslavia, Scandinavia and Iceland, accepts the SRI assessment and considers that by the year 2000 airport capacity will need to increase by 50% to meet the estimated demand of 760 million passengers. However, it concludes that existing airport development plans should be sufficient to cover this growth. Any shortfall would be expensive. An average delay of 10 minutes per aircraft at an airport with 350,000 operations per year increases aircraft direct operating costs by US $93 million (FAA, 1993). Yet past attempts to establish sites for new runways show how great the need has to be before they are readily used by the major airlines, as several of the case studies illustrate.

Lack of runway capacity results not only in increased delay but also the thwarting of additional flights by the carriers who cannot gain access to the system or who feel that the costs of delay do not justify the service. European airlines have not, in general, been keen to take advantage of the new freedoms to launch new routes. A major reason for this has been the shortage of runway capacity, combined with codesharing and frequent flyer programmes, making it difficult for a really contestable market to emerge (CAA, 1993a). Heathrow and most of the other popular European airports have been close to capacity and subject to slot controls throughout the decade. The first choice of those operators wishing to take up route freedoms has mostly been denied them. Meanwhile, the maintenance of frequency per airline in a more competitive situation only worsens the capacity problem. It appears that true competition could only occur if the major hubs had sufficient capacity to allow new low fare entrants. This was highlighted in a comparative analysis between liberal and non liberal routes between the UK and

Europe prior to the EU liberalisation (Caves and Higgins, 1995), as discussed further in chapter 11.

Lack of capacity in the stands and terminals can also have serious effects on the service offered. In evidence to the inquiry into Terminal 5 at Heathrow, British Airways argued that lack of the additional stand capacity would require that the airlines would either have to restrict or deny access to some categories of traffic on all routes, thus denying access at Heathrow to leisure traffic, or to eliminate some routes from the Heathrow network, thus removing some of the current opportunities for transfer connections (Maynard, 1995).

Even so, there has undoubtedly been more flying in terms of air transport movements (atm) as a result of liberalisation, both in the US and in Europe. Airlines have increased their frequencies, even where there is little real competition. This has resulted in a levelling off since the early 1980s of the previous increase in aircraft size, except at those airports with very serious congestion. When real competition on routes has emerged, frequencies tend to rise by even more than the extra competition stimulates traffic. This causes many observers to conclude that liberalisation must increase the pollution burden (Fergusson, 1995).

3.6.2 Environmental concerns

Despite the regulations in force to limit noise and air pollution at source through aircraft certification, society increasingly requires aviation to respond to the wave of concern for the environment. In the case of new airport development, legislation is in place to assess the impact of large changes of land use on the environment. European Community legislation (EC, 1985) through the Council Directive of 27 June 1985 requires an Environmental Impact Assessment (EIA) of proposed projects which will identify, describe and assess in an appropriate manner, in the light of each individual case, the direct and indirect effects of a project on the following factors:

- human beings, fauna and flora
- soil, water, air, climate and the landscape
- the interaction between these factors
- material assets and the cultural heritage.

Refineries, significant transport infrastructure (e.g. airports with runways greater than 2,100 metres long, ports for vessels of over 1,350 tonnes) and waste disposal installation for the incineration, chemical treatment or land fill of toxic and dangerous wastes must be assessed for their environmental consequences. Smaller transport projects and the extraction of petroleum come into a category which gives the member state the discretion to waive the need for an EIA. The information to be provided by the developer shall include at least:

- a description of the project comprising information on the site, design and size of the project

- a description of the measures envisaged in order to avoid, reduce and, if possible, remedy significant adverse effects
- the data required to identify and assess the main effects which the project is likely to have on the environment
- a non-technical summary of the information.

US legislation goes even further. Each EIA produced in the US covers four topics:

1. Brief description of proposed action, indicating states and counties particularly affected.
2. Summary of environmental impact and adverse environmental effects.
3. List of alternatives considered, including a 'do nothing' strategy.
4. For a draft statement, a list of all federal, state and local agencies and other sources from which written comments have been requested.

In order to work with their local communities, airports often have to set rules to achieve a greater control of environmental impact than that achieved through relying on aircraft certification. Several important airports are limited by daily quotas of noise (e.g. Düsseldorf), the weightings between categories of aircraft not always reflecting the relative noise nuisance. Many airports are subject to severe night noise curfews and quotas, some of which limit opportunities to serve a long haul route because the bunching which occurs when the airport opens creates the worst arrival peak of the day. Operators of the noisier Stage 2 aircraft are paying progressively more per landing, in order to support the airport's need to compensate and mitigate against noise nuisance, as well as to give an incentive to replace these aircraft before the official ban by 2002 in Europe and earlier in the US. Either way, these operators are faced with higher costs.

Attention is now firmly fixed on all other aspects of environmental impact as well as noise, as estimates are made of the total social costs of transport. The push to legislate on emissions has been led by Sweden and by California, where the Environmental Protection Agency's emissions limits are beyond the present capability of technology. It has been calculated that the full costs of Swedish domestic aviation externalities equate to 12% to 19% of the aircraft direct operating cost (compared with 12% to 15% for cars and 4% to 7% for rail) and that full recovery would reduce demand by 6% to 10% (Kageson, 1994). The Organisation of Economic Cooperation and Development (OECD) has calculated that externalities of all modes of transport equate to 5% of Gross Domestic Product (GDP), made up of 2% each for congestion and accidents, the remaining 1% being noise and air pollution (Alexandre, 1995). Levinson et al, (1998) derived values in US cents per passenger kilometre of air social costs of 0.43 for noise, 0.09 for air pollution, 0.05 for accidents and 0.17 for congestion. Their comparative values for highway social costs were 0.45, 0.37, 2.0 and 0.46 respectively. These valuations are important in the consideration of the recovery of the costs of externalities which is considered in chapter 5.

There is concerted activity by AEA, IATA and by ICAO with its Committee for Aviation Environmental Protection (Somerville, 1992) as well as many initiatives by individual airlines and airports to meet and exceed the air and water quality regulations. Pollution can be minimised by reducing movements and keeping ground manoeuvring to a minimum, and much can be done by using ground power rather than the aircraft's Auxiliary Power Unit (APU) to operate the aircraft systems during turnround and to start the engines. It is preferable, but impractical at small airfields, to pipe electrical power and air to the aircraft rather than use portable sources. However, it is possible to use electric or methanol rather than diesel-powered tractors, lift loaders and aircraft service vehicles.

However, it will be necessary to halve the movement rate in 'non-attainment' areas like Los Angeles by 2005 unless some accommodation is found. Even the use of the cleanest aircraft and minimum power takeoffs, towing to and from runways, banning reverse thrust and APU operation, using electric ground vehicles, and achieving 100% load factors would not satisfy some of the proposed regulations.

Waste disposal is the final link in the chain of effects which should be identified if a true 'cradle-to-grave' philosophy for minimising environmental impact is to be followed. The impending environmental legislation in this area can only increase costs for the industry. KLM is already spending US $50 million per year in this area (Pilling, 1991). United Airlines has a team working to apply the philosophy to the control of pollutants (Rheingold, 1991). Many other aircraft operators are similarly aware. Many airlines and airports have now been accredited by the International Standards Organisation to its ISO 14001 environmental management standard, though this is generally weaker than the British BS7750 standard or the EC's Eco-Management and Audit Scheme (EMAS). They all require an environmental policy which sets targets and objectives, EMAS also requiring external verification and a public information programme (Somerville and Meades, 1995). Environmental protection measures added 10% to the capital cost of the Munich II airport (Renz, 1996). Capital and operating costs are likely to increase further as the search continues for additional capacity in an increasingly sensitive environmental context.

Public safety may well become the next main constraint on airport capacity. Before policies are formulated, much more work is required to refine appropriate data bases on accidents and the distribution of their location, the implications for casualties and their degree of acceptability. In particular, the causal relationships with type of airline, type and size of aircraft, weather, standard of airport facilities, form of emergency and the consequences for the subsequent flight and impact path should be explored further. The accidents' locations relative to the airport may vary with factors such as air traffic congestion, quality of navigation and landing aids, weather, etc. Research by Dutch National Aerospace Laboratory (NLR) concluded that approaches to runways with precision approach aids are 5.2 times safer than those to non-precision runways (*Flight, 6 March, 1996, p 5*), but there is always doubt as to whether it is the aid or the operator using it which controls the risk. The most dangerous situation may be an inactive precision approach. Also, a resident in a developed country is not immune to risk from airlines of less developed countries: an Air Algerie freight aircraft clipped house roofs when it crashed short of Coventry airport in 1995. The mix of aircraft, weather, operators, air traffic control procedures and aids, airport facilities and oversight of national and foreign operators may all have a substantial influence on the risk. Research along these lines would allow Safety Cases to be identified that would represent the more critical situations that should be tested at any specific airport under consideration.

Most of the guidelines on acceptability of risk have been developed in association with health and safety legislation or with government enquiries into the acceptability of potentially dangerous plant. The UK approach is based on defining negligible risk and intolerable risk, allowing developments between these limits on the principle 'As Low As Reasonably Possible' (ALARP), provided that the individual at risk is aware of the risk, that sufficient benefit derives from the risk and that the maximum possible mitigation is undertaken. Official views are that a risk of one in a million per year is commonly regarded as trivial and that one in a thousand per year is just acceptable for workers. A risk of one in ten thousand per year may be the minimum acceptable for the public, compared with the risk of cancer at three in a thousand and the risk of a fatal traffic accident at one in ten thousand per year (Purdy, 1994). An alternative yardstick would be a 1% increase in natural risk: for the 10-15 age group, natural risk is one in ten thousand of dying that year, hence the acceptable involuntary additional risk would be one in a million per year (Smith, 1991). Another suggested benchmark is one in a million in a life time, against which the average risk for those on the ground from aviation in the US is too high by a factor of four (Goldstein et al, 1992).

The proposed Dutch standard for airports, which already applies to some other installations, is that the additional risk of death to an individual on the ground should be no higher than one in a million per year. This level of individual risk would be extremely difficult to satisfy near any major airport in the world (Hillestad et al, 1993). The actual policy at Schiphol is that the number of houses with greater than a one in a million risk and ten in a million risk must not increase beyond the present 3,010 and 85 residences respectively (Amsterdam Airport, 1998). The individual risk levels around several example airports are given in Figure 3.4. However, tolerability of risk must depend on the benefits to be gained from the activity by the risk-takers. The apparently greater tolerance to voluntary risk relative to involuntary risk, and the greater levels of all risks in poorer countries both reflect this fact.

Whichever basis for risk calculations are adopted, the studies indicate that large airports will have increasing difficulty in meeting emerging planning standards of involuntary individual risk. Societal risk will tend to rise with the increasing size of aircraft, though trends to greater safety per movement with time and with the scale of the airport will mitigate this effect. The indications are that serious mitigation measures may be necessary, and that the most effective mitigation is the protection of land under the approach and departure paths in the manner of the UK Public Safety Zones (PSZ) if not with the same shape. Large airports with few runways and hence high movement rates per threshold may have to adopt larger PSZ than defined in the present UK regulations, if proposed risk guidelines are to be met and the models currently available are to be believed. Meanwhile, India and the Philippines have programmes to relocate and compensate squatter settlements which encroach on their airports, creating safety and security hazards to the airport as well as to themselves (*Airport World, Winter 1996, p 40*).

Figure 3.4: Risk criteria for communities around airports

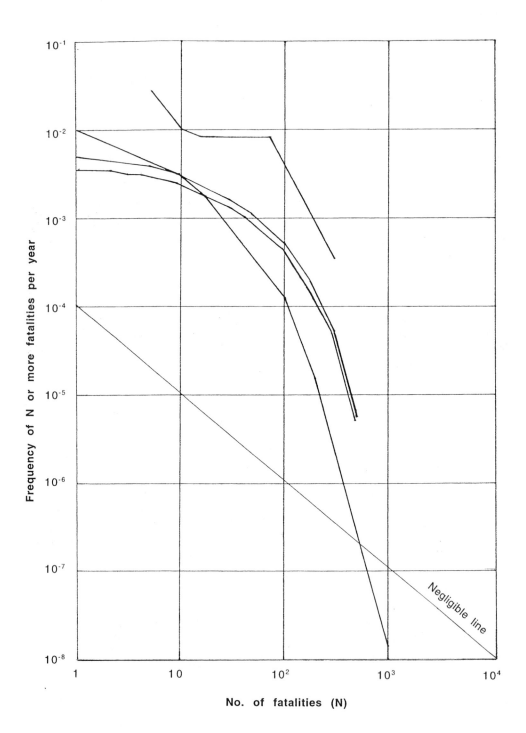

3.6.3 Financial

Strategic planning is only as good as the implementation of the plans. The plans must therefore be carefully justified in terms of their economic viability, and a realistic assessment of the availability of the necessary funding must be made. The lack of a sufficient cash flow to cover loan repayments often results in a much more incremental expansion of capacity than would be economically efficient, particularly since there is usually a temporary loss of capacity during construction.

Airports in public ownership have always had to compete with other investment priorities, while generally being seen as less socially necessary than many other activities, such as schools and hospitals, or indeed, other forms of transport. Lack of funds has therefore been a continuing reason for the failure to implement planned and necessary developments. This is true even in the most developed countries, and regardless of whether a government general fund or a dedicated fund supported by the industry's own revenues is the potential source. The US Aviation Trust Fund, administered by the FAA, has traditionally collected more in ticket taxes and other sources of revenue than it has been allowed to spend by Congress. It has resulted in an average contribution of US $2 per enplaned passenger over the last decade and accounted for 24% of the total capital funding at large hubs and 32% at medium hubs in 1994, less than that provided by new debt, with the balance coming from Passenger Facility Charges and bond issues (GAO, 1996). Admittedly, most large US airports, and others around the world, carry a AAA rating on their revenue bonds, but the bonds still have to be paid off from revenue. Rapid growth, and the lumpy and forward nature of the consequent airport investment, make this difficult. The UK local authority airport owners have the additional constraint that their borrowing is capped by the national government.

The uncertainty addressed above makes it more difficult to justify investments. It is no longer seen as appropriate to provide a single view of the future from which to calculate cost and revenue streams. Rather, a funding agency will require risk assessments to be performed, which imply judgements by experts of the probability distribution of each element of the cash streams. This has its own hazards, since all their judgements are potentially subject to bias (see chapter 7).

Private ownership usually is able to unlock more development capital, but at higher interest rates than government funding in exchange for the greater risk. Danish law allows this even for a 100% government-owned entity (Thorning, 1990). On the other hand, the Public Limited Company (plc) status of UK regional airports does not alter the restrictions of local government borrowing until a majority of the shares pass into non-government hands. Most local authorities were initially loath to take this step.

The funding situation is much more difficult in developing countries, not least because there is less prospect of the airports making a profit. Government assistance, either as a grant or as a loan guarantee, may be available on the grounds of regional and national economic benefit, but this will often be subject to the vagaries of fragile budgets. It may also be possible to obtain finance from foreign sources, at the cost of exchange reserves, unless specific aid programmes can be tapped. This can leave the developing country open to obligations to purchase other products from the aid-giving country. The international banks and funds, like the International Bank for Reconstruction and Development, are not subject to such ties, but may impose other conditions on the receiving government. A full list of the funding agencies is given elsewhere (ICAO, 1991). Commercial sources, which may arrange funding as part of a full 'turnkey' or build-operate-transfer deal, have the obvious caveat that the rest of the package must be accepted as well as the finance, and it is by no means certain that the package with the best finance will also be the best technically or economically.

4

UNDERSTANDING AVIATION SYSTEM BEHAVIOUR

The goals and objectives of the aviation system's stakeholders were examined in section 3.2. In order to plan the system so that these often conflicting goals can be brought together in a cohesive manner, it is necessary to study the behaviour of the stakeholders as they exercise whatever freedoms the system allows them, so as to anticipate the actions they may take in the future. This chapter sets out to do this, before drawing some general conclusions of the implications of that behaviour for strategic system planning.

4.1 THE ULTIMATE CONSUMERS

4.1.1 Air travel demand

Demand prediction at the system level often consists of aggregate forecasts for a complete nation, sometimes broken down by type of traveller and sometimes by region in a top-down fashion. In order to address the issues which concern strategic system planning it is necessary to know much more about travellers' needs for communication as they vary by location, activity and their socio-economic characteristics. The determinants of air travel demand are the factors which control the overall demand for travel, the extent to which it can be satisfied and the share of that traffic which can be captured by air. Demand is a function of a desire or perceived need to communicate, which depends on business and social backgrounds. The satisfaction of that demand depends on the resources available to the potential traveller and the convenience of the supply of transport. A balance has to be struck in provision of service between that which can be justified socially and economically and that which is perceived to be necessary. It is important to understand that the revealed demand as reported in the traffic data is an expression of that balance at each airport. A common but mistaken idea is to define an airport's catchment area and to presume that all those living or working or visiting that area will use that airport. In fact, not only do catchment areas often overlap, but the level of service to specific destinations influences the choice between the competing airports. The access trip itself may be a sufficient deterrent to travel, since travel will only occur if the overall benefits exceed the overall costs.

The generation of air travel will reflect the factors affecting demand and the supply alternative which gives the least disutility of travel. It may be dealt with by market category analysis techniques as used by Roskill (1971) or by econometric modelling. The econometric models will be sensitive to price and income, often with some additional factor which reflects the quality of transport supply. The implications of price and income are usually expressed in terms of elasticities, which indicate the change in demand which would be induced by a 1% change in the independent variable. The literature gives many examples of average elasticities revealed in specific markets (e.g. Button, 1993; Doganis, 1991; ICAO, 1985b). Business travel is usually quite inelastic in price but has an income elasticity approaching unity, i.e. a 1% increase in income (or gross domestic product as a surrogate) will result in a 1% increase in business travel. Leisure travel typically has much higher elasticities, that have been found to approach 2.5 for income and -2.5 for price in some markets, hence the importance of discounted air fares or the low cost carriers' entry in determining traffic levels. It is important to know the types of traveller in as much detail as possible, the travel characteristics of students being very different from retired people 'visiting friends and relatives', or that for the emergency repair of machinery from a posting to a merchant ship. With time, some individuals' desire for travel within their personal time and financial budget becomes saturated, so their elasticities diminish and hence the average elasticity also reduces, which is to be expected in maturing markets. Any further growth in the market depends on new people coming into the market as average incomes rise, prices fall, or convenience improves, so it is important to know the state of development of the markets served by the airports in a system.

In conventional transport planning the analysis of trip generation is followed by distribution models predicting the destination patterns of the traveller, usually based on some analogy to the laws of gravity. Simple models of this type ignore the important contribution of specific 'community-of-interest' between the origin and destination zones. A technique which distributes traffic as a function of relative attractiveness of the destination helps to overcome this drawback. Generation and distribution models can be applied to the air mode in isolation. For shorter haul situations, it is more appropriate to model total travel generation and distribution and then to derive the modal share of air by use of a modal choice model or by applying diversion curves that relate the modal share of air to the relative attractiveness of air compared with the other available modes. Since the relative attractiveness of air depends not only on the time saved, but also the cost difference, it is common to perform the analysis on the basis of a generalised cost that combines the money cost with the cost of the time involved, derived from an estimate of the traveller's value of time.

It may well be argued that generation and distribution are functions of the attributes of the modes serving a zone and that the travel choice is really a simultaneous decision, covering generation, distribution and modal split. Models have been developed which include mode attributes as well as the other variables used in the conventional travel demand forecasting process. Perhaps the best known example is the abstract mode model, which predicts these factors by including terms describing the attributes of both the subject mode and the best available value of each attribute. In this way, it is possible to analyse the likely patronage of a new mode if estimates can be made of its attributes. Examples of these models may be found in Ashford and Wright, (1992).

Demand depends on many things other than price, income and quality of service, but it is more difficult to model these factors quantitatively. For example, tourist demand is strongly influenced by fashion and personal security, as well as the relative total inclusive cost of the holiday. These factors are dealt with in section 7.4.

4.1.2 Passenger choices

Once a person has made the choice to communicate, a positive choice has to be made to travel instead of, or as well as, using some form of telecommunication. If travel is deemed to be necessary, there is often a choice of mode, or combination of modes, to be made. If air forms a part of the chosen trip, there is still a series of choices to be made about which airport to use, whether to use commercial airlines or general aviation, which airline to fly with and which route to take through the air transport network, though by no means all of these choices will be available for any one trip. These choices influence the supply of capacity in a competitive industry and hence the share of traffic gained by each airport, in turn determining the role of the airport in the system.

A general methodology for understanding and modelling mode, airport and route choice is presented in section 7.3 with suggestions for further improvements in the methodology, while the communication/travel and mode choices have been discussed in section 3.5.3. In each case, the general principle is always that of minimising a passenger's disutility over the total trip. The main attention in modelling is given to trip cost in terms of money and the value of time. The value of time depends on both the individual and the nature of the trip. Studies of travel choice behaviour have found values varying from perhaps three times a salary of US $150 per hour for an executive waiting for a connecting flight to a quarter of an hourly income equivalent to US $4 per hour for a person on a state pension while travelling on the main leg of the trip. The former traveller is likely to be concerned about frequency and timing of flights, while the latter will normally be primarily interested in the lowest price, regardless of the time of travel. There are, however, many other factors which influence these decisions, including return trip convenience, frequent flyer awards, type of aircraft, personal security and the ability to be productive while travelling.

4.1.3 Air cargo

Air cargo has been growing more quickly than passenger traffic in most parts of the world for several decades. Initially it was so expensive that few goods were sufficiently valuable to justify transport by air. As value per unit volume increased, with the growing market for such commodities as solid state electronic goods and out-of-season vegetables, and as the unit costs of air transport fell, it became more practical to ship by air. In addition, industry became more aware of the costs of holding inventory and that dependability, security and punctuality were also important. Tariffs fell further than average aircraft operating costs as belly hold capacity became available in the long haul wide body aircraft and tended to dictate industry-wide prices, so sustaining the high growth rates. The growth is typically 50% greater than the passenger market, particularly in the developing world where it is vital for exports and often for domestic shipments in the face of poor road conditions.

Some long haul combination carriers now earn more than 50% of their revenue from cargo operations, using a mix of dedicated freighter, combi (aircraft sharing main deck space between passengers and cargo), and belly hold capacity. Others, like British Airways, only use belly holds in aircraft scheduled around passenger operations, generally taking only the higher valued cargo. There are many all-cargo carriers offering conventional freight services, some concentrating on specialised cargo but others, like Cargolux, managing to compete directly with the major combination carriers. Much depends on the choices made by the freight forwarders who generally take responsibility for the cargo, the shipper usually not being directly involved in the decisions on carrier or route. It is common for a conventional carrier to truck the cargo between airports for some or even all the trip where it is more convenient or cheaper or necessary to meet a delivery time. These truck movements may even be assigned a flight number in order to ensure that the service appears in the schedule.

In the last two decades, the express parcel carriers, or integrated carriers as they are now called, have taken an increasing share of the smaller cargo shipments, typically weighing less than 25 kilograms. These carriers responded to the shippers' need for guaranteed service with late pick-up and early delivery, and with seamless service from door to door to support the concept of 'just-in-time' manufacturing logistics. These carriers deal with all export and import formalities, whereas the customers would normally have to claim conventional freight from customs. One of the early integrated carriers, Federal Express, bought the largest conventional cargo carrier (Flying Tiger), and competes strongly with United Parcels Service, DHL, TNT, Emery Air Freight and others with a worldwide network. The parcels, packets or envelopes are flown into hubs, sorted during the night and flown or trucked out in the early morning. The location of the hubs depends on the geographic coverage but also on the degree of independence offered to the carrier, freedom from night restrictions and the terms of the contract with an airport. Several of the integrated carriers feature in the top 20 freight airlines in terms of tonne-kilometres, and their market continues to grow faster than the conventional freight market.

Some of the conventional cargo carriers are trying to improve their service, making them more seamless and improving dependability, but this requires dedicated space to be kept available even if high yield passengers have to be refused. The biggest problem for conventional carriers, at least in the short haul market, is that freight generally needs to move at night while the passenger wants a daytime flight. One solution, adopted by Lufthansa, is to use quick change aircraft whose main deck can be converted in 30 minutes from passenger to cargo space. These factors are creating new interest in airports dedicated to cargo operations, despite the joint cost benefits which ought to accrue at airports where the passenger fleet operates by day and the cargo fleet by night, particularly for those operators with quick-change aircraft. BAA has bought a former military base at Alconbury for use in this way. Conversely, there is an equally strong trend to expand existing freight facilities at the established major hub airports. Frankfurt has the advantage that land is being released by the military, but most of the other major European airports and their carriers are also investing heavily (Gethin, 1998). There are also opportunities for some airports to meet the need for modal interchange for cargo, to rail as well as to sea. Frankfurt is again fortunate that a separate rail line runs into the former military site. In the long haul markets, sea/air trans-shipment can be attractive where this gives the right blend of tariff and inventory cost. The prospects for this traffic have led to serious competition between airports, particularly in the Persian Gulf and on the west coast of the US.

4.2 AIRLINE DECISIONS

4.2.1 Competition

Airlines behave differently in a competitive situation compared with one where they have a quasi monopoly. Competition can take many forms, from head-to-head competition between several carriers on a single busy sector, through nominal competition on duopoly routes, competition over competing hubs for indirect traffic, to the threat of being replaced as the incumbent carrier on a thin protected route. Each type of competition will trigger its own appropriate response, depending on the regulatory, capacity and financial constraints on the airline and the routes in question. Subject to those constraints, the competition will express itself in terms of network expansion, price and level of service, the latter usually being expressed as frequency, although other factors, such as the equipment used, can contribute to the level of service. The indicator over which the battle is usually fought is market share. This can be controlled by price, but airlines can, in the short term, match prices. The more important determinant of market share is frequency, following the Renard theory that a superior frequency gives an even more superior traffic share, all other things being equal. This is illustrated in the UK case study, where the Gatwick and Heathrow shares of London traffic are compared. Frequency competition results in the use of smaller aircraft, which are less economic. The direct operating cost (doc) per passenger kilometre of a 200 seat aircraft is some 30% less than a 100 seat aircraft on a short sector (Beyer, 1989). So a balance must be struck in each market between the need to compete and the need to control costs. However, given an opportunity to compete on frequency, airlines will normally tend to accept the consequent extra unit costs of smaller aircraft, leading to smaller numbers of passengers per air transport movement (p/atm).

The crux of the argument about the need for more runway capacity revolves around the expected future trend in p/atm. This depends on the number of airlines competing and their judgement of the frequency required to compete effectively on each particular route. It also depends on the airlines' own trade-off, when runway slots are limited, between frequency competition on one route and the desire to add frequency on another route. In addition, it depends on the balance of runway to terminal capacity at critical airports so that p/atm will have to rise substantially at Heathrow if full advantage is to be taken of Terminal 5 (as discussed in chapter 10).

The implications of being wrong in this area are huge, yet, compared with forecasts of passenger demand, the effort put into the forecasting of technology, market behaviour and the resulting p/atm is minuscule. When the subject of London airport capacity was being considered in the early 1960s, it was expected that the average number of seats per aircraft would rise from the current 80 to 120 in 1970 and then remain constant to the end of the century. The Boeing 747 was already on the drawing board yet the planning assumption was based on expectations of technology at least as much as on commercial pressures to increase frequency (Watson, 1965). Although a sensitivity analysis was done with aircraft sizes up to 200 seats, it was regarded as "an exercise in arithmetic". In fact, in 1992 Heathrow handled 116 p/atm, close to the 1965 "arithmetic" with the load factor of 60% assumed at the time.

Hubbing, as discussed in the next section of this chapter, encourages frequency and exacerbates daily peaking both within an airline and in competition between airlines. Contesting airlines need to compete on timing, frequency, cost and by offering better routing, so the extension of competition implies more flights per route in the peak and also more routes. Since exaggerated peaking is not economic, it is instructive to observe the outcome in the deregulated US market place, and compare it with present practice in Europe.

The average size of aircraft in the US carriers' fleet grew at four seats per year until it peaked at 154 seats in 1983, since when it has fallen back to 150 seats. Current orders suggest that it will remain at this level. The average size of aircraft operated on domestic and European service by the airlines in the Association of European Airlines (AEA) peaked at 139 seats in 1982 and has since fallen to 133 seats, despite the introduction of larger aircraft on some of the more competitive routes (e.g. BA domestic shuttle routes in competition with British Midland and rail, although these routes still have more than eight flights per day each way per carrier into slot-constrained Heathrow). In fact, even at Heathrow, seats per atm remained constant between 1985 and 1992, except for a 40% increase to 148 seats for Terminal 1 domestic flights due to the withdrawal of almost all aircraft of less than 40 seats. Meanwhile, there was a 40 seat reduction to 230 seats in the long haul Terminal 3. In contrast, European charter aircraft have continued to increase in size in the search for cost minimisation and in the face of runway and airspace slot difficulties.

The aircraft size depends on frequency per airline and the number of competing airlines per sector. Belobaba and van Acker (1994) studied the competition in overall Origin-Destination (O-D) markets in the US and found that 17% of the top 100 markets in 1991 enjoyed between five and eight carriers with more than a 5% market share and a further 25% of these markets have four carriers. The average in the top 100 markets is 3.4 carriers, varying between 3.7 at non-hubs and 3.1 at hubs, compared with an average of 2.7 in 1979. Significantly, there were the same average number of carriers (2.2) in 150 of the most dominated markets in 1990 and 1979 (i.e. the 10 top markets at each of the 15 airports at which one airline carried at least 60% or two airlines at least 85% of all enplaned passengers). However, this study does not distinguish between direct and indirect routings, and it also appears to apply to city pair O-Ds, without distinguishing between individual airports in a city.

In contrast, Europe had an average of 2.9 carriers on routes with more than 100 flights per week, with more carriers on international than domestic sectors and with Sweden and some other countries regulated to only one carrier per domestic route (Pryke, 1991). Only two routes had more than 200 flights per week, but Europe was less monopolistic than the US on routes below 40 flights per week. Pryke suggests that Europe's routes would only become slightly more competitive if it matched the US levels of competition for a given traffic density, as shown in Table 4.1.

Table 4.1: Percentage of short haul flights in western Europe and the USA by market type

Market type	Europe actual	Europe after liberalisation	USA actual
Monopoly	48.3	48.7	38.1
Duopoly	32.0	24.7	25.0
Triopoly	11.0	16.1	17.3
Quadopoly, etc.	8.7	10.6	9.6

Source: Pryke, 1991

Clearly there are barriers to entry, some of which are not caused by airline domination so much as genuine shortage of runway capacity. It is unlikely that any greater airline competition per route will emerge in Europe than in the US. In fact, the European consumer already benefits from competition with rail except on the longer and over water routes, i.e. principally routes serving the UK and peripheral countries, and the shortage of capacity is as least as great as in the US. Yet it is still common for forecasts of growth in aircraft size to be lower than those for frequency. The expectation must be that the capacity problem will somehow be solved.

In Europe, as EU liberalisation has progressed, full unrestricted fares have continued to rise, except in those few airport pair markets with at least three carriers. The major carriers at the main airports seldom attempt to match the lowest fares offered by the new entrants at the secondary airports. There is, though, a growing tendency for the major carriers to copy the US pattern of launching low cost subsidiaries, as with Alitalia Team (Jones, 1998) or BA's GO. However, it needs to be borne in mind that Europe already has a very large low fare sector in the charter carriers, so that the economies of density available to the multiplicity of scheduled carriers are limited, and would only be reduced further by more competition. This is a potent reason for mergers between the major carriers. It is also a good reason for the airlines to ensure that the low cost subsidiaries only exploit new markets rather than competing with the parent's existing services.

In the international arena, there is some evidence, from a sophisticated paired comparison of liberal and regulated routes, with assumptions of Bertrand (profit-maximising) behaviour on the former and collusive behaviour on the latter, that although normal economy fares were not affected, discounted fares fell by 35% after the liberal US bilaterals were signed (Dresner and Trethaway, 1992). It is not, however, clear if this resulted in corresponding traffic growth, any more than with the comparative analysis of early liberal Air Service Agreements in Europe reported in chapter 11.

A comparative analysis between less and more liberal regimes is valuable, because fares on all routes have fallen relative to the retail price index. The lack of a similar 'control' case for comparison makes some of the positive conclusions from the US deregulation experience doubtful, in that they do not allow sufficiently for the other changes which were occurring at the same time (Windle, 1991). Where such a comparative approach has been taken, on the longer routes between Europe and the US and between Asia and the US, it is similarly difficult to establish any clear evidence that liberalised settings are cheaper than regulated ones for any class of fare. (Feldman, 1988).

4.2.2 Route structure

Except when they are constrained by government intervention, airlines plan their network development very differently from the way a government agency, or private initiatives, would plan a system of airports. The most noticeable trend in the liberalised airline industry in the US has been the formation of a hub and spoke network based on one or more 'fortress hubs'. Hubbing can be criticised on resource grounds, in that it encourages more boardings than would be necessary for point-to-point flights. This trend is exaggerated further by 'frequent flyer' schemes designed to defend the hubs. It is extremely important for the powerful hub airports that investment in new capacity be properly justified in the light of their based carriers' future network strategies, even more so at airports which, like Manchester, are investing in order to market themselves as hubbing points. Many models, which have been reviewed elsewhere (Caves, 1993), have been developed to try to understand how the strategies might evolve (Borenstein, 1989; Brenner, 1990; Chou, 1990; Donoghue, 1988; Gillen, Oum and Trethaway, 1985; Hansen, 1990; Hansen and Kanafani, 1989; Kanafani and Ghobrial, 1985; McShan and Windle, 1989; Morrison and Winston, 1989; Phillips, 1987; Wheeler, 1989). The modellers have examined industry outcomes and, somewhat inconclusively, the correlation between hubbing and airline cost. They have shown that hubbing appears to be inelastic with respect to hub pricing. Studies of networks range from simply identifying US city pairs with no non-stop service (the largest identified potential flow was 16,000 passengers per year between San Diego and Washington), through attempts to minimise airline plus passenger time costs in a hypothetical uniform circular space centred on a hub, to cost minimisation models which derive the hub hierarchy endogenously by searching for the minimum cost for alternative minimum spanning trees rooted in each node in turn. It is possible with these models to identify levels of demand, number of nodes and distance over which hubbing is preferable, and also the extent to which adjacent zones should use direct connections rather than suffer high circuity.

Whereas passenger behaviour is presumed to have the objective of minimising the disutility of travel, airlines in a liberalised and privatised setting could be presumed to be profit-maximisers. An attempt has been made to predict the equilibrium proportion of hubbing under a profit-maximising assumption, using an 'n'-player, noncooperative game method (Hansen, 1990). Due to the many simplifying assumptions which had to be made (the use of prevailing fares, nondiscretionary aircraft size, total demand inelastic with respect to price and service level), and the *a priori* indeterminacy of the equilibrium properties of the game, the behaviour of hubs in multi-airport regions and those with relatively weak markets were not well predicted. The approach is none-the-less promising, as is the attempt to optimise the air transport network, including the effect of slot pricing, using a two-stage game theoretic approach (Hong and Harker, 1992), where again there is no variation of total demand with fare or level of service, and airlines are presumed to attempt to maximise their revenue share on each sector. The results of its application to a simplified network show clearly that oligopoly total profits are much greater when the slot allocation is endogenous, i.e. determined by the airlines' ability to pay: the airline with the larger aircraft of course gains profit relative to the others. The principles developed in these two studies could well be merged and extended to hierarchical structures and multi-airline interactions (Shaw, 1993) to study the impact of airline and regulatory policies with respect to hubs, using decision-process algorithms.

There is a clear trend for increasing concentration in the industry as airlines search for economies of scope. This leads to the major international gateways gaining an ever increasing share of country-to-country traffic (Pearson, 1997) and the major alliances in general gaining market share at those gateways. This does not, however, conflict with an equally strong trend to fragmentation of routes in a growing market, particularly since the alliance carriers are using the gateways as international hubs as well as for bridges to their partners' hubs.

It is becoming increasingly common for supply decisions to be uncoupled from revealed demand. This is sometimes to correct an under-supplied market, but it is often a result of airline or airport strategies. A number of studies have been conducted looking at the consequences for market entry of liberalised air services between the UK and other European countries (Abbot and Thompson, 1989; Button and Swann, 1989; Caves and Higgins, 1993; Johnson, 1988). These studies provide evidence of significant changes to the supply of air services in terms of destination served from London and other UK regions to Europe, increased frequency by route, improved use of jet services and an increased number of designated airlines. Recent examples are the KLM decision to markedly increase frequencies and capacity at Amsterdam and Manchester Airport's initiative to become its own hub champion and to coordinate its airlines' schedules. It must be anticipated that many other, and more extreme, initiatives will be taken as the possibilities of liberalisation of services in Europe are realised. Also, liberalisation is continuing to spread through the rest of the world.

Competition does not only exert itself due to passenger preference or to airport capability. Miami's large investment in facilities to be used largely by American Airlines is causing other airlines to react to the resulting higher charges by stepping up Latin American services from their own hubs rather than the traditional Miami gateway (*Airline Business, July 1997, p 41*).

A hub fortress helps the carrier develop a local monopoly with the opportunity to raise local fares in order to compensate for lower yield per mile for transfers across the hub. It is difficult to enforce anti-trust laws in these situations. However, the importance of hubbing to the circuity of passengers' trips, as opposed to the price they pay, should not be over emphasised. Even though hubbing is so dominant in the US, the majority of the revenue passenger kilometres (rpk) were on direct flights. This is so even on the lower density routes, largely because one end of the sector is a hub and up to half of the traffic is local. Many of the lower density sectors are only viable with frequent jet service because of the hubbing element. Similarly, hubbing helps to make the new intercontinental direct services viable. Most of the airports which can sustain direct flights are also hubbing points.

78 Strategic Airport Planning

The early hub consolidation moves by the European major airlines have now largely been consummated. Particular success in expansion of their own hubs has been achieved by BA and KLM. Despite multiple terminals and capacity constraints, BA at Heathrow has moved from a low transfer percentage to one of the biggest players. Table 4.2 shows that in 1989 BA's schedules were still effectively uncoordinated from the point of view of connecting times. Even in 1994, it did not compete well on direct non-stop links in Europe, though it had the best frequency of service and impressive long haul connections (Dennis, 1994). Yet by 1997, 40% of its passengers were transferring, induced by marketing as well as by level of service, in the form of code-sharing, aggressive over-the-hub pricing and frequent flyer awards.

Table 4.2: Comparative European hub performance

Airline	Airport	1986 [1] Transfer (%)	1989 [2] Connectivity	1994 [3] Direct links	1994 [4] Daily frequency
Lufthansa	Frankfurt	48	1.6	78	2.6
KLM	Amsterdam	30	1.9	67	2.7
Air France	C de Gaulle	19	0.9	60	2.7
BA	Heathrow	22	1.1	50	3.2
Swissair	Zürich	35	1.9	50	2.1
SAS	Copenhagen	-	-	50	2.9
Alitalia	Rome	-	1.2	44	2.7
Alitalia	Linate	-	-	38	2.3

Notes:

1. Per cent of passengers of all airlines (Carré, 1988).
2. This rates the degree of coordination of schedules for ease of transfer, where 1.0 is a rating achieved for essentially random scheduling, Vienna scored 2.2, and Madrid scored 1.0 (Doganis, 1991).
3. Non-stop routes within Europe (Bouw, 1995).
4. Source: Bouw, (1995).

Source: Caves (1997).

The hub at an airline's base serves the defensive purpose of protecting its own local traffic. In addition, the marketing advantages of hubs encourage airlines to set up other hubs. The non-base hubs serve the more aggressive purpose of raiding competitors' territory and of expanding market share to avoid capacity constraints and excessive yield dilution at the home hub. There is also added flexibility in the options for moving traffic through the network. Thus, at the height of deregulatory expansion, American Airlines had six hubs before it rationalised them during the recession in the early 1990s.

It is often possible to attract a smaller carrier to hub at a secondary airport when a major carrier elects not to serve it. Maersk Air is building up a hub at Billund to exploit the local traffic potential and to compete with SAS's operation at Copenhagen (*Airline Business, May 1996, p 14*). Air Littoral, together with the local businesses and Chamber of Commerce and Industry, has formed a subsidiary to create a hub at Nice as a 'gateway to southern Europe' (*Flight, 3 July 1996, p 14*). Crossair has invested in Basle-Mulhouse in order to establish a significant regional hub (*Air Transport World, April 1998, pp 74-75*). If the small carrier then forms an alliance with a major carrier, it may be possible to form a 'bridge'. The European carriers have been surprised by the strength of the traffic which is not local to either end of the bridges formed by their alliances with US carriers. The recent rise of low cost carriers has been another example of the use of spare capacity at secondary airports, in both the EU and the US. A further use has been for the integrated carriers with their requirement for night movements, often with rather noisy aircraft. Local opposition tends to follow the integrated carriers' arrival, so that there is a tendency for migration to another secondary airport. Further details of European network strategies are given in the paper from which this perspective is taken (Caves, 1997), and in the European case study in chapter 11.

The route structure of an airline depends on the route rights it is given. There are still many bilateral air service agreements which limit the number of gateway airports and the carriers which can serve them. The national carriers often do not wish to fragment their services and certainly do not wish to open those routes to competing carriers. Although it was in the US major airlines' interests to have their own inland bases declared as gateways so that the new long haul routes could strengthen the hub, it took a long battle before the UK regional airports managed to convince the government that, if the route restrictions were lifted, the regions and the UK as a whole would benefit, even if British Airways did not operate the routes. Some countries might be loath to increase the number of gateways because the reduced transfer onto domestic routes would make those same routes less viable.

While the industry shows a strong tendency for concentration, there has also been a complementary trend to dispersion of routes in response to the passengers' preference for direct routes, particularly with a technological advance like the extended range twin engine aircraft or the new small turbofan aircraft. UK experience suggests that an international route might operate with as few as 10,000 passengers per annum (ppa) from small airports, though at airports with one million ppa the average route size is 50,000 ppa, the airlines taking the opportunity to increase route density rather than route numbers (Morris, 1997). Airports often take the attitude that the number of direct routes is a positive marketing feature, but 25% of routes with less than 10,000 ppa failed between 1986 and 1996.

4.2.3 Strategic alliances

The real-world outcome of the application of the hubbing models described above will depend crucially on the relationships which result from mergers and alliances. A fertile field for research is that of the power relationships and processes by which an airline chooses its partners, there being no universal structural models which can guarantee the success of the partnerships (Flanagan and Marcus, 1993). The relationship is often prompted by the acquisition of slots, which changes the nature of air services when the minor partner finds itself redundant (Moorman, 1993).

Airlines have long had a tendency to form alliances. Historically these have been more for technical reasons, or due to the requirements of bilateral regulation or for interlining. Liberalisation has allowed the formation of new alliances with a number of benefits. They can enhance markets by obtaining favourable listings on computer reservation systems, by increasing network size and by increasing frequency (Oum, Park and Zhang, 1995). Network expansion increases domination of home markets, and extends international networks (de Wit, 1995; Park, Pang and Lee, 1994), where this is allowed by the respective governments. Airlines with alliance partners are also able to take advantage of the potential economies of scope by rationalising routes and fleets and by sharing sunk costs, i.e. finding possibilities to reduce costs faster than they could be reduced by a single producer when expanding output. These benefits have caused so many alliances that airlines are now having to rationalise them (*Airline Business, June 1996, pp 22-51*). Alliance mechanisms can be categorised by three main criteria (Burton and Hanlon, 1994):

- whether they are horizontal, vertical or external
- whether they are motivated more by technological factors or by market forces
- whether the mode of interfirm governance is of a relatively strong or relatively weak kind

The alliances range from specific interlining agreements to full equity participation. There has been much more resistance from governments to foreign equity participation in airlines than in other sectors of industry, but it is quite likely that a domino effect will be spawned from the EU internal aviation regulations and their implications for future 'open skies' deals with other power blocks (*Airline Business, December 1996, p 7*).

Market enhancement opportunities range from the mutual strengthening of hubs to the raiding of another major airline's markets by means of an alliance with a third airline which has permission to operate within the same territory. An example of bridging hubs is the KLM-Northwest alliance, with 201 links behind the US gateways and 107 cities beyond Amsterdam (Feldman, 1996). Raiding is exemplified by BA's purchase of TAT and Deutsche BA. Hub and frequency strengthening is exemplified by the agreement between Alitalia and Iberia to feed on their respective strengths of southeast Asia and Latin America (Flint, 1991).

Between the low key interlining and full commitment equity types of alliance are the more moderate mechanisms of codesharing and franchising. Codesharing offers most of the benefits of equity participation without the financial risks, particularly when the aircraft capacity is also shared. A franchise establishes an even greater market presence, with the use of the major carrier's livery and uniforms, while effectively eliminating the financial risk: BA is the foremost exponent of this, having relationships with several small airlines which form an integral part of the smaller hub operations.

There are also potential difficulties with alliances, as BA found out with its equity participation in USAir, TAT and Deutsche BA. The latter airlines are yet to make a positive contribution to the group profits. The cost levels of these carriers are still not under control, and the USAir relationship has been terminated. Another difficulty can be the clash of interest when a partner teams up with a carrier who is a competitor to the other partner in a different part of the world. Whichever way the alliance is formed, and despite the operational and political difficulties, the future seems to hold the prospect of ever more of them, since studies by the US General Accounting Office and others show that there are significant benefits in terms of market share, revenue and net contribution (Flint, 1996; Hannegan and Mulvey, 1995).

The formation of world alliance groups is also widely anticipated, so that a passenger will be able to start and finish any journey in the world on a single airline code. Already there are six major groupings (Burton and Hanlon, 1994). The recent Star Alliance between Lufthansa, United Airlines, Air Canada, SAS, Thai Airways International and Varig is symptomatic of the trend, while the intended alliance of BA and its partner airlines with American Airlines would make that the largest group. The KLM/Northwest alliance was reckoned to have given them an additional net profit contribution of US $26 million annually, mostly at the expense of their competitors, but also to produce a consumer surplus of US $27 million (*Airline Business, February 1995, p 67a*). It has also been estimated that the BA/AA alliance would improve BA profits by US $91 million in the year 2000, mostly from the effects of more passengers and additional yield more than countering the effects of additional competitor activity. In comparison, a simple codeshare with no open skies deal and hence no additional competitor activity would give a net profit impact of US $74 million (*Airline Business, April 1998, p 67*).

4.2.4 Non-scheduled operations

Non-scheduled operations are usually only a very small part of the total operations of the major airlines, who have traditionally been given the scheduled domestic routes or the international route rights negotiated by their governments, though there are some joint production benefits in providing charters for up to approximately 10% of total rpk (Gillen, Oum and Tretheway, 1990). Non-scheduled, or charter, operations are mostly performed by carriers who have had to resort to this method of operating in order to circumvent route entry restrictions. They are usually subject to less economic regulation than the scheduled carriers, but historically were not allowed to offer services which competed with the designated scheduled carriers. Increasingly, charters are now being allowed to compete with the incumbent carriers with pseudo-scheduled service in domestic operations. They carry significant proportions of the total traffic in countries as diverse as Canada and Nigeria. Perhaps their strongest impact has been made on the European holiday routes, carrying almost half of all passenger kilometres performed within Europe. These specialist carriers have become some of the most competent and efficient airlines anywhere, having overcome the problems of peaky demand and low yields which make them relatively immune to attack from the scheduled carriers. They have recently expanded into the long haul leisure markets. In contrast, US charter carriers largely lost out to competition from the scheduled carriers.

Charter carriers tend to operate at the price sensitive end of the market, and therefore aim to minimise their costs at the expense of optimum scheduling and seat access. Flights are often consolidated if bookings are not sufficient to guarantee a load factor in the 90% range. Aircraft utilisation regularly exceeds 4,000 hours per year, even on relatively low average sector lengths. It is therefore important for these airlines that they operate from airports with low costs and unrestricted night operations and which have runway capacity available at a convenient time for them to schedule three rotations per day. The airlines are usually vertically integrated with large tour companies, and can mostly afford to operate relatively new fleets. Unlike the scheduled carriers, they are more likely to compete on cost than on frequency, so they have tended to trade up from DC 9s through Boeing 737s and A320s to 757s and 767s and, indeed, to trijets where the traffic warrants them. The seat density is invariably higher than that used by scheduled operators, though their use of belly holds for cargo is more limited. The revenues for the airport per landing from charters therefore tends to be considerably higher than from scheduled airlines, but the spend per passenger in the terminal tends to be lower.

4.2.5 Future fleets

The earlier discussion of average aircraft size indicates the continuous assessment that the airlines must make of the balance of advantage in increasing frequency rather than controlling costs by using larger aircraft. Their decision will be influenced by the availability of slots and the possibility of circumventing slot shortage by code sharing and other forms of alliance without losing their market advantage. The UK CAA developed a model called LARAME (CAA, 1989) to predict how airlines decide to respond to increases in demand by increasing frequency and/or aircraft size. They found it necessary to use some 30 different models to describe the behaviour in the different UK markets, and still did not allow for the differing degrees of competition on the various routes. British Airways takes all the following factors into account in determining their fleet acquisition policy, in addition to the passenger and cargo demand forecasts by market:

- Infrastructure issues, e.g. terminal, apron and runway capacity, runway/taxiway separation;
- availability of new or second hand aircraft;
- aircraft economic performance;
- aircraft payload and range capability;
- regulatory policies on noise and emissions;
- investment considerations.

BA is also concerned to ensure that the fleet will be consistent with their commercial plans, will meet or better current and future environmental standards, will maximise fleet flexibility and commonality to allow a rapid response to change and take advantage of new opportunities and will minimise the lifetime costs of the fleet to BA (Acton, 1995).

A major issue is the market for a New Large Aircraft (NLA). British Airways intends to order a genuine 600 seat aircraft if it can be made available at a sufficiently lower cost than the Boeing 747-400, due to their difficulties in achieving the necessary throughput on Heathrow's two runways. A handful of other airlines are also prepared to become lead customers. However, it is not at all certain that an NLA can be produced which would give a satisfactory operating cost, with development cost estimates ranging between US $8 and US $14 billion. The requirement is to reduce direct operating cost (doc) to at least 15% below the 747-400 (Donoghue, 1997). Airbus believes that it has the technology to do this (Greff, 1996) but the programme would only make sense if it can return the investment. With an operating margin of 10%, the return on investment with a production run of 60 per year and an investment of US $10 billion would be 11.1%. This would fall to 3.7% if only 3 per year are sold and the investment were US $15 billion (*The Times of London, April 7, 1997, p 46*).

If Airbus Industrie does not go ahead with its design, BA will have to buy the stretched 747 or increase long haul frequencies, the latter being difficult because of time zone and curfew constraints. There are, of course, other options to overcome the runway slot constraints, not least the release of short haul slots, perhaps with substitution of rail service by a rail alliance partner, as done by Lufthansa at Frankfurt. Airline alliances provide another opportunity for slot rationalisation without losing hub dominance. Thus the pressure for aircraft larger than the 747 may not be as great as anticipated. Only a few years ago, the chief executive of American Airlines was adamant that his airline would never purchase one, since their philosophy was to compete on frequency, regardless of capacity constraints. It is therefore very difficult to predict the future use of an NLA. These doubts are reflected in the variety of forecasts of numbers of aircraft greater than 500 seats in service in 2014, which vary between 470 and 1,380 (*Flight, 16 October 1996, p 14*).

Airport planning requires even longer horizons than those used in fleet planning, so it is necessary to consider the likely market for a second generation supersonic transport (SST). A successful aircraft would have to conquer the problems associated with engine efficiency at both subsonic and supersonic speed, as well as with engine noise having to meet current Stage 3 requirements, with demonstrating that high altitude emissions will not accelerate global warming or further degrade the ozone layer, and with obtaining a sufficient market for the development costs to be recouped. The designs, unlike those for an NLA, are extremely sensitive to cruising altitude. The productivity should help to keep unit costs down, but will itself reduce the number of aircraft required. The best option appears to be a 200 seat Mach 2 SST (Cooke, 1995). If a genuine NLA is launched first, it is much less likely that the risk will be taken to launch an SST.

One of the main reasons for buying an NLA is lack of runway capacity, but it may not give the expected benefits, due to problems of stand and taxi space and, more seriously, due to the increased strength of the trailing vortex necessitating greater in-trail separations. An SST, on the other hand, would assist in spreading the long haul peaks with more flexible scheduling, at the expense of increased stand depth. A further serious technical challenge for the NLA is to meet the Stage 3 noise limits, given the lack of credit in the regulations for inevitable increase of noise with weight above 600,000 lb maximum all up weight. The search is on for new designs which will alleviate these problems while also being more efficient in the airborne phase of operations. The blended-wing designs being studied by McDonnell Douglas (*Flight, 30 October 1966, p 10*) and Aerospatiale could carry 800 to 1,000 passengers, but fall well short of meeting the Airport Council International requirement that wingspan should not exceed 80 metres, which they believe even airports with present ICAO Code E geometry could tolerate if a sufficient level of guidance were made available (*ACI Europe Airport Business, June/July 1996, p 47*).

Decisions on fleet mix do not only revolve around the NLA and the SST. An indication of the importance of understanding the feasibility of future technologies is the huge change in global route networks made possible by the rules permitting extended range over-water twin-engined operations. Similar implications may result from the battle now underway to introduce jets to the under 50 seat category, following the successful 'hub-busting' 50 seaters like the Bombardier Regional Jet and the EMB 145. In fact, these aircraft are being used more to extend the length of spokes into hubs than for hub-busting direct city-pair service, as with Comair's feed into the Delta hub at Cincinnati (Abbey, 1997), where 33% of Comair's passengers fly on 26 sectors greater than 400 miles, and the Continental Express hub at Newark. Lufthansa is also making similar use of this class of aircraft, though BA has delayed their planned introduction. The successes follow on the heels of the rapidly increased use of the 70 seat versions of the F 70 and Avro RJ85s. If the smaller jets do show satisfactory economics, they will bring many new airports into the air transport map, in turn demanding more runway capacity at the hubs and opening the door to even smaller jets.

4.3 OTHER AIRCRAFT OPERATORS

4.3.1 General Aviation

General Aviation (GA) may be defined as all civilian aviation activity other than that performed by commercial air transport. By far the majority of aircraft fall into this category, from private or corporate ownership of Boeing 747s to microlights. The vast majority of these aircraft fly for less than 100 hours per year, but need to be based somewhere. The number of small two and four seat aircraft was originally boosted by policies in some countries, notably the US and France, to encourage the growth of a trained pilot population for defence purposes. This class of aircraft is largely equipped for only flying in Visual Meteorological Conditions (VMC), and most of the pilots who own and fly them do not hold instrument ratings. More recently, less subsidy has been available and new aircraft have been in short supply, due partly to the increase in accident litigation in the US, so the lighter forms of conventional flying have been in decline. Growth has been seen in the new, more capable microlight aircraft for VMC use, and in the heavier end of GA. The latter sector of the market is served by turboprops and turbofan aircraft of between six and 12 seats and with impressive speed, range and weather capability, providing a viable alternative to commercial services for corporate and private business use, whether owned by the user corporations or by air taxi operators. The safety record of corporate aviation is not quite as good as for commercial air transport, but the larger companies exact very rigorous standards of their operations departments to ensure equivalent safety for their executives and their customers. The benefits of using business aircraft stem from the flexibility, productivity and security of the operations. Also, many more airports are accessible than those served by airlines, bringing the passengers much closer to their trip origin or destination.

The ultimate in accessibility is, in theory, provided by helicopters. They come into their own when servicing offshore oil rigs, as well as in the duties that they share with fixed wing aircraft. Their safety, comfort and capability have improved greatly in the last two decades, to the point that they fully complement business jets, often being used for access to them as well as for shorter point to point trips. Twin turbine helicopters form the fastest growing sector of GA.

The extent of GA activity varies a great deal from country to country. In some cases, no private flying is allowed for national security reasons, while in others it is encouraged for its ability to provide transport for business purposes and its role in providing a resource for the military and commercial industry. There are, almost always, opportunities for GA operations in industrial and social roles, from policing and inspection of power lines to ambulance duties. None of this would be possible without primary and more advanced training. Each of these activities needs specific infrastructure support, but does not want to pay for more than is required. There are, therefore, difficult decisions to be taken about joint costs and benefits, which are complicated further by the different access needs, the different environmental implications, and the tolerance of the non-flying public to these differing operations. Training is a particular problem because of the impression of lack of safety due to inexperienced pilots and the variations in throttle settings. Corporate jets are also a problem because of the excessive noise of some of the earlier types and the public sense of a 'them and us' divide. All of these factors require a formal planning response, the more so where, as in the US, there is a great deal of GA activity. Table 4.3 shows the extent of this activity in the US. In fact, US airport system planning is, to a large degree, dominated by GA issues.

The number of active general aviation pilots reached a peak around 1980 and has been declining steadily ever since. This is particularly true for student pilots, which clearly has long-term implications for the size of the general aviation pilot community. The number of pilots holding Commercial certificates, as distinct from commercial pilots holding Airline Transport certificates (who are not shown in Table 4.3), has been declining since 1972. Clearly, the factors that cause people to take up flying, and the type of flying that they do, are critical determinants of the scale and composition of general aviation activity. In spite of the importance of this to forecasts of general aviation activity, and hence planning for general aviation airports, these factors remain poorly understood.

The size of the active general aviation fleet has also declined, but not as rapidly as the size of the pilot population. The number of active aircraft peaked somewhat after the pilot population, around 1984, although the change from 1980 was relatively small and could be due to inaccuracy in the process for estimating the size of the active fleet. This slower decline in active aircraft reflects that fact that aircraft have quite long lives, while aircraft owners are less likely to give up flying than non-owners. Single-engine piston aircraft continue to form the largest part of the fleet, although the proportion has declined from 80% in 1980 to 73% in 1995.

The total number of hours flown by the general aviation fleet has declined more rapidly than the number of active aircraft. The hours flown peaked in 1979 and has declined by 41% through 1995, whereas the active aircraft fleet has only declined by 13% over the same period. While single-engine piston aircraft accounted for 70% of the hours flown in 1979, this had only barely declined to 67% by 1995. Thus not only has there been a significant decline in the average utilisation of the fleet, but this appears to have affected the higher end of the general aviation sector as much or more than the lower end. However, this finding depends very much on what part of the fleet is being considered. At the upper end of the sector, a relative small proportion of the total fleet is used very intensively indeed.

Table 4.3: General Aviation activity in the United States

	1960	1970	1980	1985	1990	1995
Active Pilots (000)						
Student	99.2	195.9	199.8	146.7	128.7	101.3
Private	138.9	303.8	357.5	311.1	299.1	261.4
Commercial	89.9	186.8	183.4	151.6	149.7	134.0
Other [1]	1.8	11.8	16.7	17.4	17.5	18.6
	329.8	**698.3**	**757.4**	**626.8**	**595.0**	**515.3**
Active Aircraft (000)						
Fixed-wing	75.6	127.9	200.1	184.7	184.5	157.7
Single-engine piston	*68.0*	*109.5*	*168.4*	*153.4*	*154.0*	*133.4*
Other	*7.5*	*18.4*	*31.7*	*31.3*	*30.5*	*24.3*
Rotorcraft	0.6	2.3	6.0	6.0	6.9	5.6
Other [2]	0.4	1.6	2.9	5.8	6.6	19.3
	76.6	**131.7**	**211.0**	**196.5**	**198.0**	**182.6**
Primary Use						
Personal [3]	34.4	66.8	96.2	96.2	112.6	109.3
Other [4]	42.2	64.9	113.9	99.5	84.2	73.3
	76.6	**131.7**	**210.1**	**195.7**	**196.8**	**182.6**
Hours Flown (000)						
Fixed-wing	11,977	25,015	38,318	29,085	29,546	22,240
Single-engine piston	*9,788*	*19,182*	*28,339*	*21,102*	*21,883*	*17,135*
Other	*2,189*	*5,833*	*9,979*	*7,983*	*7,663*	*5,115*
Rotorcraft	207	867	2,338	1,990	2,209	1,925
Other [2]	19	148	359	382	341	1,502
	12,203	**26,030**	**41,016**	**31,456**	**32,096**	**25,667**

Notes:

1. Helicopter, glider, lighter-than-air and recreational certificates.
2. Gliders and lighter-than-air craft. Includes experimental aircraft in 1995 (these were included in other categories in prior years).
3. Data for 1960 estimated from 1957 FAA General Aviation Aircraft Use survey. Data for 1970 adjusted for gliders and lighter-than-air craft.
4. Excludes commuter air taxi use from 1990.

Source: FAA Statistical Handbook of Aviation, (annual).

The proportion of the fleet used primarily for personal flying has increased from about 46% in 1980 to about 60% in 1995. Indeed, the number of such aircraft peaked in 1989, declined thereafter, and since 1994 appears to be increasing again. However, while this effect is most likely due in part to aircraft no longer being used for other purposes being purchased for personal use, the statistics are also clouded by changes from time to time in the categories of aircraft use in the surveys from which the information is derived.

The full benefits of fixed and rotary wing GA are frequently denied by lack of access to suitable facilities, despite the significant improvements in their capability. Pressure on runway capacity has meant that they are being squeezed out of the major hubs which they wish to access, either for prestige or, more seriously, to feed a commercial service. In some cases a high flat fee has replaced weight related landing charges, in other cases GA is excluded from slot allocation meetings and can only be accepted on the basis of prior permission or on an ad hoc basis when a gap opens in the schedule. The reliever airport concept, whereby equivalent facilities to those at the main hub are provided elsewhere in the same metropolitan area, is of little benefit if the intention is to connect with a scheduled service, though quicker access may be available to some parts of the city for local traffic. At the other end of the trip, airfields are often not equipped for Instrument Flight Rules (IFR) operation or with customs and immigration facilities, though in Europe these government services are often prepared to operate on a sampling basis as they do with freight. As a more realistic attitude is taken to full cost recovery in all modes of transport, GA can expect to find its costs rising and less of the necessary infrastructure being available. This will be aggravated by environmental lobbies concerned with the noise impact and use of scarce resources by a small and, as they see it, privileged sector of the population, despite the economic benefits which can spin off from a more efficient use of the time of these 'movers and shakers'. It is significant that the same aircraft, when used for purposes deemed of societal value, are accepted without question.

4.3.2 Military aviation

Historically, military activity has provided most of the technical advances, the infrastructure and the pilots used in aviation. It has also had priority in the use of airspace and airfields. Indeed, in the majority of countries, the military provide air traffic control and are fully involved in the planning of airports, as well as being joint users of both airspace and airports. It is not unusual for the airside of airports to be the sole responsibility of the military. The recent 'peace dividend' is changing this in the countries to which the dividend applies. Military flying at civil airports in the UK accounts for less than 5% of movements, and several airfields used exclusively by the military are being closed, to the detriment of the local economies. Unfortunately, they are often located in inaccessible or valued countryside, and therefore not acceptable by either the users or environmentalists for augmentation of the civilian airport system, despite their long and wide runways.

None the less, there are many military sites which could serve a useful civil role, even through joint use rather than being handed over completely. Joint use offers the opportunity to reduce defence costs as well as providing cost savings for the civilian users. Yet in the recent rounds of military base closures in the US, the decisions on which to close were taken only on the basis of military need, so that those which could have had a sound joint-use future were often overlooked. If there is to be coherent system planning, the military needs to be brought into the process at the earliest opportunity, otherwise they might feel a loss of control over the treatment of their requirements and be tempted to employ their ability to veto such plans.

The military sector has been tolerated environmentally because of the national defence needs and because of the strong environmental mitigation policies which have been put in place at the military airfields. However, low flying training, which has become more necessary with the development of terrain following tactics, has brought serious levels of complaint, as has the frequency of crashes close to populated areas. Any joint use would need to address the environmental problems of military activity, but there are many existing examples to testify to the possibility of running a joint-use facility successfully.

4.4 NATIONAL GOVERNMENT ROLES

The role of governments has been discussed in general terms earlier. The emphasis here is on how these roles affect the system planning process through the regulation and certification of the industry and the control of its infrastructure. These aspects are reviewed and then the implications for airport strategic planning are considered.

4.4.1 Airline regulation

All governments regulate the safety of their airlines, requiring them to have air operator certificates which ensure they are competent to fly and maintain their aircraft within a designated territory or route structure. They are cognisant of the need to keep a close watch on operating trends which might lead to a reduction in safety, such as leasing foreign equipment, outsourcing of maintenance or rapid expansion in the number of carriers. In contrast, many countries have now liberalised the economic regulation of their domestic routes as explained in chapter 2.

The US was always a prime candidate for deregulation because there were many significant airlines, all in private hands, operating in a large market. Most other countries have been much slower to liberalise their air service agreements and to open up their internal markets, though the European Union, the UK and some other countries started to take liberalising action in the 1980s. The major airlines in most countries are still in public ownership, still tend to be identified with carrying the flag, and often are too small to preserve a reasonable market share in open competition. It is difficult to maintain the ethos of a level playing field when there is a mixture of public and private operators. It has been suggested that the subsidies they receive should be regarded more as compensation for government interference than as providing unfair competition for those carriers who do operate independently of their governments (Doganis, 1991).

The world is, however, changing quickly as the domino effect of US and European liberalisation spreads, aided by the trend to privatise the flag carriers and to give more prominence to carriers who previously had to settle for the crumbs from the flag carriers' table. The national case studies illustrate this increasing competition, particularly in domestic markets but also with the secondary carriers being allowed a share of international routes. There is now some danger of countries with only thin traffic taking up the ethos of unbridled competition in such thin markets that instability is certain to occur.

In a liberalised setting, the carriers are able to manoeuvre so as to minimise competition on individual routes and in particular regional markets. This is made easier by shortage of capacity for new entrants, and the consequent need to supervise the allocation of runway slots and terminal gates. Few governments have taken sufficiently firm action to ensure that competition can flourish, partly due to the difficulty of providing the necessary capacity and partly because of the need to continue to ensure that their major airlines do not suffer on the global stage from excessive competition on their home territory.

There is still a need, which has not really been met, to bring the normal regulatory disciplines of free markets to bear on the airline industry. Oligopolies, and in some cases, monopolies have been allowed to flourish in markets where more competition would clearly benefit the consumer. Governments have been light-handed in their use of antitrust regulation in attempting to ensure that their countries' airlines have a strong base from which to compete in world markets. The US Department of Transportation (DOT) is only in 1998 likely to impose some limits on the predatory pricing and gate hoarding policies of incumbent hub airlines in the face of an incipient low cost new entrant (*Flight, 15 April 1998, p 4*). This follows action in 1997 to open some slots at controlled airports and to control the use of computer reservation systems. The EC is also considering action in all of these areas, but, given the strong expressed desire to encourage competition, the actions may be too late as well as too little.

There is a need in a liberalised setting to increase the safety oversight. Many studies were done in the years after deregulation in the US into the impact on safety. None were able to draw the conclusion that safety had deteriorated relative to what it would have been without deregulation, but evidence of a reduction of margins was manifest. The fear has been raised again by the Valujet accident, again without real statistical support, and defences need to be put in place.

4.4.2 Air Service Agreements

Air Service Agreements (ASA) can be bilateral or multilateral. The latter exist among some groups of countries such as trading blocks, but bilateral ASA are more common. They are designed to ensure that, regardless of the relative strength of the economies, the national carriers obtain acceptable shares of the market. This usually requires setting rules for seat capacity offered and fare levels. Often there is a limit on frequency, together with other restrictions such as the airports which can be used. The agreements are often confidential, and seriously constrain the network development of existing carriers, as well as potential new entrants.

Most bilateral agreements still limit the number of gateways, so that the networks are constrained through the existing gateways to a greater extent than the search for economies of scope would otherwise require. As the major US carriers increase long haul service to their inland hubs, the market share of the traditional gateways has been falling (Heidner, 1992). The international equivalent of deregulation is an 'open skies' policy which, in the extreme case, would give foreign carriers cabotage rights, i.e. they would be able to obtain revenue from lifting purely domestic traffic. This would usher in a whole new era of competition in a country's domestic market as well as changing the route networks around the gateways favoured by the new entrants. The network would depend considerably on how the governments treated the matter of access to the major hubs. These are almost always congested, all the slots having been guarded jealously by the incumbent carriers. If draconian policies were to be adopted for the release of a significant proportion of those slots, the route map might not change in the same way as if the only way the new entrants could express their freedom was to use existing capacity elsewhere.

4.4.3 Airport development

Airports have long been seen as national assets and a crucial part of international travel and trade. They have also had strong military significance, as well as often being seen as symbols of national and civic status. Governments have therefore historically been intimately involved in their development, and have accepted the need to fund it. The interest has, in some cases, been enhanced by the existence of national transport strategies which have defined a role for aviation and the planning of the associated land use, though aviation has more often been administered through other government agencies such as the Ministry of Defence. The changing circumstances outlined in chapter 2 have now caused many countries to consider if it is really necessary to keep such tight public ownership and control of development of the aviation infrastructure, and, indeed, of other transport infrastructure. The attractions of not having to fund the expansion, which has proved to be necessary to keep pace with the economically beneficial growth in air traffic, has caused many governments to remove themselves from ownership, or at least, management of their airports, and to encourage normal market forces and land use planning processes to apply as far as possible. The extent of this, and its consequences, are discussed in chapter 6 and also in the UK case study, where the need for strong government control of monopoly situations is identified. The implications for planning are that private ownership and funding might result in investment decisions very different from those of a government body with responsibility for a public facility.

An area which should be a government responsibility if it is to be effective, is the control of land use around airports. Time and again, governments have moved airports to remote sites in order to allow operations without large scale environmental impacts on local populations, only to allow development to follow the airport and make it difficult to increase capacity, as at Stockholm's Arlanda airport (Roenqvist and Jonforsen, 1997). Clear directives can help local authorities manage situations of encroachment, though they will always want flexibility to cope with their own priorities. The EC wishes to introduce measures along these lines. Some of the responsibilities are already taken in the rules in ICAO's Annex 14 which limit building heights, in the recommendations for noise-compatible activities and in legislation for public safety zones. There is considerable variation in government attitudes to these controls and the rigour with which they are applied, yet they are essential planks in any strategy of retaining flexibility for possible future expansion of capacity.

4.4.4 Air traffic management

The provision and development of an Air Traffic Management (ATM) system is perhaps the most important remaining input to the aviation system. The integration of airspace and airport planning is essential for the efficiency and effectiveness of the complete system. Continued development of the aviation system requires an air traffic control (atc) and management system which will be safe and effective. All types of aircraft and operations must be accepted without undue delay and restrictions on entry to the system, while maintaining safe separation between them. National sovereignty over airspace dictates that governments take responsibility for ownership of the system, which normally extends down to the runways. The system consists of the hardware, the software and the personnel for the aircraft to navigate expeditiously, and to ensure that the aircraft are adequately separated from each other. In Visual Meteorological Conditions (VMC), some countries allow controllers to devolve the responsibility for separation in controlled airspace to the pilots, which results in considerably greater movement rates without apparently increasing risk.

Flying in Instrument Meteorological Conditions (IMC) requires that separation be maintained by controllers monitoring radar returns or by receiving position reports and applying large nominal time or distance separations, the latter giving much reduced capacity. Only aircraft which are suitably equipped are allowed into the controlled airspace for which the controllers take responsibility. Thus aircraft have to demonstrate minimum navigational capability, as well as having the necessary communication equipment, which will usually include transponders to allow the aircraft to be interrogated by Secondary Surveillance Radar (SSR).

Airspace is becoming busier as airline competition increases, even though the GA sector has not grown at the expected rate. In the US, the 20 Continental Air Route Traffic Control Centers handled almost 41 million IFR operations in fiscal year 1997. The busiest Centers, Cleveland and Chicago, each handled over 2.8 million IFR operations, while Seattle Center handled the least number of operations, not quite half that. It is projected that by fiscal year 2009, the 20 Centers will have to handle over 50 million IFR operations per year, with the four busiest Centers each handling over 3 million IFR operations (FAA, 1998b). Air carrier operations are projected to account for 57% of this traffic, with air taxi/commuter operations accounting for a further 17%, general aviation for 19% and the military for the remaining 7%.

Terminal airspace capacity can be enhanced significantly by improving procedures in the airspace serving multiple sets of runways. The gains that can be achieved by any specific improvement can be estimated using computer simulation techniques, such as the FAA's Airport and Airspace Simulation Model (SIMMOD). The capacity can be enhanced further with the application of new technologies which increase the precision with which aircraft can follow complex paths. Microwave Landing Systems were originally developed to achieve this, but it is now expected in the US that the same capabilities and freedom from the constraints of the existing Instrument Landing Systems can be achieved by navigation capabilities based on the use of Global Positioning System (GPS) satellite navigation, with techniques to augment the precision of the position information. Further improvement is possible using automation to determine optimum sequencing, spacing and scheduling for arriving traffic and to help the controller in merging missed approach and 'pop-up' traffic into the final approach stream. Similar developments are being explored to increase departure efficiency. In addition to these developments, the use of enhanced Traffic Alert and Collision Avoidance Systems (TCAS) is planned to further reduce separations as well as improving safety.

In Europe, severe delays were experienced in the late 1980's, some 80% of them being caused by congestion bottlenecks in the system. A more centralised air traffic flow management (ATFM) system, termed the Central Flow Management Unit (CFMU), was developed to allow airlines to take the majority of these delays on the ground. Concurrently, a task force was set up by the European Civil Aviation Conference (ECAC) to coordinate the efforts of the individual nations, NATO, Eurocontrol and the EC to resolve the situation. The resulting strategy put the emphasis on improving atc capacity and reducing the inefficiencies of indirect routings by combining reorganisation of terminal airspace with the introduction of increasingly smart management, while preparing for the introduction of the new technologies. The efforts are being coordinated within the European ATC Harmonisation and Integration Programme (EATCHIP). Considerable improvements have already come from better communications between centres.

All high density areas were due to have en-route radar separations of five nautical miles and other areas 10 nautical miles by 1995, with automatic data communication between air traffic control centres by 1998. In the course of time, the EATCHIP programme plans to use similar technologies to those in the US, with the Mode S air/ground data link, satellite navigation and communication and the increasing use of direct routings possible with these aids to surveillance and advanced navigational capabilities. The ultimate aim is to have an integrated system which will take full advantage of capabilities defined by ICAO for the Future Air Navigation Systems (FANS) concept, and will be able to plan non-conflicting four dimensional flight profiles from gate to gate. IATA expects that as the benefits of these programmes appear in the near future, airspace capacity will cease to be a serious constraint, provided that the investments are made and that the smart technology comes on stream as planned.

FANS is also important in being able to fly direct intercontinental routes, as is the granting of right of way. In those areas of the globe where it is difficult to provide radar surveillance, Automatic Dependent Surveillance (ADS) is able to provide controllers with an equivalent picture generated by accurate position information which the aircraft are able to obtain using FANS and pass back to a control centre. Together with the use of TCAS, this allows much smaller separations between aircraft than has been possible with the alternative procedural rules.

In the long term, a crucial decision will have to be taken on whether to let an autonomous aircraft deal with its own separation requirements and fly as though it were in visual conditions because of its ability to track surrounding aircraft, or to keep the control of separation in the hands of a ground system with a lot of decision support for the controller. Either way, there will be increased sharing of information between the air and the ground, but these and other earlier decisions taken by governments as to which technology to adopt and when to install it will make dramatic differences to the airspace and runway system capacity, particularly in the need for and the siting of new runways in metropolitan areas.

4.4.5 Aircraft certification and personnel licensing

Governments are required by ICAO to take responsibility for the certification of aircraft used by the air carriers they license. This is done by a civil aviation authority, whose regulations are put into the country's law. The aircraft are given an airworthiness certificate, often only after costly modifications to take account of differing requirements between the country of manufacture and the country of operation. The majority of European countries have now become members of the Joint Airworthiness Association, which has managed to harmonise many of the safety regulations of the individual countries. Efforts are now being made to achieve harmonisation between the JAA's regulations and those of the US Federal Aviation Administration (FAA). Countries with small aircraft fleets or small manufacturing facilities have tended to use either the FAA regulations or those of the country by whom they were originally colonised.

Aircraft certification extends to noise and emissions. The requirements (ICAO, 1993) are framed around nominal flight procedures which have some relationship to actual operations at airports. The noise is measured at points under the approach and departure paths and also off to the side of the takeoff path. The main engine emissions are measured at four nominal engine settings and are then calculated for a complete approach, landing, taxi, takeoff and climb cycle. The early thinking about the need to mirror actual operations is addressed by Smith (1989a). It is interesting that, when regulation was first broached in the US in 1968, the industry felt it necessary to support it, not only because otherwise it would have been 'like being against motherhood', but because there were so many vested interests that a government leadership role was needed, because it might well affect the industry's growth, and because it would shift some of the burden from the industry's shoulders to the federal government (Stevenson, 1972).

Another duty of the responsible authority is the granting and renewing of licences to all workers in safety-critical areas of the aviation industry. This includes flight and cabin crew, including medical fitness and maximum hours of work, and maintenance personnel for all levels of the system. In the event of an accident, the government is responsible, preferably through an independent agency, for investigating the accident to determine the cause without placing blame on any individual.

4.4.6 Airport certification and regulation

Since it is the duty of the airline and ultimately the pilot in command to ensure the safety of operating into an airport, not all governments have seen fit to certificate airports. However, ICAO's Annex 14 has Standards and Recommended Practices (SARPS) which should be applied to the design and operation of the different classes of airports in order to harmonise the protection afforded to aircraft and the aids available to them at international airports. These

SARPS are interpreted for domestic airports by the national authorities, for example, the FAA issues Federal Aviation Regulations (FAR) Part 139 and the UK Civil Aviation Authority (CAA) has CAP 168. Most authorities require that these standards or their equivalents be met before a licence is granted for the relevant category of aerodrome. Most authorities also require an airport to have an approved operations manual which details, among other things, the procedure for coping with emergencies. There is also usually a system of annual inspections or audits of safety.

As airport ownership is liberalised, it becomes necessary for governments to apply residual economic regulation to them as it does to liberalised airlines. There are many options for liberalising airports, each of which calls for its own blend of regulation. The UK has charged the CAA with ensuring that prices do not rise to take advantage of monopoly status and that the BAA does not cross subsidise its airports. South Africa, in commercialising and then privatising its airport system, has regulated charges and also decided to set minimum service standards and monitor compliance (Prins and Lombard, 1995). The form of oversight provided will affect the airport's behaviour accordingly, as instanced by the BAA and Manchester airports' attempts to raise as much revenue as possible outside the activities which are subject to price control.

4.5 IMPLICATIONS FOR AIRPORT PLANNING

The variability and unpredictability of demand after deregulation in the US has made life very difficult for infrastructure operators. Traditional relationships between airlines and airports have, in some cases, been severed; some airports have been left with surplus capacity while others are under strain. The federal system of funding airside capacity means that it is often justified on the basis of transfer traffic which can fluctuate in response to airline network strategies and competition. The process of federal funding chasing the demands generated by intense hubbing tends to perpetuate the system's mode of operating, perpetuates barriers to entry and frustrates attempts to institute any changes which might be needed to reoptimise the system, perhaps by encouraging growth in regions with high traffic potential (ASRC, 1990).

Hubbing involves scheduling alternating banks of arriving and departing flights (termed a connecting complex) in order to allow the interchange of passengers and bags. This causes runway and taxiway congestion during both the arriving and the departing banks as well as congestion on the apron during the complex. In the times between complexes it is common for the apron to be completely empty. Hubs compete on frequency for transfer traffic, resulting in anything between three and ten complexes per day being performed. This has tended to keep down the average size of aircraft. The competitive forces that cause airlines to operate in this way result in serious peak loading of the airport facilities. When Manchester airport in the UK decided to construct an interline hub, it had to change its planning to construct a second runway when the traffic reached 15 mppa, whereas it had expected to exceed 20 mppa on its single runway when its business plan was based on allowing natural growth of local short and long haul scheduled and charter services to continue.

The development of hubbing in the US seems to have raised the frequency of service between major cities and increased direct connections from small cities to competing hubs without having cost the passenger a significantly higher fare. It has, however, caused congestion at the main hubs. While it appears that congestion pricing could raise the revenue for expansion without unduly inconveniencing the incumbent airlines, the system of funding for infrastructure in the US has not made it easy to adopt this solution. This reasoning does not take into account the incumbent airlines' resistance to an expansion which would simply mean that their payments would be used to allow greater competition. Bluntly, the major carriers would rather invade their competitors' territory via a new hub than open the door to a new entrant at their own base. While the FAA forecasts tend to predict further traffic at major hubs, and hence justify further expansion there, other forecasters claim that the hubbing system has distorted traffic figures and that airline decisions will determine the future distribution of transfer (or 'soft') traffic (ASRC, 1990). It is argued that the mergers which have created megacarriers will reduce competition and therefore traffic growth at major hubs, but that traffic patterns will shift to the secondary cities, reflecting demographic and economic changes and requiring extra capacity there. The soft traffic should be considered over a range of scenarios applicable to each major hub, considering the incumbent airline's strengths, commitment and potential strategies, together with the characteristics of the city and the hub's competitive position.

The peaking problem at existing hubs can only be expected to worsen as hubbing becomes more exaggerated and as competition increases. The hubs may be able to increase their capacity by the use of smart technology, and in some cases by adding runways, but not at the same rate that demand for slots is likely to increase. More airports are therefore likely to become slot constrained, and to require some form of regulatory intervention in order to avoid serious domination by the incumbent based carrier. The US General Accounting Office has suggested compulsory lotteries for slots (Walker, 1997) and the European Commission is overhauling its policy on runway slot allocation, though it may stop short of the suggestion by the UK CAA of slot auctions.

The concept of a continuous hub has been introduced by ASRC (1991) as an alternative way to reduce the peaking problem. In this, it is argued that the high unit cost of the megacarriers, caused by poor utilisation of both the aircraft and the hub airport, can be reduced by random scheduling of aircraft rather than allowing the strict 'banking' of connections to determine the schedules. Then the aircraft and airport facilities would be continuously in use, throughput would rise substantially with the same level of investment and the turnround time would become more important. Connection times would be minimised by increasing frequencies, so that the average aircraft size would fall further. Only the largest hubs would be able to operate in this way, so the implications are that there would be less of a requirement for a proliferation of smaller hubs. American Airlines is reported to be rescheduling its flights at Dallas/Fort Worth airport as a continuous hub in order to avoid some of the hub inefficiencies and to increase its capacity (Treitel and Smick, 1996).

Often the shortage of gate positions or aircraft stands is the main peak problem. Again the use of smaller aircraft is counter-productive, particularly when the problem is actually one of too few stands rather than of absolute passenger movement rates per hour. Larger aircraft use fewer stands and move more passengers per unit of apron, but may not have a greater flow of passengers per hour because of their longer turnround times (Caves, 1994a). Some alleviation of the problem is possible if the airline is willing to reschedule its flights to a pattern of smaller aircraft first in and first out, rather than bigger aircraft last in and first out (Chang and Schonfeld, 1995), if it is prepared to sacrifice some costs of aircraft delay and passenger time. This is, however, unlikely to overcome the fundamental problem, any more than is the use of larger aircraft. Apron space is becoming a very serious problem at many of the busiest airports and difficulties in locating new terminal and apron capacity between the runways results in substantial taxiing inefficiencies.

An apron which becomes used more extensively for hubbing is subject to increased pressure in several ways. There is a need to extend piers to provide more space on the air side of the security cordon for the increased boardings inherent in a hub-transfer system. This causes less space to be available for apron expansion, yet there are more boardings and fewer efficiency benefits from the use of larger aircraft. In addition, the stand use pattern, with a concentration of activity followed by a lull, reduces stand utilisation. Stands which are dedicated to airline hubbing are unoccupied for almost as long as they are occupied.

A comparison between European non-hubbing and US hubbing operations is shown in Figure 4.1, the European results referring to all operations at Heathrow and Munich by all airlines and the US operations being at Dallas/Fort Worth, Denver and Chicago O'Hare by hubbing airlines. The American Airlines operations were observed at Dallas and O'Hare, while the airline observed at Denver was Continental. The US airlines had control of their own

Figure 4.1: The effect of hubbing on narrowbody turnround times

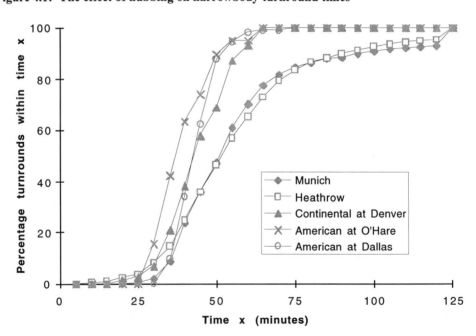

Source: Caves, 1994a

terminals and aprons. The average turnround time observed for narrow-body aircraft in the USA was 44 minutes, compared with 61 minutes in Europe. The corresponding values for wide-body twins and trijets was 51 minutes in the US and 142 minutes in Europe.

The data show that longer average times in Europe were caused predominantly by the extended turnrounds of a small percentage of the fleet rather than the excessive minimum scheduled times: indeed, the minimum times for the European fleets were shorter than those in the US.

There was a larger discrepancy between the US wide body fleet and the trijet wide body aircraft at Heathrow, caused by the preponderance of long haul international operations in the European case compared with the relatively short haul domestic operation in the US. The turnround times achieved seldom approached the inherent capability of the aircraft, even in the US hubbing context, because of the need for passengers to make connections, although there were several cases of turnrounds of both narrow and wide-body aircraft being reduced by some 50% of the scheduled time in order to make up for a late arrival. Base turnrounds of long haul aircraft need to be much longer than the minimum possible, not only to allow for delays in the turnround activities but also because the scheduling seldom allows an immediate departure, so the stand is used for storage rather than for productive movement of traffic.

Congestion has also had a serious impact on scheduled block times, and this has been exaggerated by the competition for punctuality among the airlines. A flight from Frankfurt to London Gatwick which took 85 minutes in 1982 is scheduled for 110 minutes in 1995 (Dennis, 1995). Although some of this time is spent in the air, the use of flow control tends to result in a considerable proportion of the 'slack' being taken on the ground, causing a demand for more apron and taxiway capacity.

Provision of the peak capacity demanded by the new competitive hubbing airline behaviour is a costly business, even when it can be provided at all. Yet the airlines are trying to minimise the costs of infrastructure, as they are with all their other costs, while the airports also want to ensure that they retain their market share. On the one hand, airports are offering to waive their charges for transfer passengers (Dennis, 1995) while, on the other hand they are suffering an increase in their costs per passenger compared with the non hubbing and less competitive situation.

The freedom given to airlines by deregulation was not extended to airports. They had to accept the consequences of the airlines' strategies. Some were chosen as hubs and invested in order to cope with the very large increase in boardings. In many cases the hubs continued to prosper even through the recession, due to the success of their hub airline(s) in the competition for transfer traffic, to the airport's ability to provide the infrastructure necessary for efficient operations, and to a strong local market. The impact on a city's economy of the excellent air links which a hub brings tends to reinforce the local demand further, as happened at Pittsburgh (Dennis, 1995). This cushions the exposure to the 'soft' over-the-hub traffic which is more volatile and open to diversion by other hubs.

There have also been many examples of hubs being established and then being discontinued, either because the airline failed or decided to retrench. In the case of Charlotte, traffic rose quickly to four times the original levels, only to suffer a severe cutback after a few years. In the UK, both Gatwick and Luton have seen many false starts due to successive airline failures. Even the major hub airports are not immune to the withdrawal or failure of a hub carrier, as shown by Continental Airlines at Denver and Eastern at Atlanta. Now hubs are under attack from low cost new entrants, for example US Airways' Philadelphia hub by Nation Air and Air Inter's operation at Paris Orly by Air Liberté.

In contrast, there are airports which have been bypassed by the airlines' new strategies. The UK case study shows how the smallest category of airport has suffered particularly from the tendency to consolidate service. Airports within ground access distance of a strong hub have lost service as passengers seek the utilities of greater frequency and choice of destination (Kanafani and Abbas, 1987), reinforced by the wider variety of discount fares on offer. Other airports have lost jet service in favour of turboprop feeders. It is not a unique phenomenon of deregulation for airports' roles in a system to be changed quite dramatically. Shannon was perhaps Europe's main Atlantic gateway during and after the flying boat era, only to be overflown by most carriers as soon as aircraft had sufficient range to reach the continental capitals nonstop. The South Pacific islands have similarly been the butt of political and technological change (Taylor and Kissling, 1983). Now a new threat is posed by the globalisation of the airline industry, in that rationalisation of routes and hubs may result in some other airports being neglected. One of the reasons that the Dutch government was not in favour of a KLM/BA merger may have been the fear that London might gain service at the expense of Schiphol.

The smaller airports, which are only likely to be able to grow their market share by encouraging the new lower cost carriers, often have to offer very substantial discounts. It is doubtful if the aeronautical revenue from these carriers could justify any investment in new capacity. The more established airports are not keen to agree to the terms which these carriers need, as evidenced by the stance of Aer Rianta which made Ryanair threaten to start up its own airport in Dublin, and by Amsterdam Schiphol advising Easyjet not to start a service from London Luton. At this small end of the airport spectrum, the big threat is that airlines will use regulatory freedom to downsize their operations in terms of routes and aircraft size, and ultimately pull out completely in the search for economies of scale. This can result in a downward economic spiral for the airport and, possibly, for its community. The UK case study shows how airports can improve their chances by undertaking market analyses and entering joint marketing agreements, then offering incentives for new route starts or even create new airlines themselves when the analyses are sufficiently positive. However, they need to take care not to fall into the trap of being unable to get back to average cost pricing.

It can only be concluded that liberalisation of airline regulation has brought uncertainty to the management and planning of airports. As with the airlines, there have been winners and losers. The difference is that airports have had little control over the outcome, but the uncertainty brings with it a series of opportunities and threats. It is probably easier for the opportunities to be grasped if an airport has the same blend of entrepreneurial management and freedom of action enjoyed by the privately owned airlines who are leading the reshaping of the industry. Roles as hubs or spokes are up for bidding. Publicly owned and managed airports can equally well participate in this process, although there will be concerns that this does not distort the competition. Geography, scale of operation, capacity constraints and the regulatory freedom enjoyed by the available carriers all dictate that some airports' roles in the system are more naturally competitive than others (Dennis, 1996).

It will be necessary for airports to understand the limitations of their ability to meet their ambitions as this less structured system evolves. The Dutch government has certainly understood this, as well as understanding the huge role that Schiphol plays in the country's economy. They are well aware of the benefits and the risks of their 'Mainport' strategy for the airport. Their economy is small compared with those of the countries whose airports are competing for this role in Europe, so the only way that they can enjoy the same quality of service is to encourage KLM to develop a very strong hub and thus help to keep their economy competitive as well as directly feeding more jobs into it. However, the penalty is that the environment around the airport deteriorates, while the majority of the passengers whose travel is causing the damage are not directly interested in Holland at all. The threat is that other hubs will compete even more successfully, on the basis of their stronger local demand, resulting in a greater risk that Schiphol's investment will not pay off (Veldhuis, 1992). Intensive studies of the economic and environmental impacts have been undertaken, as have studies of the implications of competition (Ashley, Hanson and Veldhuis, 1994; Veldhuis, 1997), all of which can be regarded as state of the art in their field. The planning process is itself innovative in that a form of joint planning authority comprising the government, the regions, the airport and the airline was able to agree on a plan which allowed sufficient expansion to guarantee that the airport could compete without undue constraint, while containing the environmental impacts to a tolerable level (Veldhuis, 1996). With the aid of a new runway, mainly for mitigating environmental impact, the airport should be able to grow to 44 mppa.

5

COMMUNITY RESPONSE TO AIRPORT DEVELOPMENT

The relationship between an airport and its local community is crucial to the airport's future. In many cases development of the airport will be widely viewed as necessary to support the community's economy or to expand essential transport links, in which case many in the community may consider that the benefits of the air services more than offset concerns for any inconvenience caused by them. When the airport's main role is to support a wider regional, national or international community, conflicts are likely to arise between the advantages which the airport's growth bring to the wider community and negative impacts which are perceived within the local community. In either case, the positive and negative impacts will, in general, be felt differently by different groups within the community. The differences will occur between socio-economic classes, industrial and commercial sectors of the economy, geographic districts and political jurisdictions. This chapter examines the nature and value of these benefits and disbenefits, following the arguments developed in an earlier paper (Caves, 1994b). It then describes the attempts to achieve a workable balance and the consequences for airports of adopting the 'win-win' strategies which are increasingly necessary.

5.1 AIRPORT BENEFITS

5.1.1 Economic activity

The main benefit that an airport can bring to a community is to ensure that economic growth is supported by adequate availability of commercial and private aviation. Attempts to derive a comprehensive measure of economic benefit have tended to follow a methodology advocated by the Federal Aviation Administration (FAA) (Butler and Kiernan, 1986) that measures the contribution made to an economy by aviation, by adding up the direct, indirect and induced effects. Direct impacts are the consequences of the economic activity occurring at the airport. Indirect impacts are those that are directly attributable to, but occur away from, the airport. Induced impacts are generated by the direct and indirect expenditures triggering a chain reaction through the local economy. Thus, a study for the International Air Transport Association by SRI (SRI, 1990) calculated that aviation's contribution to 22 countries in Europe was US $200 billion and 6.7 million jobs. In common with the many other studies using this methodology, nearly two-thirds of the contribution came from the induced effects. The implication drawn by SRI from this assessment was that if the traffic volume increased by 10% less than for an unconstrained forecast, there would be US $10 billion less activity per year by the year 2000, i.e. perhaps 3% less than there would have been in the unconstrained situation. A US airport advocacy group (USA-BIAS) has used a similar methodology to claim that daily service to a new US gateway from London would add at least US $268 million to the local US economy in the first year (Creedy, 1991). These estimates include stimulation of exports and foreign investment of US $139 million in addition to the impacts derived using the FAA methodology. Studies have also been carried out at many smaller airports in the US. In North Central Texas, it was estimated that each general aviation movement at airports without air carrier service was worth US $200 per operation (equivalent to US $110,000 per based aircraft per year) to the local economy, while at Dallas Love Field with its commercial traffic, the values were US $9,000 and US $4 million respectively (Dunbar, 1990). Summaries of other studies on Amsterdam, Charles de Gaulle, Copenhagen, Geneva, Glasgow, Manchester, Munich II and Zürich airports have been compiled by AACI - Europe (AACI, 1992a).

Taking the case of a typical local airport in the UK, East Midlands International Airport (EMA) contributed £20 million and 2,500 jobs directly to the local economy in 1985, when the throughput was 925,000 passengers, 15,000 tonnes of flown cargo and more than 8,000 tonnes of trucked cargo. Using an income multiplier of 1.4, a typical value for local economies in the UK, the total contribution of airport employment to the local economy was estimated at £28 million (Caves, Gillingwater and Pitfield, 1985). There is a clear difference between the total economic impact at small airports, where many of the services cannot be provided locally, and the major gateways and their associated regions. In the latter case, multipliers between 1.7 and 2.5 may be justified.

Many criticisms have been levelled at the way the impacts of airport activity have been quantified in applying the FAA methodology. Some of these are that all other sectors of the economy could claim corresponding indirect effects, that the multiplier argument can only be applied to self-contained regions providing no alternative consumption, and that the indirect effects and multiplier effects can easily be double-counted (Karyd and Brobeck, 1992). The UK Civil Aviation Authority (CAA) has noted several of the more important objections (CAA, 1990), pointing out that there are cross effects of increased air transport on other modes, both by diversion from another mode and from increased loading of the airport access systems. More importantly, it is the additional impact of investment in airports over and above the impact which would accrue from alternative investments that matters. Also, trade and tourism operate in both directions. Not only is it the net effect which should be calculated, but the result depends on how the denied passenger would have behaved. Some of these pitfalls have been avoided in a methodology being developed by European airports (AACI, 1992b).

The reality of the need to include outgoing expenditure is shown by the Japanese encouragement of overseas tourism in order to funnel back part of its huge trade surplus (Wijers, 1991). Given the preponderance of UK holidaymakers among UK regional airport passengers, the net effect on the local economy might well be negative, even though most of the holiday spending goes to UK airlines, airports and tour operators, rather than to foreign economies. Similar comments have been made by the Center for Policy Research of the US National Governors' Association (Cooper, 1990). There is also no doubt that too little account is often taken of the secondary implications of change. The SRI report discussed above estimates that some of the potentially lost traffic will be carried by accepting inconvenient timings and higher load factors, but gives no credit for the increased income from the higher fares or the increased average spend per tourist due to the poorest ones being denied the opportunity to travel. However, the CAA has studied the net impact of new long haul services at regional airports in the UK, and has shown that the region and the UK will obtain a positive net benefit even if the service is not provided by a UK carrier (CAA, 1994).

The most fundamental error associated with the use of the multiplier in assessing the aviation contribution to the economy is to imbue the resulting linkage as necessarily having a causal quality, i.e. that jobs at the airport cause jobs in the rest of the local economy. As de Neufville (1976) said: "True enough, activities at the airport are closely tied to the whole web of urban and regional activity. But as everything is connected, the argument is totally circular." The remaining direct and indirect impacts are a quite complex accounting exercise which serves simply to indicate the size of the industry relative to the rest of the economy. Thus the jobs at Toronto's Pearson airport account for 1.7% of all local jobs (Hickling, 1987) and the total impact of major US airports accounts for approximately 3% of total regional economic activity (Wijers, 1991). The Air Transport Association of America (ATA) has estimated that air transport accounts for 5.3% of GNP and it has also been estimated that, despite high leakage rates, aviation accounts for 8% of Hong Kong's economy (Nyaga, 1989). The numbers are often impressive in absolute terms and certainly indicate to policy makers the extent of the likely short term disruption to the economy which would follow a loss of confidence in the air transport industry. However, given the derived nature of the demand for transport, the large numbers may also prompt the question whether an efficient economy should not run with a lower proportional spend on intermediate goods.

This of course simply points out the fallacy of using expenditure to measure economic impact. Such an approach implies that if the price of aircraft fuel or the wages of airport workers were to double, then the contribution of aviation to the local economy would increase, which is clearly absurd. Rather, what is needed is a methodology to measure the non-aviation economic activity that occurs as a result of the availability of air services and that would not occur if restrictions were imposed on the ability of the aviation system to provide those services.

5.1.2 Stimulation of the economy

The challenge is to understand how the presence of air services actually stimulates the local or regional economy, rather than simply being bound up with it. The FAA suggests that this effect be measured by asking firms how vital an airport was to their location decision or what would happen to the firm if an airport should close. The answers to these questions are obvious in most island and some tourist economies. The Pacific island economies were severely threatened when the intercontinental trunk airlines began to overfly them on the way to Australasia (Britton and Kissling, 1984). Areas with intrinsic tourist potential usually only blossom into resorts after the provision of facilities for direct air services. This was true, for example, in Greece, Yugoslavia, Tunisia and Spain (United Nations, 1974) and in India at Goa. In remote areas, the maintenance or revival of the economy may depend entirely on the accessibility provided by aviation, whether it be by float plane in Alaska or the conversion of a run-down coal mining economy into a ski resort in Yampa Valley, Colorado (Cooper, 1990). In developing and isolated economies, air transport's more important contribution has been to enable production of specialised products like fresh fruit and flowers from Africa (ICAO, 1977), though some see airports in developing situations more as 'parasitic insertions' (Graham, 1995).

Even in these relatively clear-cut situations, the economic benefit to the local community may be small, even though the tourist industry is big. In the Pacific microstates (Britton and Kissling, 1984) and in Israel (Haitovsky, Salomon and Silman, 1987), the local economy often receives only 20% to 30% of the total holiday price. This can be improved if the country's own airline brings in foreign exchange, though the aircraft investment can be daunting, being of the same order as the hotel investment (Wheatcroft, 1994). There are other examples of attempted tourist developments where reliance has been placed on the provision of an airport to stimulate the activity, as in northeast Brazil and in East Africa at Mount Kilimanjaro, without the other necessary conditions being in place.

It is harder to sustain the same arguments unequivocally in developed economies. Certainly there is no reason to doubt the truth of the classic theory of economic geography that lower transport costs allow firms to gain new markets and then reap economies of scale, the benefits spreading to other sectors of the economy through the multiplier effect, provided that the factors of production are mobile (Evers et al, 1991), but the demonstration of this tends to rely on anecdotal evidence.

Studies have been carried out to establish the extent to which a convenient airport was instrumental to a location decision by companies. In a study of general aviation airfields in Massachusetts, 23% of firms surveyed said that an airfield was essential to their site location decision. While 19% of firms said they would relocate if their 'base' airfield closed and a further 7% said their businesses would fail, 66% said they would use the next most convenient airfield. The main effect of closure of a destination airfield would be to use an alternative mode (Weisbrod, 1990). A survey in Vancouver in 1985 indicated that the sudden loss of transborder services or all international services would reduce economic output by $669 million and $1.2 billion respectively (Cooper, 1990). A study of surveys around regional airports in the UK shows that between 13% and 26% of companies rated the presence of an airport as a very important factor in their location decision; perhaps more significantly, many would not have considered the region had the airport not been available (Newcastle Airport, 1997). This agrees with another study which concluded that almost a quarter of overseas firms in northeast England indicated they would not have come to the region without the presence of an international airport (Robertson, 1995): so more than three quarters of the firms were not conclusively influenced by the availability of the airport.

A further study (Norris and Golaszewski, 1990) used surveys to differentiate between the impact of the purchase of air transportation and the consumer surplus enjoyed by non-aviation firms through the available air services, arguing that most of the former activity only constitutes a transfer of resources within the economy while the latter reflects the net economic benefits to a region from an airport. The consumer surplus benefits are measured as the consumers' willingness to pay for the non-market benefits of a more accessible region through questions related to whether a firm would relocate, whether its operating costs would increase and what percentage of revenues would be lost if the airport ceased to exist. The surveys were performed on and around Dallas/Fort Worth (DFW) airport and also on and around a major international airport in an island economy. Some of the results are given in Table 5.1.

Table 5.1: Economic impacts of two major US airports

		DFW	Island
Purchase impact:	- total (US $ billions)	5.2	3.73
	- US $ per resident	1,570	612
Consumer surplus	(US $ billions)	1.3	11.03
Per cent of firms relocating		6	19

Source: Norris and Golaszewski, 1990

For the island economy, the net economic impact of the airport was much greater than the transportation impact, whereas the reverse was true for DFW. The difference was due mainly to the proliferation of other transportation modes and competing airports in the case of DFW.

Other surveys have also elicited similar information (Hartsfield, 1987; Mahmassani and Toft, 1985). It is clear from the results of these surveys that, in general, there is no consistent relationship between the provision of air services and the location of industry in developed economies. Although 75% of firms in Berkshire were influenced by Heathrow, 63% were influenced by the M4 motorway (Button, 1988). Efficient transport facilities were one of nine most important locational factors for more than 50% of Kansas respondents but, in Nebraska, air freight was ranked 33rd and air passenger service was ranked only 36th out of 46 location factors (Cooper, 1990). The explanation of the variation appears to be partly bound up with the richness of choice of transport available, partly the nature of the economy and partly by specific constraints and distortions. The competition among UK regions for inward investment has led to large subsidies being offered in order to capture major investors, while lack of adequate sites can lose incoming multinationals (*Sunday Times, London, 13 October 1996, Business section, p 9*).

A novel and interesting attempt to assess the economic benefits of additional airport capacity was presented in evidence to the Heathrow Terminal 5 inquiry (Coopers & Lybrand, 1995). The study wrote two optional scenarios for the case of no permission being granted for Terminal 5: either British Airways would maintain its network and favour the high value traveller, or the airlines would focus their resources on the most profitable routes. Both scenarios would increase average air fares, in addition to which the former scenario would increase the cost of travel at lower frequency and the latter would increase travel times due to the need for more interlining. The increase in direct cost of business air travel in 2010 would be some £80 per flight or nearly £1 billion, based on a value of time of £150 per hour. More importantly, the study estimated the effect on the UK economy by disaggregating the travel by industrial sector and then, using CAA statistics of trip generation rates per sector, estimating the percentage of total costs per sector attributable to air travel, and hence the percentage increase in the sectors' gross value added due to the effects presumed in the two scenarios. The study concluded that the additional costs were equivalent to between 0.3% and 1.1% of a sector's gross added value, to which extent UK industry would be disadvantaged in competition with other countries' manufacturers. In addition, the costs for foreign-owned businesses established in the UK, which account for a quarter of value added, would also increase and the cost implications would not be lost on potential new foreign investors and tourists. The lost tourism expenditure in 2010 was estimated to be some £1 billion.

Those industries with a particular need for air service, either for shipments of high value-to-weight items or to support just-in-time strategies or high-contact personnel, do seem to locate close to aviation facilities, and with good reason. Surveys of business parks near airports in the southeast of England show that tenants come mostly from the electronics, pharmaceutical, information technology and financial services sectors (Robertson, 1995). A survey in Austin, Texas, of high technology industries (Mahmassani and Toft, 1985) showed a high value of US $333/lb for surveying and drafting instruments, of which 48% were air shipped, compared with a value of US $1.3/lb for all commodities, of which only 2.8% were shipped by air. The employees' propensity to fly per month, from the same survey, are given in Table 5.2. However, the literature also quotes examples of other industrial sectors which show a negative correlation between their location decisions and the availability of air services.

Table 5.2: Monthly trips per employee in high technology industries

Type of plant	Propensity to fly
Very small R & D	15
Typical R & D	8 - 12
Mixed R & D/production	2 - 4
Large manufacturing branch	0.07
Average manufacturing branch	0.07 - 0.25

Source: *Mahmassani and Toft, 1985*

These estimates, though subject to judgement by the analysts, do not suffer from the lack of any control on the bias in the respondents' answers likely to be present in the surveys of the views of various interested parties, even though the survey results do have some internal consistency once allowance has been made for richness of air transport supply and for industrial sector.

Anecdotal evidence is often used by airports to infer causality and to combat the criticisms of stated preference methods, but such observations of the relationship between air services and the strength of the local economy suffer equally from criticism, in this case that correlation is not necessarily evidence of causality. As the poet Edwin Muir said:

> "Time's handiworks by time are haunted
> And nothing now can separate
> The corn and tares compactly sown."

Atlanta's Hartsfield International Airport indicates that direct foreign services have resulted in foreign-based firms accounting for 20% of Atlanta's businesses (Hartsfield, 1987). On the other hand, Houston has 125 Japanese companies, four Japanese government entities and the largest Japanese community in the American South, despite having no direct air service to Japan (Lopez and Wilson, 1991). In the European setting, investment in Shannon has stimulated the regional economy and foreign firms have congregated around Schiphol (IATA, 1992). However, continual investment and subsidy in Liverpool airport has done little for the Merseyside economy, and Toyota has located in the UK near East Midlands Airport despite its very limited range of international scheduled services.

More formal attempts have been made to establish causality by measuring the relationship between the economy and the supply of air transport before and after major changes in supply. One study examined the effects of deregulation on three towns and their airports in California (Kanafani and Abbas, 1987). Two airports relatively close to hubs lost much of their service, though in one case the economy grew while in the other it stagnated. A third airport, more remote, retained its local services while its economy continued to grow strongly. If anything, the economies affected the supply of air transport by influencing the judgement of carriers to serve the airport, rather than the air services affecting the economy. Convenient hub service seemed to meet any residual need to support the economic growth.

The impact of the high speed rail service (TGV) between Paris and Lyon was also examined in a before-and-after survey comparison (Bonnafous, 1991). It was found that the TGV was only one determining factor in location decisions, other factors such as government intervention being much more important. An increase in tourist visits was counteracted by a reduction in the number of nights per visit. The most positive influence was the increase in regional service industries which established branches in Paris. This was not countered by an expansion of branches in Lyon by Paris-based firms, but there was an increase in sales of products of Paris-based firms.

Neither study allows a firm conclusion that the availability of improved services stimulates the economy, even when there has been a stimulation of trips. They are, in any case, subject to the criticism that there was no study of the counter-factual situation, i.e. what would have happened if there had been no changes in the transport characteristics.

This fault can be addressed to some extent by comparing the affected routes with unaffected 'control' routes. The impact of the TGV on air traffic has been determined by this type of analysis, but unfortunately it only extends to traffic levels rather than the economic impacts. A further piece of evidence, which relies on demand as an indicator of economic activity, used time series analysis to study the relationship over time between local demand and the quantity of air transport service at three medium hubs and three spoke cities in the US over the time period 1974 to 1987 (Ndoh and Caves, 1995). A smoothed time series was fitted to the entire quarterly data set for local demand and each of six variables representing air service quality at each airport. The residuals of a pair of series were then cross-correlated for a range of lead and lag options. Causality can, of course, occur with later events affecting earlier ones if sufficient warning of the later event is available in advance, but in the more normal situation, causality can be inferred from an effect being downstream in time from an event, unless the correlation is spurious. The results show definite evidence of a supply → demand causality. At spoke airports, the influential supply characteristic was the number of weekly departures, either to non-hubs or to a nearby interchange point. The effect became apparent within a quarter of a year. At two of the hub airports, it was again the frequency to a nearby major hub which appeared to cause an induced demand, after a lag of about a year. The third hub's demand appeared to be influenced by the total number of destinations served, again after approximately a year. There was no consistent evidence of fare changes which might also have explained the changes in demand, so it is likely that they do reflect some underlying changes in the economies. There was little evidence of changes in demand causing changes in supply, but the supply indicators used were only measures of network accessibility rather than any other measures of service quality.

This rather complex methodology avoids most of the criticisms of other attempts to determine the economic benefits which can flow from improved air service, and this limited application of the methodology does suggest that air service can stimulate demand. It will increasingly require this degree of rigour to persuade policy makers of the causal reality of the relationship. An adviser to the US National Governors' Association has stated that the calculation of economic impact says little about how airports influence the type of economic development that takes place around them. He also states that, while airports are no doubt an integral part of economic development, it is hard to generalise on causality (Cooper, 1990). Similarly, in summarising a 1989 Transportation Research Board conference on Transportation and Economic Development, Drew (1990) said: "The need for a causal-based methodology has been stressed over and over again during this conference. Unquestionably, there is a relationship between transportation and economic development. .... However, transportation .... is only one of many elements. With the current state of knowledge, it is impossible for public policy makers to establish reliable, measurable, causative relationships between given levels of transportation investments and resulting economic development." The EC, on the evidence of the Regional Development Fund grants for regional airports, needs less convincing, but the response of air carriers to funded improvements at some of these airports may cause them to re-assess their views.

Care must therefore be taken when using arguments based on economic contribution or stimulation of development to use only justifiable impacts and to realise not only that any benefits will be limited to some sectors of the economy but also that some groups within the community may be losers rather than winners.

5.2 AIRPORT DISBENEFITS

There are, of course, many parts of the world where the economic benefits of air transport discussed above are seen to outweigh the disbenefits at any price. This is usually the case in the smaller isolated communities or in the developing world, and many major airports like Bangkok, Jakarta and Manila operate virtually unrestricted. It should not be presumed that this situation will last for long. Noise in some of the more affluent communities was one of the reasons for building a new airport in São Paulo, and Taiwan has recently introduced noise restrictions and compensation. Japan represents the opposite pole in Asia, with probably the world's most extensive insulation and noise control policies. It has concluded a successful mediation exercise with local opposition to expansion of Tokyo Narita, allowing a second runway only with a relatively constrained movement rate and on the understanding that there will never be a third one.

Even when in favour of the economic growth itself, people do not necessarily welcome the direct consequences of economic development. It may not comfort many English nationals to know that aviation is partly responsible for 31 Japanese banks controlling 36% of London's international banking (Crandall, 1988). Equally, the urbanisation which must inevitably accompany major airport expansion (Breheny, 1987) may be less acceptable than the direct impacts of the airport, i.e. the 'benefits' of the multiplier may be unwelcome. Estimates of the number of new dwellings required to support 50 million passengers per annum (mppa) at Stansted ranged from 20,000 to 72,000. The range for 15 mppa was 4,350 to 21,000 dwellings. Once permission for expansion was granted, but limited initially to 8 mppa and later to 15 mppa, local opinion increased from 3:1 in favour of the second phase expansion in 1985 to 5:1 in favour in 1988. This presumably reflects both a new concern for the local economy and a realisation that the reduced scale of development will be acceptable. Another instance of concern about the consequences of too much economic activity, in this case tourism, is the Hawaiian island of Maui. While two other islands decided to extend their runways, Maui rejected the idea on the grounds of higher property prices, congestion, crime and a reduced sense of social responsibility (Fujii, Im and Mak, 1992).

Communities have become much more sensitive to the whole range of environmental impacts that an airport can have on its surroundings. These were introduced in chapter 2 and have been well documented (Archer, 1993; Barrett, 1991; Pedoe et al, 1996). In addition to aircraft noise, these can include vortices lifting roof tiles, odour, visual impact from approach and apron light systems, disturbance of wetlands and other habitats, contamination of the ground water by deicing fluids and fire fighting foam, and congestion of the local roads. These aspects are investigated by means of an Environmental Impact Statement (EIS), the scope and purpose of which has been described in chapter 3. In the US, the Environmental Protection Agency (EPA) has produced a set of guidelines, requiring any Federal Agency to cover the following topic areas in any EIS:

- air quality and air pollution control
- weather modification
- environmental aspects of electric energy generation and transmission
- natural gas energy development, generation and transmission
- toxic materials
- pesticides
- herbicides
- transportation and handling of hazardous materials
- coastal areas: wetlands, estuaries, waterfowl refuges and beaches
- historic and archaeological sites
- flood plains and watersheds
- mineral land reclamation
- parks, forests and outdoor recreational areas
- soil and plant life, sedimentation, erosion, hydrologic conditions
- noise control and abatement
- chemical contamination of food products
- food additives and food sanitation
- microbiological contamination
- radiation and radiological health
- sanitation and waste systems
- shellfish sanitation
- transportation and air quality
- transportation and water quality
- congestion in urban areas, housing and building displacement
- special impacts on low-income neighbourhoods
- rodent control
- urban planning
- water quality and water pollution control
- marine pollution
- river and canal regulation
- wildlife.

In addition, it is likely that third party risk will feature more frequently in future EIS requirements (ACI Europe, 1995) as indicated in chapter 2.

The main focus of concern in the majority of cases is the noise created by aircraft, particularly where it interferes with normal home life, with classes in schools, with care in hospitals and nursing homes and with recreational activities requiring a peaceful environment. Even in 1952, Newark airport in New York was closed for several months by a concerted protest caused by increasing noise exposure and a series of crashes: ".... at 3 o'clock in the morning on that day, with no leadership, the people in that area were going to burn that airport down. They had mops, saturated with gasolene, obtained from breaking open gas stations. The airport was closed." (Stevenson, 1972). Organised pressure from those affected has now led to an increasing number of airports having to limit operations in order to make the noise impact acceptable. This is particularly the case at night, as with the night quotas at Heathrow. It was feared that Munich II would open with a night curfew and with an environmental capacity which would be fully used up if even 40% of the movements were to be by Chapter 2 aircraft (Hogan, 1990). British Airways (1996) paid £7 million in additional noise charges in 1995-1996 for using Chapter 2 aircraft which accounted for 20% of their fleet, without counting fines for noise violations. The charges were imposed at most large European airports. Outside Europe, the Korean, Japanese, Taiwanese, French territories and Australian airports were among those to whom noise charges were paid.

These restrictions are being applied despite the remarkable reductions in noise at source. The take-off Effective Perceived Noise Level (EPNdB) reduced from 112 to 87dB between 1960 and 1985 for aircraft of comparable capability (Wesler, 1988). This was due in part to the arrival of high bypass engines, but also to the rule in 1978 banning further expansion of the fleet of non-noise-certificated aircraft. The aircraft certification rules and definitions of the metrics used for the certification and the measurement of noise around airports can be found in the literature (ICAO, 1993; Smith, 1989a). The early withdrawal of the non-certificated (Stage 1) aircraft cost the operators considerable sums of money in earlier replacement and low resale value. The policy now agreed in ICAO, ECAC and the EEC to require the Stage 2 fleet to be phased out between 1995 and 2002, with some exemptions for age, size and engine type (Pilling, 1991), will also impose costs on operators. ECAC claimed that this policy will only add 0.5% to airline costs, while the Nordic Council suggested only 0.4% (Bennett, 1989). The cost to US airlines was estimated to be US $5.58 billion. Even at airports with strong growth of passengers, this has had the effect of reducing the overall noise nuisance, as shown in Figure 2.1.

The airports have pressed for these Stage 2 bans because they are suffering from environmental capacity constraints and because they are having to pay large sums for noise control. Dallas/Fort Worth proposed a noise mitigation programme with a net cost of some US $150 million in conjunction with its plans for a US $3.5 billion development of two new runways, the main expense being in buying out all 500 homes within the 70 Ldn contour, together with an avigation easement of 25% of the house value to 2,600 houses in the 65 Ldn contour (Street, 1991).

The noise alleviation from improved technology is beginning to be eroded by increased activity. It has been estimated that, even with legislation to phase out Stage 2 by the year 2001, at the average large airport the size of the noise contours may start to creep up again by 2005 despite the continued introduction of new aircraft with noise characteristics which comfortably comply with Stage 3 requirements. Complaints correlate more with maximum noise event levels rather than the energy-based metrics, so it has been suggested that environmental restrictions should be based on them. This could give even lower capacity, since the larger aircraft will always tend to be the more noisy.

Noise complaints do also increase with total integrated noise energy (e.g. Leq) and with the age of housing (Levesque, 1994), and this is to be expected. However, at John Wayne airport, it appears that local residents have been complaining at movements they could see even though they could not hear them (*Airport Forum, 1989*). This may reflect less articulated concerns, such as the safety implications of overflights. Noise complaints rose at Manchester the day after the unrelated Pan Am crash at Lockerbie and complaints from under the route of the El Al aircraft that crashed at Amsterdam were up for the following four days. There is also evidence that the irritation threshold follows the noise level down. The EU's Environment Commission tends to support this view, in that it suggests a further round of legislation beyond Chapter 3 so that all the benefits of noise control to date are not offset by traffic growth, noting that some aircraft can already outperform Chapter 3 by 12 dB(A) (Paylor, 1991). However, ICAO's Committee on Aviation Environmental Protection (CAEP), having received evidence that there is no cost-effective technical breakthrough which would allow the manufacturers to respond to a more stringent Chapter 3 (Smith, 1989b), has been unable to reach a consensus to put more stringent limitations on noise at source in line with the desires of airports and the EC (Cameron, 1997), preferring to improve guidance on operational measures and land use planning (*ICAO Journal, March 1996, pp 5-8*). In any case, the effect of the weight at which an aircraft is operated can be a more important influence on the noise impact than the difference between categories of certification (von Wrede, 1995), as can the level of thrust used and the timing of any thrust cut-back.

The continuing disenchantment of communities with aircraft noise may be explained in part by the acknowledged inability of current noise metrics to reflect all aspects of annoyance (Ollerhead, 1995). This can be particularly difficult when movements are peaky and do not produce significant levels of noise energy over the day but have a relatively small number of noisy movements. It is also true that the aircraft certification procedure is not very representative of the way that aircraft actually operate at airports. Stage 3 aircraft have been measured at the takeoff monitoring point at Heathrow making up to 10 dB more noise than their certificated values (Cadoux and Ollerhead, 1996). Unless additional limits are put on operations, there is no guarantee that a Stage 3 aircraft will not make more noise than a Stage 2 aircraft, particularly if the Stage 2 aircraft is at a low weight.

Geographic and weather factors also influence the relevance of standard methods of assessing noise nuisance (Pereira, Braaksma and Phelan, 1995). There is a gulf between the letter of the Stage 3 certification of hushkitted Boeing 727-100s and the community reaction to them, particularly when used in the night freighter role. There is also considerable doubt as to the extent to which aircraft noise affects sleep patterns (Pearsons et al, 1995), though apparently children living near Munich airport display reductions in learning ability as a result of stress (*New Scientist, 16 November, 1996, pp 14-15*). In view of the lack of confidence in the common noise metrics, the debate on the acceptability of an increase in operations often turns on the definition of a reasonable level of disruption, such as four noisy movements in two hours for a school. Whatever the improvement in noise measurements may be, it is most unlikely that complaints will decrease. An objector at Toronto airport said: "Stage 3 jets have a different sound; it penetrates through your body" (Rowan, 1991). The tensions between the regulators, the airports and the airlines are thus likely to continue into the indefinite future.

5.3 JURISDICTIONAL ISSUES

A critical issue that all too often complicates the relationship between an airport and the surrounding communities is the nature of the political jurisdiction with ultimate control over the airport development. Even where airports are public enterprises, airport management is commonly not answerable to the elected officials of those communities most directly affected by the operations of the airport. Whether the airport is owned and managed by a national authority, or simply the largest city in a region, the surrounding communities typically have little direct role in airport decisions, no say in the appointment of members of the board or commission responsible for airport management, and usually no power to approve or deny airport development proposals. Since airports are often by necessity located on the edge of the urban area, or even beyond it, they are frequently outside the geographical jurisdiction of the primary city in the region. In many cases, the city acquired control over the land at a time when the surrounding area was largely undeveloped. Subsequent development of the urban area has then resulted in the growth of the surrounding communities toward the airport boundary. This is often exacerbated when growth of the airport itself has required additional land and moved facilities, particularly runways, closer to the surrounding communities.

The situation with privately owned and operated airports can create equally complex jurisdictional issues. Should such airports be subject to the usual land-use planning controls of the local jurisdiction, and what if an airport, by virtue of its size, lies within several such jurisdictions, or the airport property is in one jurisdiction but surrounding communities under the flight paths are in other jurisdictions? Obviously, these situations will vary from country to country, depending on the legislative framework within which privately owned airports operate.

Of course, at one level these conflicts are understandable. Major airports meet an essential transportation need of a wide geographical area, and as such their development has to address broader interests that simply those of their immediate neighbours. However, when political control is thereby entrusted to those who have little or no stake in the concerns of the surrounding communities, it is also understandable that the development needs tend to drive decisions, with little real regard for the impacts that these may have on the affected communities. Even where the surrounding communities are part of the larger political entity controlling the airport, the interests of the region as a whole may still overwhelm the local concerns, and the elected representatives of those communities may simply find themselves outvoted. But at least in this situation, there is potential for serious negotiation and a forum for the advocacy of the interests of the surrounding communities. The votes of those elected representatives will be needed on other issues, and this gives them a bargaining position.

A more serious situation exists where the airport is controlled, or development must be approved by, a separate political jurisdiction from some or all of the surrounding communities. In this case, the disenfranchised communities may be left feeling totally powerless to influence the airport's decisions, or worse, the development decisions could have been made so as to deliberately shift the impacts of the airport operations from those communities owning the airport onto those that do not. In this situation, the affected communities may feel that they have little recourse but to oppose any and all development with every legal means at their disposal. While they may ultimately lose to the more powerful interests of the region as a whole, this opposition can drag out the planning and development process by many years, costing the airport and its users huge sums of money in project delays and congestion costs.

In some countries, national governments have acknowledged the concerns of airport communities and have passed regulations to require their interests to be addressed as part of the airport planning process, or created funding programs to pay for mitigation measures. The United States has been a leader in both these areas, with Part 150 of the Federal Aviation Regulations that requires airports receiving any federal funds to develop an aircraft noise reduction and mitigation program and specific funding under the Airport Development Program for noise mitigation measures. However, it is arguable that while setting standards for such studies is an appropriate role for the national government, paying for the studies and mitigation measures is not. If the benefits to a region of an airport development outweigh the disbenefits on the surrounding communities, then the users of the airport can certainly afford to pay for mitigating those disbenefits through the various fees that they pay for the use of the airport. On the other hand, if the benefits are not large enough to justify making users pay for the mitigation measures, then the net social welfare does not justify the development in the first place.

The increasing number of regions being served by more than one airport with significant levels of commercial air traffic creates another jurisdictional issue, that between the individual airport managements and the regional planning authorities. If all the airports in a region are owned by the same authority, then this jurisdiction issue tends to revolve around the consideration of airport development needs versus broader transportation and other urban and economic planning concerns. However, where the various airports are controlled by separate authorities, then the situation becomes considerably more complicated and the regional agencies not only have to resolve the conflicts between aviation and other concerns, but do so in a context of competitive, and possibly conflicting, development plans by each airport. Where airport development funds flow from national or state governments directly to the individual airports, and not through the regional transportation agency, it is inevitable that regional and broader transportation concerns will take second place to the development aspirations of the individual airports. This can cut two ways. While the most common scenario is that each airport is vying for a greater share of the traffic, it is not inconceivable that a situation could arise where none of the airports in a region is willing to incur the costs of expanding to meet future traffic needs and each expects the others to take on this responsibility. The latter scenario could become a particular concern if the noise impacts of each airport fall largely on politically influential communities within the jurisdiction owning the airport.

A particularly difficult situation arises in regions served by several existing airports, each under separate control, where a new airport is required to meet predicted traffic growth. Should one of the existing airports undertake the development of the new airport, or should it be developed by a new authority under the control of the community within which the airport will be sited? There are obvious disadvantages to both approaches. In such situations, the only effective solution may be the creation of a regional airport authority that assumes control of each of the existing airports, as well as responsibility for developing the new airport. This will of course require the jurisdictions that currently own the existing airports to give up their exclusive control of an individual airport in return for shared control of the entire regional system.

While these issues are largely beyond the scope of the technical planning studies undertaken at an airport, regional or statewide level, unless they are addressed and resolved, the recommendations of those studies may at best become impossible to implement and at worst may lead the region's airport development down a suboptimal path to a solution that imposes enormous unforeseen future costs on both the users of the system and those affected by it. Thus it would appear critical that regional and statewide airport system planning studies include an element addressing jurisdictional issues, and are free to recommend changes in the existing structure. This is of course most unlikely if the studies are funded by agencies that have the most to lose in any change of political control. Therefore national, state, and regional agencies must take the responsibility for ensuring that these issues are properly addressed.

One important factor that currently limits the ability of many regional agencies to play an effective role in airport system planning is the lack of adequate technical skills and institutional resources to address airport planning issues. Not surprisingly, most airport planners have worked for airport authorities, or the consulting firms that serve them. Regional agency staff involved in airport issues often share this responsibility with other duties and commonly lack both formal training in airport planning and experience in the more technical aspects of airport planning and operations. It is understandable, if unfortunate, that as a result these organisations are often not regarded as serious players in airport development decisions by the professionals employed by airport authorities, civil aviation agencies, and the air transport industry generally. If airport system planning is to effectively grapple with the difficult jurisdictional issues that underlie the more technical concerns, then the regional and statewide agencies that have the responsibility for this planning must also acquire the professional staff with appropriate expertise.

5.4 POLITICAL RESPONSE TO CONFLICTING INTERESTS

5.4.1 Difficulties in achieving a balance of interests

It is often suggested that the evaluation problem is one of balancing air transport benefits against environmental costs at an airport. If it were that simple, it ought to be possible to develop an index of, say:

$$\frac{\text{total unmitigated environmental disutility}}{\text{consumer surplus of airport users}}$$

where disutility measures the actual loss of amenity rather than the physical extent of the impact, such as land area affected. A maximum value of this index could be set, which legally could not be exceeded and below which a sliding scale of compensatory tax would be paid per departing passenger. Since it is estimated that less than 3% of the citizens are adversely affected by airport noise, while the vast majority of its citizens benefit from the operation of those airports (Wesler, 1983), the index might appear to be small and the problem to be overstated. Unfortunately, difficulties would arise because of the lack of accepted methods to value the impacts, differences in perception of values and the differing power bases of those holding the values, ignorance of the full societal costs, the existence of societal benefits as well as costs, and the importance of interests other than simply local passengers and local residents, e.g. regional development, other modes of transport, the wider national and international economy and broader environmental concerns.

Further, and perhaps most importantly, there is seldom an equal distribution of costs and benefits. In an ideal market economy, the price of transport would determine not only how each mode would be used but also how much travel was beneficial for society at large. In other words, the full social costs imposed by each unit of travel would be reflected in the cost of travel. However, market forces operate with a decision rule that one dollar equates to one vote, while democracy operates on the basis that one person equates to one vote (Flyvbjerg, 1984). So, even if the price of transport accurately reflected the true long term value of resources used, principles of social justice would still require additional judgements to be made. Ultimately, these judgements are made through the political process, but the judgements should be informed ones. The Environmental Impact Statement (EIS) is important in this respect, as well as proving publicly that environmental concerns are being properly addressed (Anderson and Rideout, 1992).

The most satisfactory outcome must be when a win-win result can be obtained. This becomes more likely when a 'green-gold' coalition can be formed. A perhaps fortuitous example was the development of the high bypass turbofan engine which simultaneously cut noise and increased fuel efficiency. If the tradeoffs can be correctly estimated, there must be scope for other technical and managerial win-win solutions to achieving sustainable growth in aviation, even though the result to each stakeholder may not be as advantageous as they would have been anticipating in the absence of the need for cooperation. The present capacity caps imposed on movements at airports in response to their individual sensitivity to ecology, air pollution, water pollution, noise and social issues are somewhat crude attempts to preserve a balance based more on 'do not lose too much' rather than the more palatable solutions which might be available from a system-wide assessment of the need for change. The widespread acceptance of the need for this approach requires an awareness that it is no longer possible to buy personal freedom from pollution.

At the local level, there are three basic approaches to reaching a working agreement between airports and their neighbours over increasing environmental impact. The airport can buy out the property, offer compensation in return for an easement or offer some grant towards protection from the impact. This follows from the general legal position that developers should meet the full costs of restoring each local inhabitant's position to its previous state, i.e. the cost of 'internalising the externality' (Ruddock, 1992). The UK airports are being encouraged by the government to take full responsibility for the effects of noise and to incur the costs, on the basis of 'the polluter pays' (DTp, 1993a). A similar ruling exists in the US under the Code of Federal Regulations, 49 CFR Part 24.

A prime difficulty is in establishing the deterioration in property value relative to a fair market price. It has been found that values fall initially after an increase in noise, but the process is dynamic: noise avoiders sell, some shift in land use occurs and prices eventually regain the long term trend (Crowley, 1973). Other historical evidence suggests that a noise discount of some 0.5% to 0.6% per decibel exists. A 1968 study at Heathrow gave a 0.25% to 0.30% reduction of house value per unit of Noise and Number Index (NNI), while a 1969-1973 study at Toronto gave a reduction of 0.5% to 1.3% per unit of Noise Exposure Forecast (Nelson, 1980). The Roskill Commission (Button and Barker, 1975) found something less than 1.0% per unit of NNI, depending on the value of the property: the higher the price, the more the occupiers were sensitive to noise. This result is confirmed by a later Canadian study (Uyeno et al, 1993), which also found a much higher percentage impact of noise on vacant land prices than on houses. A further Canadian study around Toronto's Pearson airport found that complaints increased by 1.92% for a 1% increase in NEF (Noise Exposure Forecast), but only a 0.7% increase for a 1% increase in the number of daytime noise events (Gillen and Levesque, 1994). However, a study at Manchester, matching noise contours to property mortgage valuations, gave only a weak and non-robust relationship. It was shown to be difficult to assign any changes of value definitively to the effects of aircraft noise (Pennington, Topham and Ward, 1990), though this has been challenged by Collins and Evans (1994) who applied neural network techniques to the same data source and found that prices fell between 0.25% and 0.5% per unit increase in NNI, depending on the size of property.

All the above results were obtained from hedonic price studies, mostly on cross-sectional data. A study using a willingness-to-pay survey found the noise premium to be 2.4% to 4.1% per unit of Ldn for home owners, and 1.8% to 3.0% for renters (Feitelson et al, 1996). Almost half of the home owners were not prepared to make any bid for a residence exposed to frequent severe noise and over which aircraft fly, over 10% of them not being prepared to consider a residence with any level of noise. The difference between the hedonic and willingness to pay results for estimating the social cost of noise has been shown to alter the sign of the net welfare contribution of deregulation (Schipper, 1997).

A clean method of approaching the problem, if a fair market price can be established, is for the airport to buy out the property in the same way as would happen in the UK if a Compulsory Purchase Order were to be obtained, with compensation for moving in compliance with the Land Compensation Act, 1973. British Rail offered to do this for properties within a 240 metres wide corridor around the channel tunnel rail link in negotiation with Kent County Council, though it appears that the Department of Transport is only prepared to allow up to £5,000 as an additional 'solace' payment for houses within 100 metres of the track compared with £30,000 for houses within 300 metres of the line of the M25 motorway (*Times of London, May 2, 1997, p 15*). In the case of Dallas/Fort Worth (DFW), residents are offered a choice between buyout, the purchase of easements, or the assurance of guaranteed sales: US $125 million has been allocated to mitigate the effects on the city of Irving, relative to a likely project budget of US $3.5 billion.

It should be noted that a buy-out programme does not necessarily mean that the land becomes valueless or that the programme need be very expensive. Homes can be resold to less noise-sensitive owners, with an appropriate avigation easement included. More severely impacted areas can be used for less noise-sensitive land uses, such as offices, warehousing or distribution centres. Such a shift in land use might not only result in recovering a large proportion of the cost of the buy-out, but generate a net increase in local property taxes.

In both the US and the UK, there are planning guidelines for the acceptable level of noise inside dwellings. The FAA can make grants from the Aviation Trust Fund to contribute 80% of the cost of programmes to ensure that houses inside the 65 Ldn contour have interior noise less than 45 dB (Shade, 1990). The cost of compliance averages US $21,000 per property. In the UK, the Department of the Environment has issued revised guidance (DoE, 1994). As well as confirming that any scheme involving a runway greater than 2,100 metres in length falls within Schedule One of the Town and Country Planning (Assessment of Environmental Effects) Regulations 1988, and therefore requires a full Environmental Assessment, it also gives guidance on the compatible use of land under given levels of aircraft noise based on a 16 hour (0700-2300) Leq measure, as follows:

- less than 57 Leq (category A), noise is not a determining factor in planning;
- 57-66 Leq (category B), noise should be taken into account for dwellings;
- 66-72 Leq (category C), there should be a presumption against planning permission for dwellings;
- above 72 Leq (category D), even industrial use is debatable.

The guidelines for other modes of transport are broadly similar. Depending on the need for the facilities being considered, the planning authorities may vary the levels up or down by 3 dB(A). The equivalent values for night time noise are 8 or 9 dB(A) lower. The US criteria given in FAR Part 150 are quite similar, but there is a more detailed breakdown of land uses and noise bands.

The UK transport industry's reaction to this and previous guidelines has been very variable. British Rail offered to finance 50% of the cost of noise barriers for houses inside a 65 Leq contour. One airport meets 80% of the cost of double glazing up to a maximum of £1,920 plus venetian blinds for houses inside 62 Leq. Another airport allows £1,880 within 66 Leq. A third airport allows £2,190 within approximately 57 Leq. In comparison, the 1970 Heathrow agreement was the full cost of double glazing within the 69 Leq contour.

The noise issue has been used in this section as an example of the difficulties of coming to a balanced view of the benefits and disbenefits of airport activity. Valuing the cost of externalities or achieving the balance is no easier for the other disbenefits. The safety debate, for example, is influenced by the perception of the risk. As shown in chapter 3, there is little consensus on the tolerability level of risk. Seven variables appear to be important: the extent to which the risk is voluntary, dread, control, knowledge, catastrophic potential, novelty, and equity (Slovic, 1987). The public's ability to control the risk may be the most important factor in their acceptance of the situation, which they can only achieve through credence in the airport owner's good faith which could be demonstrated by overt progress in reducing the threat (Apostolakis and Bell, 1995).

Perhaps the difficulty is illustrated most effectively by the debate between the staff of the Inco company who want to mine nickel in Labrador and the elders of the Innu nation. The company is trying to negotiate an Impact and Benefit Agreement, but the Innu only want to have the earth left intact for their dead and have no interest in the monetary compensation on offer (*personal communication from Mark Salvor, Davis Inlet, Labrador, February, 1998*).

5.5 THE INCREASING COSTS OF ENVIRONMENTAL MITIGATION

The result of all the work which has been done to understand the need to protect the environment around airports has been a significant increase in airports' costs, together with reductions in efficiency and capacity from the mitigation actions which have been considered necessary. It is often possible in negotiation for residents to push the airport into settling for considerably more than the legal minimum. The Port of Seattle (1990) has spent US $100 million acquiring 1,360 homes and relocating 3,900 residents. As the result of a mediation exercise incorporating an agreed noise budget, properties within the most severely impacted area (the Neighbourhood Reinforcement Area) have a guarantee of 90% of the free market value and have the choice of easements or insulation (Johnson, 1989). The easement ensures that the owners will not sue for noise, vibration, fumes or other damage resulting from reasonable increases in airport activity. Although it is usual in the West not to consider much above replacement levels of compensation for compulsory purchase, not all countries take the same view: Slovakia may compensate at up to three times the nominal value.

A survey by the UK's Airfield Environment Federation into the methods being used in Europe to mitigate environmental impacts at airports (Johnson, 1993) catalogued the following responses:

Noise:

- operational controls:
    - displaced thresholds
    - preferential runways
    - rotation of runway use
    - restriction by aircraft type, weight, category
    - restriction of annual or hourly movements
    - curfews
    - flight path and technique modification
    - ground operations
    - reliever airports
- fiscal and monitoring:
    - permitted distance limits and penalties
    - landing supplement by type, category
- planning:
    - land use
    - buy out

- airport environmental limits
- sound proofing

Air quality:

- guidelines (Annex 16; EC; WHO; national, e.g. US, Canada, Sweden)
- monitoring (e.g. Greece, Copenhagen)
- measures (car sharing; staggered hours; fixed services; waste power; fiscal, e.g. Sweden)

Water quality:

- legislation (EEC and national)
- monitoring (fuel spills and leaks, other discharges)
- measures (isolation, clearway, anti-icing).

Other measures which have been suggested to limit the effect of air or water quality are to use electric or methanol power for ground service vehicles and to cap activity.

The industry conducted its own survey (ACI, 1995). It found that the percentage of airports in the survey using the different forms of environmental control were:

- noise monitoring and penalties, 35%
- track keeping monitoring and penalties, 30%
- involvement with land use planning, 60%
- operational restrictions, 77%
- noise related operational charges, 29%
- noise insulation schemes, 25%
- air quality monitoring, 22%

61% of the airports had carried out a survey on waste, the average results being 0.68 kg per passenger at a cost of US $0.15. Almost all the airports had a bus service and 28% had a rail service, but no airports could receive cargo by public transport. Some 40% had explicit community relations programmes.

In addition to those measures given in the airport case studies later in this book (Munich II, Osaka, Gardermoen), the following examples may help to illustrate the pressure to take strong actions to mitigate environmental impacts, in some cases to the extent of seriously compromising the ability to realise the inherent physical capacity of a site.

The new Macão airport's runway was swung out of wind to avoid a city on the Chinese mainland. Landings must be from the south unless it results in a tail wind greater than five knots. If landings have to be made from the north, a curved approach must be made to an offset localiser. Bogota is to have a new parallel runway displaced 1,400 metres from the old runway which will be used for takeoffs away from the city while simultaneous landings are made towards the city on the old runway, thus relieving the city of takeoff noise on all occasions when the wind allows these operations (*Flight, 30 April 1997, p 35*). Some would argue, as they have done with respect to the use of environmentally preferred runways at Amsterdam Schiphol (*Flight, 13 May 1998, p 5*) and with the reciprocal takeoff and landing operations at Sydney airport (*Flight, 10 June 1998, p 52*), that there is a safety implication in mitigation measures like this and in those which require early power cutback.

The expansion of Zürich's terminal required an agreement to increase the percentage of passengers using public transport from 25% to 55%, to adopt rail/air containerisation of freight, to optimise aircraft parking positions, to encourage late engine starts and to build better taxiways. Considerable research was done, which established that a fee of CHF 50/kg of emitted pollutant would reduce overall emissions by 5%, and that elimination of all APUs could reduce NO_X by 78 tonnes/year (equivalent to 47,000 movements by an MD80), but only if there were a fixed system for preconditioned air as well as 400Hz (*Airport Forum 5/1992*). Charges have been introduced based on the NO_X and Volatile Organic Compounds in the certificated Landing and Takeoff Cycle (*Airport Business Communiqué, June 1998, P 7*).

The mitigation measures for the expansion of terminal facilities at San Francisco International Airport (City and County of San Francisco, 1992) included:

A Transportation System Management (TSM) program with the goal of reducing the 72% drive alone mode share of air passengers and employees to 52%, utilising flexible work hours, pricing, transit information, and employee incentives, in conjunction with:

- radio information on parking availability
- a light rail connection between the planned ground transportation centre and the terminal
- construction of a link to an adjacent station serving a commuter rail line (Caltrain) and the Bay Area Rapid Transit (BART) system, if a BART station is not constructed in the new terminal complex
- dedicated lanes for high occupancy vehicles on access ramps and the adjacent freeway
- airlines and other airport employers to contribute a share of transit operating costs on the basis of employee and air passenger patronage.

Air quality initiatives (other than the foregoing TSM measures):

- construction dust and emissions control measures
- no aircraft engine start until cleared for gate departure
- aircraft engine shut down while waiting for a gate.

Cultural resources: review by project archaeologist and procedures for reporting significant artifacts.

Geology: limitation of excavation depth and establish techniques for dewatering and erosion control.

Hazards: investigation for site contamination and application of remediation techniques; control of dust; survey for asbestos and PCBs in buildings to be demolished.

Noise:

- construction noise control measures
- continue to support the Airport/Community Roundtable
- improved flight track monitoring
- develop techniques to address backblast noise
- preferential use of Runway 19 at night
- investigate benefit of new or extended runways to allow more arrivals and departures over water
- noise barriers
- explore use of DME and/or MLS to revise landing procedures
- no trade of noise from one community to another
- noise insulation program in cities near the airport.

In Vancouver, 20 years after land acquisition, a third attempt was made to obtain a new fully segregated parallel third runway. This was shown to have an net present value of $4.2 billion relative to project cost of $110 million, yet the Community Forum for Airport Development continued their long opposition. However, the Forum was a good focus for dialogue. Under the Mandatory Federal Environment Assessment and Review Process, the Ministry of Transportation asked the Ministry for the Environment to establish a panel of four persons to scope and manage the EIS and public hearing process. The project was found to be good in terms of the total noise footprint, but it generated new noise, i.e. noise in areas not previously affected. The main mitigation strategy is to use the new runway mainly for arrivals, departures only with Chapter 3 aircraft and a curfew between 2200-0700. It is to be monitored by a Noise Management Committee (NMC) with citizen participation. The panel in fact recommended compensation, but the NMC prefers mitigation and the Minister takes the view that compensation does not cure the problem.

Manchester Airport plc adopted a sensitive attitude to the local community from the outset of the planning to provide a second runway to support the hubbing which formed the focus for its mission and consequent business plan. Three areas of search were identified as possible sites for the new parallel runway, as indicated in Figure 5.1. These were published in a draft Development Strategy. None of them satisfies the 1,525 metre separation requirement which would allow fully independent simultaneous approaches and hence a doubling of runway capacity. It had already been decided by the airport company that the physical constraints, the amount of land take and the environmental impacts to satisfy this requirement were unacceptable. Area 1 was associated with a close parallel runway with little stagger between its threshold and that of the existing runway. Area 2 would give the best increase in capacity with a relatively wide spacing and a considerable stagger. Area 3 could accommodate close parallel runways with a great deal of stagger, with either runway extended into the area; three variants of the layout were evaluated.

The three areas were subjected to an environmental review by specialist consultancies, considering agriculture, air quality, archaeology, ecology, landscape, listed buildings, noise, traffic and water quality. The comparison was in terms of the impact on people, the impact on resources, compatibility with planning policies and the potential for mitigation. Area 3 was found to be best for noise, the smaller land take and number of farms affected, the smallest disruption to the local road network, lack of archaeological importance and for minimum expansion into the countryside. Area 2 would result in the least loss of property and listed buildings. Area 1 would have the best dispersal of air pollution. On other matters the three sites were quite comparable. Area 3 was therefore preferred. A further option was raised of an angled runway cutting across from Area 3 towards Area 2. This Area 4 was subject to a further comparison with Area 3 and found to be better in terms of noise, property, air quality, road disruption, recreational rights of way, impact on water courses and natural habitats. However, Area 4 conflicted with Rostherne Mere, a wetland of national and international significance for its wildfowl. Area 3 was better for land take, particularly in the Green Belt.

Figure 5.1: Options for Manchester's new runway

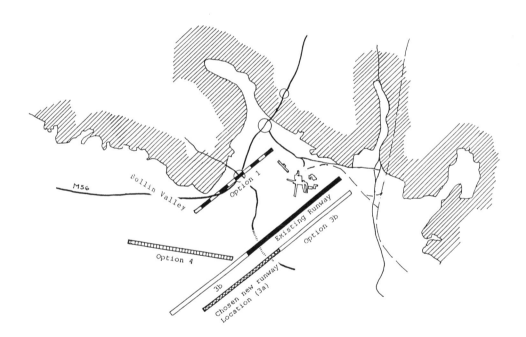

Source: Redrawn by the authors from diagrams provided by Manchester Airport

Three options were taken forward to a further round of studies and consultations. Option 3A was Area 3 with the new 3,048 metre close parallel runway extending into it by 1,850 metres. Option 3B was Area 3 with the existing runway extended into it by 2,150 metres and a new close parallel runway of 2,600 metres alongside the existing one. Option 4 was Area 4 with a second runway of 2,600 metres at an angle of 45 degrees to the existing runway. Although Option 4 was still best for noise, properties, resources, landscape, habitat and recreation, it would produce major changes in the distribution of the noise. Relief would be given to Knutsford and Mobberly but there would be significant increases in Ashley and Rostherne. Option 3A would affect fewer farms and the view from fewer houses, have the least loss of both agricultural and high quality land, have the least impact on the Green Belt, minimise urbanisation pressures, and depart least from existing noise control policies.

Option 4 would therefore be the preferred solution from an environmental standpoint, except that Rostherne Mere is a night roosting site for up to 15,000 gulls. Ornithologists, the airport company and specialist consultants studied the situation. They found that the bird hazard risk to flight was significant and that there was no way of mitigating the problem without destroying the wildlife importance of the Mere. Safety being paramount, Option 3A was selected as the scheme that should be taken forward to the public inquiry as the preferred development.

Not only did the airport company carry out that impressively transparent environmental assessment prior to the inquiry, but they also negotiated a comprehensive Planning Obligation by Agreement with Cheshire County Council and Manchester City Council under Section 106 of the Town and Country Planning Act for the regulation of the development (Cheshire County Council, 1995). The noise aspects of this agreement are given by Thomas (1996). The agreement includes:

- Continuance of the existing noise insulation policy inside the 62 Leq (24 hour) contour.

- A Community Trust Fund with an annual budget of at least £100,000 to promote, enhance, improve, protect and conserve the natural and built environment in and around those areas being directly or indirectly affected by the noise of operations at Manchester Airport.

- External independent monitoring of noise on an annual basis to ensure noise remains below the 1992 level in terms of the average of the 10% noisiest departures and the area of the 60 Leq (0700-2300) contour.

- A suitable noise monitoring system with guaranteed access for consultative groups and external auditing of quality.

- Implementation of a 'quietest operations policy', encouraging early phase out of Stage 2 aircraft, no Stage 2 night operations after 1996, 92% of scheduled operations by Stage 3 aircraft by 1998 and 96% by 2000, annual review of noise penalties, a noise element in the landing fees, annual review to identify opportunities for preferential use of the existing runway, preferential use of westerly operations.

- Adoption of a ground noise policy, with no more than 20 night time engine tests and only within the test bay, controls on daytime testing, and the increased use of fixed electrical power units.

- Designation of Preferred Noise Routes (PNR) which will be flyable by all types of aircraft, the development and implementation of systems to enable improved accuracy of track keeping and monitoring with a target of no more than 5 degrees either side of the centreline, investigation of all operations outside 1.5 km of the centreline, continued efforts to obtain the rights to fine off-track operators, regular review of the PNR noise abatement heights, and consultation with the interested committees before instituting any changes.

- Assurance that night noise will not exceed 1992 levels before 2005 and that best efforts will be made to extend this to 2011, where the impact is measured annually as the average of the 100 noisiest departures and also the modelled area of the 60 Leq night noise contour.

- Implementation of a noise points budget between 2300 and 0600 hours of 3,900 in winter and 8,750 in summer to 2005, where each operation is assigned points on the basis of its noise impact, with a doubling of points for each 3 EPNdB of certificated noise level. Also, the absolute number of night noise movements is not to exceed 7% of total movements except those producing demonstrably less than 87 EPNdB at the agreed monitoring points to 1998 with consultation thereafter.

- A lower threshold for the imposition of noise penalties between 2300 and 0700 hours, restrictions on the types of aircraft allowed to operate between those hours and minimum use of reverse thrust consistent with safe operation.

- No proposals to be brought forward for a third runway before the end of the 2011 planning horizon as set by the emerging Regional Planning Guidance and Cheshire Structure Plan Review, or other development plans submitted without prior agreement of the County Council.

- Provision of an enhanced package of environmental mitigation to minimise the impact on landscape and ecological interests, including the creation of at least twice as many ponds and six times the woodland as those lost, specific proposals with respect to bats, badgers and great crested newts, a three year aftercare period, the submission of a 15 year Landscape and Habitat Management Plan with costed one year and five year plans and the provision of a competent ecologist and landscape architect and the necessary resources to implement the plan. A steering group will be set up to assess the impact and success rate of the mitigation works, consisting of the airport company, the retained ecologists and landscape architects, representatives of the local authorities, English Nature, Cheshire Wildlife Trust, The Bollin Valley Partnership, the appointee with responsibility for the management of the area covered by the Management Agreements, and the local community.

- Provision of at least £0.5 million to support relevant road improvements.

- Support for improved public transport access by safeguarding land and partnership funding for an extension to Metrolink and also to the heavy rail network if there is a positive outcome to the feasibility studies which the airport company will also support.

- Establishing and servicing a Ground Transport Group to develop and implement strategies to maximise use of public transport by passengers and employees to meet a target of 25% of access trips by 2005. The employee initiatives include experiments with collection buses and dial-a-bus, a staff car sharing scheme, setting limits to growth in staff car parking provision. The promotion of public transport is to be assisted by devoting 10% of the annual marketing communications budget to it. There will be dedicated drop-off lanes for taxis and courtesy buses, further phases of development of the intermodal transport interchange and a new bus station. Studies will be undertaken into options for off-site road and rail check-in and satellite terminals with associated car parking.

In addition, the construction contractors will have to comply with time restrictions on the local roads and a target to import 70% of the material by rail (*ACI Europe Airport Business, April/May 1997, pp 5-9*). More than a hundred obligations of this sort were negotiated between the pre-inquiry hearings and the actual inquiry. The negotiations required the ability to get behind the obvious semantics to detect the real concerns in the community with extensive and widespread consultation. This careful attention to the needs of the local community, and the strong evidence of the creation of an additional 50,000 jobs directly and indirectly by the increased airport activity, resulted in a positive recommendation by the inquiry inspector which was accepted by the Government. It is felt by the airport staff that the inclusion of external auditing supported by airport funds and with the remit to ensure current best practice in mitigation, together with the attention given to defining development limits within the present planning horizon, contributed greatly to the favourable outcome. Even so, the airport will have to use state of the art expertise in design and operation of the airside in order to maximise the movement capacity of the chosen layout. Whereas a fully independent new runway would have given an increase of well over 40 movements per hour, it will be difficult to achieve an additional 26 movements. Despite all this effort to accommodate the environmental interests, implementation was still threatened by the same sort of environmental protests that have hampered road improvements in the UK recently, with protesters occupying tunnels dug under the location of the new runway (*The Times, 28 March 1997, p 5*). They have been removed, but not before gaining the support of many who felt that the planning system does not allow environmental opposition a fair voice.

Amsterdam's Schiphol airport has recently been the subject of an extensive study by the Dutch government to reach a balanced judgement of the environmental carrying capacity of the location relative to the economic benefits accruing to the local communities and to the whole of the Netherlands. It was understood that the Netherlands would not have sufficient economic potential to sustain a world class hub airport unless the airport and its based airline were allowed to develop a large transfer component and grow to a critical mass that would protect its competitive position relative to other major European airports. Equally, that level of activity would itself stimulate the local and national economy. Yet the last few years had seen a huge growth in noise complaints despite a very significant reduction in the number of houses affected by noise. In 1996 there were 135,000 complaints, with an average of 10 complaints per complainant. This appeared to be due to uncertainty and concern about an uncontrolled growth in air traffic, suspicion about the future plans for the airport and the region, delays in decision making by the government and much adverse publicity (Boer, 1997). It will also have been influenced by an El Al Boeing 747 freighter crashing into a block of flats.

An estimate was made of the carrying capacity, taking account of air and ground traffic noise, air pollution from aircraft and airside and road traffic, third party risks from aircraft activity, water and soil pollution, waste management, energy consumption, 'green' management and landscaping. The noise rules in force allow only 15,100 houses to be within the 35 Kosten (Ke, a Dutch version of Leq, which contains a penalty factor applied to each flight, varying from one in daytime to ten between 2300 and 0600 hours). The four main runways are used to minimise noise impact. At night, runway 01R-19L is not used at all, runway 09-27 is only used when other runways have more than 25 knot cross winds or 10 knot tail winds, special standard instrument departure routes are used, approaches to runway 06 use continuous descents, all approaches should start from 3,000 feet altitude and reverse thrust is banned. Altogether there are 67 standard instrument departure routes.

A fifth runway is to be constructed by 2003 at a cost of Dfl 700 million parallel to the two existing north-south runways to mitigate the environmental impact rather than to augment the physical capacity, with an additional Dfl 680 million being spent on noise insulation of 25,300 houses, paid for from noise charges which have been collected since 1981. It will be used for preference, resulting in only 10,000 houses being inside the 35 Ke contours. A large variety of new procedures are being investigated to allow traffic to continue to grow in the face of these severe rules. It should prove possible to do this, but the Dutch Parliament appeared not to trust the effectiveness of the noise reduction policy and put on an additional safeguard that the passengers per annum should not exceed 44 million. The politicians appear to prefer to consider the possibility of constructing a new offshore airport at a cost of some $12-16 billion (*Airline Business, January 1998, pp 22-23*), rather than maximise the use of Schiphol within the carefully defined environmental carrying capacity. Meanwhile, Schiphol is to charge landing fees on the basis of noise and time of day rather than landing weight. The airport believes that to solve the equation with the community in the long run it needs a new social contract with its community neighbours, based on the concept of no involuntary exposure to environmental impacts. This will mean that those people with high exposure and who do not work at the airport should be offered compensation or a subsidised move, together with a greater feeling of control through public panels (Kuijt, 1997).

Dallas/Fort Worth is in danger of having a 'serious non-attainment' rating relative to Texas regulations for ozone concentrations. It has therefore instituted strict policies, including stopping all non-essential engine use, maximising use of mass transit, consolidating bus services, using clean diesel heavy vehicles, and zero emission vehicles on aprons. It has been difficult to improve the emissions from light vehicles (Ryan, 1998). Compressed natural gas is mostly methane which is very clean environmentally, but requires heavy tanks and has given reliability problems, while liquid petroleum gas (LPG) suffers from 25% evaporation per hour. However, all the taxis in Tokyo run on it, and many Brazilian road vehicles use alcohol derived from sugar cane. There is obvious potential for large reductions in emissions without losing the flexibility or mobility of conventional vehicles, though the move to the Brazilian type of sustainable energy source will not always be cheap. Munich airport plans to experiment with the use of hydrogen power for operational ground vehicles (*Flight, 10 December 1997, p 26*).

These examples show the essential difficulty in balancing the pros and cons of airport expansion caused by the difference between economic and political analyses. Not only is there a difference in outcome due to differences in voting power, depending on whether the basis for evaluation is economic or political, but there are also distortions in each voting system. In economics, market prices often do not reflect true costs of resources, so the principle of 'the polluter pays' does not function efficiently. Even when some of the more tangible pollutants are properly priced, it is still difficult to draw firm conclusions from the analysis. The London case study shows that full compensation to 4,000 affected households for relocation in the event of a new full length runway being built at Heathrow would be amortised over the 40 million per annum additional passengers at the rate of only some £3 per passenger compared with the £50 premium which British Airways claims to receive at Heathrow over the yield at Gatwick. However, it could not be claimed that building the runway would be in the best interests of society at large without some judgement of the implications for, among other things, the changing balance of economic activity and transport infrastructure in the south of England, for the loss of amenity and national heritage associated with historic churches and buildings, and for the changing behaviour of the air transport operators. Meanwhile, the UK House of Commons Transport Committee (1996) feels that Heathrow charges should rise to reflect the environmental impact, the proceeds being allocated to mitigation. Clearly some form of political decision is required which reflects the feelings of a wide cross-section of the British public on these issues. Yet the political process is itself open to abuse from the lobbying of power groups, from 'regulatory capture' and ultimately from a lack of transparency in decision making.

6

ECONOMICS OF AIRPORT DEVELOPMENT

The behaviour of the airport's users and their communities has been examined in the previous chapters. In order to plan the aviation system effectively, it is also necessary to consider the behaviour of the airports themselves. In one way or another, the airports need to be managed as a business, whatever the political and economic context. The goals and objectives of the business will vary with the context, as will the options for meeting them, but the framework for managing them will be largely common. They all are influenced by their balance sheets, their needs for investment and their options for obtaining it, their efficiency and effectiveness and the attitude of government in prioritising investment in aviation rather than other activities. Furthermore, they all have an interest in balancing the management of demand with investment in new technology to get more out of existing capacity. This chapter addresses these issues.

6.1 THE AIRPORT AS A BUSINESS

The transformation of a growing number of airports from government owned service providers to fully private enterprises has been traced in chapter 2. It is necessary to delineate the exact nature of privatisation and its consequences for management autonomy in terms of land use, employment and funding before it is possible to understand how this might affect the adoption of normal business practices. There will be many different motives for privatisation, serving different agendas. These range from purely ideological beliefs that public ownership and regulations are inherently inefficient and that individuals should own a country's assets, to the need to raise public funds or to access capital markets. The sale of the three largest Australian airports (other than Sydney airport, which is to be sold later) raised approximately US $2.5 billion, though this equates to only $1.5 billion if allowance is made for that loss of corporate tax payments over the next 20 years which a public floatation would have ensured

(*Jane's Airport Review, December 1997, p 11*). This is equivalent to 16-21 times earnings before depreciation, interest and tax, compared to 9-10 times for Vienna, 7-8 times for Copenhagen and 7 times for BAA (*Jane's Airport Review, June 1997, p 3*). Privatisation can even be seen as a catalyst to resolve longstanding and controversial planning issues (Zeverijn, 1997).

It is by no means certain that these privatisation expectations will always be fulfilled, given that they are often contradictory, nor that privatisation is a necessary step to fulfilling the expectations. It certainly overcame government funding restrictions for UK airports, which particularly helped Birmingham's Euro-Hub development, allowed East Midlands to buy some land to ensure long term expansion capability despite having to justify the investment to the National Express board of directors and shareholders, and similarly helped several of the smaller airports. East Midlands staff also felt that it reduced the short term local political input to the planning equation, but at the same time made dialogue with the local planning agency more difficult (Froggatt, 1998). Elsewhere in the UK it is hard to find evidence that the change in status to a private limited company (plc) has itself produced the expected benefits: BAA's own performance indicators show no real discontinuity before and after privatisation, and the heavy post-privatisation spending was already under way (Poole, 1992). Equally, it could not be shown that the changes produced any adverse direct effects. The tax status of privatised airports has a considerable bearing on their subsequent competitiveness and hence their value at the time of sale. At the smaller scale, management contracts for British Airports International (BAI) to run airports previously run by local government have not managed to rescue their fortunes. BAI sacked all 93 employees at Southend at the beginning of its £100,000 per year contract in 1984, in an attempt to recoup a £500,000 loss. After a spell of increased services the airport was once again virtually without service (*Airline World, December 6, 1984*). At Liverpool, British Aerospace's plans for a major inter-continental hub came to nothing, and it was not able to retain British Midland's Heathrow service. There are, in fact, signs that potential buyers are beginning to take a more cautious approach to the opportunities being offered, and that the relationships among consortia may be difficult to manage, as in the case of the bidding for Berlin Schonefeld airport (*Jane's Airport Review, July/August 1998, pp 6-7*). Privatisation is clearly not a panacea. It may increase an airport's options, but can also increase risk and uncertainty.

Privatisation is also not an essential prerequisite for the imposition of a business ethic. Manchester airport is corporatised under the UK Airports Act of 1986 so that it is an autonomous legal entity, subject to the rules and regulations of private enterprises, including accounts which clearly separate the airport's costs and revenues from the general accounts of the owning authorities. It responded to competitive pressures and the perceived need of its local authority owners to enter the world stage without relinquishing even partial ownership. The airport management followed normal corporate practice in developing a corporate vision to be:

- a hub airport in a deregulated market
- in the world's top 10 list of airports
- a leader in service quality
- a place where people want to work

- part of a quality partnership with employees and customers
- a world leader in environmental excellence.

From this they developed core strategies in aviation activities, commercial activities and service quality, together with a range of enabling strategies for pricing, infrastructure provision, ground transport, environment, human resources and funding. This approach has allowed them to compete with the BAA London airports, not only for traffic but also in the global market place for airport management expertise. The exact implications of the different forms of ownership differ according to national law. In the UK, corporatisation does not exempt a local government owned airport from the borrowing limits imposed by the Public Sector Borrowing Requirements, these setting limits on borrowing as a result of the government's annual spending review. Amsterdam's Schiphol airport is similarly corporatised, but its government ownership exempts it from corporation tax on its earnings (Zeverijn, 1997).

As airports become more like conventional businesses, they have to analyse their position in the marketplace. They may use conventional tools such as Ansoff matrices, the Boston box and SWOT (Strengths, Weaknesses, Opportunities, Threats) analysis (Shaw, 1990) to provide a basis for the development of strategic plans. Risks have to be assessed in all areas of the business: technical, economic, commercial, political, human resources, operational and environmental. Strategic thinking about its privatised future is leading Schiphol to consider how, as it copes with capacity constraints, it might develop into an 'interconnector', with the ability to enter into alliances and to adopt an innovative mindset (Zeverijn, 1997). Manchester, meanwhile, is promoting activities whose revenues will not be controlled under the UK CAA's powers to intervene in the setting of charges.

The strategies need to be operationalised through business plans (Ashford and Moore, 1992), which set targets over, typically, five years, together with enabling policies and the resources to achieve them. There has to be a close relationship between the business plan and the airport's master plan to ensure that each supports the other. For many European airports, the policies focus on redefining the customer, competing for transfer traffic, promoting the airport and its region, developing retail activities and improving ground access (Feldman, 1996). The corporatised Malaysian Airports Board has formed subsidiary companies to develop retail and leisure activities, including a Formula One racetrack and a nature park within the confines of the 10 sq km of the reserve site for the new Sepang airport (*Airport World, ACI International, April-May 1997, p 22*).

An emerging trend is for airports to realise that they can benefit from multinational status in the same way as other enterprises. It is said that: "multinationals have grown confident and become indispensable .... they are rapidly internationalising and try to have no territorial definition .... they have learnt to let governments add one egg to the mix, so relieving criticism. .... They have also learned to roll over, becoming socially responsible citizens. .... The quickest way to technology, capital and productivity for a poor country is through local subsidiaries of global corporations" (Madsen, 1981). British Airways is taking the message to heart in downplaying nationality in its livery (*Times of London, 14 June 1997*). Airports are rapidly following. BAA has a stake in Naples airport and, with other members of a consortium, has won a 50 year lease on Melbourne airport in addition to its US interests. Schiphol is a partner in renovating and operating the international arrivals terminal at New York Kennedy and also has a share in the 50 year lease on Brisbane airport. Both Brisbane and Melbourne airports are expecting to make use of their owners' connections with airlines to reach new customers (Ballantyne, 1997). Aer Rianta has bought 48% of the Birmingham (UK) airport (Feldman, 1996). Miami airport has signed a marketing agreement with Madrid and Frankfurt airports so that the interests and opportunities of each can be promoted by the others (*Airline Business, July 1997, p 42*). Milan Airport Company has bought 30% stake in the complete Argentinian airport system, while Rome Airport Company, whose privatisation will be completed in 1998, has bought into the terminal at Santiago and into several regional Italian airports (Feldman, 1998). These moves should not only protect shareholders from any local difficulties and provide strong income streams, but also improve possibilities for financing further capital investments and could influence airline decisions on airline routings and hence airport roles.

Airports used to be at the mercy of the airlines' decisions on the routes which are operated, the passengers generated and the impacts on the local communities. This latter aspect is already changing with environmentally driven operating restrictions. Although it is still believed, at least by some in the UK CAA, that airports do not generally compete, but only the airlines serving them (UK House of Commons, 1996), it may not be long before airport consortiums are influencing the routes served and hence the individual airport roles with results similar to those achieved by government regulation of route rights, or traffic distribution policies. Frankfurt, Düsseldorf and Cologne/Bonn airports are all preparing for federal and regional privatisation, and are considering the possibility of operating as a single unit tied by high speed rail. Munich, Leipzig, Stuttgart and Dresden airports have an agreement to cooperate in purchasing, training, data collection and marketing. Munich airport can also see the benefit in a partnership with Manchester airport to share the development of long haul routes. Amsterdam airport took stakes in nearby Rotterdam and Lelystad airports some time ago for additional capacity (van Leeuwen, 1992), and is now considering a joint hub operation with Brussels and Charleroi airports (*Flight, 4 June 1997, p 22*). Amsterdam airport also has a 1% interest in Vienna airport, which in turn is to build and operate a new terminal at Istanbul (*Association of European Regional Airlines, Regional Report, March 1998, p 38*). If National Express had been able to buy a controlling share in Birmingham airport together with its ownership of the neighbouring East Midlands airport, it could have influenced the traffic split between them. However, it must be noted that BAA had very little success in tempting airlines to move from Heathrow to Gatwick or Stansted (see the UK case study), and such airport groupings will always be subject to examination by anti-trust authorities. Indeed, many have seen the privatisation of BAA as a set of airports which dominate London and Scotland as a missed opportunity to promote competition (Poole, 1992).

It has been claimed that larger airport structures are more profitable than individual hub airports (*ACI Europe press release, 5 December, 1997*). To the extent this is true, it points up on the one hand the benefits of monopoly status and, on the other hand, the incentive for airports to accelerate the exploration of alliance opportunities. A good business case for alliances can be made on the strength of group brands, on scale economies in purchasing, training, marketing and administration, as well as the conventional advantages of any multinational enterprise mentioned earlier.

At the other end of the scale, the smaller regional airports, whether privatised or not, are trying to develop policies to restore or improve air service to their small communities. This involves not only improving customer service and understanding their markets, but also entering into joint agreements with airlines and tourist boards, offering incentives to airlines and even creating their own airlines.

6.2 AIRPORT REVENUES AND COSTS

The components of costs and revenues are well documented (Ashford and Moore, 1992; Doganis, 1992) and the details are not repeated here. The focus of this chapter is on the way in which the contributions can be managed in order to improve an airport's balance sheet. This has implications for the way an airport may behave with respect to fulfilling social objectives and as part of an airport system, and the consequences of planning decisions on the airports in a planned system.

6.2.1 Costs

Capital is involved with the initial development of an airport, when the costs are mostly associated with airside development which is largely independent of throughput, due to the indivisibilities of runway construction. It does, however, vary with the type of role the airport is to fulfil. Considerably more capital is required as the scale of operations increases over time, so that the total investment may be as small as £10,000 for a rural STOL strip or several billion dollars for an airport the size of Dallas/Fort Worth. A large part of the total capital may be required before the airport can operate at all, particularly if a large amount of earth moving is necessary: the land reclamation costs at Hong Kong's Chep Lap Kok airport were US $1.15 billion (*Airport World, April-May 1997, p 10*). Expansion is also subject to large indivisibilities; expenditure may have to increase by between 80% and 300% in order to double airside throughput. The landside (i.e. from the access system to the aircraft gate) may eventually cost considerably more than the airside, but the initial investment can be much less and expenditure can keep more in step with throughput. Also, the BAA has decided that unit costs of new terminals can be cut substantially by reducing construction time and other initiatives. The phasing of investment can be crucial to the economic viability. There may be diseconomies associated with keeping the indivisibilities too small, in that congestion costs may arise from capacity effects and inefficiency due to construction. The cost of construction of new capacity causes a disproportional increase in cost compared to the depreciated investment in existing facilities, and usually costs more per unit of capacity due to the need to use increasingly more difficult or remote sites. United Airlines raised its ticket prices by US $40 per round trip to offset the charges at the new Denver airport (*Flight, 26 April 1995, p 34*) which were set at approximately twice the levels of those at Stapleton airport, and there has been a long battle with IATA over the scale of charges to cover costs at Kansai (as discussed in the Japan case study later in this book). This will become an increasing problem as sites become more difficult to find, and there appear to be no easy cost control measures for building on offshore sites. Japan is even considering floating airports at a cost of US $30 billion.

134   *Strategic Airport Planning*

Operating costs increase with the size of an airport, particularly if the expansion involves landside facilities, though unit costs ought to fall. The landside unit costs per terminal passenger or tonne throughput depend on:

- traffic mix, with facilities for charter and domestic passengers usually less expensive than for scheduled international traffic
- sharing of facilities, leading to a greater utilisation of space
- policy on space standards, depending on level of service
- peaking of traffic, affecting airport utilisation.

The airside unit costs depend on:

- scale of aircraft used: the effect on cost is hard to generalise because large aircraft generally require a higher standard of provision of facilities, a higher regulatory classification, a longer runway, a much larger area to be maintained and amortised and a lower maximum movement rate, but these factors must be balanced against the greater throughput per aircraft movement;
- apron layout: e.g. remote stands and shuttle buses versus nose-in parking and air bridges.

The economies of scale which should be associated with larger buildings and their servicing are hard to realise because other factors are almost always increasing as an airport's throughput increases, namely:

- aircraft size
- number of operators
- international operations
- capital borrowing
- opening hours
- unionisation
- construction on the airport
- ancillary services.

The UK provides a good opportunity to examine these effects, because the airports' accounts have had to be free standing and transparent since the 1986 Airports Act. An idea of how these factors combine to affect unit costs is given in Figure 6.1, where a work load unit (wlu) is one passenger or 100 kg of cargo. The wlu is commonly used to attempt to account for the cargo contribution to expenses and revenue. There is no intention to imply that 100 kg of cargo is equivalent to one passenger plus bags, except in terms of weight and therefore the certificated weight of the aircraft which carries them.

The accounts of UK airports have been gathered together annually by the Chartered Institute of Public Finance and Accountancy (CIPFA), now the Centre for the Study of Regulated Industries, though comparisons within the local authority grouping and between this group and the BAA airports have been made difficult not just by the differing nature of the airport but also by differing accounting principles and reporting practices, particularly with regard to

Figure 6.1: Operating costs of UK airports, 1994/95
a) **Operating expenditure (£ million)**

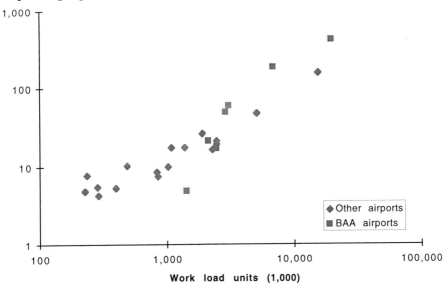

b) **Operating expenditure per work load unit (£)**

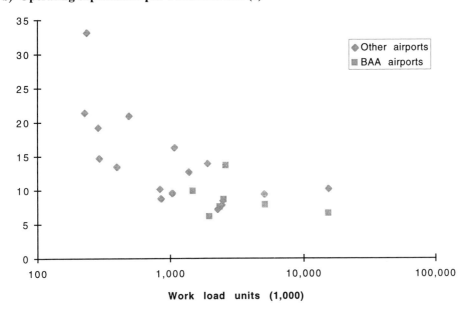

Source: CIPFA, 1995, modified by the authors

depreciation. Figure 6.1a plots the total operating cost (excluding debt charges) from CIPFA (1995) against wlu. There is generally a fairly linear relationship between cost and wlu, particularly if it is recognised that some airports were over-provided with facilities relative to their traffic levels. These data suggest that there may be some level of fixed costs, which make smaller airports relatively more expensive. This is shown more clearly in Figure 6.1b, which shows expenditure per wlu. It can be seen that, even before debt charges are considered, no airport was achieving less than £6 per wlu in 1994/95. Airports in the 100,000 to 500,000 passenger per annum bracket were costing in the range of £12-£35 per wlu, while those with 5,000 to 30,000 passengers (not shown in the figure) were costing up to £300 per wlu. The airports in the Spanish and Australian systems show much the same characteristics (Graham, 1997).

The major item of operating cost in the UK data was always staff, which never accounted for less than 30% (with the exception of Heathrow). The average staff cost for the local authority airports was 55% of the total operating cost before the addition of debt charges. The 13,590 staff were employed to handle £5,201 million of fixed assets, i.e. £383,000 per employee, which is much higher than the average for airlines, i.e. airports are capital-intensive in this sense. The variation in the employee cost per airport largely depends on the functions performed in-house rather than put out to contract, e.g. Luton airport did virtually everything in-house whereas Birmingham airport let many activities out as concessions. Airports' costs for supplies and services have historically been high if they employed the CAA to provide air traffic services and where services (and the associated staff) such as security and the fire service were being provided by the local authority on behalf of the airport. The staff cost also depended on the cost per employee. The variation here was considerable: although some of the variation was due to the range of functions performed and the opening hours, there were also locational and other factors at work. The BAA costs per employee would have been less if corporate and central service staff had been allocated to each airport to reflect the way these central costs have been allocated.

The second largest item was almost always the depreciation and interest on loans. There was a general trend towards relatively low debt charges in airports with a low throughput. The variations about the mean cost per wlu, after taking account of economies of scale and the above factors, were due to the differing levels of utilisation of the facilities, to investment policies which emphasise regional advantage rather than airport finances, and to the differing mix of traffic. The latter effect can often be reflected also in the income levels, e.g. an international airport has to provide customs and immigration facilities but can usually benefit from the income of a duty-free shop.

These costs have not been normalised to account for the lack of homogeneity in the airports. Attempts were made to carry out a normalised analysis in the 1970s (Doganis et al, 1978). These concluded that there were substantial economies of scale, particularly below a throughput of three million wlu per annum, and that the costs were strongly dependent on the ratio of international to domestic passengers, on the way in which atc was provided (CAA or in-house) and on the degree to which an airport had undertaken a recent development programme. These studies were later extended to include those European airports whose accounts were sufficiently self-contained to be analysed (Doganis and Nuutinen, 1983). The larger average size of these airports may have obscured any economies of scale; certainly none were found, and neither development programmes nor the proportion of international passengers appeared to affect unit costs. An update of the European study confirms these earlier results and also the large labour content in airport costs: the average result across 16 airports was that staff accounted for 42% of the total annual costs, while capital amortisation was responsible for only 24% (Graham, 1992). The European studies made an attempt to allow for differing degrees of capacity utilisation between the airports, but they were not very successful. It is therefore not easy to use these results to guide investment decisions or verify economies of scale, though they provide a useful basis for the generation of performance indicators to monitor the management of productivity.

Returns to scale (or economies of scale) describe how output increases as the factors of production are increased proportionately. In practice, this is often measured by whether long run average total costs increase or decrease with increasing output. A value less than unity implies that as output increases, long run average total cost decreases. Returns to density describe how short run variable costs change with increasing utilisation of some fixed factor of production, in the case of an airport the airside and terminal facilities.

The distinction between short run and long run costs, and return to density and return to scale, deserve some discussion for the reader not familiar with these terms. The short run cost function describes how costs increase with increasing output (traffic in this case) if facilities are not expanded. In contrast, the long run cost function takes account of the higher costs of capital but the offsetting reduction in operating costs if facilities are expanded as traffic grows. The average total cost (short run or long run, as the case may be) is simply the total cost divided by the output (in this case measured in wlu), while the marginal cost is the increase in total cost for each unit increment in output (i.e. the slope of the total cost function). By definition, the short run total cost (and hence average cost) is higher than the long run total cost, except at the point where the optimal amount of facilities are provided, where they are equal.

If an airport has over-invested in facilities or has embarked on a major expansion in advance of being able to utilise the full capacity of the new facilities, as is inevitable given the indivisibilities and "lumpiness" of airport facilities, then the short run marginal cost is generally lower than the long run marginal cost. On the other hand, if the airport has delayed investing in facilities beyond the point at which further expansion is required to achieve the long run total costs, then the short run marginal cost is higher than the long run marginal cost. The timing and scale of investment should depend on the trade-off between the amortised costs of expansion and of congestion, which, in turn, depend on the initial capacity and the time path of growth of demand (Oum and Zhang, 1990), and are thus sensitive to the quality of forecasts.

6.2.2 Revenues

Revenue generation comes from a variety of sources. Aeronautical charges are the traditional mainstay of revenue and, for small commercial airports, may be the only source. At some types of airport, the emphasis is on landing fees supplemented by passenger charges, while at other types revenue will come mainly from rents and, particularly in the case of general aviation airfields, levies on fuel sales. As passenger traffic grows, and particularly if it contains a long haul component, more of the revenue will come from non-aeronautical sources such as concessions or car parking. At a typical large hub airport in the US, 43% of revenue comes from non-airline sources, half of that from parking and rental cars. The relative importance of revenue from various forms of road transport access is even greater at medium hubs (Ernico and Walsh, 1997). The precise nature of the agreements between the terminal landlord and the concession holder varies from a percentage of sales to a fixed charge per square metre of floor space, or some combination. The length of lease and the conditions of sale are also normally specified. Revenue generation per passenger varies considerably with the types of passenger handled by the airport and the quality of the service provider.

In a liberalised and competitive setting, particularly where there is an element of airport privatisation, the profit motive is likely to be a primary determinant of income and pricing policy. Maximum profit depends on the difference between revenue and cost, and may well not be made at minimum unit cost, even in a competitive situation. BAA quite deliberately provides more space for passengers than is strictly necessary for processing in order to improve concessions revenue, and is planning to provide approximately 1,000 sq metres of retail space per million annual passengers (Donald, 1996). The BAA earned £654 million from European concession operations in 1997/98, of which 22% came from perfume, 15% from gifts, 15% from liquor, 13% from tobacco and 12% from car parking (BAA, 1998). This not only produces a high marginal profit contribution but also helps to keep airline fees down. On average, 16 European airports have obtained 45% of their revenue from non-aeronautical sources (Graham, 1992). Airports which give this a high priority appear to be able to generate about US $10 of non-aeronautical income per passenger (O'Toole, 1997). At Schiphol, 55% of the revenue was non-aeronautical in 1996 (Zeverijn, 1997). Atlanta airport also generates almost half of its revenue from concessions (Nelms, 1997). A quarter of the non-airline revenue at Las Vegas airport comes from gaming (*Jane's Airport Review, September 1996, p 69*). These types of non-aeronautical revenue generation are not possible when, as in the duty free shops at the Airports Authority of India's airports, the airport has no rights in the operation of the concessions (*Airport World, Winter 1996, p 37*). Also, the more competitive the situation, the more it may be necessary to maintain a high and costly level of service, in order to assure the airlines that their passengers are satisfied. Little research appears to have been done into the ability to influence passengers' choice of airport by level of service, though some airports (e.g. Dubai, Manchester) believe it can be affected by the price of duty free, particularly for transfers. However, airlines' decisions on which airports to serve certainly appear to be based partly on the level of service offered by each airport.

At the margin, i.e. for those air transport operations which are financially fragile, airlines are also swayed by the landing fees, passenger charges and other infrastructure costs which have to be incorporated in the ticket price. A considerable amount of attention therefore needs to be given to these charges.

Aeronautical charges can be used as a competitive tool in an underutilised situation, but also, in theory, as a method of controlling demand and maximising cost recovery in a congested situation, hence facilitating the funding of additional capacity. In fact, "under the assumption of perfect divisibility of capacity expansion, the well-known cost recovery theorem states that congestion toll revenues just cover the amortised costs of capacity expansion if there are constant returns to scale in capacity construction. .... We have shown that when capacity investment is lumpy, the cost recovery ratio depends on the time path of traffic growth, and thus the conventional cost recovery theorem no longer holds". (Oum and Zhang, 1990). The same authors point out that marginal social cost pricing (i.e. pricing based on marginal costs that include both congestion costs and externalities), although it gives the most efficient allocation of resources, implies increases in user charges prior to expansion and a rapid reduction immediately after expansion, which is difficult to manage. A constant optimal price is, however, seen to result in too high a welfare loss. IATA favours a minimum of forward funding from a development fund raised through congestion fees, preferring that the capital be raised in other ways and amortised over the life of the project.

Since congestion tends to occur only at peak periods, some airports, notably BAA at Heathrow and Gatwick, have for many years differentiated their landing and passenger handling charges between carefully defined peak hours and those in off-peak periods. In 1996, Heathrow's peak landing fee for all Chapter 3 fixed wing aircraft was £400 per aircraft in addition to £5.95 per domestic passenger and £10.95 per international passenger. This compares with off-peak charges of £100 per aircraft up to 16 tonnes and £162 per aircraft between 16 and 50 tonnes in addition to £2.95 or £4.20 per domestic or international passenger respectively, the definition of the peak hours varying between domestic and international movements. BAA found little ability to control the schedules of the major operators with this pricing mechanism, though marginal operations are sensitive both to this and to the flat fee, the latter being particularly onerous for short haul operations by small aircraft. Operators who retain their peak slots typically do not pass these extra costs specifically to their peak passengers. In the matter of maximising revenue from scarce runway capacity, IATA tends to favour a flat fee rather than one varying in the traditional fashion with aircraft weight, whereas ICAO prefers this method only when the capacity problem is severe (Vandyk, 1993). It has been suggested that passengers themselves ought to be charged directly through a peak surcharge, thus creating additional indirect pressure on airlines to shift to off-peak slots (Hardaway, 1986).

Control over pricing of aeronautical services has normally been in government hands, either as owners or as the authority charged with policing monopoly behaviour. International influence is exerted by ICAO and IATA. The UK CAA has a duty to ensure that the BAA airports are not being cross-subsidised. Also, it is CAA and ICAO policy that charges should be cost-related, ICAO now allowing costs to include a reasonable rate of return (Clayton, 1997). So the prices paid by airlines to the airports should be influenced by the airports' costs, but attempts to relate charges to costs are often made difficult by the lack of discrete cost information. The EC also wishes to harmonise the relationship of fees to cost, but is under pressure from airlines to harmonise the charges themselves. This is clearly inconsistent with a competitive market place.

In the UK, most regional airports have always been in a competitive and uncongested environment, (as discussed in the UK case study later in this book), despite the regulated nature of the airline industry. The temptation was always to keep fees low in order to retain airline service, despite the nominal uniform pricing policy of the Aerodrome Owners Association (AOA) and the Joint Airports Committee of Local Authorities (JACOLA). The poor joint profitability of these airports (Doganis et al, 1978) led to a five year plan of phased and coordinated increases in fees, particularly the passenger load supplement. These were agreed between AOA, JACOLA and the airlines (Bowers, 1979). This would, of course, have been against the principle of cost-related pricing, given the apparent economies of scale, though perhaps not to the same extent as the principle of 'what the market would bear' which had been firmly ingrained in the industry. The Director of Fair Trading ruled that, under the Restrictive Practices Act which had been made applicable to the service industries in 1976, the agreement would have to go to a court ruling, but the essence of the agreement was carried through.

At the other extreme, in a profitable but potentially monopolistic situation, as is the case with the BAA airports in southeast England, and has also been deemed to be the case with Manchester airport, the CAA is required under the 1986 Act to rule on the allowable fees after consultation with the Monopolies and Mergers Commission (MMC). A 'Retail Price Index - X' formula has always been used to ensure the consumer benefits from the expected improvements in efficiency, though this method has been criticised on the grounds that the charges should have been reset to appropriate levels prior to privatisation (Poole, 1992). Manchester airport's changes in charges are limited to 5% below the retail price index (RPI) from 1998 for five years (Field, 1998). In its review of the BAA (CAA, 1991), the CAA ruled for a very marked reduction in fees, which was modified after support came from the airlines for BAA's argument that their investment plans would be put in jeopardy. The situation was resolved by the CAA tapering off the fee reductions (in real terms) to a varying amount below the RPI of 8%, 8%, 4%, 1% and 1% over the five years from 1992, and the BAA agreeing not to cut back on necessary investment (Nuutinen, 1992). It is not certain that the CAA has been equally vigilant in ensuring that Stansted is not subsidised by revenues from Heathrow and Gatwick. Indeed, Luton airport has accused BAA of predatory pricing (Betts, 1993). For the CAA, it is a matter of balancing competition policy against expansion funding.

These two cases of regulatory intervention are examples of the difficulties that airport companies have in behaving like a profit-maximising private enterprise. They also exemplify the importance of a clear analysis of the power relationships in the industry if workable strategies are to be developed and implemented successfully. This will certainly be affected by the UK and EC regulators' view of monopoly pricing, predatory pricing and cartels in a privatised airport industry, just as they are in the airline industry. It will be necessary to predict likely initiatives by legislators with respect to charges and their harmonisation. There are examples in Germany, Australia, Canada and the US of reserving some peak hour slots for general aviation and third level scheduled services, with the implication of reduced fees for these slots. It is also likely that fees will have to increase in real terms to cover the costs of expansion, even if after the event, and also for levies to be raised for the control of noise, emissions and energy use (Toms, 1993). There is a general feeling in the industry that any taxes to address externalities should be harmonised worldwide through ICAO, but the basis for harmonisation is far from being agreed. Also, some legislators feel that it is wrong to insist on the airports being able to earmark such taxes for the alleviation of the nuisance, on the basis that pollution knows no boundaries (Vaagen, 1993).

The EC legislators appear to be against the recognition of the monetary value of a scarce runway slot, despite the fact that trading in these slots occurs legitimately under FAA regulation in the US, and the US airlines' valuation of their predecessors' routes to London were clearly influenced by the high value of slots at Heathrow. It appears that Heathrow slots might have changed hands for US $3 million (*Flight, 8 July 1998, p* 27) and to be worth US $1.7-3 million at Narita (*Air Transport World, July 1998, p* 154). It has been argued that the real owners of the slots are the airports, and that sales or leases to the highest bidder would give the most appropriate method of funding capacity expansion (Cameron, 1991).

6.3 IMPLICATIONS FOR INVESTMENT STRATEGIES

6.3.1 Performance objectives

All airports will wish to obtain operational improvements from their investments. The benefits may arise from reduced delays, improved schedule predictability, more efficient traffic flows, the ability to accept larger aircraft, better compliance with safety or environmental standards or operational efficiency (FAA, 1996). However, from the point of view of the airport as a business, a healthy operating margin is a prime concern, and investment strategies will be chosen to ensure it. Certainly privatised airports are expected to be profit maximisers. Others will increasingly have to demonstrate their economic viability, though they may have to carry other obligations. It has been usual for airports to be service providers with no need or desire to make a profit. This is reflected in the residual cost approach in the US, where airline fees are adjusted to cover the difference between costs and non-airline revenues. This means the airlines do not encourage investment in advance of the need, nor the generation of an operating surplus to fund it. It gives no incentive for cost efficiency, since the costs cannot be retained (Trethaway, 1996). The system of federal funding through the Airport Improvement Program, which pays for the majority of airside investment if the need can be justified, in any case relieves the airport of the need for large cash reserves to fund airside improvements. Similarly, the widespread use of revenue bonds provides additional capital which can be repaid from revenues after the expansion has occurred.

While the large established airports may be viewed as an almost limitless source of profits, even allowing for regulatory intervention, this will only be true to the extent that they manage their resources effectively. If they opt for a 'cash cow' policy and defer investment in expanded capacity, there is every danger that the consequent congestion and inability to accommodate traffic growth would favour their competitors and the airlines which use them. If they do invest, the absolute level of investment to gain a reasonable percentage increase in capacity and hence a reasonable uncongested time horizon may be so great that the economies of scale no longer apply. Non-linear costs of large scale environmental disturbance may occur. The UK case study later in this book indicates how non-linear these may be with respect to an extra runway at Heathrow, though it also shows that even the high costs of a full third runway could be acceptably amortised over the large number of extra passengers that could be accommodated.

It should be recognised that the competitive dynamics of a fully privatised airport system are not well understood. The combination of the geographical advantage that many existing

142 *Strategic Airport Planning*

airports enjoy with respect to the market they serve and the extreme difficulty in many situations that a competitor would experience getting approval to construct (or expand) a nearby airport means that the potential to extract monopoly profits in the absence of price controls is very real. This may be viewed as simply obtaining a location rent, which if it has been properly accounted for in the original purchase of the airport may not harm overall social welfare. However, the difficulty of doing this in practice, given the uncertainty over future levels of demand, airline strategies, and the actions of competing airports, should not be underestimated.

The greater danger lies with those small airports who feel the need to expand or die. Figure 6.2 shows the reported operating surplus for the UK airports, i.e. before covering debt charges, in the 1994/95 financial year. There were some anomalies in the results. London City was losing over £6 per wlu and Stansted over £3 per wlu, while Norwich made an operating profit of more than £4 per wlu. Generally, however, the effects of scale appear to be strong, with a breakeven on operations at approximately 0.5 million passengers per year and an operating surplus of £1 million at 1 million wlu per year. The addition of debt charges requires perhaps a further 0.5 million passengers to ensure net profits. It is clear from Figure 6.2 that it is very easy to lose money if traffic does not materialise to cover an investment. Below 50,000 passengers it is very difficult to avoid losses - though some succeed in staying in the black by not investing, by maintaining almost exclusively a domestic operation (e.g. weekend flights to the Channel Isles) and by only opening 10 hours per day. In the US the pattern is similar. The operating ratios are: large hubs, 2.05; medium hubs, 1.95; small hubs, 1.7; non-hubs, 1.0; other commercial, 1.15; GA airports do not in general break even (GAO, 1998). A rule of thumb is that an

Figure 6.2: Operating surplus per work load unit (£)

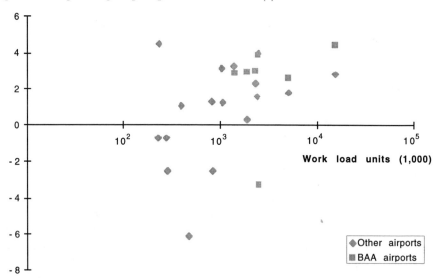

Source: *CIPFA, 1995, modified by the authors*

operating ratio of 1.25 is the minimum necessary to earn a good bond rating in order to be able to borrow for further investment. One third of the members of Airports Council International do not have enough income to cover their costs, yet still have to meet ICAO operating standards (*Jane' airport Review, July/August 1998, p 9*).

The concept of escapable costs can be applied to airports, (Doganis et al, 1978). Thus, while salaries are inescapable, maintenance and repair cost could be postponed for a year, and depreciation and renewal costs might be postponed for five years, subject to licensing requirements, and interest and development charges could theoretically be postponed indefinitely. The implication of deferring expenditure is that the airport's traffic would at best stagnate and the airport would lose market share. Some have survived for a long time using this approach. The really difficult choice is whether to accept declining share of the overall market, with perhaps a niche function whose market is close to saturation (e.g. ferry routes to offshore islands or oil rigs), or to attempt supply-led expansion in the hope of reclaiming prior market share.

More generally, there has been a growing realisation that financial stability depends primarily on position in the market, i.e. on the strength of the local air demand rather than on the mere presence of an airline. This has meant a move away from the residual cost approach of balancing the books, from long term agreements which limit the possibility to reflect changes in value in levels of charges, and from the weaker 'majority in interest' clauses at US airports. More airports have moved towards the compensatory approach which is more common in Europe. There is more competitive bidding, new sources of revenue have been identified, and Passenger Facility Charges (PFC) have been levied where this is legal. Investments now need more careful justification.

6.3.2 Competitive position

Subject to capacity constraints, airports will wish to track their market share as a leading indicator of their competitiveness. Examples of competitive situations and their consequences are given in chapter 13. Airports might fall into the trap, which has proved fatal for some airlines, of chasing market share regardless of the yield. It may not be too difficult for an airport to encourage a low cost carrier and so increase its market share, but the landing fee may have to be so low and the concessions revenue per passenger also so low that only marginal costs are covered. The strategy may pay off if the traffic growth convinces other carriers that a market exists, but investing in new capacity without sound attention to yield would not be sensible. However, it may provide the means to maintain cash flow when a change in airline strategy leads to a reduction in traffic. Geneva airport has adopted this strategy as a response to Swissair's withdrawal of most long haul services to Zürich (*Airline Business, December, 1996, p 21*), but Russian airports simply put up their fees (Duffy, 1996) while Shannon responded to the Irish government's decision to allow half of Aer Lingus's transatlantic flights to overfly it by strengthening its role as a hub for Russian transatlantic services (*Flight, 6 March 1996, p 10*) and by waiving all charges for the first three years of a new service (*Jane's Airport Review, May 1997, pp 8-9*).

Similar risks are attached to a large growth in the percentage of transfer passengers. The consumer's preference is strongly for direct service, with US passengers being shown to be indifferent between one direct flight per day and 15 flights via a hub (Hansen, 1990). The comparable result between the UK Midlands and Europe seems to be 1:7 (Ndoh, Pitfield and Caves, 1990). However, some caution is appropriate in interpreting these figures because of the effect of multiple airlines serving a market. A situation with one airline offering direct flights while several competitors offer connecting flights via different hubs is clearly very different from the same airline offering one direct flight per day and fifteen via a hub. Not only will passengers usually take a direct flight where one is offered, but they will be largely indifferent to the choice of hub if they have to transfer (although perhaps not to the choice of airline). Further, they normally contribute little to revenue inside the terminal building, because the airlines wish to offer rapid connections. Their costs therefore have to be covered by the aeronautical charges, which will be subject to regulation. Schiphol reduces the passenger charge for transfers to some 70% of those for terminal passengers, which must assist KLM in its competition with Lufthansa and British Airways for hubbing passengers, since Frankfurt and Heathrow make no such reductions. Olympic Airways, Sabena and Air France are in the happy position that their base airports exempt their transfer passengers completely (Stockman, 1996).

There is a long tradition in Europe of differential fees, particularly for domestic versus international flights, passenger versus freight flights and for start-up services. The European Commission is set on eliminating these differences, but any business needs to be free to use some degree of pricing to what the market will bear and to what it takes to compete. The Australian Federal Airport Corporation have lowered charges at Sydney airport for Qantas and Ansett airlines so as to compete with the newly privatised airports, while increasing minimum landing charges (*Jane's Airport Review, July/August 1998, p.*4). Similar distortions occur in the US through the federal funding mechanisms. Minneapolis/St Paul gets back in grants only 6% of the US $200 million its travellers pay in ticket taxes (*Air Transport World, September 1995, p 93*), while its based carrier is faced with including a US $3 Passenger Facility Charge (PFC) in the ticket price for each embarking passenger, which its competitors at other hubbing airports with no PFC do not have to face (*ITA Press 171-172, August 1992, p 18*).

However, it must be recognised that such comparisons can very easily get twisted into narrow self-serving arguments. If the airport did not levy a PFC, it would fund its capital development some other way, which would end up getting paid by the traveller (and perhaps the airline) through some other fee. Likewise, who is to say what proportion of the ticket tax paid by passengers starting their trip or changing flights at a particular airport should be returned to that airport in grants. Presumably the entire point of a federal airport development programme is to provide a mechanism to selectively allocate funds within the system. If every airport gets back exactly what its travellers contribute, development might as well be funded through landing fees or some other local revenue source.

Schiphol anticipates that its impending full privatisation may expose it to the threat that a KLM alliance could dilute its transfer traffic in favour of a partner's hub, as happened at Copenhagen after SAS's alliance with Lufthansa. This is another example of the dependence of an airport on its main based carrier, and therefore on the competitive performance of that carrier. The investment message is that some form of control of risk needs to be engineered by way of airline guarantees or, if airport policy is to avoid diluting its autonomy, a viable alternative investment strategy should be formulated which could cope with the weaker outcome.

6.4 COMPARATIVE PERFORMANCE MEASURES

Measures of financial performance are important to any business. In a publicly owned airport system, there will normally be no requirement to publish separate airport system accounts or to identify the financial performance of individual airports, though the airlines will wish to have transparent evidence of the costs which their fees are being required to cover. In an open market setting, at least the performance of any commonly owned set of airports will have to be made available so that shareholders know what they are buying. Often, the antitrust authorities will also require evidence that there is no cross-subsidy within a privatised system which would assist a weaker airport to compete with rivals. It therefore becomes easier to adopt normal business accounting methods to establish the financial performance of airports. Some of the common ones are:

- Operating Ratio - operating and maintenance expenses divided by operating revenue. This ratio measures the share of revenues absorbed by operating and maintenance costs. A relatively low operating ratio indicates financial strength, since it signifies that only a small share of revenue is required to satisfy operating requirements. A high ratio (close to 1) indicates that relatively little additional revenue is available for capital spending.

- Net Take-Down Ratio - gross revenue minus operating and maintenance expenses, divided by gross revenues. This similar to the operating ratio, but it also includes non-operating revenues (e.g. interest income). It is a slightly broader measure of the share of airport revenues remaining after payment of operating expenses.

- Debt-to-Asset (or equity) Ratio - gross debt minus bond principal reserves, divided by net fixed assets plus working capital. This measures the fraction of total assets provided by creditors. Creditors prefer low debt ratios because each dollar of debt is secured by more dollars of assets.

- Debt Service Safety Margin - gross revenues less operating and maintenance expenses and annual debt service, divided by gross revenues. This ratio measures both the percentage of revenues available to service new debt and the financial cushion to protect against unexpectedly low revenues.

ICAO suggests the following additional indicators for measuring performance:

- income per passenger
- expenditure per passenger
- aeronautical income per passenger

- non-aeronautical income per passenger
- passengers per employee
- income per employee
- value added per employee
- capital expenditure per passenger
- net assets per employee

with the caveat that it is important to note any internal or external factors which might cause distortions to them (ICAO, 1991).

If proper allowance is made for the effects of scale and type of operation, these partial measures can be used for benchmarking to give an idea of comparative performance, though there is no unique way of allowing for all the individual characteristics of each airport. Ideally, with a sufficiently large and disaggregated data base, some form of total measure of productivity could overcome the deficiencies of partial analyses. Total factor productivity and data envelopment analysis have both been used to analyse other public transport systems (Graham, 1997), and are beginning to be applied to airports (Hooper and Hensher, 1997; Gillen and Lall, 1997). In lieu of this, some quite elaborate attempts have been made to normalise data across international sets of airports. The firm of Symonds Travers Morgan used a selected set of core activities corrected by the Special Discount Rate (based on the trade-weighted values of seven major currencies), and showed, among other things, that Heathrow made the largest operating profit per passenger of the 28 airports studied (Mackenzie-Williams, 1997). A recent study of European airports has also attempted to allow for some of these differences and to normalise the currencies with Purchasing Power Parities (Lobbenberg and Graham, 1995). In general, it confirms the influence of scale, type of traffic and the utilisation of capacity as the main determinants of productivity, but also shows that the three highest earners of non-aeronautical revenue are all privatised.

It is, however, safer to use these and other non-financial indicators to track changes of performance over time than to compare operators, because airports generally operate in relatively unique settings, not least in terms of the inherited expectation of the roles they should play. It is, in any case, unlikely that airports will want to share most information about their quality, unless they are part of a commonly owned or operated system. The most direct way of checking service quality to the ultimate consumers is to survey achievement against standards. BAA measures its service delivery against its standards (e.g. queue length) and by surveying 150,000 passengers per year to establish the level of satisfaction with cleanliness, mechanical assistance, procedures, comfort, congestion, airport staff behaviour and value for money on a scale of 1 to 5 (BAA, 1998). These results should confirm that design standards on space and service times are being met. These latter indicators will give the clearest indications of the need for more capacity, either landside or airside. Similarly, Amsterdam monitors its environmental performance against its objectives for 30 or so indicators (Amsterdam Airport, 1997), to ensure that it complies with the law and the agreements which will have been made with the local authorities. It is also sensible to allow consumers to comment on service and to monitor the complaints. It is, in addition, necessary for management to monitor its relationship with all other interested parties: the airlines, freight forwarders, concessionaires, the community.

Benchmarking on quality depends primarily on third party surveys. A 1995 survey of IATA's long haul passengers ranked 43 airports for overall convenience, as well as separately for ground transportation, availability of baggage carts, shopping facilities, eating facilities and speed of baggage delivery. Orlando was the top North American airport, while Manchester was the best in the rest of the world (*Airport World, 1996, pp 50-52*). A similar study in 1997 of the views of 78,000 passengers on 23 service categories put Singapore first of 62 airports, Helsinki second and Manchester third (*Airport World, Vol. 13, Issue 2, 1998, pp 30-33*).

6.5 PUBLIC INVESTMENT PRIORITIES

Regardless of the degree of privatisation, there will clearly be a public interest in the operation of an airport. At a minimum, this will be a concern for the proper management of the site for its designated land use, and its conformity with local and national regulations. In addition, an airport will be expected to provide some form of public service. Public sector sellers of airports often retain a stake as an assurance that they will be able to influence the airports' policies, even if only in the form of a 'golden share' which allows them a veto if they feel the airport is not fulfilling its proper function. The BAA management has not yet tested the UK government's resolve to use their golden share, having yet to run counter to an official government policy.

6.5.1 Justification for public sector role

Governments have a duty to use public funds wisely. They therefore have to decide not only which airports they would like to support, but also whether they should spend resources on airports at all, rather than on other public goods, such as hospitals or defence. Those charged with managing the aviation system thus need a methodology for generating a case for investment which will carry weight with elected representatives. To the extent that they see fit to support aviation, governments will wish to ensure that adequate facilities are in place for the satisfaction of reasonable demands, so that common rights of passage and the process of national administration, industry and commerce may continue. The public sector will therefore need to monitor the service, and where the service is in jeopardy, invest to a level necessary for the maintenance of a minimum standard of service or, alternatively, decide that the public interest does not require the facility. Again, this requires a process for evaluating the performance of the system and the projects brought forward for improving it.

It has been argued that the public policy of serving the transportation industry and travelling public is most effectively implemented through government ownership, while regulation and franchise conditions applied to private sector enterprises are less effective since control of development and asset utilisation is indirect. On the other hand, government ownership is held to be, by definition, less efficient because of the intentional limits on flexibility imposed by procurement rules, employment regulation and investment rules to make certain that the public's resources are properly used (Turbeville, 1996). The balance of public advantage between these alternatives is in the value of the sale, which is determined by the risk for the investor and varies dramatically according to the terms of sale.

The UK government effectively decided, in the 1978 Airports Act, that the national interest was only concerned with the gateway airports, the maintenance of all others being devolved to local authority policy and, where necessary, support. Some would argue that an airport system is only as strong as its weakest member, in which case all airports would be important to the nation, but this will only be true if the domestic air links between the airports are maintained. The UK government position has been that the local communities are best able to take such decisions. That philosophy, of a system not being more that the sum of its parts, clearly breaks down when a community wishes to support its airport so as to provide a link to a national hub, only for the hub to deny access due to lack of runway slots. It has been argued that a market-driven system can cope with that difficulty by allowing the remote community to buy hub slots, even starting an airline in order to being allowed to enter the bidding. A more logical process would be for national government to take the regional case to heart and influence either the rationing of capacity or the provision of additional capacity. There is theoretical evidence that, at least in hub and spoke networks, there is a welfare gain to be made by switching from a regime of pricing every airport independently of other airports in the network to the regime of pricing the airports jointly (Oum, Zhang and Zhang, 1996).

A further argument for public sector involvement, apart from the ideological considerations of selling national assets, is the theory that liberalisation and privatisation lead naturally to monopolies, and that the consequent pricing regulation may encourage under-investment (Helm and Thompson, 1991). A suggested solution to this is more explicit regulatory contracts on the investment programme, though the specific terms of pricing regulation can also affect the issue. Those privatisation processes which only intend to regulate aeronautical charges will allow a higher starting capital value (Zeverijn, 1997), and will also give more headroom for the generation of profit to plough back in investment. However, the BAA privatisation is of this form, and has been criticised because it would be possible to raise charges in the non-regulated activities to choke demand in the regulated sector 'to produce even less output than a monopolist' (Gellman, 1990).

In fact, one of the main reasons for considering privatisation is that, in many countries, public investment priorities have resulted in a proportionate fall in airport investment. The appropriation from the US Aviation Trust Fund has been roughly constant during the 1990s, so the industry has had to rely more on the Passenger Facility Charges (PFC) and bond issues to fund the necessary investments. However, this demonstrates that if the principal concern is public borrowing or taxation limits, there are alternatives to privatisation that can provide airports with investment funds without recourse to restructuring the ownership of a major component of the transportation system. The two central questions that need to be answered are: which is more effective, public or private ownership; and how is effectiveness to be measured?

6.5.2 Programmatic considerations

If the funding is to come partly or wholly from public funds, as has been the case in the US, a process has to be developed to collect and distribute the funds. Criteria have to be established for eligibility of projects and for the ordering of priorities, given that the identified needs will almost always exceed the available budget. The FAA at the federal level, and the individual states where there are state funding programmes, use a process of weighting proposed projects between classes of airports and project purposes. In California, a pool of projects for airports is formed into a 10 year Capital Improvement Program (CIP) from an unconstrained needs list. The California Department of Transportation (Caltrans) then applies a weighted evaluation matrix to tailor a proposed programme to the budget for presentation to the California Transportation Commission. Although the State only funds general aviation needs, the CIP includes all airport needs, so as to accurately display the aeronautical needs of the whole system to the public and the policy makers. The evaluation matrix is re-evaluated every two years by a Caltrans advisory committee composed of aviation system users and operators and transportation planning agencies. The Commission then decides if it is necessary to adjust the weights to reflect changing capital improvement priorities (Oldham, 1998).

The FAA's funding is partly by apportionment and partly discretionary, in each case reflecting priorities, the apportionment weightings being embedded in agreed formulas which relate to legislative mandates. Proposed projects for inclusion in the Airport Improvement Program (AIP) are generated by a bottom-up five year capital improvement plan, synthesised into a top-down national programme and delineated by region for a current year (TRB, 1994). An internally generated weightings matrix gives a ranking between 1 and 12 to each type of work at each type of airport, as shown in Table 6.1. It is proposed to move to a more structured method of identifying priorities. Benefit Cost Analysis (BCA) has been suggested for projects of more than US $5 million (FAA, 1997), the FAA interpreting its goal as encouraging what is best for the society of aviation providers and users. Their suggestion is that other priorities should be dealt with outside their evaluation system, but others believe that a blend of quantified economic criteria and softer analysis techniques is required to capture all the multidimensional objectives of a series of projects (TRB, 1994). Economic analysis does not deal with the problem of providing a level playing field, since small airport projects will always have difficulty demonstrating net present values of the same order as those for large projects, and BCA does not deal with the distribution of costs and benefits, with the relative importance of them to stakeholders, or with the long term issues which are discounted too heavily by conventional discount rates. The optional methods of evaluation are shown in Table 6.2, in addition to which it is suggested that a form of peer review among representatives of all the airports of a given class or in a given region would have a useful complementary role in testing a proposal's merits and fine tuning it.

All authorities with a responsibility for funding as well as planning a system of airports have to face these problems of allocation of limited funds. The discussions currently going on in the US are pertinent to the development of programme methodologies which can be seen to reflect society's view of the priorities.

Table 6.1: FAA project appraisal weightings

Airport Class	W	X	Y	Z
PLANNING CATEGORIES				
– Initial study for existing airport	1	2	3	4
– Study for new airport	1	2	3	4
– Complete/continue phased projects	1	2	3	3
– Periodic update	2	3	4	4
– Supplemental grant for ongoing study	2	2	2	2
DEVELOPMENT CATEGORIES				
– Special programs (e.g., Safety)	1	1	1	1
– Reconstruction	2	2	3	7
– Standards (includes Noise mitigation)	2	3	4	9
– Upgrade	3	4	5	10
– New capacity	3	4	5	12
– New airport capacity	3	5	7	12
– New airport community	5	6	7	12

Class W is: Primary in large or medium hub and its relievers or noncommercial 100 or more based aircraft or 40,000 or more itinerant operations
Class X is: Primary outside large or medium hub and its relievers or noncommercial 50-100 based aircraft or 20-40,000 itinerant operations
Class Y is: Commercial service other than primary or noncommercial 20-50 based aircraft or 8-20,000 itinerant operations
Class Z is: Noncommercial less than 20 based aircraft or less than 8,000 itinerant operations

ADD-ON FACTORS (No Add-on Factors for SPECIAL PROGRAMS):
+1 = Primary landing surface and associated taxiway, approaches
+2 = Aprons, secondary landing surface and associated taxiway, approaches
+3 = Fundamental configuration or for noise compatibility in DNL 75 dB
+4 = CFR maintenance facilities, electronic navaids, AWOS, snow removal equipment/storage buildings
+5 = Primary access roads, noise compatibility (DNL 65-74), terminal buildings
+6 = Snow abrasive/chemical storage buildings
+7 = Other (such as service roads, secondary access roads, noise compatibility projects outside DNL 65 dB, fencing, etc.)

NOTE: The Priority System conforms to the following hierarchy of general goals:
1) Support airport safety and security;
2) Carry out statutory policy and regulatory direction;
3) Encourage airport/planning agencies to plan for improvements;
4) Preserve existing infrastructure;
5) Bring airports into compliance with FAA design criteria; and
6) Add new capacity.

Source: TRB (1994), adapted by the authors

Table 6.2: Criteria for evaluating projects

Name of Criterion	Advantages	Disadvantages
Net Present Value	1. Most accurate, according to theory	1. Confusing: value reported cannot be placed in accounts 2. No notion of scale, so cannot easily be used to compare projects 3. Ranking of projects sensitive to discount rate
Benefit-Cost Ratio	1. Easy to explain as a concept	1. Wrongly biased against projects with operating costs: misleading 2. Ranking of projects sensitive to discount rate
Internal Rate of Return	1. Ranking of projects not sensitive to choice of discount rate	1. Ambiguous, wrong results possible when project has large final costs
Payback Period	1. Easiest to calculate and explain 2. Ranking of projects not sensitive to choice of discount rate 3. Focuses attention on immediate, more certain benefits	1. No allowance for discounting of value over time, and thus 2. Biased for short-term projects and against those with more lasting benefits
Debt Service Coverage Ratio	1. Market test of values ascribed to project	1. Benefits and costs not borne by users of facility not included 2. Scarcity of capital not taken into account
Cost-Effectiveness	1. Objective measure of relative performance of projects	1. No clear guidance on whether performance justifies the cost
Social Cost-Benefit or Consumers' Surplus	1. Theoretically correct way to calculate social value of projects	1. Major elements of calculations are speculative, and thus 2. Approach impractical 3. Ignores biases of distribution of income
Impact-Incidence Matrix	1. Focuses clearly on distribution of benefits and costs 2. Provides guidance for mitigation efforts	1. Emphasises local interests, possibly to detriment of regional or national interests

Source: TRB (1994), adapted by the authors

6.6 MANAGING CAPACITY CONSTRAINTS

Most airport systems suffer from excess capacity in some areas and a lack in others. The problem of excess capacity can be addressed through a combination of marketing and asset management. It is often possible to find creative aeronautical uses for underutilised facilities, in addition to benefiting from the overspill from congested airports, particularly in the cost-sensitive sections of the market. When this is not possible, some of an airport's assets can be put to use in aeronautically related activities in ways which do not close off the options for eventual aeronautical use. In the case of seriously uneconomic situations, the only route to retention of an airfield is some form of subsidy based on government policies on social integration. Strategic planning should address the systems issues to assess the value of retention of facilities and the possibilities of making economic use of the available capacity.

Capacity constraints will always exist somewhere in an economically efficient system, whether the limit is derived from physical or environmental shortcomings. This is not a subject which can be considered further here, but has been addressed by Brander and Cook (1986), who argue that, if capacity is not constrained, the overall market will settle at zero economic rent and that social surplus will be maximised when peak prices are based on estimates of the social costs of congestion. This will be so regardless of the ownership of the facility in question. The management of demand requires either that policies of intervention in the market be acceptable or that the market be fully functional, to the extent of pricing in the peak above the level of average cost. Demand at any particular facility can be reduced by intervening through route licence regulation or by traffic distribution rules or by some method of coordinated rationing, all of which would have consequences for the development of competitive services. Another intervention option, where it is available, is to shift traffic to other modes of transport. However, once all modes are competing on a level playing field with full social cost recovery, further restriction on air travel demand would imply an uneconomical distortion of societal choice. The important question for the efficient strategic planning of the infrastructure is to judge the appropriate level of demand management, as opposed to the expansion of capacity, as it will be decided by the future decision mechanisms.

Shortage of capacity may present itself at any part of an airport, and the sharing of the available capacity is a problem which must be faced wherever it occurs, but the allocation of runway slots is usually the focus of most attention.

6.6.1 Slot allocation

Some of the most attractive airport markets for airlines will continue to be short of capacity, even without allowing for the need to inject more competition by reducing this barrier to entry. The FAA in the US has long operated a slot allocation policy at its four most constrained airports which grants slots automatically for foreign airlines, another small percentage are reserved for general aviation, and the rest are open to trading, the initial allocation having been made on grandfather rights. A 'use it or lose it' provision has recently been added. Other countries have relied on the scheduling committees of individual airports to fulfil the requirements resulting from the twice yearly IATA scheduling conferences plus the needs of charter and general aviation operators. The US and the IATA methods both pose increasing barriers to entry as activity levels approach capacity in the peak hours, either because the larger incumbent airlines can afford to cross-subsidise slot prices at the few pressure points across a large system or because they have undue influence in the local scheduling committees and can continue to protect their grandfather rights, especially if they are a subsidised flag carrier. All the present methods effectively presume ownership of slots by airlines, as part and parcel of their historic route rights. This is now open to challenge, particularly from airports, and the EC is as yet undecided on this issue.

The EC has taken a first step in regulating slots by requiring an independent company to run the local scheduling committees and by requiring all slots not used for 80% of the time to be put into a pool for reallocation, 5% of which is to be available to new entrants. It decided that it would be better to modify the IATA system in this way, rather than attempt to dismantle it. It is doubtful if much impact will be made on improving entry at the busiest airports, compared with the much more draconian measures first proposed. The EC had suggested a maximum limit on frequencies per route, so encouraging larger aircraft. Currently, 17% of capacity is devoted to one route at Madrid airport, 6% to one route at Heathrow, while the average seats per aircraft on short haul in Europe are 95 at Brussels airport and 145 at Heathrow, compared with 175 at Bangkok airport and 260 at Narita. There are actually discounts for higher frequency at Brussels and Madrid airports, and for small aircraft at Amsterdam, Athens, Copenhagen, Paris Charles de Gaulle and Lisbon airports. In fact, frequency per se is not the main issue when physical capacity is the constraint, since it is always peak capacity that is in short supply, and concentration of the demand into the peaks with larger aircraft may actually make the peak more severe, both on the runway due to the effect of large aircraft on the runway capacity, but also on terminal and stand capacity.

Even though they are acknowledged as models of their kind, the Heathrow and Gatwick interpretations of the EC slot allocation method still have operational difficulties. There are problems with definitions, such as the time period relative to the assigned slot during which it must actually be used before being declared unused, as well as problems with monitoring the reasons for non-compliance and with unrealistic multiple bidding for slots. The EC regards slot swapping as anticompetitive, but many airlines have gained entry to Heathrow by substitution, others by black market trading. The EC also regards runway slot allocation as separate from traffic rights, but many of the difficulties surrounding the US-UK bilateral renegotiations relate to access to Heathrow. A further difficulty, particularly at some airports in less liberal settings, is the assessment and declaration of capacity, where too low a level acts as an artificial barrier to entry.

The operational difficulties with slot allocation and the extreme shortage of slots clearly create barriers to competition at Heathrow. British Midland holds the second largest number of slots at Heathrow, but still ceased domestic services to East Midlands, Birmingham and Liverpool airports in order to compete successfully on the Brussels and Frankfurt routes. Virgin managed to obtain 40 slots to move some services from Gatwick, but had to delay the planned start of a service to Johannesburg. Canadian Airlines International obtained slots after four years to move its Toronto services from Gatwick to Heathrow, but only for departures between 1730 and 2030 hours. Meanwhile, United Airlines was able to start a fourth daily frequency to New York Kennedy.

This limited attempt to increase access can certainly be improved by better monitoring, by a system of deposits to deter unreal bids and by penalties for late returns of slots which turn out not to be needed (Harding, 1994). If the authorities are serious about improving competition on a route by route basis, and if they respect the airline view that access to the dominant airport in a metropolitan area is the only effective way to compete in a way that will bring market control to full price fares, then sooner or later the problem of extremely high frequencies per carrier per route will have to be tackled. The pros and cons of the peak pricing, auction, trading and administrative reallocation of slots is well rehearsed in the literature (e.g. Banister et al, 1993). All have their drawbacks. Peak pricing has done little except to bar entry to small aircraft at Heathrow. The Swedish government recently delayed a recommendation of a Commission into domestic deregulation to allow competition on routes which have 300,000 passengers per year until Arlanda's extra runway became available, arguing that competition would not happen until capacity becomes available.

The CAA has recently analysed the impact of existing EC regulation on runway slots (CAA, 1995a). It concluded from this analysis that the EC regulation is not achieving its competition objectives, at least at Heathrow and Gatwick. It recommended that:

- Nearly all slots created by declaration of greater capacity, by returns or by withdrawal of service, should go into the pool for new second and third carrier entrants rather than half as in the present regulations.

- Requests for slots should be prioritised to allow the most promising new carrier on each of the denser routes the opportunity to build up an effective level of route entry. This would still be unlikely in the first year of application, so slots once allocated should be held in trust for the prioritised carrier and route but available short term for other uses.

- The priority rules should be set at supranational level. They must be clear, unambiguous, blind to nationality and tightly drawn. Some flexibility is seen to be necessary, but the process must be transparent.

The UK CAA can see little advantage from the point of view of aviation policy in reserving slots to protect access for smaller operators and thin routes. The recommendations are set against the background of the CAA's opposition to the concept of slot confiscation or capping, and a belief that slot trading is necessary and inevitable. The exchange processes should therefore be registered with the slot coordination company. Some of the assumptions underlying the analysis are open to question. It may be that more competitive results could be obtained by encouraging a fourth, low cost, carrier on existing three carrier routes - little thought seems to have been given to the post 1997 fully liberalised situation in Europe. Experience also shows that the main beneficiaries of competition have been leisure passengers, whereas the report aims primarily at enhancing competition on business routes. It could be argued that many potential new entrants have been put off from applying for slots because of the difficulty of obtaining at least two arrival and two departure slots in the peak hours, so that analysing the ratio of bids made to bids satisfied tends to underestimate slot demand.

There are areas about which the report has little to say. It notes that there have been significant increases in declared capacity at Heathrow and Gatwick, in part due to an acceptance of a ten minute delay standard rather than five minutes. If the process is to operate fairly throughout Europe, it is essential that common ground be established for the declaration of capacity. Meetings between airport coordinators might further this. Beyond admitting the existence of slot trading, little is said about how to manage the process in view of the acknowledgement that the free market may well suppress competition. There is only passing reference as to how to tighten up the operating rules to reduce problems of the 'paper airline' and the late return of slots to the pool. The former might benefit from a system of refundable deposits, the latter by monetary or exclusion penalties. Either way, the proposed measures do nothing to further the right of airport operators to manage this most crucial of airport assets, since the issue of who really owns the slots is not addressed. The common view that airlines have prior rights to slots is only an assumption that remains to be tested in law (*ACI Europe Communiqué No 81, February 1997, pp 6-7*).

6.6.2 The role of yield management

Airlines have invested heavily in yield management systems which tend to maximise load factors and increase their share of traffic, often cross-subsidising discount fare passengers from full fare passengers. These systems are now seen more correctly as revenue maximising tools. When faced with a capacity shortage, the airlines therefore have a self-compensating tool which will tend to raise average yields and increase profitability, or at least compensate for the additional delay costs. They will then be less inclined to push for new capacity, so making it harder for new entrants to break into the market. While it may be true that some of the present attitude to encouraging the marginal passenger through the use of these systems has been due to inadequate mechanisms for allocation of scarce capacity, this ability of incumbent airlines to take advantage of capacity constraints to achieve higher yields adds further to the lack of contestability in the present system.

This effect does offer another possibility for managing capacity constraints rather than adding capacity. However, it also shows that the need for more capacity can be greater than implied by delay data, particularly if a real attempt is to be made to achieve competition.

6.6.3 Technological solutions

The prime goal of the system planners must be the continued availability of appropriate capacity. The strategic options for responding to a shortage of capacity are well known (Hamsawi, 1992). Capacity can often be increased by bringing an airport up to current best practice and by expanding existing facilities, to the extent that the airport has remaining economies of scale. Ultimately, in a free market, congestion costs ration the capacity and make the costs of developing a new site more attractive, despite the substantial costs of air service fragmentation, of ground access and of compensating for the creation of new environmental impacts. Yet past attempts to establish new sites show how great the need has to be before they are readily used. Strategic planning should assess the anticipated best practice, should develop methodologies for determining the potential physical and environmental capacity of existing facilities, and should be able to guide decision makers in the consequences for the air transport system and society of the provision of new airports.

The remaining way to respond to the shortage of capacity is to develop innovative technologies and operational techniques to increase the throughput of the existing system. These innovations should be aimed at the most severe anticipated constraints. For some airports the worst constraints will be environmental, for others they will be aircraft flow rates or ground access congestion. In most cases the innovations will require readjustments across the whole system, rather than isolated action by an individual stakeholder or, indeed, even action by any one set of stakeholders. The challenges for strategic planning are to identify the innovations which are likely to make a worthwhile impact and to incorporate in the planning framework the necessary motives and opportunities for the synergistic modifications across the system to take place.

Runway capacity is likely to be the most serious constraint to large increases in system capacity. In the realisation that putting more physical capacity in place will never again be easy, the FAA's approach is to make the maximum use of technology to increase the capacity, thus restricting the extent to which new building is required. This approach has also been adopted in Europe by the ECAC/APATSI initiative (ECAC/APATSI, 1994; ECAC/APATSI, 1995).

Enhancements to the capacity of existing runways are being achieved by new approach procedures:

- reduced longitudinal separation on wet as well as dry runways from 3 to 2.5 nautical miles;
- simultaneous (independent) parallel IFR approaches using the Precision Runway Monitor to runways separated by as little as 3,400 feet, by allowing the controller to have almost instant knowledge of aircraft track deviation as at Raleigh-Durham and Sydney (Herbert and Custance, 1994);
- improved dependent parallel approaches to runway separated by between 2,500 and 4,300 feet that reduce the required diagonal separation from 2.0 to 1.5 nautical miles;
- dependent converging instrument approaches using the Converging Runway Display Aid;
- extending simultaneous operations on intersecting runways to wet runways;
- use of flight management system computers to transition an aircraft from the en-route phase onto existing charted visual flight procedures and ILS approaches.

- Separate Access and Landing System (SALS) procedures to take advantage of the superior manoeuvring capability of regional aircraft (as demonstrated at London City Airport) for simultaneous operations with conventional aircraft.

It is expected that lateral and in-trail separations on approach and departure will continue to improve as the navigational capabilities of aircraft improve with the use of the Global Positioning System (GPS), subject to restrictions due to vortex wakes. The FAA plans to use differential GPS rather than the Microwave Landing System (MLS) to fly the approaches. System error will be much lower than Category I ILS and the infrastructure required will be less expensive. The poor weather capability will be further improved by enhanced vision systems in the cockpit, again lowering the infrastructure requirement.

The benefits of the new technologies are very site-dependent, but could increase capacity by 50% in existing situations, and may make it possible to add 100% rather than 50% when incorporated in new expansion plans. Frankfurt airport plans to increase its declared runway capacity from 70 movements per hour in 1995 to 80 in the year 2010, using many of the above techniques and also laser detection of vortices. In addition to these technologies, new runways are planned for 25 of the 33 most congested airports in the US. Short parallel runways have been on the agenda since it was shown that significant capacity gains could be made at some key airports in the US (Sinha and Rehrig, 1980), but little progress has been made in installing them, so the plans for all these new runways may not be fulfilled.

<u>Stand and apron capacity</u> is now one of the most serious areas of concern in capacity provision. As aircraft become larger, the runway traffic throughput tends to grow in direct proportion without additional cost or construction, but the larger aircraft need more parking space per stand, which is often not available without extensive reconstruction. In the process, taxilanes and taxiways become more constrained. The introduction of the planned New Large Aircraft will cause serious difficulties at most of the world's congested major airports, even if Frankfurt's new terminal and the planned fifth terminal at Heathrow will be able to accept these aircraft.

If more air transport movements (atm) occur, as well as the aircraft becoming larger, the constrained sites of many busy airports will make it difficult to provide the land for the expansion of aprons. Already some airports are having to displace maintenance activities or develop remote aprons. Los Angeles airport had to park the 747-400 at special remote stands, with a 2 km airside bus transfer. At Heathrow, aircraft often have to leave the stand and find temporary parking while waiting for ATC clearance, which creates problems for the ground controllers. If the criterion developed by Chevallier (1992), that apron congestion becomes critical with activity levels above 1,600 turnrounds per stand per year is accepted as a guideline, then eight European airports are under severe pressure. The new Munich airport is, indeed, almost at capacity on the apron during the peak hours, while the terminal is only at some 73% of its declared capacity. An Air Transport Action Group study shows Sydney and Bangkok airports to have apron capacity problems now. There will be increasing competition for space on airports, which should eventually find its way to increased aircraft parking fees.

The availability of stands is the most serious barrier to entry, even if the totality of apron space is sufficient. There is a significant difference in level of service to the consumer between being taken by bus to a remote open apron and having an airbridge stand within a short walk of the terminal facilities. Heathrow could use some stands more intensively by changing the mix of airlines in the terminals, and could develop further stands, but they would not be in the most convenient locations and, in the limit, would involve closing the cross runway (Poole, 1995). In the US, the shortage of stands is reflected in inter-airline lease rates, the sub-lease sometimes costing an order of magnitude more than the original one. When the stands are all common use, which is normal practice in Europe, this situation does not arise, but there is little point in an airline gaining a runway slot if no stand is available. It may pay an incumbent airline to extend its scheduled turnround time to ensure continuous use of stands near its own terminal and aircraft handling facilities. There are no easy solutions to this problem, which is likely to worsen as more delays are taken on stand. Novel suggestions have been made for improving apron operations. Recently, Boeing has proposed a drive-through single-stop dock with fixed servicing facilities on either side of the aircraft (Fotos, 1986). This would improve stand efficiency at the expense of additional manoeuvring space, not to mention the capital costs involved.

Productive capacity can, in general, be increased by speeding up the movement of traffic through a given space or by moving the same traffic through a smaller space. On an apron, space is limited by some combination of the numbers of gates, numbers of stands, apron depth or terminal frontage. The aircraft flow rate is controlled by the turnround time of the aircraft and the interval between a departure and a subsequent arrival. Attempts to improve any of these aspects of productivity must be based on a knowledge of current levels of productivity and the constraints which are limiting it.

Any trend towards increased hubbing is, as explained in chapter 4, likely to negate attempts by European airlines to improve traffic flow rates per unit of apron space by reducing turnround times. The spatial efficiency of aprons would become even more important, because the overriding criterion will be to increase traffic throughput per complex rather than traffic flow per hour. Efforts would then need to be concentrated on the spatial aspects of apron productivity, initially with the present aircraft fleet and later by designing aircraft which are more spatially efficient on the apron.

It is by no means certain, however, that airlines in Europe will be able to justify the low utilisation of aircraft and stands implicit in conventional hubbing indicated in chapter 4. Even in the US, there are already some indications that airlines may adopt operating strategies which avoid conventional hubbing complexes in order to increase aircraft utilisation. Southwest Airlines achieve substantially higher aircraft utilisation, with their multistop routings into secondary airports and with no emphasis on hubbing, than do the conventional hubbing carriers.

It can be deduced from Table 6.3 that, when only spatial effects are considered, the use of larger aircraft results in a more productive use of apron space: the required clearances are proportionally smaller, the aircraft volumetric efficiency is greater and the average load factors should be higher due to the lower stochastic variability at higher traffic volumes. The traffic flow per complex could therefore be doubled for only a 50% increase in apron area by switching from narrow to wide-body aircraft. A similar result applies for freight aprons dedicated to express parcels operations with only a single hubbing cycle per 24 hours: a situation which is now occurring in Europe. A small turbo-prop requires some 600 sq metres per tonne uplifted compared with 250 sq metres for a small narrow body and 150 sq metres for a wide body. This, of course, counts heavily against commuter operations. In the US, the effect is mitigated by multiple parking of small turboprops with perhaps six per jet stand, when the assumptions implicit in Table 6.3 of rectangular stands and ICAO clearances break down. Similar but smaller savings can also be made with jets by using staggered nose-in parking to circular frontages and by overlapping wingtips, but such solutions are easier to implement when an airline controls its own apron.

Table 6.3: Stand productivity per turnround

Aircraft type	Pax per gate	Pax per metre of terminal frontage	Pax per 100 m^2 of apron
Widebody jet	170	5	6
Narrowbody jet	70	2	4.5
Turboprop	15	0.5	2.5

Source: Caves, 1994a

The situation is different when the productivity of the apron over time also needs to be considered. The data in Caves (1994a) show that, with operations representative of Europe in 1987, maximum traffic flow rate per hour per stand occurs with the use of the largest aircraft, the best rate per metre of terminal frontage occurs with medium size jet aircraft and the best rate per square metre of apron with small jet aircraft. These results stem from the fact that the larger and longer haul aircraft are taking relatively longer to turn round, while turboprops have large wing spans per seat. In comparison, if all turnrounds could be made in the times indicated in manufacturers' data, all jet aircraft irrespective of size would be able to achieve approximately 12 passengers per 100 sq metres of apron per hour, i.e. some 250% better than current practice.

If aircraft size were to increase directly with growth in traffic, the gain in apron spatial efficiency would reduce the extra apron needed by approximately 50% compared with a situation where all the traffic growth were achieved through increases in frequency. The apron throughput efficiency would increase by the same amount. Further increases in throughput per hour could only be achieved by reducing the slack time between turnrounds or by reducing the turnround times.

The slack time could only be reduced if the European airports operate a stand allocation policy which results in an even better stand occupancy than that achieved now with common user allocation. This would be very difficult to achieve in a hubbing context. Even in conventional operations, inefficiencies tend to arise from the mix of airlines and aircraft sharing an apron frontage, from the incompatibility of aircraft with available stands and from conflict on taxiways between arriving and departing aircraft.

The turnround productivity is influenced not so much by aircraft size per se as by the long sectors flown by the larger aircraft. The stand productivity will only improve significantly with a concerted effort to reduce the turnround time at base of long haul aircraft in the cases where operating windows and other constraints would allow advantage to be gained from this. The trend to increase sector length can only exacerbate the problem. It may be that part of the solution rests with passenger behaviour rather than with technology: Southwest Airlines cabin staff and passengers combine to clean up their aircraft during descent. Certainly new turnround operational concepts are likely to be necessary if the introduction of the NLA is not to compound the problem.

A more certain way of increasing apron productivity is to improve the spatial efficiency. The spatial efficiency benefits which might arise from increasing the average capacity of conventional geometry aircraft would be offset to some extent by the need to modify existing apron/taxiway layouts, so losing operational flexibility or increasing delays and effectively diluting the spatial efficiency benefits which ought to flow from the use of the larger aircraft. These pressures can be only marginally alleviated by reducing the present clearance standards between aircraft wingtips with more accurate tracking (Wilson, 1988) or video monitoring of wingtip location, since the collision risk will rise in association with the greater percentage of movements by the larger aircraft.

It would be possible to avoid these problems to some extent if the dimensions of future aircraft could be reduced by adopting less conventional geometry, limited ultimately by the necessary living space per passenger. If this proved feasible, it would then be possible to further increase apron productivity as well as reducing the space required on the other parts of the manoeuvring area. The fuselage length for a given payload could be reduced by using horizontal or vertical double bubble fuselages as in the current proposals for the NLA, though the consequent increases in tail height might give other parking and hangar problems, and it may be difficult to evacuate passengers from these configurations and to fit the necessary servicing activities within the reduced stand envelope. Reductions in fuselage length would certainly improve apron area productivity as indicated in the industry's preliminary evaluations (IWG, 1990), though doing nothing for gate or terminal frontage problems. Their main benefits would be seen in situations where apron depth is a problem, as where more taxiways are required to alleviate airside congestion or where taxiways must be moved closer to terminals in order to maintain separation standards with larger wingspans.

Terminal capacity can be made to go further by spreading peaks, but there is a limit to how much the airlines can accommodate these changes. Some alleviation could also come from reduction in airside delays, since passengers would then have shorter dwell times in the terminal. If traffic is to grow, and if most of the traffic is to be handled on contact stands, the terminal frontage will need to be increased in one way or another. However, the space given over to passenger processing could be used more efficiently by the use of new technology. The most promising advance is likely to be through information technology leading to machine readable passports, and self-ticketing at the gate or at home via the internet. These innovations will undoubtedly speed up the processes and bring some space efficiency benefits unless further additional processes are required at the airport, as happened with security in this decade. However, the baggage problem is unlikely to improve in the same proportion, and, though the greater passenger processing efficiency may delay the need for investment, it will not generate profits. The space gained may well be used for more concession space unless congestion levels become intolerable.

Landside access may become a serious cause of capacity constraint, particularly when the air pollution caused by road traffic can be seen to be a major contributor to that in the local metropolitan area. Many, however, argue that the additional loading on metropolitan transport systems caused by airports is sufficiently small that the overall responsibility for improvement should rest with the local government. While this may be true at the level of the region as a whole, major airports are one of the largest trip generators in a metropolitan region and contribute a significant proportion of the traffic on nearby streets and highways. Thus they are likely to be increasingly expected to take actions to reduce their impacts, particularly as part of the environmental review of plans for further expansion. Aside from such pressures, the airport and airlines must also be concerned to ensure that they do not lose competitive advantage relative to other airports. Much of the local traffic generated by an airport is associated with the workers rather than the passengers, or with freight which moves mainly off peak, but the competitive nature of airports requires attention to the passengers' quality of service.

For medium-sized airports, improvements in access will normally mean bearing some part of the necessary costs of improvements to the local road network. The aim for the largest airports should be a progressive shift towards public transport. It has been suggested that a high public transport share of access trips by passengers and workers requires an integrated rail network, baggage assistance and an integrated multimodal airport terminal (Coogan, 1995), supported by through ticketing. Some authorities, e.g. Oslo, are expecting to be able to achieve a 50% use of rail access by passengers. However, for this strategy to be successful, it is necessary to have a good rail network which covers the whole metropolitan area as well as serving the city centre. This will clearly not be the case in many cities, particularly in North America, for the foreseeable future. Therefore a broader set of strategies to support the use of high occupancy modes will be required (Gosling, 1997). These could include off-airport terminals with remote parking linked to the airport by express bus service, shared-ride vans providing door-to-door service, and the use of advanced traveller information systems to ensure that air passengers are aware of the options available.

Where it proves feasible to incorporate the airport into the intercity rail network, it may be possible to take advantage of convenient transfer facilities to replicate the Lufthansa experiment at Frankfurt airport that has substituted rail services for some shorter feeder services, which are in any case expensive for airlines to provide. Manchester airport is progressively expanding its direct regional rail feeder services. Swissair has already developed a system that allows passengers to check baggage at train and postbus stations for rail access to Basle, Geneva, and Zürich airports (Jud, 1994). In many cases, however, it will prove to be very expensive and institutionally difficult to provide convenient direct interchange to high speed rail. For this to happen, either the airport or the airline would be faced with high costs for the infrastructure and services involved, unless they could be covered by supranational policies and funding instruments associated with, say, sustainable mobility as envisaged in the European Airport Network Plan. The more the airlines take Lufthansa-style initiatives, the more they could exert their economies of scope.

Landside capacity can also be expanded by the use of off-airport parking and consolidation of car rental facilities away from the terminal. In the US, it has been found that the very high level of service typically provided by courtesy buses has meant that terminal curbfront congestion and air pollution can actually be worsened. Some airports are now requiring that these bus services be consolidated and use low-emission vehicles. Schiphol and Eindhoven are studying the setting up of remote check-in sites with free bus transport to the airport, though city centre terminals served by bus have had a chequered history. In most cases, when these were provided they were well-used (Gosling et al, 1977) but fell victim to a complex set of factors, including airline desires to reduce costs, shifts in the distribution of trip ends within the metropolitan area, and failure of the airport authorities and regional planning agencies to address the difficult institutional issues involved. On the other hand, the airline check-in facilities at London Victoria rail station for Gatwick are still well used. As smart technology is progressively applied to the ticketing, immigration control and seat assignment, it may be more necessary to guarantee access journey times if these systems are to be fully utilised in reducing terminal congestion.

7

REGIONAL AIRPORT SYSTEM PLANNING

The previous chapters have introduced the concept of strategic planning and reviewed the behaviour of the stakeholders in civil aviation. Strategic planning needs to consider all the actors whose interactions create an aviation system, as well as the context within which the system exists. The airports in an aviation system are separated spatially but can be grouped by jurisdiction or function. Responsibility for planning, and indeed owning and operating, them is usually taken by an agency representing one or more of the jurisdictions. This is necessary because of the need to relate the airport activities to general land use, transportation and other issues for which the jurisdictions are also responsible. Difficulties are still likely to arise in dealing with strong interactions between airports in different jurisdictions and within any one joint planning agency where the costs and benefits are not distributed evenly across the members. The responsible jurisdictions are often nested hierarchically, in which case there may be mechanisms for dealing with the wider interactions.

The planning of airport systems has been institutionalised in the US, where the hierarchy consists of local, regional, state and federal tiers of responsibility. The FAA directly administers the federal plan and the airspace planning aspects of the other tiers. It also provides guidance on how to prepare individual airport master plans (FAA, 1985), metropolitan systems plans (FAA, 1996) and state system plans (FAA, 1989). While there is no legislative requirement that airports, regions or states conform to this guidance, airports, metropolitan planning organisations, and states are expected to follow this guidance if they wish to receive federal funding and have their airport development projects included in the federal plans. Considerable use is made of the latter two publications in this chapter and the next one because there is less formal guidance available in other contexts, but it should be appreciated that the US process is tailored to its needs and context. This is particularly so in its emphasis on providing for a very substantial general aviation activity. Not all airports in any one tier are necessarily included in the system formed in the higher tiers. Most nations have a similar arrangement, though often lacking the top tier unless they are federations like Germany, Brazil, or increasingly, the European Union.

The upper tiers of the system are usually more concerned with national issues like cohesion, setting and maintaining standards for design and operating practices, and the equitable use of central funds. The local tiers tend to be more interested in land use issues. The commonly used terminology tends to vary from country to country, reflecting the varying political and institutional structure, so it is necessary to define the meaning of the terms used here to describe the system's tiers.

The primary distinction between a regional airport system and a state airport system is that the former is defined on the basis of a geographical grouping of airports while the latter is defined on the basis of political control over funding and planning. As discussed in chapter 1, for the purposes of this book, a state is considered to be the next tier of government below the national government with responsibility for planning transportation facilities, including airports, within its jurisdiction. Obviously, if states (or provinces, or whatever they may be termed) do not have responsibility for planning or funding airport development, then the issue of the need for state airport system plans is moot.

A regional system comprises a collection of airports within a defined geographical area for which there is a need to plan their development in a coordinated way. It will usually include at least one major airport, together with some smaller commercial airports and several general aviation airfields. There will usually be several local jurisdictions within the region, only some of which will have airports. The most common regional system occurs where more than one airport serves a major metropolitan area. However, other regional systems could include several adjacent metropolitan areas or other geographical groupings where there is both a perceived need to plan the airport infrastructure in a coordinated way and the institutional framework to do so. Sometimes these regions will cross state, or even national, boundaries. This occurs, for example, with the Port Authority of New York and New Jersey, set up to manage the ports and airports in the tri-state metropolitan area that includes New York City, northern New Jersey, and parts of Connecticut. Sometimes such regions do not have a single representative agency and have to work under a system of quasi-bilateral agreements between the jurisdictions. On the other hand, there often is a planning agency for large metropolitan areas, which may well have resources equivalent to many entire nations and which can form a focus for planning the airport system within that jurisdiction.

This chapter and the following one explore the generic processes and issues involved in the planning of regional and state airport systems. Depending on the scale and context of the jurisdiction, many of the concepts and issues dealt with in one chapter may be relevant to the other. Chapters 10 to 13 then illustrate the processes by means of case studies of a wide variety of airport systems and contexts, before an attempt is made in chapter 14 to match these cases to the generic processes and to identify some ways of improving them.

7.1 METROPOLITAN AIRPORT SYSTEMS

7.1.1 Evolution of multi-airport systems

Many larger cities are now served by multiple commercial airports, a full list being available elsewhere (de Neufville, 1995). Traffic forecasts indicate that many more major cities will have to face this problem during the next decade as their present principal airport reaches physical or environmental capacity (Hansen and Weidner, 1995), growth often being exaggerated by airline hubbing. Even with the best use of new methods to increase capacity and control environmental pollution, it is eventually necessary to find a site for a replacement or an additional airport. In the US, where much of the activity is general aviation or turboprop commercial schedules, cities of more than 250,000 people typically need more than one airport for general aviation, while above about 10 million annual originating passengers they may well need a second airport for commercial traffic (FAA, 1996).

Decisions on the development of supplementary or replacement airports are notorious for vacillation caused by inadequate decision mechanisms, changes in political power base and genuine changes in circumstances. London, Oslo, Sydney, Lisbon and Manila are all examples of long drawn-out sagas, of which only Oslo has been categorically resolved (see chapter 13). London is examined in depth in chapter 10. It now has two main airports, both owned by BAA plc, plus Luton and Stansted. It has already seen problems in fragmenting services between Heathrow and Gatwick and now needs to fragment further because land use and airspace limitations preclude any new large close-in replacement airports.

The US National Plan for Integrated Airport Systems (NPIAS) suggests several solutions to the extra capacity still needed in large metropolitan areas, when all possible expansions have been made and all smart management has been exhausted. The first suggestion is new commercial service at existing airports near the congested airports. This presumes that the alternate airports can off-load the growing traffic at the principal airport. However, these movements are growing because the airlines want to serve that airport rather than the alternatives. If not, then the forecast for the principal airport is presumably wrong. However, nearby airports which do not have the potential capacity or local market to become the primary airport for the region can still be developed if airlines can be persuaded to fragment their services. Supplemental airports of this sort are being studied for Dallas/Fort Worth, Chicago, Seattle-Tacoma, Boston, Atlanta, Eastern Virginia and San Diego, as well as Stewart Airport for the New York region (see chapter 13).

The NPIAS also includes the concept of reliever airports, designed to provide full facilities for general aviation and possibly some regional airline services, so offloading the major airports and releasing more capacity for the commercial carriers. The smaller operators do not pay too high a penalty, provided those operations performing a feeder function either are retained at the main airport or given excellent ground connection facilities between the reliever and main airports. In most countries, the general aviation content of traffic is much smaller than in the US and it is more acceptable politically to use pricing or legislation to reduce its access to congested airports.

It is hard to convince the commercial carriers to operate from secondary airports. As de Neufville (1984) said: "Competition also has a remarkable feature: a slight advantage may be decisive .... Any attempt to model the details of the competition between airlines is quixotic. The airlines have too many different kinds of moves to make. The dynamics are too sensitive. Fortunately, a detailed model is quite unnecessary as it is in fact relatively easy to anticipate the final outcome of the competition. And that is all that matters .... Concentration is the end result of the sequence of choices both airlines and passengers make in a multi-airport system. The choices made by each group, as well as the choices made by each airline, influence that of everyone else. These interactions inevitably lead to concentration as the only stable outcome .... The competition will lead to concentration in each market, but this concentration may focus on different locations. It is thus possible to have substantial airports in a multi-airport system .... A planning effort that seeks to develop a second airport should therefore try to identify particular markets for the facility right from the start. These markets will be essential to its success. If a special market cannot be identified, the prospects for a second airport may be bleak". Current experience suggests that multi-airport systems only arise in situations of less than 10 million boardings per year if there are overwhelming technical or political reasons (de Neufville, 1995).

The market for a second airport was easier to identify when route entry was firmly regulated, although even then the system managed to find a way round inconvenient regulation. Sector length restrictions were put on Washington National airport in order to encourage the use of Dulles airport, but the airlines still served it from further away with transit stops. Even in emerging economies the same effect happened. The designated intercontinental airport for São Paulo was Viracopos, at some 100 km from the city, but most airlines transferred their passengers in Rio onto domestic services into the downtown São Paulo Congonhas airport. Case studies of the later system developments in São Paulo and the somewhat similar interactions affecting the Montreal and Glasgow airports are presented in chapter 13. At Lae in Papua New Guinea the airport was originally located close to the port, for easy transhipment of mining equipment for delivery to the gold mines. When Air Niugini wanted to serve Lae with jets, a new airport was built on flat land some 50 km from town. Talair continued to fly into the old city centre site. Within a year Air Niugini had to shift back to the crowded city centre airport due to passenger preference. Models of passenger choice indicate that passengers will be lost to the metropolitan systems of London (CAA, 1989) and Los Angeles (Wellman, 1998) if the necessary new capacity is not made available at Heathrow and Los Angeles International respectively. It will be interesting to follow the behaviour of airlines serving Seoul as the new US $12 billion Inchon International Airport 52 km west of the city opens in the year 2000, while the existing Kimpo airport will accept only short haul traffic (*Passenger Terminal World, July-September 1997, pp 20-24*).

The options for the airlines are much more limited when the original airport is forced to close, as happened at Munich and Denver, and is planned to happen at Athens, with Berlin's Tegel and Tempelhof airports and with Oslo's Fornebu airport.

The decision of whether it is feasible to plan and operate a fragmented hub system is difficult to make, depending on transfers, international politics, the availability of cross-subsidy and the way the costs fall on the operating airlines. Despite the apparent diseconomies, fragmentation appears to be occurring naturally in the air transport system, but planners must foresee the trends and use them. The main transatlantic carriers saw fit to fragment their services to Newark in New York as well as continuing to serve Kennedy, without affecting their traffic levels at Kennedy (*Air Transport World, September 1990, pp 95-97*), and British Airways has recently reversed its earlier policy of concentration at Heathrow by developing a major hub operation at Gatwick to supplement that at Heathrow. Route fragmentation also occurs if a carrier can see the opportunity to exploit a new market. The low cost carriers seem to prefer secondary airports, as with Ryanair's use of Charleroi rather than Brussels.

The choices will always be made primarily by the airlines, but they can be influenced by the investment decisions of the airports, which will themselves depend on the nature of the cooperation or competition between the airports. The primary determinant of this relationship is likely to be ownership. There is a strong feeling that the UK CAA should re-examine whether BAA should continue to own the three major London airports, not only because the common ownership denies the opportunity for competition between the airports (House of Commons, 1996) but also because, despite the CAA's duty to monitor the situation, there are fears that BAA can distort the market by cross-subsidy.

7.1.2 Opportunities for new airports

Military bases have been generally regarded as off-limits for civil use until recently in most countries which have separate civil airports. Now several military air bases are being released for other uses in western countries, and make feasible sites for new civil airports when they are conveniently located. This solution has been adopted at Austin, Texas for one of the few successfully implemented major airport projects in the US in the last decade. The existing 288 hectare airport is only 10 km from the centre of the city and has 29,000 residents within the 65 Ldn noise contour. The Bergstrom Air Force base, 19 km from the city centre, became available in 1990 and the city decided by public referendum to replace the downtown airport with an upgrade of the military site. The plan is shown in Figure 7.1. The airport is due to open in 1999, and is on budget at US $630 million, despite delays in clearing the contamination from the military occupation and the inability to reuse most of the existing facilities because of poor location or non-conformance with civilian codes (Rosenberg, 1997). Increased costs due to noise mitigation, land acquisition, demolition and the elimination of toxic waste were recovered by savings in the management of construction, which has helped the project team win an award for quality. The airport will have a terminal with 20 stands, greatly expanded cargo facilities and a new 9,000 foot runway parallel to the existing 12,250 foot runway. It will have only 1,502 residents in the Ldn 65 contour compared with over 10,000 when it was an active military base, which went a long way to ensure a positive result in the referendum that was held to approve the project. At least on noise grounds it was a win-win situation, with almost nobody suffering new noise impact. In many ways, this is a similar solution to Oslo, with the use of a more remote military site at Gardermoen to replace a constrained site closer to the downtown, as explained in chapter 13. Other examples of the use of military sites include Guarulhos in São Paulo and the proposed reuse of Clark Air Force base in the Philippines (*Air Transport World, February 1997, p 75*).

Figure 7.1: Conversion of Bergstrom Air Force Base, Austin, Texas

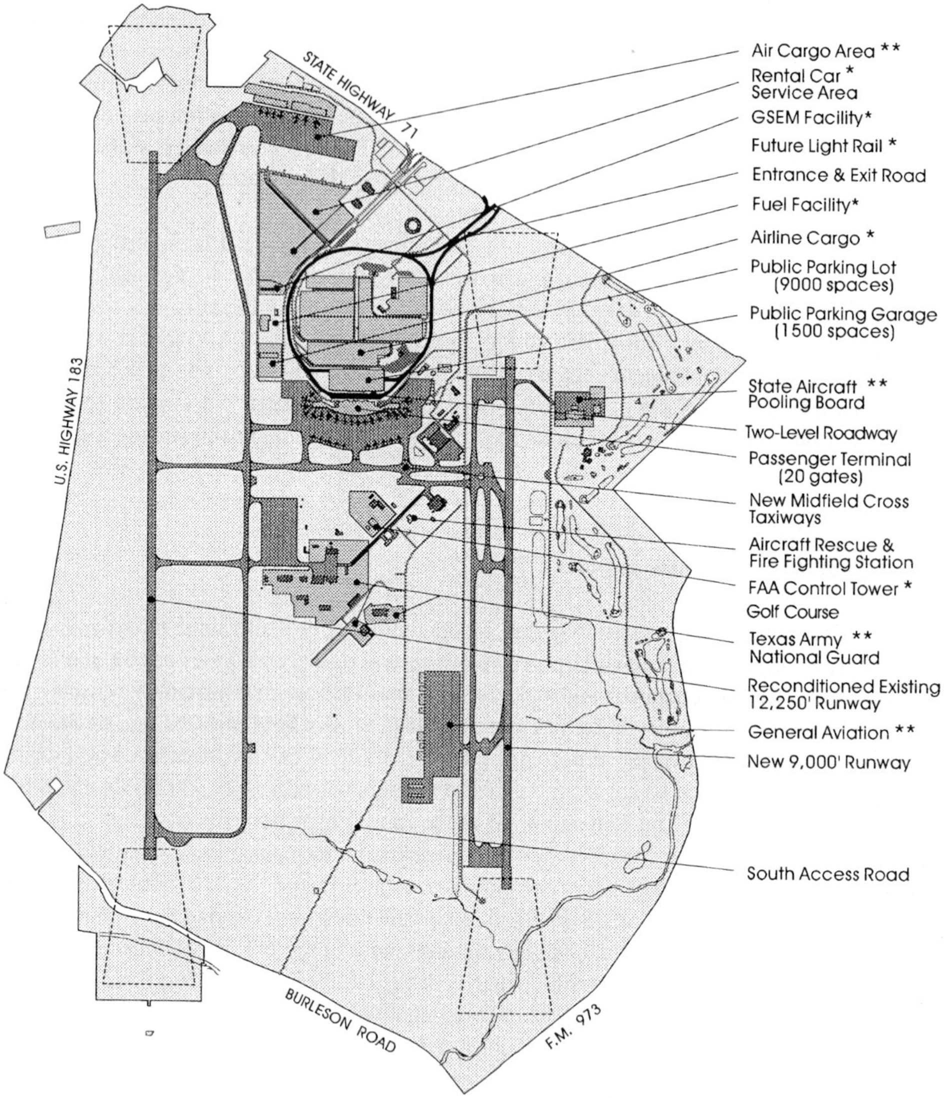
Air Cargo Area **
Rental Car *
Service Area
GSEM Facility*
Future Light Rail *
Entrance & Exit Road
Fuel Facility*
Airline Cargo *
Public Parking Lot
(9000 spaces)
Public Parking Garage
(1500 spaces)
State Aircraft **
Pooling Board
Two-Level Roadway
Passenger Terminal
(20 gates)
New Midfield Cross
Taxiways
Aircraft Rescue &
Fire Fighting Station
FAA Control Tower *
Golf Course
Texas Army **
National Guard
Reconditioned Existing
12,250' Runway
General Aviation **
New 9,000' Runway
South Access Road
STATE HIGHWAY 71
U.S. HIGHWAY 183
BURLESON ROAD
F.M. 973

Source: Parsons Brinckerhoff Construction Services Inc., courtesy of New Airport Project Team

Greenfield sites still offer the potential to provide the most efficient solution to the search for new capacity from the point of view of the air transport industry. The idea is particularly attractive when it would provide the opportunity to concentrate existing dispersed low levels of activity onto a single site which can gain economies of scope. Such solutions have been considered for most of the UK regions and shown to be efficient from the points of view of the industry and the consumers, but were all rejected (Caves, 1980). The UK government has now all but ruled out the idea of a greenfield site as an inappropriate use of land which would have no chance of gaining approval at a public inquiry, though a private consortium has brought forward a scheme for a major airport in the Thames estuary. This could only succeed if the government were to champion it.

The new Denver airport is a tantalising example of the almost infinite expansion possibilities available if the right site can be found, though even this is having difficulty in meeting its environmental agreements. The other major new airport development at Munich serves as a potent reminder that most cities do not have the same opportunities: despite choosing a site with minimal environmental impact, Munich airport still has serious restrictions on its available capacity. This suggests that any solution which causes adverse impacts to people previously unaffected could only be adopted if the most extreme necessity could be demonstrated. Similar judgements would be made in most developed economies, while some, like Japan, have additional topographical constraints which make it even more difficult to find a suitable site.

Offshore airports provide a further option for cities near large stretches of water. This has already been used for Kansai airport (see the Japan case study later in this book) and has, indeed, been a partial solution for other earlier airports like Nice and New York La Guardia. It has now been adopted for Macão airport and Hong Kong's Chep Lap Kok, and is being considered for airports serving Buenos Aires (*Air Transport World, May 1997, p 45*), Tokyo and Amsterdam. There are significant problems to be overcome in adopting this solution. Costs for Kansai and Nice airports were inflated very substantially by both expected and unexpected settlement of the ground. Fog can be a problem, even with Category 3 operation, not least in hampering ground access. Access times and costs are inevitably high. The environmental consequences of silting and disturbance of sea bird habitats can inhibit development. Many of these problems came to light in the consideration of the Maplin Sands as the site for a new London airport (Bromhead, 1973). The problems of offshore airports built on fill have caused Japan to consider floating airports (*Flight, 10 January 1996, p 27*), which were first considered as staging posts for transatlantic landplane flights. They have their own problems of mooring, corrosion and the effect of large masses of iron on navigation aids as well as access difficulties, but at least they might survive tidal waves which would swamp a filled site. San Diego is said to be also considering this solution.

7.2 SYSTEM PLANNING PROCESS

There is a generic process for any system planning, the steps of which have been introduced in chapter 3. They are described in the FAA advisory circulars and illustrated in Figures 3.2, 3.3 and 3.4. This section expands on the steps involved and tailors them to the regional level of planning. It is an iterative process with increasing detail where necessary, and may require even bringing earlier options back into the process. The planning goals are to provide an orderly and timely development of a system of airports which will be adequate for the present and future needs of the area (FAA, 1996), while placing aviation in perspective, protecting and enhancing the environment, encouraging aviation planning expertise, promoting effective governmental processes for implementation, providing a framework for individual airport plans and ensuring compatibility with wider plans. Related goals, such as identified in the Boston Regional

Airport System Study (Flight Transportation Associates, 1989), include "defining appropriate roles for each airport".

7.2.1 Policy objectives and study design

The stakeholders in a system will all have their own policy objectives. The underlying aim of planning should be to get to a win-win situation where every stakeholder takes something positive away from the table. Schiphol airport believes that this aim should apply right down to the level of the individual. The nature of the gain achieved by each stakeholder can take many forms. Vancouver airport has managed to achieve permission for its new runway by mitigation rather than financial compensation, the Canadian government not believing that the latter provides a lasting solution.

Each situation will have its own unique needs. Policies to satisfy the needs within the legal and financial constraints should be developed with the full participation of the community. This is not easy to achieve, even with assiduous attention to public meetings, informational material, and focus groups. The main problem has been that, though there are four identifiable levels of participation:

- recognition of the problem
- recognition of the solution
- willingness to act as an individual
- willingness to participate in collective action,

most processes are structured only to reach the third and fourth of these levels (Brogan, 1993).

The keys to the success of system planning are good information and the construction of a working methodology which will maximise the possibility of consensus. It is essential that the group developing policy should be able to speak for the many political jurisdictions within the region. A fundamental difficulty is in getting all the various agencies within and outside the region to dovetail their efforts. It is necessary to understand the real needs of present and future users of the system, and the real concerns of the relevant communities. This is likely to be an iterative procedure, as in the Manchester example described in chapter 5.

The planning process is getting longer. Good study design should start with a scoping exercise to identify those factors which are going to be critical to the outcome, either because of importance or due to the time needed to address it satisfactorily. The environmental impact assessment should be done largely in parallel with the physical system planning, since the evaluation of options will have to depend partly on its outcome

7.2.2 Inventory of facilities

As with master planning, it is necessary to take an accurate inventory of all facilities in the system, even those outside the responsibility of the planning authority and possibly even those in adjacent regions or countries if they have a functional relationship with the system under study. In some countries this may prove almost impossible, particularly if there is significant military involvement. It is also necessary to collect data on the utilisation of the facilities and the efficiency with which they are being used. It is entirely possible that changes in operating practice could avoid the need for enhancement of capacity.

The inventory should extend to airspace structure and navigation aids, land use and environmental considerations, and ground transportation. Local and state laws and ordinances and inter-jurisdictional compacts should be studied. Potential sources of finance should be established. An understanding of the regional, social and economic characteristics should be obtained, together with the attitudes in the community towards airport development. It is essential to understand the makeup of the aviation activity. Not only is it necessary to count the actual activity by type and location, but to assess latent demand and to know where the traffic is coming from. This may well require expensive surveys of travellers.

7.2.3 Forecasts of future demand

Forecasts are needed of air passenger and freight traffic, the aircraft movements by category and the associated ground traffic by mode generated by the users and by the direct and indirect employees. These are needed at annual levels to determine impacts and, if there is likely to be a capacity problem, for design hours to determine capacity requirements. Although it is possible to develop forecasts based on time series data describing historic growth in throughput of the facilities by traffic category, it is difficult to have any confidence in such estimates if the traffic may redistribute itself in the future as a result of possible reallocation of capacity. Passenger preferences for particular airports would then change, and future traffic at individual airports can best be predicted by the allocation models described in section 7.3. An equally serious limitation of time-series models is that they contain no information about the underlying factors that drive the growth in demand, and therefore provide no basis for predicting the consequences of alternative policy scenarios or assumptions about future trends in the underlying causal factors. Ideally, trip generation models should be constructed which can feed distribution and modal split models, as described in chapter 4 as well as the allocation models. Some attempt to model the future behaviour of the airlines and GA operators should also be made, since they are unlikely to have produced such long term forecasts themselves and, if asked, may wish to paint a picture which would produce a favourable outcome for them. Other far-reaching changes in the system environment also need to be considered, as explained later in the chapter.

Uncertainty should be recognised. One way of dealing with this is by scenario writing as explained in section 7.4. The Boston regional study (Flight Transportation Associates, 1989) evaluated plans based around five possible future sets of airline behaviour, in which expert judgement played a vital role. It may be necessary to collect primary data if these models are to have a firm basis. The collection of the data, and the subsequent modelling, should be seen as an opportunity to bring together the diverse interests. If common forecasts can be agreed, this will help considerably to resolve conflict among local groups and with state and federal interests. This approach makes it difficult simply to use any one ready-made forecast, whether from the master plans of airports in the system or top down ones from a higher tier of government.

The need to coordinate the forecast process between the various levels of system planning is becoming increasingly recognised. Apart from the potential damage to the credibility of the entire aviation system planning process when one agency bases its plans on forecasts that differ significantly from those used by another agency, and the resulting disputes over which forecast is more credible, it clearly makes no sense for one part of the system to be making airport development decisions on the basis of one set of assumptions while other parts of the same system are making decisions on the basis of entirely different assumptions. This issue is addressed in more detail in section 7.4.

7.2.4 Definition of alternative plans

The regulations for environmental impact assessment require the development of alternative solutions, but prudent planning would, in any case, want to explore a range of options that encompass different approaches to the problem at hand. These might range through options on sites, layouts, multimodal opportunities, life-cycle costs, demand management and mitigation. These multidimensional options should be focused into a manageable number of alternative plans with real-time participation of all the stakeholders. It may be expected that they will change as the implications become clearer, given that economic, environmental and supply-side analyses will be proceeding in parallel.

It is important that the definition of alternatives to be considered include some that address the need to accommodate future growth through non-aviation strategies or through efforts that only meet part of the demand. It is as important for the evaluation of alternatives to be able to articulate and quantify the foregone opportunities that result from not being able to satisfy the unconstrained demand as it is to estimate the benefits that flow from meeting that demand. This is emphasised in the recent draft FAA benefit-cost analysis guidelines for airport capacity projects (FAA, 1997a).

In a formal planning environment there will be a defined process to be followed, and it is important that the process does not take over the content by becoming the prime concern of the planners. Though the FAA has adapted its planning process to the legal requirements for Environmental Impact Statements (EIS), fears have been expressed that both the FAA and the courts have focussed on the use of the EIS for internal decisions on options rather than for the ends of ensuring full disclosure. Since the preparation of the EIS relies heavily on information generated by extensive technical studies, many of which are sponsored by the applicant, there is an almost inevitable bias towards those schemes which will be financially favourable to the applicant. The courts tend to stress procedures over substance, since the National Environmental Policy Act of 1969 only requires 'consideration' of the environment rather than 'protection and enhancement of the human environment'. The EIS preparation and review process has therefore taken on the role of resolving conflict, following the "theory of desirable omission". In this, "the sponsor prepares as incomplete a draft as it feels it can get away with, thorough enough to avoid total embarrassment .... Once comments are received within the prescribed time limits, the agency (FAA) can conduct studies (if necessary) only on the issues raised.... Should the 'watchdogs' be asleep, disorganised, or lacking enough political influence to have their objections acted upon, ... the Draft EIS may become .... the basis for an 'informed' ultimate decision." (Orlick, 1978).

The danger for the planning system, in a situation where the initiative is from an entrepreneurial investor, is that options will be chosen for analysis which weigh profit more than net societal benefit. The owner of a project always wants to retain a maximum of power, but this can result in planning difficulties because it will be necessary for the entrepreneur to cover the same planning ground as a public sector entity in order to garner support from the local or regional planning authority and the community in general. The danger for the entrepreneur is that the chosen alternative will either give too much away or will be diluted in exchange for planning consent to the point where the project will not be profitable. The London case study in chapter 10 highlights this latter problem for the BAA with respect to Stansted.

Comparison with best planning practice suggests that, strong as it is, the metropolitan planning process as it operates in the US falls short in some areas. There has been little use of formal goal-setting processes employing citizen input. Community views about aviation and the environment have tended to be either ignored or intuitively estimated or responded to only after litigation. Only a small number of alternative aviation solutions have received initial attention. Evaluations have tended to be made on the basis of subjective listing of benefits and costs by those persons conducting the analysis. Little or no monitoring of the impacts of completed projects appears to have been incorporated into the continuous planning practices of operating agencies. The new draft version of the FAA advisory circular (FAA, 1996) attempts to repair some of these problems, particularly in terms of citizen input to goals and researching community views, but notably not in the methodology of evaluation of alternatives.

Once system objectives have been agreed and the required facilities have been derived, the normal master planning process can be followed to size and cost the options and to support applications for strategic authorisation. Broad design criteria should be used rather than too much attention to the detail of the design, data on which can be found elsewhere (Ashford and Wright, 1992; Blow, 1996; Horonjeff and McKelvey, 1994). It can be assumed that an independent runway working with a mix of takeoffs and landings will be able to handle continuous peak flows of between 40 and 50 air transport movements (atm) per hour in Instrument Flight Rules (IFR) operations, but this can be exceeded for short periods or when Visual Flight Rules (VFR) are in operation. The declared runway capacities at the major European airports are given in Table 11.1. They are often well below the best practice levels, for political and environmental reasons. The additional capacity of a new runway reduces if it has to be located close to the first runway or at an angle to it. There is hope that advanced technology will allow perhaps another five atm per hour in the next decade, and will also reduce the minimum separation of parallel runways which will give the maximum additional capacity.

For a strategic study, terminals can normally be sized on the basis of 15 sq metres and 25 sq metres per domestic and international design hour passenger respectively. The level of service and hence the spatial requirement will be higher when there is a large proportion of business passengers in the mix of traffic, partly due to the requirement for each airline to provide an executive lounge, but the average dwell time in the terminal will normally be lower than for leisure passengers, so the typical spatial requirements per hour are still appropriate.

The definition of the design hour, often referred to as the Standard Busy Rate (SBR), varies with the airport owner. The British Airports Authority adopted the 30th busy hour, the FAA suggests the use of the peak hour of the average day of the peak month, and another measure is the traffic level at which 5% of the annual traffic is subject to greater flow rates. Typical ratios of SBR to annual flows are given in Figure 7.2, together with Typical Peak Hour Ratios (TPHR) as used by the FAA and the Port Authority of New York and New Jersey. It is also usually adequate at larger airports to assume the need for three aircraft stands per million passengers per annum (mppa), with some 4,000 sq metres per stand. Maintenance and remote parking may require an additional 30% above these estimates. It will be necessary to provide between one and three car parking spaces per 1,000 annual passengers, depending on average length of stay and mode of access, at three hectares per 1,000 spaces. In addition, there will need to be parking for car rental. There will be some 1,000 jobs on the airport per mppa, or perhaps twice as many if there is a large amount of freight activity or a based airline, with staff and freight vehicles also requiring parking.

Figure 7.2: Typical ratios of annual to peak hour traffic

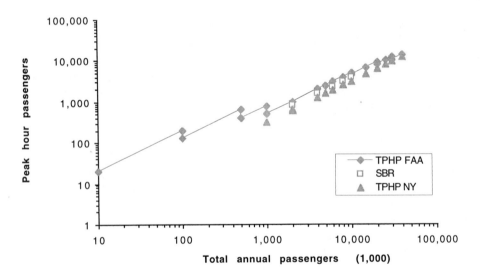

Source: Garcia, (1996)

Particular attention needs to be given in strategic studies to airport layouts, so that a balance is struck between short term efficiency and retaining the long term capability to provide an optimal layout for the ultimate site capacity. The layout will also be sensitive to the airspace in the region. It is necessary to avoid conflicts between airports, with obstacles, and with airspace reserved for military or security reasons. The environmental implication of each alternative system must be assessed. The FAA's Integrated Noise Model and air quality models are helpful in this regard. Before comparative evaluation of the alternatives, every effort should be made to mitigate the environmental impacts.

7.2.5 Evaluation of alternative plans

The criteria on which the alternatives should be ranked will reflect the planning objectives, but will inevitably include costs. It is important that the evaluation should not be driven only by the capital cost, but that due weight be given to life cycle costs. On the other hand, an emphasis on operating resource costs, as often happens in simplistic comparisons of energy use, ignores the inevitable consequence of capitally intensive projects, namely that a large proportion of the total resources will be consumed in construction. This is particularly the case with new rail systems, where it is common for more than half of the life cycle energy budget to be consumed during construction.

It has become recognised that a broader basis is necessary for evaluation than Cost Benefit Analysis (CBA). There are very fundamental questions to be asked of proposals for airport expansion projects, which formal analysis can help to answer but which ultimately require political judgement. A community which is attempting to decide on the net value of a proposal wants to know just how much a new facility is needed, and also, if it decides to go along with the proposal, whether it is then opening the door for larger schemes in the future.

A particularly narrow CBA (really cost effectiveness) was used by the UK CAA to analyse airport requirements in the southwest of Britain, which suggested that no airports were needed (Caves, 1980). Twenty years later there are still six airports operating with scheduled services. It is not possible to say whether inefficient pricing has led to an uneconomic system or whether the wider benefits are judged by consumers and communities to outweigh the costs. In either case the planning methodology could have included the effects.

System costs of course include not only the costs of constructing and operating the facilities, but the costs incurred by the users of the facilities. Since airport development is often undertaken in response to increasing or anticipated congestion, estimates of future delays form an important part of the evaluation process. The technical aspects of capacity and delay modelling are beyond the scope of this book, but can be found in the standard references on airport planning (e.g. Ashford and Wright, 1992; Horonjeff and McKelvey, 1994). However, there are two important issues that deserve some discussion in the context of evaluating alternative plans.

The first issue is how to evaluate the no-action alternative, or indeed any alternatives that do not provide the necessary capacity to satisfy the unconstrained demand. It is common to apply the full forecast demand to these alternatives and calculate the resulting delays. Because of the nonlinear relationship between traffic and delay, particularly as peak period traffic levels exceed capacity and the duration of saturated periods when the airport is operating at capacity increases, the resulting calculated delays are often ludicrously high. In reality, the airlines would not operate schedules that incur such levels of delay. Rather, the yield management systems would raise the average fares during peak periods by limiting the number of tickets sold at discounted fare levels, which would tend to reduce demand and redistribute it to other periods, other airports, or even other modes. During prolonged periods of bad weather, when capacity is reduced and which typically contribute most of the delay, some flights will be cancelled to prevent escalating delays from crippling the entire network. These consequences of insufficient capacity are not costless, particularly to the air travellers. However, these costs may be quite different from the calculated cost of delays had the number of flights simply increased in proportion to forecast demand. This need to explicitly address and evaluate the likely airline response to a capacity shortfall is recognised in the draft FAA guidelines for performing benefit/cost analysis for airport capacity projects (FAA. 1997).

The second issue is the need to recognise that there will usually be tradeoffs between delay and other costs and impacts. Delay can be reduced, but at the cost of building and operating more facilities (particularly runways), which in turn create additional noise and other environmental consequences. Addition of capacity at different airports in the system will not only result in different levels of total cost and impact, including different access costs to air travellers and air cargo shippers, but the distribution of those costs and impacts will be different. The evaluation process needs to identify these tradeoffs and distributional consequences for each alternative.

Some weighting scheme is often adopted to allow a comparative analysis of the alternatives. Any such approach should reflect the system's goals, and include capital cost, social cost, environmental impact, air traffic delay, consistency with other sectors' plans, ground access costs, airspace utilisation, ability to satisfy demand and regional economic impact. However, it must be recognised that any attempt to reduce as complex a decision process as airport system planning to a single number for each alternative is doomed to fail, except in the happy but unlikely circumstance that one alternative is clearly superior on all counts. Any weighting system necessarily reflects the values of those creating it, and it is most unlikely that all the stakeholders will have the same values. The most useful contribution of such an exercise is to help decision makers better articulate the nature of the tradeoffs inherent in the various alternatives, and how those tradeoffs are viewed by the different stakeholder groups. This in turn provides the basis for the inevitable compromises or mitigation measures that can move the process to a politically acceptable decision.

The evaluation methodology may have to change at each clearance point in a hierarchy of decisions. In a system where most of the planning issues are settled at the local level before being brought forward to a bid within an overall system budget, as is sometimes the case with metropolitan system planning, the local decision would be best supported by a comprehensive multi-objective evaluation (Hill, 1986) before perhaps sifting the options with a cost-benefit analysis. Thus any application of multi-objective evaluation methods which have been developed to deal with the need for the resolution of conflict between the various affected parties, e.g. planning balance sheet (Litchfield, et al, 1975), goals achievement matrix (Hill, 1968), the minimum requirements approach (Hill and Lomovasky, 1980), will have been needed only at lower tiers of the system planning process.

The number as well as type of alternatives evaluated should not be reduced prematurely. Computing power is cheap, particularly if data for sequential tasks are carried through the analysis (as discussed in section 8.5.3). Eliminating alternatives too early in the process not only can leave stakeholders feeling that their preferred alternative was unfairly eliminated in order to favour a less desirable outcome, but could deprive the project sponsor of an option that might end up looking quite attractive if unresolvable difficulties emerge with the sponsor's initially preferred alternative.

Central to the evaluation of alternative strategies in a regional context is the ability to predict how the regional demand for air service, or for access to general aviation facilities, will be distributed among the airports in a region under each alternative. This is necessary both in order to be able to assess how well each alternative meets the goals of the planning process as well as to determine the magnitude of the environmental impacts of each alternative. Due both to the importance of this issue and its technical complexity, this is treated in greater detail in section 7.3.

7.2.6 Selection of the preferred alternative

Up to this point, the process is largely technical and the analysis is in the hands of the technical team members. The actual decision will clearly be in the hands of public representatives and will involve political judgement. The results of the technical analysis should be designed to inform this decision and the technical analysis itself should be, in turn, shaped by the decision process. Further iteration may be necessary to obtain a plan containing revised policies.

The selection should be as fully transparent as the earlier stages of the planning process, so that suspicion of the way that the planning authority made the final decision does not undermine the entire process and infect other initiatives. The process of selecting evaluation factors, scoring each factor for every alternative solution and then balancing the tradeoffs to arrive at a preferred alternative, can only lead to an implementable outcome if the representation of policy goals is not only complete but if their consideration is commensurate with their influence on the outcome. Of course, the objective is to derive a neutral assessment of the alternatives, but many failures of implementation are due to an inability by the planning team to understand the influence of key professional and political players and build this into the evaluation.

It is quite likely that the most suitable option for a metropolitan area will be to build up the capabilities of the primary airport in a system while safeguarding the possibility of developing major additions to capacity in the future by adding more runways or building additional airports (de Neufville, 1995). However, many problems may stand in the way of this aviation-oriented solution. There may be overwhelming environmental or safety issues at the primary airport, and the cost to society of foregone opportunities at the reserved site, such as residential development or recreational uses, may be judged too great. Joint-use airports and all-cargo airports should be considered as alternative ways to redistribute demand.

The decision should be made in the light of potential futures rather than in a present day setting. Probabilistic methods of assessing the future only congeal existing relationships into the future, tending to make the future essentially the same as today: this has been described as 'colonising the future' (Rosenhead, 1989). Some weighting scheme should be adopted to allow a comparative analysis of the alternatives. It should reflect the system's goals, and include capital cost, social cost, environmental impact, air traffic delay, consistency with other sectors' plans, ground access costs, airspace, utilisation, ability to satisfy demand and regional economic impact.

7.2.7 Implementation of the plan

The thoroughness with which the previous phases of the study have addressed both the technical and institutional issues will largely determine the success in implementing the system plan through the master plans of individual airports, though the inevitable changes of circumstances over time will also affect the outcome. It is therefore important that the planning timescale attempts to achieve a balance between what can be reasonably foreseen and the need to establish a planning framework that allows the stakeholders to make long term commitments of resources. Given the lead time involved in the development of major new airport facilities, such as runways or terminals, the system plan should take a longer term perspective, perhaps looking 20 or 30 years into the future, while recognising that the chosen option should allow a flexible approach to investment commitments as events unfold. This requires tight linkages between the development of the overall system plan with the development of the other elements of the planning process. This will be relatively simple if all the airports are in common ownership or management, though they will each have their own challenges in complying with local land use and multimodal transportation issues. The plan implementation will need to be managed over time as additional phases commence and flexibility has to be exerted. The periodic updating of both system plans and airport master plans allows the process to respond to changing circumstances in a coordinated way.

The functional definition of the system may well mean that parts of the system have to meet inconsistent environmental or planning standards. Differing ownership will add an additional complication, even where collaboration is seen to be desirable. The more difficult the coordination, the more it is necessary for there to be a 'champion' organisation for the system plan with the authority and the resources to manage the study through to implementation. Since implementation is an ongoing process, the champion needs the authority to respond to changing circumstances and take appropriate action. This could be the main metropolitan planning authority of a region, which normally has primary responsibility for surface transportation but which is also increasingly charged with responsibility for multimodal issues.

In smaller countries, the principal metropolitan area will typically be the site of the main gateway airport and a government agency will be the coordinating and enabling 'champion'. In this case, within the functional system of international airports, the government will tend to behave in the same way as would an entrepreneur in a liberalised domestic situation, so that any system planning is then internalised and for the benefit of that nation only, unless funding aid is required and the World Bank, say, insists on an international overview. Macão's planning was essentially a master planning exercise, with no attempt to integrate the plans into those for Hong Kong or Shenzhen. Holland took a great interest in the competitive situation in planning Schiphol as a 'mainport', but without any collaboration with other countries to agree on appropriate roles. In both these instances, the planning authorities took independent decisions which gave them the opportunity to impose their own preferred future onto the system. This is also possible in the UK's regions, because the metropolitan authorities are relatively independent and the regional planning agencies have few teeth. In all these competitive situations, where the administrative responsibilities do not coincide with the functional system, the key to driving the system to individual advantage is to establish an early market lead.

No plan can be implemented unless a funding stream can be established. In the competitive situations considered above, an independent funding stream will help to ensure dominance. Where funds are required from central national or international sources, their availability will generally depend on some vetting of the overall system benefits. In general, also, there will be some element of cross-subsidy to be faced within the system, unless the chosen option is based on a criterion that entry to the system be confined to those airports which demonstrate their financial independence. This would defeat one of the primary aims of most system plans, in that they recognise some system benefits which are greater than the sum of the parts, e.g. the reservation of expansion capability, though a sufficiently complicated internal accounting exercise could probably be devised to allow for this if the objective were made sufficiently explicit. The decision may have been taken that supra-regional funds should not be used, in order to avoid the obligations which generally are attached, e.g. by the use of federal Airport Improvement Program grants in the US rather than funding from local bond issues. In any case, the plan should contain some prioritisation for funding. The funding for the implementation of most system plans will come from many sources and will require specific management, particularly of the multimodal aspects of the plan, in order to ensure an equitable participation and the timely availability of the funds. Sometimes the purpose of system planning is simply to allocate funds across the systems administered by the planning agency. In these cases the plan implementation is the allocation itself but this also needs to be managed in a continuous manner.

7.2.8 Monitoring

Monitoring of the implementation and achievement of airport planning has been almost nonexistent, despite the exhortations of the guidelines for best planning practice. This is largely because there have been few explicit objectives against which to compare the outcomes, as is shown by the system case studies in chapters 10, 11 and 12. Certainly a check is kept on the overall system delays, but little is done to monitor the total factor productivity of the complete air transport system or to assess the difference that a change in infrastructure investment would make to that productivity. The criteria themselves, i.e. the objectives which drive the planning, are seldom monitored to confirm their continuing applicability, the US Essential Air Services Program being something of an exception. The progressive commercialisation of the industry is making it easier to derive financial indices, but there is still a great deal of opaque accounting. There should at least be indicators of efficiency, effectiveness and economy, as indicated for individual airports in chapter 6.

There has been almost no attempt to assess the full societal balance sheet for air transport, which would allow firmer conclusions to be drawn about whether and where society should support further expansion of the system. It is therefore not possible to judge the performance of airport strategic planning, in terms of the output of each airport. The extremely political nature of the process even makes it difficult to apply best planning practice to airport system planning, so it would not be surprising to find the outcome to be non-optimal from the perspective of the system as a whole. There are, in any case, few agreed indicators of system performance, the level of service indicators given in the FAA's metropolitan system planning guidance being appropriate to the individual airports but not reflecting the system content of the planning, such as regional development or cross-subsidy. The allocation of delay between local and system-wide elements is itself difficult.

7.3 ACTIVITY ALLOCATION MODELS

The general discussion of demand models in chapter 4 indicated that a crucial element in planning airport systems is understanding the way in which passengers, airlines and other users of the system make their decisions as to how to use the available facilities. Many attempts have been made to model the revealed behaviour of passengers. Most of these attempts have used logit models, because of their specific properties (Fischer, Nijkamp and Papageorgiou, 1990; Hensher and Johnson 1981). The logit model expresses the logarithm of the ratio of the probabilities of choosing any two alternatives as the difference in the perceived utility of each alternative, where the perceived utility is a function of the attributes of the alternative and of the traveller. This formulation results in an intuitively reasonable S-shaped choice function, but is constrained to give equal cross-elasticities between alternatives for each attribute. Also the result is only valid if all alternatives are genuinely different and the choice between any two is unaffected by the attributes of the others (the so-called independence from irrelevant alternatives). This constraint can be lifted either by use of the less computationally tractable multinomial probit model, or, more easily, by nesting the choices and using logit models within each level of the resulting hierarchy. Almost always, these models have been found to best explain behaviour in choosing between airports, between routes or between carriers with some combination of access time, trip cost and frequency offered.

7.3.1 Air passenger airport choice

It is widely recognised that in broad terms the choice of airport by air travellers is determined by the air service offered at each airport on the one hand and the accessibility of those airports for the traveller on the other (Lunsford and Gosling, 1994). The relevant air service variables obviously are those for the market of interest to the traveller and include flight frequency and air fares (where these differ across airports). Accessibility is typically measured by highway travel time, although a more sophisticated approach would consider the full range of access modes available and derive the access disutility from the ground access mode choice process (Harvey, 1987). The advent of yield management systems and hub-and-spoke networks has significantly complicated the treatment of air fare and service frequency. The fact that a particular fare is offered in a market does not mean that it is available for a given traveller, particularly at busy periods. Higher frequencies may be available to travellers willing to take a connecting flight through a hub at the cost of a somewhat greater travel time.

There is evidence that the specification of the frequency function on travel choice models has a profound effect on the utilities at both low and high frequencies, and also on the implied value of time for schedule delay relative to time on mode, where schedule delay measures the disutility of the interval between departures. In spite of the recognition of the importance of frequency to the choice of service, there is no agreement on how best to reflect this in the form of the utility function. Table 7.1 shows a normalised comparison of calibrations of several choice models, together with the form of frequency term in the utility function adopted (Aroesty et al, 1990). The implied premium for schedule delay varied from 1.0 to 1.8, and the implied value of main mode travel time varied from US $7.1 to US $44.6 per hour, even when the two most extreme results are ignored. Another comparative study also shows that, while some models give a coefficient for the natural logarithm of frequency difference greater than 1.0 (implying a greater market share than frequency share), not all models even display this characteristic (Alamdari and Black, 1992). It is suggested that the models might have given more consistent results if they had also considered the mutual interaction between demand and frequency offered, the influences of computer reservation systems and frequent flyer programmes, and also the relationship between the flight departure time and the desired departure time. The assumptions on travellers' value of time are also important: a log-normal distribution gives a lower price elasticity than assuming average values (Ben-Akiva et al, 1993).

Table 7.1: Travel choice model parameter comparison

Model	Travel choice	Source data year	Adj. cost parameter (1987 US $)	Travel time parameter (hour)	Sched. delay parameter (hour)
Grayson	Intercity mode	1977	-0.018	-0.200*Y	0.0244*Y
Morrison & Winston		1977	-0.0020	-0.051	-0.276
Kanafani & Ghobrial	Airline route	1980	-0.0201	-0.897	0.239*F
Hansen		1985	-0.0043	-1.45	1.29*ln(F)
Kanafani et al	Airport	1970	-0.014	-0.10	0.021*F
Harvey		1985	n/a	-9.96	(0.94-0.55*F)F
Gosling	Airport access	1980	-1.74/PCI	-1.10	same

Notes:

1. Y = Household income/2000 ("wage rate")
   PCI = Household income per capita/1000 (children counted as 0.5 adults) = 0.7*Y (approx.)
   F = Daily departures

2. Morrison & Winston included party size variable (auto mode only) = 1.366*N

3. Morrison & Winston based schedule delay on average time in minutes between departures over
   24-hour day. Parameter adjustment for schedule delay in hours for a 15-hour day:
   $P_{adj}*15*7/(2*D) = P_{org}*10080/D$, where D = weekly departures

4. Schedule delay parameter for Harvey model based on (15 + F) daily departures in market from
   all airports and series expansion of relative frequency ratio

5. Gosling & Harvey model parameters for residents of region.

Source: Aroesty et al, 1990 (corrected by authors)

Other areas under further investigation are the nationality mix of the airlines, the aircraft technology and the importance of fares, using new data on fare actually paid. The effect of replacing turboprops with jets on competing routes from airports in the midland region of the UK was estimated by comparing turboprop predictions with jet outcomes (Brooke et al, 1994). The improvement in market share due to jet service appeared to be less than 10%. Indeed, it may have been less, because no attempt was made to disassociate the changes in type from the increase in aircraft size of the jets (85 seats, rather than 50-65 seats) and many in the airline industry believe that it is capacity rather then frequency which defines market share. However, other evidence suggests that a balanced market with both operators using turboprops would change to an 80:20 split if one operator used jets at the same frequency (Aroesty et al, 1990). The fundamental difficulty with both the effect of frequency (Kaemmerle, 1991) and of aircraft size (Russon and Hollingshead, 1989) is that, without lagged variables or some other method of inferring causality, it is not possible to distinguish between these effects and the tautological relationship of demand with capacity.

More work is required in the specification of the choice models and the appropriate scope of their application. The UK CAA's models (CAA, 1989), which were used in the RUCATSE studies described in the UK case study, presume relative airport attractiveness to be directly related to market share, thus perpetuating current trends in changes in market share provided that runway capacity remains available. Nesting of decisions on access choice, mode choice, destination choice and generation (i.e. choice of whether to travel), with simultaneous estimation of model coefficients has been used in Sweden to build a complete integrated model for the whole country (Algers, 1993), but it is not clear how well the model copes with dominant hub frequencies. It may be that the heuristic learning approach of neural network models (McNally and Lo, 1993) may overcome some of these difficulties, as well as the assumption of random utility maximisation implicit in logit models, but they are far from transparent and not easily capable of sensitivity analysis.

The choice models require a prior assumption of level of service offered. This is very different from predicting the likelihood of the service actually being offered. There is much historic evidence of the choices airlines make in offering service in airport systems with unequal market shares from which some understanding could be gained, as attempted by the UK CAA (1989). However, airline strategies are continually evolving, as are the marketing efforts of the airports, so that much more needs to be understood about airlines' attitudes to the risk of failure, their response to financial and other incentives, and their overall route development strategies before predictions of service and the subsequent continuity of the service can be made with any confidence.

Finally, it is important that the market being shared is also understood, if correct traffic predictions are to be made. This requires taking a rigorous disaggregate approach to the modelling of trip generation, including long term saturation effects (Shaw, 1979), and of the multi-modal nature of transport. Unfortunately, the latter analysis awaits a surface mode data base commensurate with that provided for air transport, at least in the UK.

There are hints in the development of the air transport demand and choice models described above, that the specification of the models leaves much to be desired. They may calibrate quite well, but they are less secure when used for prediction, particularly of the effects of changes in supply. Aircraft technology, airport characteristics, the effect of capacity per se and the adjustment of travellers' time budgets to available flight timings may be more important than the traditional variables, though harder to isolate. The impression is quite intuitive, as it is with other authors: "These considerations lead me to suspect that current mode choice models as well as models of other travel choices are seriously mis-specified in ways that have nothing to do with the random effects and covariance parameters of multinomial probit" (Horowitz, 1991). The feeling is that not only are comfort, reliability and safety important, but also that choices are 'state dependent' (i.e. they depend on past choices) and that perceptions of service quality may be very different from the objective measurements normally used in the models. Other work suggests that serial correlation is even more important than state dependence, again indicating that previously ignored variables may be strong determinants of trip generation (Kitamura, 1989). Also recent work has shown that "different individuals use different decision-making processes in choosing travel modes", pointing to the need for really effective segmentation of demand (Chou, 1992).

It may be that the concept of 'satisficing' or of making decisions which are 'good enough' for the situation as understood (Simon, 1955) needs to be developed. Another approach could build on the idea of 'value stretch' (Mansfield, 1992). The traveller is presumed to have a highest level of preference, defining realistic maximum goals, as well as a lowest level of tolerance, reflecting minimum basic travel needs. The level of the last experience will be situated within this band of preference or 'value stretch'. The range between the 'tolerated' and 'last experience' levels will indicate the 'satisfaction gain' and that between the 'last experience' and the 'highest preference' levels will indicate the 'reconciliation gap'.

A third approach would be to distinguish between cognitive, affective and conative behaviour (Michie, 1986), where the latter is the overt activity of making reservations and travelling, i.e. the revealed demand which is usually modelled. Cognitive behaviour is the acquisition of the information which will be used to formulate beliefs and patterns of psychological activity. Affective behaviour evaluates the activity in the light of the cognitive structures to develop an attitude towards its desirability.

All of these approaches are relevant to improving the understanding of the behaviour which has been investigated by the travel generation models (e.g. Brooke et al, 1994) and the airport choice models described above. Some combination of revealed preference and stated preference surveys (Fowkes and Preston, 1991) may be required to test the validity of model specifications, particularly in the areas of demand-constrained leisure travel and of new travel opportunities.

7.3.2 Airline service choice

Whereas passenger behaviour is presumed to have the objective of minimising the disutility of travel, airlines in a liberalised and privatised setting could be presumed to be profit-maximisers. An attempt has been made to predict the equilibrium proportion of hubbing under a profit-maximising assumption, using a 'n'-player, noncooperative game method (Hansen, 1990). Due to the many simplifying assumptions which had to be made (the use of prevailing fares, nondiscretionary aircraft size, total demand inelastic with respect to price and service level), and the *a priori* indeterminacy of the equilibrium properties of the game, the behaviour of hubs in multi-airport regions and those with relatively weak markets were not well-predicted. The approach is none-the-less promising, as is the attempt to identify an optimal air transport network, including the effect of slot pricing, using a two-stage game theoretic approach (Hong and Harker, 1992), where again there is no variation of total demand with fare or level of service, and airlines are presumed to attempt to maximise their revenue share on each sector. The results of its application to a simplified network show clearly that oligopoly total profits are much greater when the slot allocation is endogenous, i.e. determined by the airlines' ability to pay, with the same fee for each airline, rather than by prior allocation. The airline with the larger aircraft of course gains profit relative to the others. The principles developed in these two studies could well be merged and extended to hierarchical structures and multi-airline interactions (Shaw, 1993) to study the impact of airline and regulatory policies with respect to hubs, using decision-process algorithms.

The real-world outcome of the application of the models will depend crucially on the relationships which result from mergers and alliances. A fertile field for research is that of the power relationships and processes by which an airline chooses its partners, there being no universal structural models which can guarantee the success of the partnerships (Flanagan and Marcus, 1993). The relationship is often prompted by the acquisition of slots, which changes the nature of air services when the minor partner finds itself redundant (Moorman, 1993).

There is a clear trend for increasing concentration in the industry as airlines search for economies of scope. This leads to the major international gateways gaining an ever increasing share of country-to-country traffic (Pearson, 1997) and the major alliances in general gaining market share at those gateways. This does not, however, conflict with an equally strong trend to fragmentation of routes in a growing market, particularly since the alliance carriers are using the gateways as international hubs as well as for bridges to their partners' hubs.

It is becoming increasingly common for supply decisions to be uncoupled from revealed demand. This is sometimes to correct an under supplied market, but it is often a result of airline or airport strategies. A number of studies have been conducted looking at the consequences for market entry of liberalised air services between the UK and other European countries (Button and Swann, 1989; CAA, 1995; Caves and Higgins, 1993; EC, 1996). These studies provide evidence of significant changes to the supply of air services in terms of destination served from London and other UK regions to Europe, increased frequency by route, improved use of jet services and an increased number of designated airlines. Recent examples are the KLM decision to markedly increase frequencies and capacity at Amsterdam airport and Manchester airport's initiative to become its own hub champion and to coordinate its airlines' schedules. It must be anticipated that many other, and more extreme initiatives will be taken as the possibilities of liberalisation of services in Europe are realised. Also, liberalisation is continuing to spread through the rest of the world.

While the foregoing studies may have contributed significantly to a better understanding of the dynamics of the industry, this understanding has not yet been sufficiently developed or operationalised in accepted modelling tools to be much help to a system planning study trying to determine, for example, how traffic will redistribute in a region if a new airport is constructed or if the primary airport does not expand its capacity. Past efforts to use airport choice models, such as the Multiple Airport Demand Allocation Model (Campbell, 1977) or the ACCESS model (Harvey, 1988), have implicitly assumed that the service frequency will follow demand, in order to maintain load factors at an acceptable level. However, this assumption fails to address such factors as the role of yield management systems, the effect of competition, and alternative route options presented by a hub-and-spoke network. There is an on-going need for more research into how airline service strategies respond to capacity constraints or the provision of additional capacity in an airport system.

7.3.3 Based aircraft airport choice

Where a region is served by several general aviation airports, aircraft owners have a choice of airport at which to base their aircraft. This decision is likely to be influenced by a range of factors, including the travel time to reach the airport from their home or business, the airport fees, and the facilities available at the airport, including such issues as lighting and instrument approach aids, and the presence of a control tower. For larger or higher performance aircraft, the length of the runways may also be a concern. The way that these factors are weighed by different owners will reflect the use to which the aircraft is put. Corporate owners or those using their aircraft for business flying are likely to find all-weather and night capability important, as well as the higher level of perceived safety offered by the presence of air traffic control services. The availability of fixed base operators with the technical capability to service the aircraft is also likely to be important. On the other hand, owners of aircraft used primarily for recreation may be willing to forego these services in order to take advantage of lower airport fees at smaller, less well equipped airports. Some may even prefer the greater sense of freedom that comes from operating from an airport without a control tower.

Another factor that may well influence the choice of where to base aircraft is the availability of hangar space. An inventory of the 23 general aviation airports in the San Francisco Bay Area (Metropolitan Transportation Commission, 1994) found that every airport in the region had vacant tie-down spaces while none had empty hangar space. Thus it could be expected that the construction of additional hangar facilities at some airports would attract some owners to relocate their aircraft to those airports.

Understanding this choice process is important for two reasons. The first is that the number of aircraft operations at a general aviation airport depends on both the number of aircraft based at that airport and the composition of that fleet. An aircraft operated by a flight training school will generate far more operations per month than an aircraft owned by an individual for recreational flying. Without an understanding of how to predict changes in the number and composition of based aircraft at each airport in response to actions taken both at that airport and others in the region (for example changes in fees or construction of hangar facilities), forecasts of future aircraft operations at each airport become largely guesswork. The second reason is that this information can be used to guide investment policies. With limited funds available to support capital needs of general aviation airports, it makes sense to focus those funds where they do most good. However, this will require a way to measure the consequences of different funding allocation decisions. For example, is it a higher priority to resurface the runway at one airport or provide an instrument landing system at another? In the extreme case, it may be better to allow some airports to close, so that their based aircraft redistribute to other airports and help improve their economic viability.

These decisions can be modelled on a discrete choice basis, using similar models to those for air passenger airport choice. Aircraft owners can be assumed to select the airport that maximises their perceived utility, where the utility of each airport is measured as a function of the access distance or travel time, the airport fees, and the services available. In order to calibrate such models, it is of course necessary to have data on the existing distribution of based aircraft and their owners. In California, this information can be obtained from the County Assessors, who maintain records of each aircraft based in the county for property tax purposes. These records include where the aircraft is based and the address of the registered owner. There are a number of operational difficulties that need to be overcome in working with these data, including such situations as flying clubs that may register all their aircraft at a single address (which may even be the airport itself) or corporations that register their aircraft at their head office, which may not even be in the same state. There are also concerns over misreporting of aircraft location by owners trying to minimise (or avoid) property taxes. Even so, an attempt to build such a model as part of the update of the Bay Area Regional Airport System Plan (Metropolitan Transportation Commission, 1994) resulted in a model with statistically significant parameters that was able to explain the existing distribution of based aircraft, and was used to predict the expected distribution of based aircraft among the regional airports under a range of system development alternatives.

7.4 FORECASTING IN A SYSTEM PLANNING CONTEXT

Air transport is facing a future which conventional wisdom suggests will be characterised by a falling growth rate as the market matures, the cost per unit of technological progress increases, modal competition increases and environmental constraints increase the cost of expanding capacity at least in the developed world (Thomas, 1993; Boeing, 1997). This prediction is in a sense self-fulfilling, since it will determine the strategies which profit-motivated airlines will adopt in an increasingly free market. Even so, the future clearly is not cast in stone. Among other things, it depends on the continuance of liberalisation policies, on the degree of competition which results, on economic growth, on public attitudes toward the environmental impacts of aviation, on the way in which airline and airport entrepreneurs (together with their support industries) innovate, and on the degree of control which governments deem to be necessary.

7.4.1 Effect of supply on demand

The effect of supply on demand is important to the preparation of traffic forecasts, as indicated in chapter 4. Yet there is confusion as to its existence. It has been pointed out elsewhere (Grigson, 1978) how the UK government view varies: when deciding national policy the effect definitely does not exist, for regional policy it probably does not exist, yet it definitely does exist in the autonomous term of econometric forecasting models. In view of the trend, noted earlier in this chapter, for supply decisions to be uncoupled from revealed demand, it is of interest from the point of view of forecasting demand per se, as well as for the wider implications for stimulating the economy, for airport policy and for bilateral negotiations, to establish the causality, the scale and rate of response of the relationship between supply and demand, rather than simply establishing the correlation. Yet the effects of changes in the quality of service or in network accessibility on demand are commonly ignored, unless they are subsumed within some autonomous trend term (e.g. as discussed by Ashford and Wright, 1992), though the most recent models of the UK Civil Aviation Authority (CAA) make some allowance for demand generated by additional service by applying elasticities to changes in generalised cost (CAA, 1990). Smaller airports might therefore reasonably expect their traffic will grow faster than average as more routes are added and the generalised cost of reaching destinations by air falls. This expectation of a shift along the demand curve is exemplified by the claim that the lower propensity to fly in the rest of the country compared to the southeast of England is due to a relative lack of opportunity to fly from regional airports as much as to the lack of a need to fly (Caves, 1986). Traffic models which do not include the effect of changes in supply effectively deny the expression of airport-based competition among the airlines. The supply characteristics which affect demand are primarily availability, price and quality. These are mostly incorporated in liberalisation scenarios though it should be noted that the changes (particularly in availability) may be mostly autonomous rather than induced by liberalisation. However, availability and price (through congestion costs) will be very dependent on the provision of infrastructure and policies with respect to cost recovery, as evidenced by Ryanair's continual search for airports with minimal charges and minimal congestion.

The demand at an airport may also be related to supply through the in-migration and stimulation of economic activity as firms find the accessibility attractive enough to influence their location decisions. The UK CAA's domestic traffic model does allow propensity to fly to increase with airport traffic, but the effect is discounted in other traffic categories. This implied shift in the demand curve reflects the estimates of the economic impact which are used to help justify airport expansion, as discussed in chapter 5.

7.4.2 Technology

Technology has historically been the main determinant of aviation's rate of progress, as explained elsewhere (e.g. Caves, 1995). The technological improvements have been adopted because of their intrinsic value, with their affordability being strongly influenced by investments in military aviation technology during the two World Wars and the subsequent Cold War. This conservative path to the present was not anticipated by many forecasters in the 1970s. Most were prepared to expect large numbers of supersonic transports, 1,000-seat aircraft, a high probability of hypersonic hydrogen-powered aircraft and intercity vertical takeoff and landing (VTOL) aircraft by the year 2000. Only those who attempted to relate their forecasts to capacity requirements, retirement rates and replacement cost/benefits produced forecasts which partially stood the test of the unfolding future. Even now, there is no shortage of forecasters prepared to envisage extreme technologies based on experimental vehicles, paper studies or normative but unknown solutions to the industry's problems.

History, then, suggests that the industry is maturing and becoming ever more careful in its approach to the future. Yet history also suggests that it was innovative technology which allowed aviation to achieve its present status, and that only a new infusion of new technology will be able to allow it to overcome a highly constrained future. One dilemma is that the longer the industry leaves the search for the appropriate solution, the harder it will be to introduce it in the face of ever more entrenched incremental preferences. Another dilemma is that true innovation tends to occur in a haphazard way, rather than in response to a specification.

It has been suggested that the overall industry growth curve is actually composed of multiple 'S' curves, each reflecting the impact of a major structural change. Gifford (1993) identifies three curves associated with the DC3 era, the first generation jet era and the wide body/deregulated era. Moonen and Schaper (1994) suggest that it is possible to consider even the smoother progress within each era as composed of a series of 'S' curves associated with more incremental changes of technology or management style. The introduction of the overwater twin engine aircraft and code sharing are two recent examples of incremental change for the industry as a whole which have large implications for specific parts of it. At any given point in time, the decision makers will have had their perception of the ultimate saturation of demand coloured by where they perceive they are on the current 'S' curve. Examples of the next incremental changes may include the incorporation of noise cancelling devices with other developments on engines (*Flight, 24 August, 1994, p 20*) or the use of steeper approaches (Caves et al, 1998), while the next step change may be the use of hydrogen as a fuel to overcome the emissions problems at high altitude if not in response to high oil prices (*Flight, 17 December, 1997, p 24*).

More often, with the deregulation era maturing in the western economies and predictions that the overall economic growth will also continue to mature being increasingly seen as an inevitable consequence, the adaptive view is being interpreted as having to 'do more with less' (Fuller, 1983). Gillen (1993) suggests that this means introducing smart management and technology, these involving full social cost pricing and dynamic information systems respectively. With the appropriate institutional changes encouraging free market forces, it is expected that these measures will produce the best match of supply with demand (from the point of view of community welfare). The increasing need to respond to environmental pressures for a sustainable level of mobility, that is now widely foreseen, will be satisfied with a self-fulfilling outcome of early maturity. This does not, of course, imply no further expansion of the infrastructure. Even with low growth rates, the absolute increases in traffic in a mature industry can be very large (Oster, 1993).

7.4.3 Timescale

A common problem in forecasting is the lead time in the more developed countries between initial planning of major works and the commissioning of the project, so that the necessary time horizon for master planning is unfortunately extended. Even 25 years may not cover the expected mid-life of the project if the planning process requires 7-10 years. One of the most serious criticisms of the demand analysis in the 1978 UK White Paper on airports policy was its failure to plan for more than a 15 year time horizon, when it was obvious that the intransigent capacity problem of the London hub system would be bound to worsen after that date. It is difficult to excuse the lack of consideration of a longer time horizon on the basis of the uncertainty of the forecasting process.

The forecasts themselves are dependent on the level of technology, average aircraft size, income per capita, population, regional planning and a number of 'external' effects; in other words, the forecasts are usually implicitly 'locked-in' to the time frame for which they have been developed. Furthermore, unless the time frame is reasonably firm, the ability to coordinate planning between the various interested parties may be severely jeopardised. Also, the estimates of the factors that depend on the forecasts, such as the capital requirements, economic viability, environmental impacts, and the sizing of facilities, may well alter in scale if they are dislocated in time. Thus, when the UK Department of Trade estimated that Stansted would need a second runway at 25 mppa, these capacity requirements were estimated for the most likely timescale of traffic growth and they will be wrong if that timescale does not occur.

7.4.4 Judgement

Many of these difficulties are dealt with as a matter for judgement rather than analysis, and it is often felt appropriate to seek out the accepted experts in the field to make the judgements. However, research shows that errors of judgement are systematic rather than random, manifesting bias rather than confusion. Many errors of judgement are shared by experts and laymen alike. "Erroneous intuitions resemble visual illusions in an important respect: the error remains compelling even when one is fully aware of its nature" (Kahneman and Tversky, 1979). Most predictions contain an irreducible intuitive component, for example in the assessment of confidence intervals. Research shows these to be assessed too optimistically, particularly before repetition provides the opportunity for feedback on predictive performance, because judgements are often made on minimal samples and assumptions are too restrictive of possibilities. A classic example of this is the tendency for long term forecasts to be unduly influenced by short term trends. Another case is the European carriers' small concern for market saturation relative to the Asian carriers, as revealed by an IATA survey of airline forecasting techniques (*Newsletter 34, Transport Statistics Users Group, March 1995, pp 6-9*), despite the greater growth potential in Asia. Where econometric modelling is used, the judgement component is usually in the prediction of the independent variables, which is a field in which the acknowledged experts are notoriously wrong even in such fundamentals as population and wealth per capita, and also in the future change in the elasticities.

7.4.5 Scenario writing

This is a well-known approach to exploring potential futures. Often it is limited to exploring options within the system under consideration with a view to assigning probabilities to each of several potential futures. Thus, a recent panel of experts confined themselves to mentioning only noise legislation, fuel prices and video-conferencing as issues outside the direct control of aviation managers which were regarded as having important or special significance in shaping the industry (Thomas, 1993). However, a more productive use is to explore the range of feasible potential futures and their consequences in terms of the needs to which a system might be asked to respond and the steps it would need to take in order to respond effectively. It should be emphasised that the objective is not to predict the future, even by assessing probabilities of the various potential futures and hence take a view on the most likely future. Scenarios explore potential futures so that some light can be thrown on the scope and flexibility which needs to be designed into the system, on the extent to which this might be accomplished autonomously by changes in the system's capabilities which would be embedded in any specific scenario, and on the consequences for the system's performance of not being able to meet the needs of some scenarios, as well as the identification of those futures which the system should not be designed to accommodate.

It can be taken for granted that the most likely expected future is the one which is presently being projected by in-house planners and system designers. It can also be taken for granted that, except in so far as the system's future evolution is predetermined by the closing off of future options, this expected future is most unlikely to coincide with any actual future state. G. K. Chesterton (1946) described a game called 'Cheat the Prophet', where people listen to all the prophets, wait for them to die and then live out a future different from any prophesies. He remarks how difficult the game was becoming at the beginning of the twentieth century due to the extreme profusion of prophesies, though the game was not yet pointless, because all the prophets were using the same technique to derive their different scenarios. They had each noticed a trend in society and extrapolated it as far as their imagination would take them. In fact, it is extraordinarily difficult even to predict live births, which is the most fundamental of all the factors which might influence future demand for travel (Makridakis and Wheelwright, 1979), despite the apparent underlying logic behind the very long term trends (Lenon, 1992). Yet accepting the 'myth of predictability' powerfully influences the provision of infrastructure and hence the ability to respond to change.

Scenarios for future air traffic can be described in terms of socio-political, economic and technological characteristics. Alternatively, they can be considered in terms of future demand and supply possibilities. Either way requires the incorporation of interactions between the descriptive groups, as well as the need to speculate on the implications of currently unknown initiatives in the management of processes and innovations in system capability.

Increasing maturity is usually reflected in the demand models by a slackening of the reduction in the real cost of air travel and also a decreasing income elasticity over time. The former assumption rests on projections of the costs of fuel and other factor inputs, together with projections of improved productivity. It takes no account of major changes in management practice. The latter assumption reflects the apparent increasing saturation of the desire for travel, yet seldom analysing how demographic changes and migration patterns might alter, nor how the discretionary income of the ethnic and age cohorts in the population may change (Evans, 1989). If there is some underlying determinant of market maturity, is it related to the mode's capabilities, to environmental and physical capacity constraints on total trip making, or to some inherent budgetary constraint on the individual? If so, the constraint may be some combination of percentage of income (Airbus 1994), percentage of time (Hagerstrand, 1987) even an energy (or discomfort) limit.

However, great caution is needed because of the tendency with large and complex systems, such as the air transportation system, to base predictions or analysis on aggregate data. Yet there is at most only one "average" traveller, and most likely none at all. Rather, the market for air travel is composed of myriad submarkets, each of which is responding to different forces in different ways. In developed countries there are many individuals who make dozens of air trips per year, but far more who make only one or two. A shift in demographics or the economy that moves significant numbers of people into occupations or income strata that make greater use of air travel could fuel continuing growth in demand for decades to come.

Some feel that the information revolution, or 'fourth logistic revolution' (Anderson and Batten, 1989), will result in fast and volatile flows of commodities, people and information, leading to a preference for the fastest and most direct mode of travel. This type of future, where a new hierarchy of cities evolves with the most powerful being those based in the four Cs of culture, competence, communication and creativity, is already visible. It should be put alongside other demographic trends, which include decentralisation of metropolitan areas (*New Scientist, 15 June 1996, p 11*), the labour migration flows which eventually form the basis for air trips involving visiting friends and relatives (*International Labour Rights, Second Quarter, 1993*), and changes in fertility and mortality. These population changes are far from easy to predict (*The Economist, 4 June 1977*), nor are the implications of the changes: only a few years ago, the increasing number of pensioners was expected to allow more discretionary spending, but the difficulties that many countries now face in funding pension schemes may well alter that (Evans, 1989). It is likely that present estimates of income elasticity have been wrongly specified, since they have been derived without consideration of the contribution of wealth to travel demand (Alpetovich and Machnes, 1994), particularly among the retired population.

Economic factors which require a scenario approach include exchange rate differentials which drive all types of travel, as exemplified by the changes in North Atlantic traffic as the dollar weakened in the late 1980s (*Avmark Aviation Economist, March 1994, p 11*), but make a major difference in the competition between holiday destinations, as does the threat of terrorism (*Times of London, 29 February 1996, p 23*). The changing structure of industry and commerce could affect air trip generation rates. It was seen in chapter 5 that propensity to fly varies greatly with the type and size of firms. There is evidence of a reversal of domination by the top 500 Fortune companies in the US (Makridakis, 1995), whereas the European economy appears to be increasingly dominated by multinational enterprises who are preferring to locate in the advanced regions (Amin et al, 1992). Whichever trend dominates in a particular setting will have a strong bearing on the future propensity to fly.

The past evidence of liberalised services does not indicate any large changes in traffic, partly due to capacity constraints and partly because the recent activity by low cost new entrants had not surfaced. Yet liberalised airlines will reorganise their networks and alliances, abetted by profit-seeking airports. Some will make mistakes, and survive by taking advantage of free route exit rules; some will fail completely. This will create much greater volatility in the traffic levels at individual airports and on individual routes (de Neufville and Barber, 1991). If models are to offer valid guidance on the distribution of demand, they need to be "capable of simulating, for a very heterogeneous market, the expected growth under assumed economic scenarios while explicitly taking account of increased competition between airports, airlines and air routes, increased frequencies and changed network structures, and competition with an emerging European High Speed Rail Network" (Kroes, Bradley and Veldhuis, 1994). The FAA's models for predicting the distribution of traffic in the US have been criticised for taking insufficient account of the increasing economic activity of the medium size cities (ASRC, 1990).

The impact of technology and smart management on future capacity (or simply the assumption of future capacity) is often the most neglected area of supply prediction. Heathrow's growth is a case in point, early predictions being hampered by miscalculations of the maximum feasible size of aircraft and more recent analyses by under-prediction of annual runway capacity. Despite the highlighting of these broader system implications in the literature (e.g. TRB, 1990), most studies only perform sensitivity analysis involving changes in the single variable which is closest to the immediate focus of the study.

Scenarios can be used either in a normative sense to choose the most appropriate technology or to aid in predicting the success of a given technology. Either way, it is necessary to consider the influence of the socio-political-economic setting rather than only the technical and economic qualities of the technology itself. Brooks (1971) says that prior to 1970, it may well have been correct to assume that technology was an exogenous variable. The earlier failures of technological forecasting, i.e. optimistic in the short term and pessimistic in the long term, were due to failures of technological or economic imagination such as not appreciating the likely progress in ancillary technologies. However, even in 1971, he felt that these lessons had been learned and that the problems of technological forecasting were shifting to the social and political arenas: "rationality demands some centralisation of responsibility, but centralisation is increasingly incompatible with participation, while participation is increasingly incompatible with action or progress." Roy (1976) also concluded that technical-determinism often does not explain why some technologies are not adopted, though pointing out that none of the alternative economic, political or ideological determinism models are any more successful on their own.

Future scenarios cannot be complete without the inclusion of the implications for the alternative provision of infrastructure within each composite scenario. This was the central focus of the UK RUCATSE studies (see chapter 10), where it could be argued that the options examined were unrealistic in terms of environmental acceptability and conservative in terms of the potential for technology to augment capacity.

There are many options for writing scenarios to include whichever of the above travel determinants are thought to be appropriate. The important characteristics are that the scenarios should be cohesive and should show a feasible route from the present state to the potential future state. The latter requirement is made difficult because the dynamics of change are not well understood (Keynes quoted in Wills, (1972), p.165). The scenarios could consider the relative rise and fall of the percentages of the population holding various ideologies in order to explore their implications on the underlying socio-politics (Post Office, 1976). Condom (1993) uses two levels of the economy and three levels of regulation, deleting the combinations of high growth with status quo regulation and low growth with strong regulation as infeasible scenarios. Miles, Cole and Gershuny (1978) use a three dimensional matrix with axes of the economy, egalitarianism and world views to develop 12 scenarios, the three world views being conservative, reformist and radical, equating approximately to the Adam Smith *laissez faire*, the Keynes interventionist and the Marx philosophies respectively.

The Dutch have used a consolidated version of these scenarios to inform their planning for Schiphol airport (Ashley, Hanson and Veldhuis, 1994). The macroeconomic scenarios were:

- 'balanced growth' (BG), in which free markets, the rapid diffusion of new technologies and international cooperation lead to balanced world economic growth.

- 'global shift' (GS), where variations in the abilities of individual economies to exploit new technologies lead to high economic growth in the Asian-Pacific region and low growth in west and central Europe and Africa.

- 'European renaissance' (ER), which leads to high growth in the European economies and relatively poor US growth.

These macroeconomic scenarios were then linked to three air transport liberalisation scenarios:

- 'business as usual' (bau), reflecting 1992 Europe
- 'European liberalisation (el)
- 'global liberalisation' (gl)

the latter two allowing competition and the end of protectionism to lead to lower prices, higher frequencies, smaller aircraft and greater development of hub-and-spoke systems across, respectively, Europe and the globe. Of the potential nine combinations, three scenarios were judged to be feasible:

- BG - gl
- ER - el
- GS - bau.

It was predicted that Schiphol's passengers would increase by 120% between 1990 and 2015 under GS-bau or ER-el, but would increase by 250% under BG-gl. Aircraft movements were predicted to increase by 90%, 110% and 190% respectively under the three scenarios, reflecting the greater growth in larger aircraft in all cases.

Obviously there is a loss of variability in the use of such a simplified set of economic scenarios, and the air transport scenarios also do not reflect the full range of possible changes which may result from liberalisation: in particular, they ignore the impact of new low cost non-hubbing airlines and the privatised airports' roles in the market-making. The scenarios were, however, augmented by further scenarios of possible government environmental policies with respect to Schiphol. The study remains the only real attempt in European aviation planning to recognise that policy for infrastructure provision needs to be informed by an exploration of the future rather than by forecasts based on the 'myth of predictability'.

This example illustrates two key issues that need to be considered in developing and using such scenarios. The first is the need to be able to articulate in a coherent and consistent way how a given set of macroeconomic and policy assumptions translate into demands for air travel and how the airline industry would respond to these demands within the regulatory context that would exist. For example, it is one thing to say that a more liberalised environment would lead to more competition, lower prices, higher frequencies, the use of smaller aircraft, and greater use of hub-and-spoke networks, but quite another to quantify those effects and determine what changes would occur at a specific airport or group of airports. If these effects are assessed on the basis of expert judgement, then there is a risk that the higher level assumptions are merely window-dressing to give a set of fairly arbitrary assumptions about the evolution of air service some apparent methodological validity. Rather, what is needed are analytical models that can be shown empirically to reflect airline behaviour, and that can then be applied to the scenario assumptions to predict potential outcomes. Since the underlying process is not deterministic, and the models are not likely to be perfect anyway, it must be recognised that there is a very large set of possible outcomes, and the models need to be able to generate a large number of different outcomes together with their relative likelihood of resulting from the scenario assumptions.

This emphasises the second key issue, that not all scenarios are equally likely. While it may be useful to simply identify the range of outcomes that conceivably could happen, so that the potential vulnerabilities of alternative strategies to different futures can be assessed, in the real world decisions need to be made and decision makers will make judgements based on their assessment of the likelihood of different outcomes, with or without the help of the analysts. Thus any information that the analysis can provide regarding the relative likelihood of different outcomes should lead to better decisions. For this to occur, it is important that the analysts are honest about the reliability of their models, and that the variability of the outcomes reflects not only the uncertainty in the behaviour of the system, but also the uncertainty in the ability to model that behaviour. Fortunately, as a practical matter the distinction between the two issues is largely academic, with one important exception. That is the stability of behaviour over time. A model calibrated on past behaviour (as they all must be) may give very good explanation of that behaviour. However, if the behaviour subsequently changes, then the model may give a very poor prediction of future outcomes. Thus the stability of the behaviour of components of the system over time is an important research area that needs to be given more attention than it has received in the past.

8

NATIONAL AIRPORT SYSTEM PLANNING

The previous chapter described the general process of system planning introduced in chapter 3, in the context of multiple airports serving a single region. Those airports are likely to have considerable influence beyond the geographical boundaries of the region, as shown by some of the case studies in chapter 13, and problems will arise in identifying appropriate roles for the airports within the system. However, the goals for the regional system are likely to be relatively clear.

This chapter considers national level perspectives on a system of airports. The process, and the concerns of the planners, are likely to be similar if a state within a federation is attempting the task. The model for the process is once again to be found in the US, where the FAA has produced guidelines (FAA, 1989) for states to produce a State Airport System Plan (SASP). However, the process and its products will vary greatly, depending on the context. Chapter 12 gives examples of widely diverse contexts, and the UK and European contexts described in chapters 10 and 11 provide further examples of this. The US context, discussed in chapter 9, is unique in many ways, one of which is the emphasis given in SASPs to general aviation and the need to allocate small amounts of state funding. The issues are usually quite specific, with a need to rationalise the system while maintaining access in the face of threats of closure. Some states, like California, are including their large airports in the SASP in order to demonstrate the total system need, though the funding for these comes mostly from a combination of federal and local sources, including the federal Airport Improvement Program (AIP). The National Plan for Integrated Airport Systems (NPIAS) serves to identify funding needs at a national level. The main drivers in the US system are the 'bottom-up' expression of the metropolitan areas' needs for capacity, together with efforts to maintain a balance between the needs of all users of the system.

Despite the value of having clear documentation of the system planning task for the NPIAS and SASPs, the uniqueness of the US context makes it necessary for this chapter to take a more generic approach. It attempts to concentrate on those issues which will be present at national level in most countries, leaving the case studies to sharpen the focus in a variety of specific situations.

8.1 NATIONAL PLANNING CONCERNS

The FAA (1989) advice on SASPs gives the purpose of system planning to be: "to determine the extent, type, nature, location and timing of airport development in the state to establish a viable, balanced and integrated system of airports." If the state has a highly developed aviation programme, the plan may include elements addressing:

- goals and objectives with respect to airport development and its relation to the economy, development, transport infrastructure, land use, and environmental concerns.
- aviation objectives with respect to safety, efficiency, level of service, and economic self-sufficiency.
- resource requirements, timing and priorities for the state budgetary process
- information for legislation consistent with state aviation goals
- identification of the airports' present and future roles, giving direction for improvement needed to satisfy them
- policy and technical direction for master planning by individual airport sponsors
- management and coordination resources for metropolitan and regional planning
- special studies and activities associated with the planning process
- support for a continuing airport planning process
- the organisation, structure, authority and responsibility for implementation
- implementation measures, e.g. revenue generation, land use controls and legislative recommendations.

If strategic system planning is to be a viable and useful activity it should weld together the inevitable bottom-up planning with a perspective that the more focussed local planning cannot take. Specifically, it can assist in:

- resolving conflicts by providing a forum for discussion
- allocating resources efficiently

- promoting national policy goals, e.g. equal opportunity
- providing the tools to make societal trade-offs between the economy and the need for sustainability, between regions and between local, regional and national priorities
- harmonising analysis techniques, to promote consistency in approach
- establishing best practice, to promote efficiency
- identifying appropriate public/private roles, questioning market place solutions
- addressing global issues to save wasteful and inconsistent considerations of these issues at regional and local levels of planning
- putting necessary enabling processes in place
- improving the planning process.

Some of the issues that relate to these system level tasks are now addressed. The implications for improving strategic system planning are discussed in chapter 14, after the issues have been revisited in the specific case studies.

8.1.1 Role of air transport

In the early days of aviation, there were powerful political lobbies in favour of aviation which almost guaranteed that capacity would be made available in response to revealed demand, if not in advance of it. Nowadays, when the feasibility of an airport expansion programme has been established technically and financially and brought forward for local public examination, the debate turns on the issues of the environmental disbenefits and the economic benefits - assuming that the expansion is not in an area where the economy is overheated. However, the eventual decision will also take account of national and international ramifications of the decision.

Since aviation growth has substantial national policy implications, the industry must therefore direct itself not only to the ultimate user but also to policy makers. Society as a whole must see the need for growth if the policy makers are to be persuaded to allocate the resources required to support further growth. Furthermore, society, as well as the local communities around airports, must be persuaded that the benefits are worth the full social costs which will accrue from the further growth.

There is a natural acceptance that there is a positive mutually supportive relationship between aviation and the economy, while at the same time there being a general feeling that aviation is socially divisive. However, in many ways aviation allows similar social benefits to those from other modes and is the only mode which can provide them over long distances. Social benefits are self-evident when aviation is the only way to respond to disaster relief, medical evacuation, law enforcement or the protection of the environment (FAA, 1978). So are some of the other leisure benefits which are claimed by the FAA. Another example of the benefit to society, over and above the benefits to the individual, is the statement by Pope John Paul II that the world is becoming a global village in which people from different countries are made to feel like next door neighbours. Tourism has become a real force for peace in the world (Edgell, 1990). The type of inter-cultural tourism to which the Pope was referring can only be accomplished in practice by air.

The social advantages of aviation are more readily apparent in developing countries, in promoting cultural unity within a country and allowing cultural, ethnic and educational links with the developed world. The direct benefits from the support of industrial and social service activities are readily apparent in low density situations and in the pre-industrial phase of an economy. However, it is perhaps the less direct benefits, flowing from the transfer of technology, which have the greater beneficial effect on a developing economy. The transfer process can be formalised as occurring at four levels of sophistication: appropriating, disseminating, utilising knowledge and ongoing and interactive communication (Williams and Gibson, 1990). It would be very difficult to provide this latter most effective form of technology transfer without international aviation.

The establishment of air services brings with it the need for technical support, which extends from governmental administration through to qualified mechanics, perhaps ending with a country constructing aircraft of its own design. The training and quality control requirements result in a higher total industrial capability. However, if the dependence on foreign expertise is not to be greater than before, the transfer must set its own pace. Aircraft can be bought quickly and building an airfield does not take that long, "but these represent only the foliage of the tree. The trunk and roots of aviation - the aeronautical training, the navigation and meteorological skills, the concepts, rules and procedures - must be nursed along slowly and carefully over many years" (Heymann, 1962). The levels of quality required can only be maintained through a system of specifications and quality controls which educates a broad base of industry into this level of production (Ozires, 1976).

The technological spin-off has a particular part to play around airports in developing regions, where the airport can form a base for local industry. Indeed, the airport is often the only place where clean water and a consistent supply of electrical power can be found. The effect of an airport can then be beneficial both technically and through increased job opportunities, and real meaning can be attached to the multiplier effect. The principal benefits are, however, unlikely to show up in the short term in calculations of economic contribution, input-output or cost-benefit analyses, since they are embodied in the changes of attitude brought out in the local people. The effect can be quantified to some degree by ranking projects by the effect they have on attitudes (Wilson, 1967). At one end of a ranked continuum, the projects would only influence attitudes, while at the other end they would only produce directly productive activity. Transport projects appear to be somewhere in the centre of the continuum.

These issues relate primarily to the countries with developing economies, but even the most advanced economies also have other goals than economic growth to which aviation can make a positive contribution. Japan has a relatively explicit statement of its goals and the role of transportation in meeting them, as shown in the case study in chapter 12. Norway has adopted similar goals of social interaction which has resulted in a string of Short TakeOff and Landing (STOL) ports (Strand, 1983). The social benefits of these STOL ports have been well documented (Wilson and Neff, 1983). However, in most cases, national planners have to work with much less explicit social goals. Only a well informed political process can evaluate the weighting which these effects deserve within the general debate on national priorities, and hence on the formation of airport system planning goals.

In the most advanced economies, society derives its aviation benefits primarily from the consumer surplus generated by users who choose aviation in free competition with other available modes of transport, though even in these countries the case for competitive advantage in the stimulation of the economy through access to air transport is often also made. The evidence for these economic benefits has been examined in chapter 5 with the conclusion that they cannot be assumed without specific studies. There are some additional developmental aspects which should however be addressed in a national context, particularly the tourism and other routes to development. They centre on the provision of access to markets and for high level communication.

For many countries, tourism is a vital component of the economy and air transport is an essential part of the industry. Wheatcroft (1994) develops a methodology to compare the relative benefits of policies favouring tourism with those favouring the interests of the national airline. Notional calculations are made for an example island economy and a larger, more developed country. These show that there are usually more benefits to a country from investing in tourism than in the national airline. The calculations include allowances for leakage (of benefits outside the local economy) but not for the costs of externalities, though the environmental disbenefits of tourism can be substantial.

Countries where there have been tourism gains can be grouped into those where the gains come from liberalised air transport policies, those where national airlines are still protected, those whose tourism has boomed through charter airline freedoms and those who continue to operate charter bans in order to control against a low spend per tourist. Tourism is sometimes shown by the statistics to grow strongly despite restrictive airline policies. Where tourism growth has been slow, the reasons often have little to do with airline policy. The tourism demand equation is multivariate, airline policy being but one of many important factors. Others include exchange rates, personal security, originating countries' policies with respect to foreign travel, the provision of hotel accommodation, the strengths of competing destinations, and policies on airport infrastructure, all of which can influence the apparent relationship between tourism and transport policy. Brazil, for example, had economic and security problems which were the predominant causes of the decline in tourism in the late 1980s.

The relationship between economic development and provision of transport infrastructure is not straight-forward (Caves, 1993), but the main issues are summarised here. Goals of mobility, and of social and spatial equity call for much the same transport solutions as does the economic goal of regional development, since they both imply improvements in relative, as well as absolute accessibility. Spatial inequality is normally defined in terms of income or employment (Slater, 1974), which may be related in some way to accessibility. The inequality can occur both in an absolute sense in that some regions are geographically remote, and also in a relative sense, in that some towns happen to be located on main arteries while others lie between these main routes. The former was the main focus of the analysis in Action 33 (OECD, 1977), which showed that air transport provided an effective way for an integrated Europe to meet cohesion objectives.

There is evidence to support the intuition that economic growth is inversely related to isolation. Firstly there is the theoretical notion of 'impure' public goods, i.e. centrally located public services which are less useful to some of the community because more of their real income is taken up in accessing them (Breheny, 1974). Other studies also infer strong mutual interaction between communication potential and regional economies. One piece of research identifies personnel in high contact employment as the principal determining factor on both sides of the interaction (Tornquist, 1973). Another study notes that more than 50% of jobs in large metropolitan centres of the US controlled from headquarters are not in the immediate metropolis or its hinterland (Pred, 1976). This study also finds that such linkages are not gravity-based and that they frequently involve smaller towns at a considerable distance. The inference is that growth transmission depends on the job-control and decision making activities reflected by these linkages and that only between 3% and 21% of this activity falls within the subject firm's hinterland. This evidence for the role of communication stimulating the economy contrasts with the consequences in advanced economics of relatively poor availability of transport. Several studies show little relationship between transport and regional growth (Gwilliam and Judge, 1974; Chisholm, 1976). It seems that only the most remote locations are at a serious disadvantage in respect of transport costs, at least in the UK, and that the underlying determinant is whether there are external economies of scale resulting from the spatial concentration of activities.

Spatial inequality tends to be reinforced by the transport planning process unless specific intervention counters this natural tendency. The consequent 'spiral of disadvantage', which triggers the provision of new transport capacity to enhance the comparative accessibility of the growing and congested locations, has been noted at the European scale (OECD, 1977) and is clearly visible in the US federal airport planning. The EC is clearly keen to promote inter-regional accessibility and to encourage aviation in that role, and the SRI (1990) report points up the value of this, though it has still be determined whether any of the EU's regional policy efforts will be able to counter the increasing comparative trade advantage of advanced regions with their power to attract the large multinational companies (Amin et al, 1992).

If accessibility is to form a plank of development policy, suitable indicators need to be adopted. It is suggested in the literature that the indicator should be technically feasible and operationally simple, easy to interpret and preferably be intelligible to the layman (Morris et al, 1979) and that only the accessibility to relevant opportunities should be measured. This tends to suggest that some compromise between relative and integral indicators should be adopted, where relative accessibility is the ease of reaching a single destination and integral accessibility is the mean case of reaching all relevant destinations. Jones (1981) concludes that network measures are generally insensitive to types of travel and location, and that contours may be the most useful compromise measure, given that each of the alternatives violate some of the criteria at least some of the time. With contours it is possible to disaggregate attraction and deterrence with different aims for different travel types.

The market place may be expected to take care of transport provision to support activities among the central place hierarchy of cities (Christaller, 1966), and low density demand is almost certain to be too small to expect other than a network treatment. But if the important economic connections are the community-of-interest ones, no blanket measure of accessibility is adequate and no policies based on it can expect satisfaction.

8.1.2 Investment strategies

Governments clearly have to prioritise their investments to satisfy national goals and international responsibilities across a broad spectrum of activity. Historically, aviation has been seen almost as an advertisement for a country's development status as well as a source of foreign exchange and public sector employment. With the realisation that, as a derived demand, aviation and other modes of transport must be efficient if a country is to be competitive, and also that it is not necessary to spend scarce government funds in order to have a working air transport system, governments are reassessing their previous commitments. It is becoming more common to expect aviation to cover the full cost of its infrastructure, the debate turning more to methods of cost recovery and the extent to which government should be involved in the process.

The discussions on finance in the earlier chapters are also relevant to national strategy. There is a wide range of options for funding, depending on the context. In the US, aviation is almost completely self-funded as a stand-alone mode of transport. At the federal level a Trust Fund is administered by the FAA and provides grants to the operators of the facilities being funded. The users contribute to the Trust Fund through ticket and other taxes, but the government can intervene as the controller of public sector spending.

The General Accounting Office has recently undertaken a review of airport financing options within the US context (GAO, 1998). It has compared the optional sources against five key characteristics:

- economic efficiency as reflected by the extent to which beneficiaries pay in proportion to the benefits received
- the extent to which airports' capital costs are equitably distributed among those who benefit
- the ease or cheapness of managing funds
- the extent to which the federal government can control the choice of project and amount of funds provided
- the extent of substitution of federally authorised funding for state and local funding.

The funding options considered were the Airport Improvement Program (AIP) for airports incorporated in the NPIAS, Passenger Facility Charges (PFC), tax-exempt bonds, state and local funds and airport revenue. The largest source in the US has been tax-exempt bonds, while the AIP has reduced as PFC have grown since their introduction in 1992. PFC funding is seen as having a strong linkage between those who pay and those who benefit, but some benefit without payment. Tax-exempt bonds cause non-users to pay for airport services through the tax revenue foregone (US $560 million per year). Federal control of the AIP is strong and it has to approve an airport's PFC application, but it can only influence the distribution of tax-exempt bond funds if they are used to meet the matching share requirements for federally funded projects. In contrast, the UK government now takes no part in the funding of the aviation infrastructure, and there are severe limits on the borrowing by airports in the public sector. The expectation in emerging, as well as in developed economies, is that the privatisation approach adopted in the UK is the funding panacea for the future. However, the consequences may be not only a reduction in employment but also a smaller and leaner airport system in which the weaker members will not survive.

8.1.3 Development of gateways

Historic gateways developed as a natural function of national geography and the technical capability of aircraft. They shaped the air transport network, providing natural hubs for the designated carriers. Those hubs were protected by the restriction of new traffic rights to those gateways and insisting on an equal capacity share. This has distorted the network development and denied many provincial cities the opportunity of direct international services. It is now becoming clear, as explained in chapter 5, that protection of the flag carrier may do less for a country, and certainly for the peripheral regions of that country, than the encouragement of new air gateways. It is still important for a country that its main hub should not lose its competitiveness. Not only does a strong gateway provide an opportunity for transfer, or sixth freedom, traffic but it also strengthens the domestic links into the main hub. Olympic Airways has lost a lot of this feed as many more Greek islands have developed runways which allow direct flights by charter carriers, so that the passengers do not have to make the difficult transfer at the congested Athens airport. Tourism development can be severely hampered by the need for domestic connections or long bus transfers to the popular resorts.

As the main hub reaches a world scale of activity and becomes progressively congested, it becomes more interesting for the airlines to begin to fragment air service, to the extent that local markets can support it. The relatively high growth rates of traffic at second city airports in Europe shown in Table 8.1 indicates that there is latent demand in many provincial cities.

There is often no equivalent of the US situation where competition between the inland hubs of major carriers causes the carriers themselves to lobby for new international links, as the UK case study in chapter 10 makes clear. Mostly, airports and their cities have had to do the lobbying, leading to the BIAS studies mentioned in chapter 5. Similar situations occur in Japan and many other countries. If the expected changes in demography come to pass, with the high propensities to fly being taken up by the large provincial cities and some degree of political self-determination in the regions of some countries, the arguments for a liberalisation of gateway status would become stronger. The most interesting test case is perhaps in the EU, where the EC believes that it should do the negotiating of Air Service Agreements with third countries on behalf of the EU member countries, but most countries prefer to apply the principle of subsidiarity, which embodies the concept that nations should control their own affairs except where it is against the larger interest of the Union. If the EC position prevails, it is difficult to see how they could satisfy all aspirations. In any case the airlines will exert their own judgement and find some way of expressing their need to satisfy markets.

Table 8.1: Passenger growth (%) at European airports, 1994/1993

Country	Main city		Second city		Ratio
England	London	6.2	Manchester	10.7	1.73
			Birmingham	17.6	2.84
France	Paris	7.8	Lyon	6.1	0.78
Spain	Madrid	5.0	Barcelona	6.5	1.30
Portugal	Lisbon	5.4	Oporto	8.7	1.61
Italy	Rome	5.3	Milan	6.1	1.15
Sweden	Stockholm	8.7	Gothenburg	12.9	1.48
Norway	Oslo	13.7	Bergen	6.1	0.45
Turkey	Istanbul	7.11	Ankara	13.4	1.89
Greece	Athens	6.3	Thessaloniki	10.9	1.73

Source: Derived by the authors from ACI Airport Traffic Report, 21 March 1995

8.1.4 Modal priorities

Aviation has usually been planned as an independent mode, but a system planner must think on a broader canvas about where it would be best to spend the marginal dollar. Not all countries have passed specific legislation similar to the Intermodal Surface Transportation Efficiency Act of 1991 (ISTEA) in the US, but there is much pressure from the sustainability movement in Europe and the new socialist government in the UK for an integrated transport solution to congestion and pollution. The key to a successful intermodal policy is to create a level playing field so that the market can express the value of the modes. This certainly requires enabling actions by governments, because some modes have become unavailable in important markets and because the playing field is not level with respect to subsidies and environmental charges. There is a very strong trend to favour the high capacity modes, particularly the various rail options, but this can lead to very low vehicle utilisation in lower density settings unless the level of service is to be allowed to drop well below that which is available with privately owned vehicles.

There are encouraging signs that entrepreneurs are once again leading the way to an intermodal future, as reported in chapter 6, at least as far as air transport interchanges with ground modes is concerned. An interesting situation is developing in the UK with the rail routes being franchised out, in several cases to bus operators despite the anti-competitive possibilities, but also to airlines and airport owners. This certainly improves the chances of through ticketing and a seamless travel product, though the large differences in service standards still have to be tackled.

8.1.5 System capacity

Airspace and airport capacity issues and the government role in providing it have been discussed at length in chapter 4. It was seen that there was a shortage of airspace, runway, apron terminal and ground access capacity at many major airports. Governments have a choice between the provision of capacity on demand and intervention to manage the system's operations so as to reduce demand in the worst affected parts of the system. One example is to reduce runway capacity requirements by limiting frequencies per route at the expense of allowing free expression of airline competition. To the extent that additional capacity is provided, governments can also choose the degree to which they wish to be involved in planning and funding, rather than leaving the decisions to the private sector. Left to the private sector, there may be too little investment if subsequent monopoly regulation reduces profit potential or if there is not profit potential at small but necessary airports. Equally, there may be excess land take if many entrepreneurs chase a market for which there is not enough traffic.

The government is the final arbiter of the balance which must be drawn between international commitments to provide access, conformance with land use planning guidelines, economic development, defence needs, and the health of the aviation industry. Government can enable the release of military facilities and dictate the form of any runway or airspace slot rationing. In those countries where government is directly involved in the provision of capacity, it is capable of influencing the distribution of that capacity by varying the criteria for provision. Thus government could act to decrease the spiral of differential investment which results from a criterion based in reducing delay. The balance to be achieved in this case is to manage demand and/or to improve the capacity at the congested points so as to alleviate the environmental consequences of congestion without taking away the discipline which comes from a certain amount of rationing. It should also manage interactions between airports where the expansion of one would deny a role for the other, whether because of too little traffic or because of airspace interference.

8.1.6 National standards

Most governments have agreed to apply international standards of safety from ICAO and control of global environmental pollution from the Rio Earth Summit, or have filed their disagreement. Some states have gone well beyond these standards, e.g. with public safety zones in the UK and emission standards in California. The latter will require substantial changes in technology if transport modes are to comply with them. Normally the international standards are accepted by writing them into national law, so that the relevant operators then risk penalties for non-compliance. As standards change, the onus is on the owners or operators of the facilities to invest to bring them up to the new standards. The government then needs to have in place an auditing mechanism to ensure compliance.

There will also be national standards for the design and construction of aviation facilities usually tailored to the differing function needs. It is therefore necessary to categorise airports in a plan on the basis of the defined role. The US uses five 'service level' categories to reflect the type of public service (primary commercial service, other commercial service, reliever airport with commercial service, reliever airport, general aviation airport). These also reflect the AIP funding categories. There are also nine categories which reflect an airports' role within the NPIAS, in so far as its design influences the aircraft it can accommodate. The categories distinguish between utility and transport operations, short and long haul, and also cover helicopters, STOL and seaplane facilities. The design standards associated with these categories cover not only airport design and operating requirements similar to those found in ICAO's Annex 14, but also, for the smaller utility airports, give guidelines for capacity calculations and for terminal sizing.

Part of the continuous system planning process is the monitoring of these standards and classifications to ensure that they remain relevant. Examples of the need for this are the changing capability of the business jet element of the general aviation fleet, and also the emergence of the tilt-rotor aircraft.

8.2 IMPLEMENTATION STRATEGIES

8.2.1 System operation and management

A planned system of airports which cannot be implemented is no better than a wish list. Implementation will require that all necessary resources must be in place, both within the airport system itself and in supporting areas like air traffic control which form part of the overall aviation system, for the airports to be operated as planned. In the developing world, this implies not only that the physical facilities must be in place, but that personnel resources need to be prepared by training and fostered so that they will invest their talents in the industry rather than using their skills to find more lucrative employment. Continuity of training will probably require that a training establishment be incorporated in the system plan.

The planning and the subsequent operation and management of the system must be mutually compatible. Put another way, the combined management of the system must be committed to the planning process. An example might be a system designed around the assumption that the airlines would operate via regional hubs, as was expected in the north of Canada and also was found to be the most efficient solution for the Spanish domestic air transport system. If the airlines later decide to bypass the regional hubs, the whole pattern of required airport capacity would change. The communities would be faced with a very different economic outcome from the one which perhaps persuaded them to invest in the airport facilities.

This points up the fact that, if planning is to be useful, there must be an 'ownership' of the plan which can ensure implementation. The inference is that the initiation of the planning effort should be by those same owners. There is little dispute about this ownership if the regional or national government is, in fact, the owner of the airports and the author of the plans, and is also able to influence the behaviour of the airlines through a regulatory process, though many would doubt the efficiency of such a system. As explained in chapter 4, it is not easy to predict or control airline behaviour, even in a regulated setting. Increasing liberalisation and privatisation gives the stakeholders many more options, and makes it more difficult to identify any one owner with enough influence to drive the implementation in the direction of expectations. This would, in any case, not necessarily lead to a more efficient system than a fully regulated one as it would tend to favour one of the players and fail to respond to justifiable pressures for the needs of all stakeholders to be met.

Difficult though it may be to achieve, the preferred solution appears to be one where there is consensus ownership of the planned strategy, wherever the actual ownership may rest, together with a realisation of the need for continuous review of the relevance of the plan's goals.

8.2.2 Financial support

If plans are to be successfully implemented, sufficient financial resources must be available, not only for the construction of facilities, but also for the operations to be sustained and developed as necessary, and for the planning itself to be a continuous process. Too many projects fail downstream of the initial implementation because of a lack of maintenance of the physical and human resources. A good system plan will ensure that a cohesive and sustainable financial plan is included. If a goal of the system is to cover its own costs, careful phasing of development should be employed to ensure that the cash flow can support the operations and allow access to additional sources of capital so that subsequent phases can be constructed as needed. Few funding organisations will assist in meeting operating costs. If profits are not expected, a secure source of subsidy must be identified before any commitment is made to start construction.

One option that should always be considered before embarking on costly new development is to improve the efficiency of the system as it stands. This may take some urgency out of the development needs, and can result in the appearance of potential investors prepared to share in the risks and rewards of committing to the planners' strategy for the system. The benefits of this funding source will have to weighed against the desire of the new owners to stamp their own identity and ideas on the plan. The correct way forward is for the planners and the owners to see that a plan will only succeed in being implemented if it fulfils to the best possible extent the requirements of all the stakeholders, rather than just one set of interests.

8.2.3 Regulation

A plan will only remain relevant if it is continuously updated, including a review of its goals. However, this is unlikely to be sufficient to keep the system aligned to the interests of all stakeholders, particularly those outside the main management structure. The global and local environmental interests will require protection beyond that provided by national and supranational regulation, even if they are addressed on an on-going basis through a monitoring function. Legally enforceable agreements will be necessary at local and regional level which will pay particular attention to specific local concerns, as described in chapter 5 in regard to the approval for a new runway at Manchester.

Depending on the nature of the system, it may also be necessary to safeguard the system's users against abuse of a monopoly position by some form of price control as practised for the privatised airports in the UK or South Africa. Other areas where it has been found that residual regulation is necessary are in allowing competition for ground handling (including self-handling) and in the allocation of runway slots at congested airports. A challenge to anti-trust regulation may come from attempts by airlines to integrate vertically by buying into airports, particularly since the profit margins are perceived as more attractive than those in the airline business (see chapter 6). Another aspect of regulatory control is the limit imposed by governments on foreign ownership which, as with airline privatisation, limits the efficiency gains in return for some perception of retaining control within the country. Yet some large degree of outside investment may be necessary if plans intended to achieve efficiency improvements are to come to fruition.

8.3 INSTITUTIONAL CONSIDERATIONS

Clearly these will vary greatly depending on the national setting of system planning. It is therefore a task of the case studies to give a flavour of this variety. However, some general remarks which may be helpful are given here.

8.3.1 Departmental responsibilities

If there is to be a national airport or aviation system plan, an organisation will have to be assigned to design the study. The staff will need to ensure they have interpreted government policy correctly, given the inherent conservatism of their profession, and that all the above relationships are in place and effective. They should also be aware that, where government agencies are primary owners of the system, there is a risk of the ownership bias being pronounced and should have a responsibility to balance this with independent inputs of research and advice. The process will have to identify issues, develop a work schedule and budget, organise the work programme and the use of consultants and create or adopt a set of procedures. Detailed guidance on these study management topics is given by the FAA (1989). The organisation carrying out the process will ultimately have to formulate implementable policy which is acceptable across the system, to institute a process for coordinating regional and metropolitan plans, to provide a framework for individual master plans and to assure a technically sound and dynamic planning process.

To do this effectively requires the selective use of consultants. It also requires close relationships with the regional and local planning agencies, with the system's stakeholders, with the other departments representing the interests of the stakeholders and the wider societal concerns, and with the Treasury department or whichever agency is the ultimate paymaster. Ideally, the agency or government department which has the budget for implementation should assign priorities for the allocation of funds. There is often a department with overall responsibility for land use planning which will have some form of joint ownership of the plan and will share the responsibility for its implementation and the ongoing operation of the system (Rydin, 1993). This latter responsibility implies a requirement to set up a formal monitoring system to ensure that goals are met and remain relevant.

8.3.2 Regional and local governments

It will be the main responsibility of local or regional government to ensure that general planning law is followed and that equity is preserved among the local stakeholders. Their executive departments should have the expertise to ensure this and also to understand the role and needs of the aviation industry, if necessary being supplemented by independent expertise from similar but larger agencies. It is important to ensure that representation afforded by the planning system is converted to real effect rather than being only a token due to lack of information or time or money.

It is an equally important responsibility for local government to represent the local case accurately and forcefully to central government. Where there is likely to be conflict with adjacent jurisdictions, efforts should be made to draw these into the planning process at an early stage in an attempt to present a common case to the central planners.

8.3.3 Supra-national issues

In most airport planning situations, the legal responsibility for the implementation of internationally agreed regulations or standards rests in national law, with the execution being in the hands of the same agency which would normally create the system plan, so that these legal issues can be effectively internalised. The main supra-national issue which could be influenced by government is the competitive cross-border aspects of the wider airport system. This is particularly evident in the European gateway competition explored in chapter 11, where the attitudes taken by the respective governments towards the expansion of their major hub airports are influencing the shape of the overall European system. Where there is an overarching supra-national authority with the power to influence the plan, the national planners are faced with a larger scale version of the problems faced by local governments in representing their competitive case to central government. Frequently, in the international arena, this is not the case, and governments need to assess their behaviour in an open market situation. They then need to bear in mind that what is best for their national airline is not necessarily best for the ultimate consumers or for the economy.

8.4 ROLE OF ANALYSIS IN POLICY FORMULATION

8.4.1 Scale and complexity

Some situations demand a great deal of study while others require no more than an affirmation of the goals and the availability of enabling funds. The latter may apply when a strategic tourist development policy requires an airport system upgrade to accept long haul aircraft, provided that the environmental consequences are minimal. The opposite extreme may be when the socio-political setting expects a full public debate on the merits of further aviation development and the issue is in the balance, as has recently been the case in Holland. It is usually better to do more analysis than is judged to be the minimum necessary, since ad hoc judgements about a system are often made with an unjustified belief that the system has been understood, and a little time and money spent on analysis can avoid expensive mistakes being made. Furthermore, if issues arise later that have not been adequately addressed in the prior analysis, or stakeholders reject the conclusions on the grounds that the analysis is inadequate to support them, and as a result work has to be redone, the resulting delays may be more costly than a more comprehensive analysis would have been in the first place. Many planners feel that the move towards softer methods of evaluation has reduced the ability to engage in the ongoing learning process which provides the essential background to the judgements they face in allocating study resources and advising the decision makers. However, budget is always important, and it is hard to justify carrying analysis further than is necessary to support the required decisions.

In some countries, like the US, the requirements for a system study are laid down in quite some detail, though the analyst is still faced with decisions as to the necessary complexity. This is often dictated by time constraints and the lack of 'off the shelf' data, which emphasises the need to take every opportunity to increase the data base. Most countries will rely on external consultants to undertake the analysis and to justify the resources included in the bid for the study. The need to win the contract usually dictates the amount of analysis that can be done, so the consultant with the greatest relevant existing data and local expertise would normally be chosen. However, this presumes that the consultants bidding for the work have comparable technical analysis capabilities, or at least sufficient for the study. For complex technical issues, prior experience and analysis capabilities may be more important than local knowledge.

8.4.2 Stakeholder acceptance

The scope of the analysis needs to be agreed with the stakeholders if they are to contribute data and, indeed, if they are to accept the outcome. There are always different ways of looking at data even when the data are agreed, as in the interpretation of the inclusion of aircraft which just comply with Stage 3 noise certification requirements in the list of allowable aircraft for night operations. When models of airport choice come to be considered, past calibrations can always be challenged by stakeholders on the grounds that subsequent changes in facilities or methods of operation have not been taken into account. Perhaps the most serious problems of acceptance stem from misunderstandings about what each of the stakeholders really wishes to take away from the table. It may well be that residents really want an assurance of a finite limit to the growth of aircraft movements when they are only articulating concerns about individual noise events, in which case they will not be satisfied with an agreement based on the size of noise contours and the banning of non-Stage 3 aircraft. Before airlines accept noise quotas they will want to be assured that they reflect community views of acceptability. The only satisfactory answer to these concerns is to involve all the stakeholders in the study right from the time of identifying planning goals and of the study methodology which is proposed to develop potential solutions. Unfortunately, all too often, the clients are unwilling to participate in this type of procedure, expecting that they are paying for exogenous advice rather than that they have to put work into the study.

If any of the stakeholders are left out, it is easy to make erroneous assumptions as to their expectations and the way they will make subsequent use of the planned system. This is particularly true of the airlines and the GA users, whose individual decisions may be very different from those implied by generalised modelling. If the future is explored by scenarios, and hence the alternative plans are evaluated against their ability to cope with the full variety of possible futures, it is important that the stakeholders are able to identify with those futures. This will be a painful learning process for some of them, who will not have faced the possibility of futures which are not of their own construction. Good strategic planning is an experiential learning process, where the analyst is primarily a facilitator and catalyst.

At least in the US (FAA, 1989), public participation is seen as both essential and feasible, even at the system level. It must be recognised that there are multiple publics: travelling, aviation, neighbouring, commercial, elderly, young adult, rural, urban, etc., each with their own interests and priorities. Ultimately, the appropriateness of the system is determined by these publics, and not by the planners. These publics already understand the parts of the system they need to understand and contribute to its development, albeit in an unstructured way. Furthermore, if denied participation in its development, "they are perfectly capable of impeding the progress" (FAA, 1989). If they are to be satisfied, all the publics must perceive that the process is open, that opportunity for participation exists, and that the study is designed to accept and consider all inputs.

If this is to happen, the study design has to grapple with the fact that "even when a policy is crafted in an open and broadly representative process, political adjustments during the implementation process 'often are narrowly based, typically are achieved covertly, and therefore encourage self-serving behavior'" (Gifford et al, 1994). It is therefore important that the study designers understand the competing interests, and the power and salience of their deeply held commitments and beliefs.

8.4.3 National models

Analysis implies modelling, even if the models are relatively soft, i.e. based in scenarios, threshold analysis and decision algorithms. Any form of modelling requires a knowledge of the present situation and the history leading up to it. There is no substitute for an accurate and well defined data base. Those countries with ongoing collection of trip making data, like the UK Origin-Destination surveys or the US 10% ticket sample, have a natural advantage in understanding the way in which the system is used. Even then, similar data for other modes, particularly for intercity car and truck traffic, often do not exist. The best that can normally be expected are traffic counts across screen lines, at least where national borders are involved, together with customs and immigration records, if they can be released. Few countries do systematic surveys of the type of air freight commodities, or how much moves by air rather than in trucks with flight numbers. Another gap in the data in almost all countries is the real fare paid as opposed to the nominal tariff, so that the only price indicator available is aggregate yield per passenger, when the analyst really wishes to know how much each segment of the market is being stimulated by airlines' discounted fare offers.

Historic data on aggregate flows are often available, but of doubtful quality and with unspecified changes in definition over time. It is usually much more difficult to obtain data on the supply of flights and other modes of transport. Most of the scheduled flight guides have only recently converted to electronic format, and hard copy of historic schedules are hard to find. The situation is worse for charter seats offered, not least because they are frequently different from the seats flown. It is more difficult still to obtain historic data on price (rather than nominal fares) and on freight capacity, particularly in terms of belly-hold and combi capacity.

When it is necessary to model traffic between third countries, e.g. to assess the potential for sixth freedom traffic, the ICAO publishes route flow and 'origin-destination' (O-D) data, but not only are data missing or incomplete for many routes, but the O-D data are on a flight basis rather than the true itinerary, and so are of limited use for airport and route choice analysis.

If econometric or behavioural models are to be used, rather than simple time series extrapolations of traffic, past and future data on independent variables are needed for model calibration and its subsequent use for prediction. Historic population and economic data are usually available, but not at a sufficiently disaggregate level to allow good models to be constructed for each market segment or to allow modelling of the airport access aspects of trip generation, airport choice or modal choice. The analyst is then left with the need to make assumptions on the market segmentation or use aggregate data which clearly lacks the richness to represent real travel choices. It is also necessary to borrow experience, including elasticities, from other work. Extreme care should be taken that the models from which the coefficients are to be borrowed have the same format as those to which they are to be applied.

Future estimates of the independent variables which are available should not be taken on trust. It is preferable to treat them as a subset of the potential futures, and put them into a set of scenarios together with a range of other possible futures. The most difficult problem is posed by the client with an existing set of normative socio-economic forecasts built into a series of national plans, and unwilling to accept the inevitable uncertainties.

It is often possible to work with an existing set of national forecasts in performing a regional system study, but an independent review of those forecasts should be done. It is easy to be misled by the statistically good models which can be obtained when working with aggregate data into believing that the future is easily predictable at that level and that a 'step-down' procedure will adequately predict traffic throughout the system.

8.5 INFORMATION DISSEMINATION

Airport system planning at both a regional and national level involves a complex web of intersecting interests and agencies. Quite apart from the jurisdictional issues discussed in chapter 5, the range of issues that need to be considered requires the active participation of an array of government departments from aviation authorities to wildlife agencies, not to mention the range of private sector organisations and citizen groups. For all these parties to be able to participate in the process in a meaningful way, considerable thought and effort needs to be given to how information generated by the planning process is made available and distributed.

Although these issues apply at all levels in the airport system planning process, the national government can exert leadership in two important ways. The first is in making its own planning process as open as possible, and facilitating the dissemination of information at a national level, as well as the results of state and regional studies funded from national resources. The second way is in providing technical support and establishing recommended or required procedures to promote information dissemination at the local, regional and state level. While the process of information dissemination will need to be tailored to the circumstances of each planning study, the more consistent the approach, the easier it will be for the diverse stakeholders to know how to effectively participate in the process. This is particularly true for those stakeholders, such as airlines or other governmental agencies, that need to become involved in planning studies in many different locations.

8.5.1 Closed versus open processes

A fundamental issue that determines the nature of the planning process, and most likely its success, is the extent to which information is made available to all interested parties. This is naturally very expensive and time consuming in any process as complex as airport system planning, and it is tempting for the professionals and managers involved in the planning process to restrict access to much of the technical analysis, whether intentionally or otherwise. In some societies, there is also the tendency to view airport development decisions as the rightful prerogative of a small class of technical professionals and senior aviation administrators. In the extreme case, these decisionmakers may not only view much of their analysis as no business of outside parties, but in fact as protected by national security or similar restrictions.

Such attitudes are the antithesis of the planning philosophy espoused in this book. So complex are the issues that underlie sound airport system planning that no single individual, or even one agency, can hope to have a full and complete grasp on all the issues and subtleties. No matter how well intentioned, attempts by those involved in the conduct of airport planning to conceal information from other interested parties are likely to lead to poor decisions due to a failure to properly address relevant issues. Human nature being what it is, restricting the flow of information can also become a means for incompetent managers to conceal their inadequacies, or hide the pursuit of narrow vested interests from wider public accountability. In such a climate, senior managers may come to rely heavily on their intuition or prejudices, in place of careful analysis, and their subordinates may fear to put their jobs at risk by challenging unsound ideas.

An equally important concern lies in the tremendous economic stakes involved in a large modern airport. It would be naive to ignore the fact that individuals can become enormously wealthy through tapping the cash flows associated with airport development and operations, and that those with the decisionmaking authority to affect these flows wield considerable power. If this power is not being used in the best interests of the public good, it is hardly surprising that those in control of the process might wish to keep the details hidden from public scrutiny.

In contrast, an open process holds all the parties accountable for their actions by enabling the full disclosure of the basis for each decision. If arguments are weak or fallacious, or decisions not supported by the evidence, or made on arbitrary or self-interested grounds, this will become rapidly apparent to the other parties in the process. Likewise, where one party has information that materially affects the outcome of decisions being made by another, it is more likely that these factors will be properly considered if that information is readily available to all concerned.

Naturally, there will be some information that is necessarily sensitive or confidential, such as security procedures or the assessed value of some land that an airport is negotiating to acquire. However, the sensitivity of this type of information is usually sufficiently obvious that there is little difficulty protecting it.

8.5.2 Availability of study results

Timely public distribution of study results is critical to keeping stakeholders informed of the progress of the planning process and allowing them to make meaningful input. While the planners involved in the study may well be devoting full time to the issues, most of the other stakeholders have other duties, and may even be participating in their free time. A major national, state or regional airport system planning study may well result in a final report several centimetres thick, backed up with working papers or study elements that in total may occupy a foot or more of shelf space. While not all stakeholders will need, or want, to read all this, it can be expected that every part will be of interest to at least some. The worst scenario is for the technical experts to work behind closed doors for a year or more, then release all this documentation at once and give stakeholders 30 days to submit comments.

Not only should stakeholders be given adequate time to review interim study results and formulate their responses and concerns, but this process should happen while there are still time and resources remaining to adequately address issues raised in the comments. The dismissive or evasive repetition of earlier statements that often passes for a response to comments that is legally required in the US environmental impact statement process is not only likely to exacerbate any hostility on the part of the stakeholders, but can prevent important issues that have been overlooked in the study from getting a balanced consideration. This suggests that the process needs to be designed to release interim work products on a progressive timetable that allows for some debate and iteration.

An equally important consideration is the level of detail presented in the reports, particularly those pertaining to technical issues, such as traffic forecasts, capacity and delay analysis, or noise projections. The reports should be written to allow a moderately well-informed, but not technically trained, stakeholder to understand what was done and what the results were. At the same time, they should also allow a technical specialist to determine whether the analysis was competently performed, and to understand all the assumptions that were made in performing the analysis. The technical details may well be presented in appendices, or even supplementary working papers, but they should be sufficient to permit the traditional scientific test of repeatability. Sufficient information should be provided to allow an appropriately experienced person with access to the relevant tools to repeat the analysis and come up with the same results. This not only helps ensure a level of objectivity in the technical analysis, but can help minimise the risk that errors overlooked in earlier work subsequently come to light and compromise the results of later analysis, resulting in wasted time and effort.

8.5.3 Availability of data and analysis tools

The advent of powerful personal computers has changed the nature of airport planning in many ways. One of these is the ability of stakeholder groups to perform their own analysis, to satisfy themselves that their concerns have been addressed or to formulate more considered and technical input into the process. While this also requires some expertise, many groups have such expertise among their members, or know where to get technical advice. Rather than being viewed as little more than distractions to the primary analysis underlying the study, such efforts can contribute significantly to the success of the study by allowing a broader range of analyses to be considered, and involving the stakeholders more actively in the conduct of the study.

This can be facilitated by making data and appropriate analysis tools available to those groups that wish to use them, and have the resources to do so. Much of the data involved in airport planning becomes available eventually anyway, particularly in the US where there is a strong tradition of public access to government data. Organising the study so that data gathered in the course of the study becomes available in a timely way for access by interested stakeholders facilitates this process and strengthens the value of any such analysis.

Beyond the use of data and analysis tools by the stakeholders themselves, the ready availability of inexpensive computing power allows a much broader examination of alternatives. In the traditional process, the technical analysts determine what alternatives to examine, perform the analysis, and present the results. If stakeholders are interested in looking at other alternatives, or variations on the alternatives analysed, this is often resisted on the grounds of the additional time and cost involved. However, by analysing a wide range of alternatives, even if only a selection of the results are initially presented, such questions can be more easily addressed without delaying the process. An even more responsive approach, now well within the technical capabilities of most analytical tools, would be to provide a capability for stakeholders to choose the combination of inputs and assumptions that interest them, run the models, and view the results. This could be done on a dedicated computer to which the stakeholders have access, or via the Internet, as discussed below. In addition to being more responsive to the concerns of the stakeholders, this can also serve as an educational tool for both the stakeholders and decision-makers, allowing them to see how the results of the analysis vary with the assumptions made.

8.5.4 Opportunity to comment

It is becoming increasingly recognised that any major social decision, including airport planning issues, should be made in a way that affords those affected with the opportunity to make their views and concerns known. This can take the form of releasing draft documents with an invitation to submit written comments, or holding public hearings where stakeholders can present their views as well as question those involved in the technical analysis. Airport system planning studies will commonly have a technical advisory committee that will oversee the study as it proceeds and represent a broad cross-section of stakeholder interests. Consideration should be given to making the meetings of this committee open to the public, although not necessarily to allow public input. It is common in the US for such groups to set some time aside at each meeting for limited public comment or questions.

Naturally, any such forum will attract those who wish to pursue a narrow, self-interested agenda. In a sense, an airport authority itself is no different, although it may view its own agenda as being somehow superior to those of the surrounding communities or business interests. Certainly, most airlines have a very clear, very narrow, and decidedly self-interested agenda. By fostering a public debate between these various interests, it is to be hoped that not only will the final decisions be made in the best interests of the broader community, but that these decisions will receive broad political support that will allow them to be implemented.

Since airport system planning decisions take place on a wide stage, involving an entire region, or even a state or nation, many of the stakeholders are very dispersed, and may not have any history of involvement in the planning process. Special efforts will almost certainly be necessary to involve the relevant stakeholder groups. These may include newsletters or organising a speakers panel that is available to give talks at regular meetings of stakeholder groups. Special efforts to involving the media in the process can prove an effective way to reach a large number of people.

8.5.5 Role of information technology

Recent developments in information technology present new opportunities for information dissemination, and ways to involve stakeholders in the planning process. These techniques not only offer the potential to provide access to information in a more timely way, but to do so at considerably reduced cost and waste of resources.

Electronic publishing provides a means to provide access to documents on an as-needed basis. Standards now exist to allow users to download, browse, and print documents from Internet sites using personal computer software. Users can be sure that they are working with the latest version of the document, and by scanning the list of available documents can be made aware of other information that may be of interest. Since almost all documents are created in electronic form anyway, the additional work involved in making them available in this format is relatively modest, and in fact may be considerably less costly than responding to requests for copies of publications in printed form.

Internet web sites not only provide an effective means to distribute electronic documents, but can provide a wide range of other information in a timely way. Announcements of forthcoming meetings and the minutes of past meetings can keep stakeholders involved in the process, particularly those who are unable to attend some or all of the meetings. Increasing amounts of statistical information about the aviation system are available on a wide range of Worldwide Web sites. The ability to link web pages to information on other web sites both avoids the

216 *Strategic Airport Planning*

need for duplicative information, as well as ensures the currency of the information. This information not only helps stakeholders become informed participants in the process, but can greatly simplify the work of the technical professionals involved in the studies.

One of the best examples of this process is the availability of the US FAA Terminal Area Forecasts for each airport in the nation at *http://api.hq.data.faa.gov/apo_pubs.htm/*. This allows both aviation planners and the broader public to have immediate access to the latest updates of the forecasts as soon as they are available. The underlying traffic data for every airport that is used in developing the forecasts is also available from a related web site at *http://www.apo.data.faa.gov/*. This web site takes advantage of another feature of the web, that allows two-way communication between the user and the web site. Rather than download a vast amount of data, the site allows the user to specify the data of interest and only receive that. A large number of other documents relating to national aviation planning are available on FAA supported web sites, and can be found using the search capabilities on the FAA home page at *http://www.faa.gov/*. Extensive information on statistical data on the US aviation system is also available from the US Department of Transportation Bureau of Transportation Statistics web site at *http//www.bts.gov*. The site includes the FAA Statistical Handbook of Aviation and access to airline on-time statistics, together with information on other aviation data sources.

The California Department of Transportation has been funding the development of an Internet-based California Aviation Database, that is planned to include a large amount of statistical data on the state aviation system. A preliminary version of the database can be found at *http://www.its.berkeley.edu/nextor/cavd/*. Other state aviation programmes with extensive information available on their web sites include Virginia (*http://www.doav.state.va.us/*) and Washington (*http://www.wsdot.wa.gov/aviation/*). Some regional planning agencies in the US are beginning to make use of their web sites to support their aviation system planning programmes, such as the Puget Sound Regional Council in Seattle (*http://www.psrc.org/air.htm*).

CD-ROM disks offer yet another technology to distribute large amounts of information in a cost-effective way. The FAA publishes its annual Aviation Capacity Enhancement Plan on a CD-ROM, that not only includes the latest issue of the plan, but prior versions, a range of Airport Capacity Enhancement Plans that have been prepared over the years for various airports, and an extensive database of statistical information. Commercial vendors also use CD-ROM to distribute large amounts of aviation statistical data, including FAA pilot and aircraft registration data, airport physical characteristics, and airline financial and traffic information reported to the US Department of Transportation.

8.5.6 Summary

As experience is gained with the foregoing techniques, they will become better integrated into accepted planning practice. Expectations and standards will become established, as much by emulation as by fiat. However, this will require some thought and appropriate budgeting to ensure that adequate resources are available to establish and maintain these procedures. The design and implementation of an effective information dissemination programme requires professional skill, and is not something that should be done as a sideline when one of the planners has some time free, or left to a summer intern. Initially, there may be some resistance to devoting resources to what may be viewed as only "window dressing". Addressing these concerns will require attention to both a clear definition of what the information dissemination programme is attempting to accomplish and to developing ways to measure its effectiveness. This is consistent with recent efforts in the US, stimulated in part by the 1993 Government Performance and Results Act, for government agencies to begin asking how they can measure their effectiveness in performing their various functions, including planning.

9

UNITED STATES EXPERIENCE

As befits a nation with one of the largest airport systems in the world, as well as one that has played a leading role in both institutional and technical aspects of air transportation, the United States was among the first countries to establish a formal airport system planning process, and has the most experience with implementing such a process. Even so, as those involved in this process would be the first to admit, there is much that it not yet well addressed or even understood, and much needs to be done to improve the process.

This chapter provides an overview of the US airport system planning process and presents three case studies of specific system planning studies, one at the state level and two at the regional level. The intent is both to illustrate the structure and content of typical system planning studies as well as to provide a context in which to discuss a number of issues that represent some of the latest thinking on how to perform airport system planning. Thus in an important sense, these studies are far from typical. Sadly, most such studies are undertaken in a very pro forma way, that steps lightly over the more difficult issues and fails to grapple effectively with the underlying concerns. The three selected case studies attempted to face some of these issues head-on, and in so doing have tried to advance the state of the art in various ways. For this alone, they are worth studying. However, they are important for another reason. In each case, the planning process has encountered political pressures that have constrained the outcome, in ways that have compromised the ability to achieve the original objective. Thus they also present important object lessons in the pitfalls of the airport system planning process, and real-world conflicts with which it has to contend. Hopefully, by understanding what each of these studies has been able to achieve, and what it has not, future studies will both build on their accomplishments and find ways to overcome the difficulties that they encountered.

In selecting these studies, the authors faced the dilemma of what to leave out. With fifty states and hundreds of metropolitan regions, the number of airport system planning studies that could have been included is overwhelming. Many include interesting technical analyses or address critical issues, and their exclusion was motivated largely by the need to keep the chapter to a manageable length. Another real constraint was the difficulty of obtaining copies of the studies or supporting documents. In spite of the establishment of a formal system planning process at a national level, there is no central repository of system planning studies. While each state and regional office of the Federal Aviation Administration (FAA) will generally receive copies of system planning documents prepared within its jurisdiction, there is no common procedure to catalogue these reports, much less provide a means for distributing copies.

This last point has ramifications that go beyond considerations of preparing case studies. If those involved in airport system planning have no easy way to compare their proposed approach with what has been done elsewhere, the state of the art will advance slowly, mistakes and failures will be repeated over and over again, and there will be no real consensus on what constitutes "best practice" standards of study conduct. This is obviously deeply troubling if the results of those studies are then to be used to prepare higher level assessment of airport development needs. Of course, comparing study reports is not the only way that knowledge is shared. The larger consulting firms perform system planning studies for clients on a nationwide basis, and carry their expertise and learning from study to study. Also, planning methodology and sometimes results are presented from time to time at professional meetings and through articles and papers in appropriate journals. However, there are obvious limitations to reliance on consulting firms as the "institutional memory" of the process (not the least of which is that their personnel turn over quite rapidly and their role in any study is heavily circumscribed by their scope of work and budget), and system planning issues have generally taken a very distant second place in the literature to discussion of airport-specific developments.

Notwithstanding these concerns, over the past 30 years the FAA has developed an airport system planning process that defines procedures for conducting regional and statewide system planning studies, and has provided financial support through its Planning Grant Program to enable state and regional agencies to undertake those studies. The results of these studies, together with airport master planning studies, are then used by the FAA to provide input to the National Plan of Integrated Airport Systems (NPIAS), which attempts to summarise the airport development needs at a national level.

9.1 THE US AIRPORT SYSTEM PLANNING PROCESS

The evolution of the US airport system planning process was described in chapter 1. In its current form, system planning takes place at three distinct levels:

- national
- state
- regional.

This division of responsibility reflects both the geographical extent of the country, as well as the respective roles of different levels of government. In particular, while the Federal government traditionally has been responsible for regulating and facilitating interstate commerce, including regulation of most airline activity prior to the deregulation of the industry, the states have largely been responsible for both economic development and provision of the primary transportation network within the state. The growing recognition of the need for coordinated planning within the larger metropolitan areas, most of which are composed of multiple counties, cities and townships (the San Francisco Bay Area metropolitan region, for example, includes nine counties and over one hundred separate cities, ranging in population from over half a million to less than 10,000), has led to the creation of Metropolitan Planning Organizations with broad responsibilities for regional planning issues, including transportation planning.

This division of jurisdiction is reinforced by the flow of funding, particularly for much of the transportation system. Taxes are collected at the federal, state and local levels, but much of the federal spending on transportation programmes is accomplished by passing the funds down to the state and local governments through a combination of predefined allocation according to formulae in the enabling legislation and discretionary awards within specified programmatic guidelines.

The complex details of transportation policy and funding in the US are well beyond the scope of this book. However, some background is necessary in order to understand many of the factors driving the airport system planning process. One such factor is the distinction that is made between surface transportation programmes and air transportation. The creation of the US Department of Transportation in 1967 was intended to provide a coordinated approach to transportation policy and programmes, and indeed the Federal Aviation Administration has been an agency of the US Department of Transportation ever since, and technically reports to the Secretary of Transportation. However, aviation legislation and funding is controlled by different congressional committees from surface transportation, and thus in practice the coordination between air and surface transportation policy is virtually non-existent. This is reinforced by a funding process by which transportation tax revenues, such as fuel taxes and airline ticket taxes, flow into separate modal "trust funds" (such as the Highway Trust Fund or Aviation Trust Fund), and can only be used for the purposes specified in the authorising legislation. The passage of the landmark Intermodal Surface Transportation Efficiency Act in 1991 introduced a much greater degree of flexibility in funding transportation projects, but as the name of the act implies, this was restricted to surface transportation modes. This approach has been preserved in the 1998 reauthorisation of surface transportation programmes under the Transportation Equity Act for the 21st Century. However, not only have the congressional aviation committees shown no support for a similar approach to developing the air transportation system, but from time to time proposals surface to make the FAA independent from the US Department of Transportation.

9.1.1 National airport system planning

The Federal Aviation Administration performs two critical functions that shape its role in airport system planning: it operates the nation's air traffic control system, including most of the airport control towers, and it distributes the airport development funding from the Aviation Trust Fund. As the agency responsible for managing the flow of air traffic through a network of air routes linking every airport in the nation, it must cope with the congestion that results from inadequate airport capacity on a daily basis. At the same time, in its role of distributing capital development grants from the Airport Improvement Program (AIP), it is in a position to influence where additional airport capacity is provided.

However, its ability to do so is severely constrained by two factors. The first is that the allocation of a large proportion of the AIP funds available each year is predetermined by the enabling legislation. The second is that the FAA does not operate any of the airports itself. Most US airports are owned and operated by municipal airport authorities, while the rest are operated by regional, state or bi-state agencies (such as the Maryland Aviation Administration or the Port Authority of New York and New Jersey). Until 1987, the FAA operated Washington National and Dulles airports, but even these are now operated by the Metropolitan Washington Airports Authority, a regional agency with federal, state and local representation. Since the FAA does not own or operate the airports, it cannot decide whether to proceed with any particular capital expansion programme. It can only agree to provide funding for eligible projects that the airport authorities decide to initiate.

As discussed elsewhere in this book, the factors that determine whether an airport authority can embark on a capital development programme, particularly where this involves constructing new runway capacity, much less new airports, go far beyond whether financing is available for the projects. Even where an expansion programme has made it through the multiple steps of evaluation and consent and has been approved for construction, the constraints on available AIP funding are such that the federal funding is likely to only contribute to part of the cost, and the airport authority will have to raise much of the required funding from other sources.

These two constraints therefore shape the nature of the national airport system planning process, which comprises two separate activities. The first is an ongoing process to identify the future capital development requirements for the nation's airport system, ideally so that Congress can determine appropriate taxation levels to ensure that sufficient funds are available in the AIP to meet the needs. This has been formalised into the National Plan of Integrated Airport Systems (NPIAS), which is updated on a regular basis, and is discussed in more detail in the next section of this book. The second activity consists of a series of planning studies that are designed to enable the FAA to both anticipate the future demands that will be placed on the system, identify potential opportunities to expand the capacity of system, and assess the implications of these projects for the overall performance of the system. This second activity results in several documents that are updated on an annual basis, including the FAA Aviation Forecasts, the FAA Terminal Area Forecasts, and the FAA Aviation Capacity Enhancement Plan. The relationship between these studies and the NPIAS is also examined in the following section.

9.1.2 State aviation system planning

The state aviation system planning process is defined in FAA Advisory Circular 150/5050-3B (FAA, 1989). This advisory circular discusses the purpose of state airport system planning and describes the relationships between the various agencies involved in the process, as well as the need for public participation, and organisational aspects of undertaking a system planning study. The planning process is considered to comprise seven steps:

- Identify issues, goals and objectives
- Assess existing system
- Prepare unconstrained forecasts of demand
- Identify system deficiencies
- Identify and analyse alternatives
- Define a recommended system
- Prepare an implementation plan.

The analysis activities performed in the course of the planning study include developing an inventory of existing facilities and activity data and establishing a data base, preparing forecasts of aviation demand, and analysing the future performance of both the existing system and alternative system development plan against the forecasts of demand. The results of this analysis then serve to select the recommended system. The advisory circular discusses sources of data and analysis techniques, although in fairly broad terms and with more emphasis on process than technical details. It emphasises the need for continuous review and update of the state system plan.

One key function of the state airport system plan is to identify the existing and expected future role of each airport in the system, so that airport development can be undertaken to appropriate standards and adequate provision made for future expansion. While this has traditionally focussed on airport design standards, the need for compatible land use planning in the vicinity of airports is being increasingly recognised. This will become particularly critical where the role of the airport is expected to change, such as the introduction of jet service at airports that had formerly only received turboprop service or where a general aviation airport will be expanded to provide commercial air service. Another functional change can occur where former military airfields are being decommissioned and will become available for civil uses.

The recent trends in the general aviation sector in the US pose a significant challenge for the state airport system planning process. The existing planning process was largely defined when the level of general aviation activity was increasing and the principal issue was whether to expand existing airports or develop new ones. However, with general aviation activity in steady decline, capacity is no longer an issue at most general aviation airports and the issues have shifted to how many airports to try to preserve in order to ensure an economically viable system. New airports may still be required, particularly if existing airports are expected to close for other reasons or where growth in traffic at commercial service airports is creating pressures to relocate the general aviation aircraft based at those airports.

Another aspect that may need to be addressed at the statewide level is the relationship between airport system planning and surface intercity modes. Several states are actively considering the development of high speed rail systems, and others are upgrading or expanding conventional rail services. There are also prospects for new technology systems, such as tiltrotor aircraft or automated highways. Whether any of these new systems ever get built remains to be seen. However, their potential impact on the demand for air travel gets brought up in the airport planning process at all levels from master planning to state system planning. In order to address these issues in a coordinated way and to avoid duplicative analysis in every study, it may be helpful for the state airport system plan to address these issues in a way that can then be incorporated in other studies.

9.1.3 Regional airport system planning

Regional airport system planning addresses both metropolitan and non-metropolitan regions. The planning process in both cases broadly follows that defined for state airport system planning. Indeed, what is considered a region in one state may be larger than some other entire states. However, airport system planning in metropolitan areas introduces two factors that require particular consideration. The first is that these regions often include the large hub commercial airports and may be served by multiple air carrier airports. Since the larger hub airports are also usually the most congested, and expansion opportunities are limited, the provision of adequate airport capacity for future demand for air service can become a central focus of the system planning studies. The other factor is the relationship between airport system planning and urban transportation planning. Not only do large airports impose significant traffic loads on the regional highway system, but access to the airport is frequently constrained by highway congestion. Therefore airports need to address their intermodal ground access needs and the metropolitan planning agencies are concerned to mitigate the impacts of increasing levels of air travel. In many metropolitan regions there are also concerns about air quality, and airport system planning needs to address emissions by both aircraft and ground vehicles.

The airport system planning process for metropolitan regions is described in FAA Advisory Circular 150/5070-5 (FAA, 1970). A draft update of this advisory circular was released for comment in April 1996, but as of September 1998 had not been finalised. The planning process described in the advisory circular follows that described in the state airport system planning advisory circular, but the draft update provides considerably more discussion on the issues that need to be addressed at each step of the process, with particular reference to those aspects that arise in metropolitan regions. Even so, the advisory circular provides limited information on the technical aspects of performing the necessary analysis.

A critical issue that arises in many metropolitan airport system plans, and is discussed in the case studies later in this chapter, is the choice between continued expansion of the existing air carrier airports and the development of new airports to serve the region. The expansion of existing airports is often constrained by adjacent land uses or topography, such as rivers, shoreline, or terrain elevation. Thus while each increment of capacity expansion is more cost-effective at the existing airports than by developing a new airport, the marginal cost of expanding the airport system keeps rising until eventually the cumulative cost of fairly modest increases in capacity would have paid for an entirely new airport. However, determining the point at which a new airport can be economically viable, and convincing the airlines to provide service at that airport, presents a major analytical and institutional challenge. This is further compounded when the primary airport in a region is an airline hub and the air carrier airports in the region are not all operated by the same airport authority.

A second major issue that can arise in metropolitan airport system planning is the provision of rail access links to the air carrier airports. While there is often strong political pressure to provide rail access to these airports, partly as a way to reduce the amount of highway traffic generated by the airports and partly to improve the reliability of access travel times for air passengers, these projects are usually extremely expensive and require a major commitment of regional transit capital improvement funds. At the same time, the development of new airports on greenfield sites outside the urban area will almost certainly raise demands for the provision of a high speed surface transportation link. Thus analysis of airport ground access issues will form a significant component of metropolitan airport system planning studies.

9.2 THE NATIONAL PLAN OF INTEGRATED AIRPORT SYSTEMS

The National Plan of Integrated Airport Systems (NPIAS) is prepared by the Federal Aviation Administration to identify airports that are eligible to receive grants under the Airport Improvement Program, and to estimate the future airport development costs that are eligible for federal funding under the AIP over the subsequent five year period. The most recent update of the NPIAS was published in April 1995 (FAA, 1995) and covered the period 1993 to 1997. Thus by the time the update was published, there were less than three years remaining in the five year period. As of September 1998, a year after the five year period ended, the FAA was still working on the next update.

As of January 1993 there were 18,233 airports in the US, of which 5,534 were open to the public. Of the public use airports, 4,196 were publicly owned. The 1993-1997 NPIAS included 3,294 existing airports, of which all but 199 were open to public use, and 366 proposed public use airports.

9.2.1 Airport development funding requirements

The information on future airport development requirements is drawn from airport master plans and airport system plans prepared for state and local agencies. The estimated development requirements for the period 1993 to 1997 came to over US $30 billion, of which 79% was accounted for by airports with significant commercial service. Of the total, the 29 large hub airports accounted for 44%, the 39 medium hub airports accounted for 13%, and the

347 small hub and non-hub primary airports accounted for 20%. Projects were divided into eight categories:

- Safety and Security
- Reconstruction
- Standards
- Environment
- Terminal Building
- Access
- Airfield Capacity
- New Airports.

Airfield capacity enhancement projects accounted for 31% of the total, followed by projects to bring airports up to design standards, which accounted for 20%. Terminal building projects were the next largest category at 15%. However, while 49% of the airfield capacity projects and 77% of the terminal building projects were at large hub airports, 47% of the projects to bring airports up to design standards were at reliever or general aviation airports. Funding for new airports accounted for 11% of the total, of which 71% was for large hub airports.

The estimate of funding requirements for new large hub airports illustrates the difficulty of projecting future airport development capital requirements even five years into the future. Of the US $2.25 billion projected requirements, 20% was accounted for by the New Denver Airport, which was still under construction at the time, 49% was allocated for a new supplemental airport in the Chicago region, and 31% was allocated to a new air carrier airport for the San Diego region. As of 1998, neither of the two latter regions had even reached a decision on sites for these airports, much less acquired approval to start airport development. Given the uncertainties involved in the airport development process, a five year horizon is simply too short to be either meaningful or useful. Indeed, airport development projects can be thought of as falling into two categories. The first category comprises a large number of fairly small projects that can generally proceed without too much controversy. The funding requirements for these projects can be predicted fairly well. The second category consists of a relatively small number of large and frequently controversial projects: new major airports, new runways at major hubs, and major new terminal building expansion at primary airports. While the funding requirements for each of these projects can be estimated reasonably well, although the relative error is likely to be greater due to the small number of projects, the timing for those funds cannot. This suggests a need to treat these two categories of project differently. Although a rolling five year time frame may be adequate for the first category, a much longer time frame would be appropriate for the second category, at least 10 years and perhaps even 20 years. Of course, this would require an airport system planning process that would have an even longer time horizon, to allow for the lag between the planning studies and the resulting cost estimates becoming available for inclusion in the NPIAS. Unfortunately, this may require a planning process that goes well beyond the time frame that political decision-makers are willing to consider.

9.2.2 Airport system performance

The 1993-1997 NPIAS includes a section on the condition and performance of the airport system, addressing six aspects:

- Capacity
- Safety
- Aircraft Noise
- Pavement Condition
- Accessibility
- Financial Performance.

The treatment of each of these aspects varies in level of detail, reflecting the varying attention that it has received in past studies and the availability of comparable data. Aircraft delay is routinely tracked by both the FAA and the US Department of Transportation. The NPIAS considers an airport to be severely congested when average delays exceed 9 minutes per operation. In 1992, 7 airports were in this category, and this was expected to increase to 17 airports by 2002 if no new runways were constructed at those airports. The average delay per aircraft operation systemwide was estimated to be 7.1 minutes in 1992, and projected to increase to between 7.7 and 8.4 minutes by 2002.

The NPIAS notes the generally declining trend in aviation accident rates and states that it has not been possible to develop a statistically significant relationship between safety and airport capital investment levels, although it suggests that an increased emphasis that was given to the adequacy of airport marking, lighting and signage in airport inspections beginning in 1991 may have contributed to the subsequent reduction in the rate of runway incursions.

As of January 1993, 208 airports were participating in airport noise measurement and reduction programmes under Part 150 of the Federal Aviation Regulations, of which 155 had Noise Exposure Maps in compliance with the requirements of the programme and 135 had approved Airport Noise Compatibility Programs. The population exposed to high noise levels (presumably levels above 65 dB Day-Night Level) was reported as declining from about 7 million in 1975 to about 2.4 million by 1992. Due principally to the phase-out of Stage 2 aircraft, this was projected to decline to about 0.4 million by 2000.

Pavement condition information is collected as part of the FAA annual inspection of public-use airports. Runway pavements are classified as good, fair or poor, depending on the extent of unsealed cracks and joints, surface and edge spalling, and vegetation growing through cracks and joints. In 1993 some 68% of all runways at NPIAS airports were rated good and only 7% were rated poor. Runways at commercial service airports were in better condition, with only 3% rated poor. This represents an improvement in pavement condition over 1986.

The NPIAS measures airport accessibility in terms of the percentage of the population residing within 20 miles of a NPIAS airport. Using 1990 census data, there is a commercial service airport within 20 miles of 70% of the population, while 98% of the population live within 20 miles of some category of airport included in the NPIAS. Of course, the level of air service available at these commercial service airports varies widely, as does the use of air travel across the population. The 1993-1997 NPIAS also presents data for the distribution of air passenger origins and destinations with respect to travel time to the airport by highway and transit for three large metropolitan areas. Not surprisingly, a much higher proportion of air passengers can reach the airport by highway in a given time than by transit. However, the source of the data is not cited and it is unclear how transit is defined. Public transportation services at large airports typically include a range of public and private services, including door-to-door shared-ride vans and express buses to hotels and remote parking facilities. For many of these services, travel time estimates need to reflect service frequency, access time to the stop used, and any en-route stops or circuity to pick up other passengers. On the other hand, there are cost differences between different access modes that may make some public modes appear much more attractive to the traveller than is suggested by a simplistic comparison of travel times.

Financial information for different categories of airport was estimated from the results of a survey of airport revenues and expenditures conducted by the American Association of Airport Executives. Survey responses were obtained from 196 airports, including 81% of the large, medium and small hub airports. The data appear to suggest that expenditures exceed revenues for most categories of airport, although the results are distorted by the inclusion by some airports of construction costs rather than debt service while others included depreciation as an operating cost. Overall, in 1992 the 529 commercial service airports were estimated to have incurred capital expenditures of US $4.8 billion and operating expenditures of US $3.9 billion, and had revenues of US $8.6 billion, including federal and state grants. The 2,932 reliever and general aviation airports included in the NPIAS were estimated to have incurred capital expenditures of US $601 million and operating expenditures of US $420 million, and had revenues of US $1.05 billion, including US $487 million in grants. Since it is unclear whether the apparent shortfall in revenue at commercial service airports is an artefact of the survey methodology or the accounting conventions used by the airports, and there is no attempt in the NPIAS to link the financial data to any operational performance data, it is not clear what useful conclusions can be drawn from this information. It is evident that the AIP grants play a major role in funding capital development, particularly at smaller airports. However, whether these airports would be able to fund their capital development needs some other way if the AIP funds were not available, or even whether they would incur those development costs in the first place, cannot be determined from the information in the NPIAS.

Projections of the future performance of the airport system, and indeed of the larger aviation system of which it is part, clearly depend on two factors. One is the growth in traffic and the other is the investment in airport facilities and related programmes. Estimates of future delay levels are prepared as part of the annual FAA Aviation Capacity Enhancement (ACE) Plan (FAA, 1997d). These estimates are based on forecasts of traffic demand prepared as part of the annual FAA aviation forecasts (FAA, 1998a) that project the growth in aviation activity at a national level for the coming 12 year period. These forecasts in turn serve as the basis for the FAA Terminal Area Forecasts (FAA, 1997c) that allocate the national activity to individual airports.

In principle, Terminal Area Forecasts (TAF) identify the future demands at each airport, the ACE Plan identifies the capacity benefits of specific airport improvements, and the NPIAS determines whether adequate funds will be available to implement these improvements. There are clearly feedback effects that should exist in this process. The traffic levels at a given airport will depend on the capacity enhancements that are implemented and the priorities assigned to these projects should reflect their relative benefit in terms of delay reduction. In practice, the three activities are undertaken independently by three different offices within the FAA, with no explicit feedback between the activities. In part this reflects the fact that the timing of most capacity enhancement projects depends on factors largely outside the control of the FAA, and in part the fact that the NPIAS only considers the total capital development needs over a five year period. Not only is the timing of these projects not addressed, but the time frame of the NPIAS is too short to be of much use in developing either the ACE Plan or the TAF.

9.2.3 Airport Capital Improvement Plan

In parallel with the NPIAS, in 1990 the FAA began to develop an airport capital improvement planning process (FAA, 1997b). The goal was to identify the highest needs and priorities for airport development, and to move AIP funding decisions from a process based on the historical distribution of funds to one based on needs as determined from a national priority rating system. This process received additional stimulus from the Government Performance and Results Act of 1993, which required federal agencies to establish objective measures of the effectiveness of their programmes, and Executive Order 12893 in January 1994 that addressed Principles for Federal Infrastructure Investments. The Airport Capital Improvement Plan (ACIP) is intended to provide a comprehensive documentation of airport development needs and funding plans. FAA regional offices and Airports District Offices (ADOs) compile ACIP information from states, airport sponsors, and other sources into a regional ACIP. The regional offices then review funding requests and make initial AIP funding decisions based on national guidance, timing of the need for funds, region-wide demand for funds, and other considerations, including projects of extraordinary need that require special funding consideration. The regional ACIPs are then analysed by the FAA Office of Airport Planning and Programming with respect to set-asides and other requirements established by AIP legislation, the purpose, goals, and objectives of the AIP, and annual funding appropriations. The results of this analysis are used to make the AIP funding allocation decisions.

Central to this process is the use of the National Priority System (NPS) to provide a priority rating for projects. This priority rating is given by the following formula (FAA, 1997b):

$$\text{Priority Rating} = 0.25\ P * (APT + 1.4\ P + C + 1.2\ T)$$

where APT, P, C, and T are codes assigned to airport type, project purpose, airport component, and project type respectively. The airport code is assigned between 2 and 5 points depending on the airport type and size. The other three codes are assigned points between 0 and 10 depending on the nature of the project. Project priorities are developed in a two-stage screening process. The first stage uses the NPS rating to determine a set of candidate projects. Some projects that do not make candidate status on this basis can be considered based on a justification using qualitative factors. The second stage uses such qualitative factors as benefit-cost analysis, risk assessment, system impact, and state and local priorities to determine the final priorities.

The idea that benefit-cost analysis and risk assessment are considered "qualitative" factors, while assigning points based on a classification of the type of project and the size of the airport is somehow "quantitative," demonstrates both the difficulty of defining a process that is able to assess the relative contribution of specific projects to the overall goals of the AIP, as well as the desire to retain a large element of funding discretion in deciding the final priorities.

The points assigned to different categories of airport component and project type also contain some rather surprising implied values and categories of project. New airports are only assigned 4 points in the component code, whereas runways (presumably at existing airports) are assigned 10 points. Funding for "construction" is assigned 10 points for project type, while funding for a "people mover" project is only assigned 3 points. Apparently provision of a people mover is not considered construction.

9.3 THE CALIFORNIA AVIATION SYSTEM PLAN

With a 1996 population of 32 million, California is the most populous state in the US and extends 700 miles from Mexico to the Oregon border and between 200 and 250 miles from the Pacific coast to the Nevada border. It includes two of the largest metropolitan areas in the US, the Los Angeles region with a population over 15 million and the San Francisco Bay Area with a population of more than 6 million, and two more with populations over a million, San Diego with a population well over 2 million and Sacramento. The Los Angeles region is served by five air carrier airports and the San Francisco Bay Area is served by three. As of 1997 there were 256 public use airports in the state, of which over 30 had commercial air service. The so-called California Corridor between the Bay Area and Southern California is one of the densest air travel markets in the world, while Los Angeles International and San Francisco International airports are the two principal international gateways for air travel between the US and the Pacific Rim countries in the Far East, Southeast Asia and Australasia. Thus aviation plays an essential role in the state's transportation system.

Statewide planning for aviation in California began in 1968 and the first California Aviation System Plan was completed by the California Department of Transportation (Caltrans) in 1981 (Caltrans, 1998a). This was updated in two phases between 1987 and 1991. The Phase I update was completed in 1987. Phase II of the update addressed airspace, air cargo and ground access, reports on which were completed in August 1991. A Policy Element, updated in coordination with staff of the California Transportation Commission (CTC) and the CTC's Technical Advisory Committee on Aeronautics, was adopted by the CTC in October 1991. During the preparation of the 1987 update, the Caltrans Division of Aeronautics (later renamed the Aeronautics Program) began to develop a process to coordinate the statewide aviation system planning with input from the state's 43 Regional Transportation Planning Agencies (RTPAs).

In 1990 the California Legislature passed Senate Bill 707 that modified the state's Public Utilities Code to require the California Aviation System Plan (CASP) to be updated every five years and contain the following elements (Caltrans, 1998a):

- Background and Introduction
- Statewide System
- Air Transportation Issues
- Capital Improvement Program
- Regional Plan Alternative
- State Plan Alternative
- Regional/Statewide Comparison
- Summary and Conclusion
- Any other elements chosen by Caltrans and the RTPAs.

The initial versions of the CASP were prepared by Caltrans Aeronautics staff with support of consultants. However, the process of updating the CASP has evolved to include input from the RTPAs, and eventually to coordinate the preparation of the Regional Airport System Plans with the CASP process, so that the CASP can directly incorporate the results of the regional airport system planning process. This coordination is facilitated by two activities. The first is the establishment of the RTPA Aviation System Planning Committee (RASPC) that includes representatives of Caltrans, the RTPAs, the FAA, the CTC, the military, and other stakeholders in the aviation system planning process. This committee meets several times a year, and serves to keep the RTPAs informed of aviation system planning issues and activities at the state level, keep Caltrans informed about activities at the regional level, and provide a vehicle for regional feedback on issues affecting the CASP. Caltrans Aeronautics staff serve as the secretariat to the RASPC and the minutes of the committee meetings provide a continuous record of the status of the statewide aviation system planning process.

The second activity is a coordinated approach to programming FAA aviation system planning grant funds under the federal Airport Improvement Program. Formerly, each regional agency and Caltrans itself submitted applications to the FAA for these funds, and the FAA regional office was left to decide between the competing applications. The resulting allocation of funds was subject to a certain amount of lobbying by the various agencies and was largely shaped by the FAA's own view of the relative priorities. Under the current process, which has come to be termed the "Plan for Planning," regional agencies are encouraged to submit their proposed planning studies to the RASPC in advance of submitting applications to the FAA. These are then incorporated into a multi-year rolling plan based on the priorities determined by the committee. The FAA regional office has largely followed this plan in allocating funds for system planning studies, while reserving the right to fund other applications for studies that the FAA believes to have a higher priority. This provides some flexibility to respond to critical issues that emerge at short notice, or address issues where federal and state priorities differ. The process provides the regional agencies with some assurance of when their studies will be funded, which allows them to develop their work programmes several years in advance, and prevents wasted effort submitting grant applications year after year without knowing what chance they have of being funded.

Thus the current vision of the CASP is that it is a bottom-up process, in which the regional airport system plans provide a framework within which individual airports prepare and update their airport master plans, and merge the results of those master plans into a regional plan. These regional plans then form the building blocks of the statewide plan. Through appropriate coordination, the CASP becomes in effect the sum of all the regional plans. In practice, information flows both ways. The timing of individual airport planning studies, regional system planning studies, and work on elements of the CASP make this a continuously evolving process. The elements of the CASP allow the regional agencies to put their own airport system planning studies into a statewide context, as well as avoid duplicative work defining aviation system planning issues and assembling technical information to support the planning process.

9.3.1 Technical support

The state's 43 RTPAs vary widely in size and their ability to undertake aviation system planning. The larger Metropolitan Planning Organizations (MPOs), which serve as the RTPAs for the metropolitan areas, have dedicated airport planning staff, although often only one person who may also have other duties. However, the smaller RTPAs may have no staff with any technical experience in aviation planning. Because of these staff constraints, the larger agencies rely heavily on consultants to perform system planning studies, and the smaller agencies have difficulty obtaining funds to perform any aviation system planning at all. Thus an important role of Caltrans Aeronautics staff and the CASP process is to provide technical support to the regional agencies.

One area where this has been addressed is in the development of a coordinated approach to aviation forecasting methodology. Forecasts of future demand for air travel and general aviation activity are central to the aviation system planning process. These forecasts establish the adequacy of existing facilities, the need for new facilities, and provide the basis for the evaluation of alternative plans. If the forecasts fail to anticipate likely future demand, then the other elements of the system plan are at best of limited value. There is also the concern that in a bottom-up process, in which the statewide system plan builds on the regional system plans, inconsistent forecasting methodology at a regional level could result in gross underestimates or overestimates of demand at a statewide level.

Therefore Caltrans sponsored a study of forecasting methodology by the Institute of Transportation Studies (ITS) at the University of California, that proposed a framework that could be adopted by the regional agencies to develop aviation forecasts on a consistent basis (Gosling, 1994). This study included a survey of various Caltrans offices, RTPAs, and airports to determine the needs of the users of the aviation forecasts contained in the aviation system plans in terms of the information that these users required. The survey responses were then used to define the level of detail that these requirements impose on the forecasting methodology. It was found that a considerable mismatch existed between the user requirements and the sophistication of the forecasting techniques typically used.

For example, one application of aviation demand forecasts is to estimate future traffic on the surface transportation system generated by the airports in a region. This not only requires estimates of modal shares, but some idea of the pattern of origins and destinations within the region. Forecast techniques commonly used simply project total enplaned passengers, and provide no information on the composition of the traffic (such as visitors to the region versus travel by residents) which could be used to estimate changes in access mode use or the distribution of trip ends. The forecasting methodology study proposed a demand modelling framework to address these issues. However, it was also recognised that the provision of appropriate analytical tools, and training in their use, would be required to enable most of the RTPAs to apply these techniques.

9.3.2 The Central California Aviation System Plan

The first application of the shift in philosophy of the state role in aviation system planning toward the provision of technical support to regional agencies occurred with the preparation of an aviation system plan for the Central California region. This region stretches down the Central Valley of California from the Sacramento metropolitan area in the north to the Bakersfield metropolitan area in the south, and includes 13 counties, with a combined population in 1996 of 4.75 million. In 1995, the region was served by eight airports with commercial air service, the busiest of which, Sacramento Metropolitan, enplaned over 3 million passengers. The next busiest, Fresno Air Terminal, enplaned just over 500,000 passengers, while the third busiest, Bakersfield Meadows Field, enplaned a little over 100,000 passengers. All the others each enplaned less than 25,000 passengers, mostly on regional airline service to the state's major airports. The region also contained a further 60 public use general aviation airports, of which four had no based aircraft.

The ability of the ten RTPAs in the region to perform aviation system planning varied widely. Therefore Caltrans Aeronautics Program obtained an FAA planning grant to prepare a joint regional aviation system plan in conjunction with the RTPAs. A Technical Committee to undertake the Central California Aviation System Plan (CCASP) was formed with representatives from each of the RTPAs, selected airports, and other agencies in the region, and supported by Caltrans Aeronautics staff. Funds were provided to each of the RTPAs to prepare selected CCASP elements for their respective areas, following a model developed by the CCASP Technical Committee. Forecasts for each of the airports in the region were prepared by a consultant under contract to Caltrans, who applied a modified version of the forecasting methodology defined by the earlier ITS study (ICF Kaiser, 1996).

Over the four year duration of the study, each of the RTPAs prepared a Regional Aviation System Plan for their county or region that contained the following elements:

- Background and Introduction
- Inventory
- Forecast
- System Requirements
- Financial Plan
- Action Plan.

Caltrans Aeronautics Program staff then integrated the information from these regional plans into a summary Central California Aviation System Plan.

9.3.3 Multimodal transportation planning

One important function of the state and regional aviation system plans is to provide a means to link airport planning in the state to planning for the broader transportation system. This is complicated by the separate federal legislation governing airport development and that governing surface transportation. Under state law, each of the RTPAs is responsible for preparing a Regional Transportation Plan (RTP) that identifies regional goals and transportation improvements to be implemented over the next 20 years. These RTPs are updated periodically and are required to include three elements: a policy element, an action element, and a financial element. There is no requirement to address airports as such, but many of the RTPs include an aviation element, typically based on a regional airport system plan where this exists.

The RTPs are implemented through Regional Transportation Improvement Programs (RTIPs), which are updated every two years and define a specific programme of projects. These projects are programmed by the CTC and incorporated into the State Transportation Improvement Program (STIP), that in turn forms the basis of the Federal Transportation Improvement Program (FTIP) for the state and identifies all the federally funded surface transportation projects at the state and regional level. There is no requirement that federally funded aviation projects be incorporated in the FTIP. However, surface transportation projects for improving airport ground access must be included in the FTIP if they are to receive federal funds, and in the STIP if they require state funds.

State legislation (Senate Bill 45) passed in May 1998 changed the project programming process, consolidating nine categories of project into two, and shortening the programming of projects from seven years to four. While the legislation did not change the allocation process for funds in the State Aeronautics Account, it did add a requirement that all RTPs should address the coordination of aviation facilities and services with other elements of the transportation system. In addition, the RTPs in any region that contains a primary air carrier airport shall include an airport ground access improvement programme. The legislation requires that these programmes give highest consideration to mass transit projects and authorises the RTPAs to recover the costs of preparing or updating these programmes from the airport operators if federal funds are not available to cover these costs.

9.3.4 The Capital Improvement Program

Following the creation of the California Transportation Commission and the STIP process in 1978, airport projects to be funded with state funds were to be included in the STIP. In 1990, changes in state law removed the Aeronautics Capital Program from the STIP and established it as a separate four year programme. Other legislation (Senate Bill 707) that established requirements for the CASP also directed that the CASP include a 10 year Capital Improvement

Program (CIP), based on adopted airport master plans with the information prepared by the RTPAs for submission to Caltrans Aeronautics Program for inclusion in the CASP. The CIP was to be updated biennially. In November 1993, the CTC changed the Aeronautics Capital Program (ACP) to a three year programme, to be developed from projects included in the CIP using a project priority methodology adopted by the CTC at that time and revised in April 1997.

The project priorities are determined through the use of a "Project Evaluation Matrix." This assigns points to each project on the basis of two criteria (Caltrans, 1998b). The first criterion defines the Project Type, using 25 categories such as "New Pavement" or "Security Fence." These categories are divided into eight groups:

- Primary Runway
- Other Runway/Taxiway
- Ramp/Apron
- Navaids
- Safety
- Acquire Land
- Planning
- Other.

Each category is assigned a number of points that range from 3 for Automated Weather Reporting to 10 for Seal/Overlay/Rehabilitate Primary Runway.

The second criterion defines the Project Purpose, and has five categories:

- Safety
- Planning Documents
- Reconstruction/Standards
- Upgrade
- Miscellaneous.

Each of these categories is also assigned a number of points that range from 5 for Miscellaneous projects to 20 for Safety projects. Project points are then determined by multiplying the project type points by the project purpose points. Finally additional points are added based on the number of based aircraft and annual aircraft operations at the airport. Both based aircraft points and annual operations points range from 5 to 30 in six steps.

It is clear that this prioritisation process takes no account of the cost effectiveness of different projects. The project points depend only on the classification of project and the activity at the airport. It also tends to favour projects at busier airports. While this reflects the source of the funding, since airports with more activity generate more aviation tax revenues, it has the effect of using a large proportion of state funds to pay for projects at those airports that are in the best position to generate their own revenues. This raises important questions about the objectives of the ACP.

The CIP includes projects for which airports have applied for federal funds, whether or not they have also applied for state funds. In some cases, airports apply for state funds to provide the non-federal match required to receive federal funds. Although the intent of the CIP is to identify capital requirements for a 10 year period, it is clear from the projects submitted that many airports have only defined their project requirements a year or two into the future. Also, most airports have only identified projects for which either state or federal funding has been requested. Thus many projects that airports intend to fund entirely with local funding sources, whether directly from revenues or by issuing bonds, are not included. In the 1998 CIP update (Caltrans, 1998b), neither Los Angeles International nor San Francisco International airports show any locally funded projects, although at the time San Francisco International was in the middle of a two billion dollar expansion programme! Thus the CIP provides a somewhat distorted view of future capital requirements of the airport system in the state.

9.3.5 Evolution of the state role

The state aviation system planning process performs three broad functions. The first is to provide an analytical basis to guide the allocation of state and federal funds for airport development, particularly in addressing the needs of the state's general aviation airports. The second function is to assess the future needs of the state for airport facilities and ensure that these needs are being adequately addressed. The third function is to provide a framework to incorporate aviation within the state's multimodal transportation planning process, which is primarily oriented to surface transportation issues over which the state has much more control.

The CASP process appears to be working fairly well in performing the first role, although the bottom-up process tends to result in a disconnect between the system planning activities at a regional level and the project priorities in the CIP at a state level. The success of the CASP process in addressing the second and third functions is less clear. The principal airport development needs facing the state concern the provision of air carrier airport capacity in the three largest metropolitan areas, all of which are facing severe congestion at the primary airport. Yet the airport authorities in each region are pursuing their own development plans with very little heed to the regional airport system planning process, which is largely forced into a position of simply documenting and facilitating these plans. The bottom-up approach to the CASP means that the state is in no better position than the MPOs to either anticipate or influence these decisions. Finally, efforts to integrate aviation into statewide transportation planning are still limited by the lack of an effective mechanism to link decision-making across the modes. With airport and surface transportation funding coming from entirely separate sources, there is little motivation to do much more than pay lip-service to the concept, while continuing to take decisions on a modal basis. Thus while the state is a major player in the development of the surface transportation system, it plays almost no role in development decisions concerning the air carrier airports in its largest metropolitan areas. Whether this is a cause for concern will depend on how effective these regions are at meeting their own airport capacity needs without state intervention.

9.4 THE MINNEAPOLIS/ST PAUL DUAL-TRACK PROCESS

The Minneapolis/St Paul International Airport (MSP) is not only the primary airport serving the twin cities of Minneapolis and St Paul, Minnesota, but is the headquarters and principal hub of Northwest Airlines. It occupies a fairly constrained site, surrounded by residential and commercial development to the north, west and south, and the Minnesota River to the east (Figure 9.1). Aircraft noise affects large numbers of homes in the communities to the northwest and southwest of the airport, as well as lower density areas to southeast of the airport beyond the Minnesota River. Expansion of the existing main terminal complex is constrained by its location between the two parallel runways, the crosswind runway to the northwest and the State Route 5 freeway to the southeast (Figure 9.2). Development options are further constrained by the Fort Snelling National Cemetery on the south side of the airport and Fort Snelling State Park on the northeast side.

Following airline deregulation and the resultant restructuring of the US airline industry, with Northwest Airlines (like most of the major carriers) adopting a hub-and-spoke network, the airport experienced a rapid growth of both passenger traffic and aircraft operations through the first half of the 1980's. By 1985 the airport was the 14th busiest airport in the US in terms of passenger enplanements, with traffic growth outpacing national trends and delays becoming an increasing concern. In 1987 the Minnesota Legislature directed the Metropolitan Council of the Twin Cities Area (the metropolitan planning organisation for the Minneapolis/St Paul region) to asses the long term adequacy of the airport to meet the aviation needs of the Twin Cities area through the year 2020 (Metropolitan Council, 1988d). In April 1987 the Council appointed a 35-member Advisory Task Force to undertake an Airport Adequacy Study. The task force included representatives of the aviation industry, federal, state and regional agencies, the business community, affected communities, and the general public.

The Airport Adequacy Study lasted a year and a half, and included the preparation of background reports and working papers by a consultant team and several panels of national and local experts (Metropolitan Council, 1988c). The task force concluded that its analysis indicated a need to take decisive action to plan for future airport capacity for the region. However it faced the dilemma common to many such situations: should the region continue to invest in the existing airport with all its constraints or should it plan to develop a new airport on a less constrained site. This decision was complicated by two factors. First, any new airport would take many years to develop and additional capacity was urgently needed in the meantime. Second, the need for a new airport would depend on future traffic growth, the extent of which would remain unclear for some time. While aircraft operations at the airport had grown rapidly from 1981 to 1985, they had subsequently declined. Was this a minor irregularity in a generally steady growth trend, or did it indicate that the surge of growth following airline deregulation was essentially over? The task force recognised that continuing to invest in the existing airport if steady traffic growth were to continue was not a viable long term option. However, constructing a new airport could have serious financial consequences if in fact the traffic growth did not materialise.

238 *Strategic Airport Planning*

Figure 9.1: MSP 2005 noise contours - No action alternative

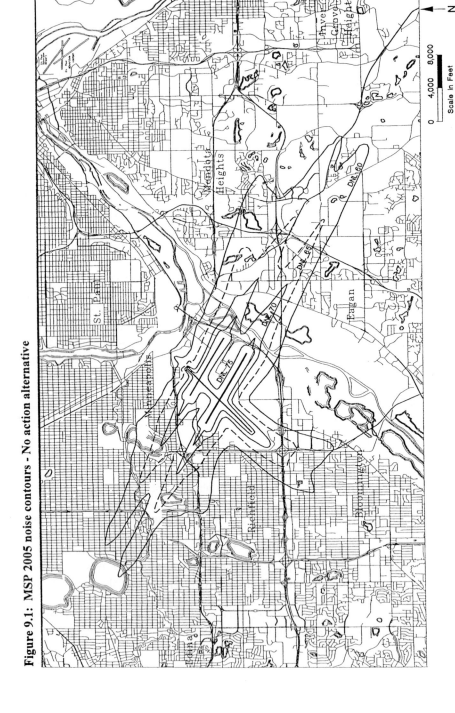

Source *Metropolitan Airports Commission/Federal Aviation Administration, Dual Track Airport Planning Process - Final Environmental Impact Statement and Section 4(f) Evaluation, May 1998.*

Figure 9.2: MSP airport configuration - No action alternative

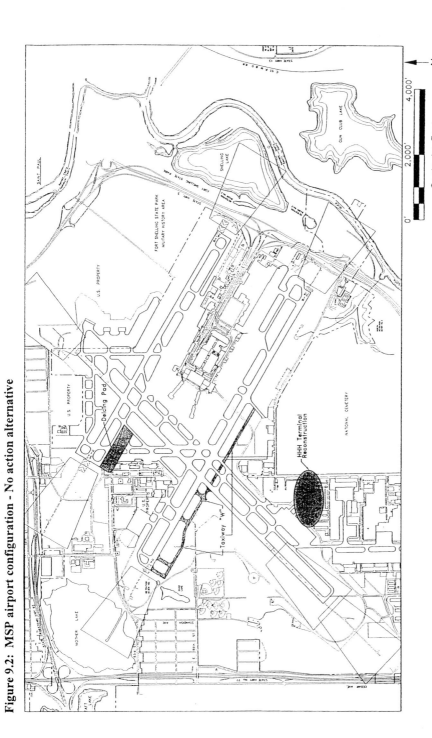

Source: Metropolitan Airports Commission/Federal Aviation Administration, Dual Track Airport Planning Process - Final Environmental Impact Statement and Section 4(f) Evaluation, May 1998.

Therefore the task force recommended what it termed a dual-track strategy, that involved provision of near-term capacity enhancements at MSP while keeping the new airport option open by selecting a site and banking the necessary land (Metropolitan Council, 1988a). It was recognised that the process of selecting a site and acquiring the land would take many years, by which time the need for a new airport might be clearer. If at that point it was apparent that further capacity enhancement at MSP could allow the airport to serve the traffic adequately for some time, the banked land would continue to be held for long term future airport development. On the other hand, if traffic growth was such that adequate expansion of MSP was no longer viable, the region would be in a position to develop the new airport and transfer the activity from MSP.

9.4.1 The dual-track strategy

The cornerstone of the dual-track strategy was to be an annual contingency planning process to analyse timing, magnitude and cost-effectiveness of public investment for both tracks of the strategy (Metropolitan Council, 1988d). The first track (Track A) would involve making capital and operational improvements at MSP. The second track (Track B) would involve identifying a search area, then locating and acquiring a site for a new major replacement airport for the metropolitan area. The Metropolitan Council would prepare an annual assessment, with participation of the Metropolitan Airports Council (MAC), the operator of MSP and several other smaller airports in the region, to ensure a balance between the tracks and to avoid either overinvestment at MSP or a premature move to a new airport.

The Metropolitan Council accepted the recommendations of the Airport Adequacy Study task force that expansion of the capacity of MSP (Track A) should begin immediately, with the implementation of the current airport capital improvement programme. This included an extension of one of the existing runways (Runway 4-22), various ground access and terminal improvements, and implementation of a Part 150 noise abatement programme. Track A also included enhancement of the regional system of reliever airports to accommodate corporate and general aviation traffic diverted from MSP, and implementation of demand management strategies to maximise the capacity of MSP. It was also proposed that if the current traffic growth projections were to materialise, a new north-south runway would be constructed in the 5 to 10 year time frame.

This latter project was clearly the most controversial element of Track A, representing not only a major investment in its own right, but allowing the decision to construct a new airport to be postponed further into the future. The proposed new runway would be located on the west side of the existing airport site, and would be restricted to operations to and from the south. This restriction would be necessary to prevent operations on the new runway from reducing the capacity of the existing runway system, but would also have the effect of avoiding the noise impacts from overflights affecting residential communities to the north of the airport.

9.4.2 Addressing uncertainty in the planning process

Apart from explicitly recognising the tradeoffs between continued expansion of MSP and development of a new airport, the Airport Adequacy Study was noteworthy for another aspect of its analysis: the approach to uncertainty in forecasting future traffic levels. It was recognised that decisions on whether to expand MSP or develop a new airport depended not only on the extent of future traffic growth, but when that growth occurred. On the one hand continued investment in expanded capacity at MSP would be wasted if rapid traffic growth forced the need to develop the new airport earlier than expected. On the other hand, constructing a new airport could be a financial disaster if the traffic growth did not materialise.

Therefore the traffic forecasts were developed using an approach that explicitly recognised the inherent uncertainty in the forecasts, which were expressed not as a "best guess" projection, but as a *probability distribution* for a given future year (Metropolitan Council, 1988a,b). This approach not only provided a direct measure of the uncertainty in the forecast, but allowed such statements as:

- Twenty years from now, there is a one-in-three chance that the number of passengers will have doubled (exceeding 32 million each year) but an 84% chance that they will have doubled within 30 years.

- Building the north-south runway offers about a 70% chance of meeting the expected demand in 10 years, but only an 18% chance of meeting demand in 20 years.

    (Metropolitan Council, 1988b).

The probability distributions were developed through a two stage process. In the first stage a set of baseline forecast models were developed that related the traffic measures to be forecast to socioeconomic factors and airline service variables, such as average aircraft size. The Airport Adequacy Study based these models on the traffic forecast that had been prepared as part of an ongoing master planning study being performed for MSP at the time. These models were revised in subsequent forecast analysis performed during the dual-track planning process (MAC/Metropolitan Council, 1992) and a series of forecast models were developed to predict various components of the forecast, comprising:

- Air carrier originations, enplanements, and operations;
- Regional air carrier enplanements and operations;
- Scheduled international and non-scheduled activity;
- Air cargo operations;
- General aviation and military operations.

The structure of these forecast models followed conventional practice. The forecast of domestic scheduled passenger originations on certificated air carriers was based on an econometric relationship between historical traffic and regional employment, per capita income, and national average airline yield. Enplanements were forecast by assuming a continuation of the historic trend in the ratio of enplanements to originations, to the average level experienced at other hub airports, after which it would remain constant. Aircraft departures were then projected based on assumptions of the future trends in average aircraft size (in seats) and load factor.

Regional carrier enplanements were forecast based on an econometric relationship between enplanements and regional employment, with a dummy variable to account for the effect of code-sharing between Northwest Airlines and its regional airline partners. Aircraft operations were forecast on the basis of assumed future trends in aircraft size and load factor. No attempt appears to have been made to separate the effects of local traffic to and from the Twin Cities and feed traffic connecting to Northwest Airlines or the other large air carriers.

The forecast of scheduled international traffic was based on a separate study of the potential market for international service from MSP. Non-scheduled traffic and air cargo were forecast on the basis of empirical relationships involving metropolitan income. foreign exchange rates, and regional employment. General aviation operations were forecast on the basis of assumptions about the change in based aircraft, with the number decreasing at MSP but increasing if the new airport is constructed, and the use of FAA national forecasts of the ratio of operations to based aircraft. Military operations were assumed to remain constant.

In the second stage, a risk analysis process was applied to these forecast components, in which a series of expert panels provided information on the expected probability distributions of the key input variables. These distributions were then used in a Monte Carlo simulation to estimate the distributions of the forecast output variables.

The expert panels were asked to review preliminary forecasts of the independent variables that were prepared by the forecast consultant, and provide their assessments of the values in future years in terms of the median value, together with the values that they thought there was a 10% chance the actual traffic would be less than or would exceed (the lower and upper deciles of the distribution). These estimates were then used to create probability distributions for each variable. It was not necessary for the experts to reach agreement, since their different estimates could all be incorporated in the distribution. The forecast distribution for the various traffic components in a given future year was then obtained by performing a large number of projections, sampling at random from the distributions of the input variables.

Although this process generates a distribution for forecast traffic levels, rather than assuming an arbitrary set of "high" and "low" assumptions, this distribution is only as good as the underlying models and the estimates of the distributions of the input variables provided by the expert panel. Even for experts, it is hard enough predicting what the median value of a variable like per capita income or airline yield will be in 10 or 20 years, much less estimating the values of the 10th and 90th percentile of the distribution. The other source of uncertainty is the forecast models used in the simulation. While these models may have provided a reasonably good fit to the historical data, that is a very different thing from saying that their independent variables are the only factors that influence air travel, or even that the relationships between air travel demand and those variables that existed during the period on which the models were calibrated will remain unchanged for the next 30 years. Unfortunately, the only way to find out is to wait and see.

9.4.3 Implementing the dual-track strategy

In 1989 the Minnesota legislature enacted the Metropolitan Airport Planning Act that defined the Dual Track Airport Planning Process and directed the Metropolitan Council and the MAC to determine whether the long term air transportation needs of the metropolitan area and state could best be met by expanding MSP or developing a replacement airport. The Act specified a seven year timetable that would result in the Metropolitan Council and the MAC submitting recommendations to the Legislature by July 1, 1996. The Act also defined a series of intermediate studies and reports that should be prepared by the Metropolitan Council or the MAC, including:

- A definition of long term aviation goals for the major airport facility in the region;
- A study of facilities requirements and a conceptual design plan for a major new airport, including an analysis of estimated costs and potential funding methods;
- Designation of a search area for the new airport site;
- Development of a Long-Term Comprehensive Plan for MSP to satisfy the air transportation needs of the region for a 30 year planning period;
- A study of policies on the reuse of MSP should a new airport be developed;
- Selection of a site for a new airport and preparation of a comprehensive plan for the development of the new airport.

The New Airport Conceptual Design Study and Plan was completed by the MAC in December 1990, and the Long-Term Comprehensive Plan for MSP was adopted by the MAC in November 1991. In December 1991 the Metropolitan Council designated a search area in Dakota County for planning the development of a new airport, and the Council completed the MSP Reuse Study in December 1992. The selection of the site within the search area and the development of the comprehensive plan for the new airport were handled through the preparation of an Alternative Environmental Document (AED) under the procedures of the Minnesota Environmental Quality Board. This required the publication of a Draft AED, public hearings, and a Final AED, with periods for public and agency review and comment on both the Draft and Final AED before a decision by the MAC. The Commission selected the site for the new airport in March 1994 and selected the New Airport Comprehensive Plan in April 1995. A similar process was followed for an update of Long-Term Comprehensive Plan for MSP, which was completed in February 1995. The MAC and FAA then prepared a Draft Environmental Impact Statement (DEIS) for the entire Dual Track Airport Planning Process, which was issued in December 1995 (MAC/FAA, 1995).

The MSP Long-Term Comprehensive Plan included a major reconstruction of the terminal facilities and construction of the proposed new north-south runway, as shown in Figure 9.3. The new terminal facilities would comprise a new landside building in the northwest corner of the airport linked to two new concourses on the site of the existing terminal by an underground people-mover. There would also be a remote parking facility at the southeast end of the terminal complex, served by the existing terminal access road. The new West Terminal would require entirely new access roads to connect with the existing freeway to the north of the airport.

The site selected for the new airport was located some 20 miles southeast of MSP in an area mostly of farmland southwest of Hastings (Figure 9.4). The New Airport Comprehensive Plan included provision for up to six runways, four aligned north-south and two east-west, with a terminal complex in the centre.

9.4.4 Abandoning the new airport option: the legislature intervenes

As could be expected, a wide array of organisations and individuals submitted comments in response to the December 1995 DEIS. Predictably, the communities in the area of the proposed new airport opposed the new airport option, arguing that the construction of a new airport would result in extensive loss of high quality farmland and induce significant development in a largely rural area outside the Metropolitan Urban Service Area boundary. Indeed, as documented in the DEIS, there was a well established legislative history in the Minneapolis/ St Paul metropolitan region of limiting urban growth and preserving existing farmland.

Perhaps more significantly, the comments submitted by Northwest Airlines not only supported continued expansion of MSP, but argued that the proposed new West Terminal was unnecessary and presented a concept study for an expansion of the existing terminal area that was termed Concept 6A (the MAC preferred alternative described in the DEIS had been designated Concept 6 in the prior Alternatives Evaluation). This plan was based on the same traffic forecast projections used in the DEIS and proposed adding 15 air carrier gates to the existing terminal concourses, resulting in the same number of gates as proposed for the new West Terminal in the DEIS. The plan also included a new 50,000 sq ft regional/commuter airline concourse and an automated people mover along the landside face of the Green Concourse that forms the north side of the terminal area. Although the plan discussed renovating the existing terminal buildings and some modest expansion of facilities, it did not present any operational analysis of the proposed concept nor explain how a projected increase in enplaned passengers of over 55% from 1992 to 2020 could be handled with a relatively modest increase in facilities. The proposed terminal expansion would expand the number of air carrier gates by about 20%, but increase the terminal building floor area by only about 10%. The proposed regional/commuter concourse would add another 4%, but this would not be usable by air carrier passengers. The number of parking spaces in the public parking structure would also be expanded by only about 20%. The two most significant features of this new concept were that is was estimated to cost only about US $145 million, compared to US $1.37 billion for the proposed West Terminal, and the fact that the expansion would predominantly serve the needs of the Northwest Airlines hub operation.

United States Experience 245

Figure 9.3: Proposed MSP development alternative

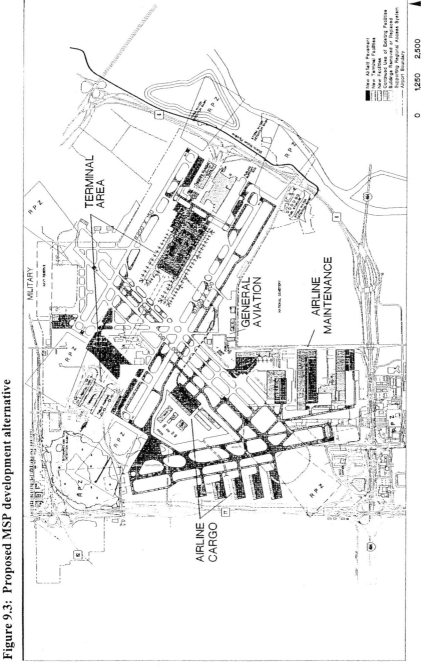

Source   *Metropolitan Airports Commission/Federal Aviation Administration, Dual Track Airport Planning Process - Draft Environmental Impact Statement, December 1995.*

246 *Strategic Airport Planning*

Figure 9.4: Location of new airport alternative

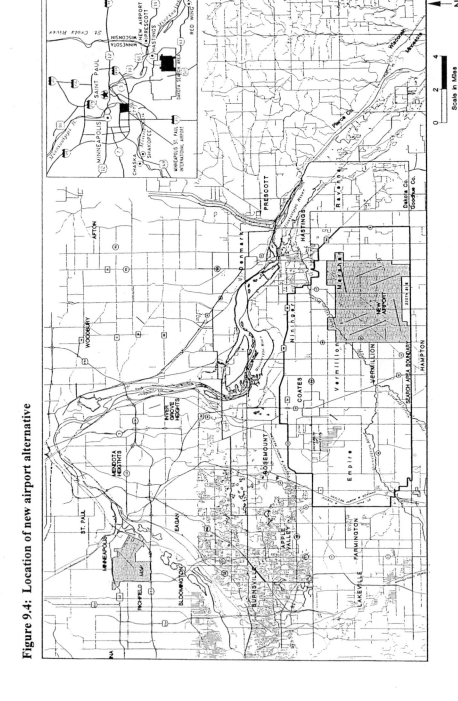

Source: Metropolitan Airports Commission/Federal Aviation Administration, *Dual Track Airport Planning Process - Final Environmental Impact Statement and Section 4(f) Evaluation*, May 1998.

In April 1996, before the Final Environmental Impact Statement (FEIS) could be issued responding to the comments on the DEIS, the Minnesota legislature passed a bill requiring the MAC to implement the 2010 Long-Term Comprehensive Plan (the preferred alternative in the DEIS) and eliminating the requirement to provide environmental or technical analysis of the new airport alternative in the FEIS for the dual track planning process. Since the whole point of the dual track process was to compare the merits of expanding the existing airport with those of developing a new airport, this was a remarkable decision to say the least. Moreover, the legislation prohibited the MAC from acquiring land for a major new airport or constructing a new passenger terminal on the west side of the airport without further legislative approval. It also prohibited the MAC from constructing a third parallel runway without first obtaining the approval of each affected city, defined as one that would experience an increase in the area located within the 60 Ldn noise contour as a results of operations using the runway. Interestingly, this requirement was not applied to cities affected by the proposed new north-south runway proposed as part of the preferred alternative.

9.4.5 Traffic growth overtakes the process

In May 1998, over two and half years after the release of the DEIS and two years after the legislation directing the MAC to pursue the Long-Term Comprehensive Plan as the preferred alternative, the MAC and FAA released the Final Environmental Impact Statement (FEIS) for the dual track planning process (MAC/FAA, 1998). The FEIS divided the proposed airport development projects into two phases: the 2010 Long-Term Comprehensive Plan (LTCP) and the 2020 Concept Plan. The new north-south runway and associated facilities, including some modest expansion of the existing passenger terminals formed the 2010 LTCP (Figure 9.5). The new West Terminal formed the basis of the 2020 Concept Plan (Figure 9.6), which the FEIS noted would require legislative approval. The FEIS presented estimates of the environmental impacts of both plans.

As part of the development of the Long-Term Comprehensive Plan, revised forecasts of the expected growth in air traffic through the year 2020 had been prepared in December 1993. These consisted of a detailed Base Case forecast and a number of sensitivity analyses that examined various scenarios in which forecast assumptions were varied from the Base Case. The Base Case forecasts were subsequently used in performing the analysis presented in the DEIS.

However, by the time the FEIS was released in May 1998, it had become clear that traffic growth at the airport since 1993 had significantly exceeded the rate projected in the Base Case forecast. Total passenger enplanements in 1996 exceeded 14 million, a level that was not projected to occur until some time between 2005 and 2010. In 1997 the FAA published a Terminal Area Forecast (TAF) for MSP that projected total enplanements increasing to over 22 million by 2010. Therefore the FEIS included a Forecast Sensitivity Analysis, which examined the effect on the projected environmental impacts from the proposed airport development that would result from assuming the highest of the forecast scenarios analysed in the 1993 Revised Activity Forecasts (termed the MAC High Forecast in the FEIS).

248 *Strategic Airport Planning*

Figure 9.5: MSP 2010 Long-Term Comprehensive Plan

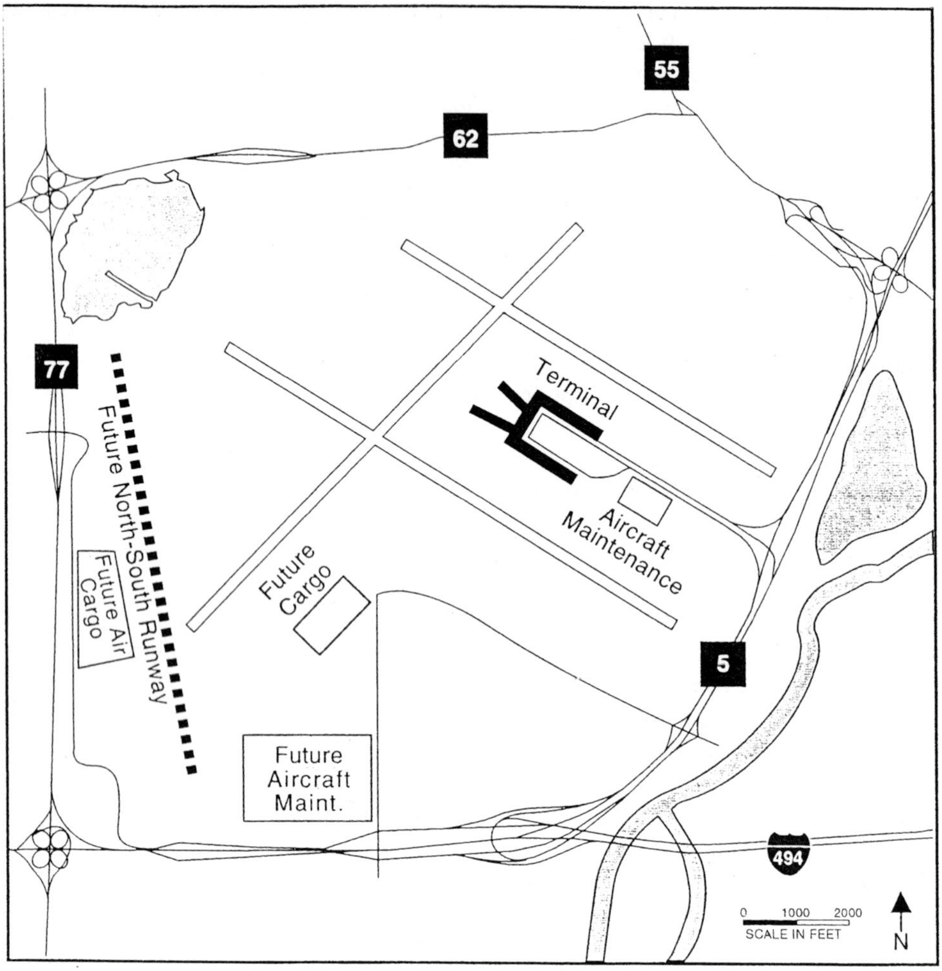

Source: Metropolitan Airports Commission/Federal Aviation Administration, *Dual Track Airport Planning Process - Final Environmental Impact Statement and Section 4(f) Evaluation*, May 1998.

However, the 1997 FAA TAF projected over 2 million more enplaned passengers in 2010 than the MAC High Forecast. Furthermore, although the TAF projections did not extend to 2020, the MAC High Forecast enplanements projected for 2020 exceeded the 1997 FAA TAF projected enplanements for 2010 by less than a million passengers. While future traffic levels are of course inevitably speculative, particularly more than 20 years in the future, the discrepancies between the MAC High Forecast and the 1997 FAA TAF, together with the recent growth rates, suggested the possibility that even the MAC High Forecast may have considerably underestimated the potential long term growth in air traffic at MSP. This of course would have important implications for the need for expanded terminal facilities, and their consequent environmental impacts.

Figure 9.6: MSP 2020 Concept Plan

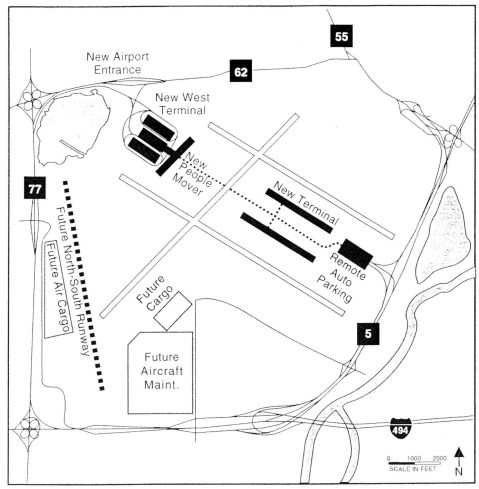

Source   Metropolitan Airports Commission/Federal Aviation Administration, *Dual Track Airport Planning Process - Final Environmental Impact Statement and Section 4(f) Evaluation*, May 1998.

The Forecast Sensitivity Analysis included in the FEIS reviewed each of the environmental impact categories in the FEIS to determine whether an increase in traffic levels at the airport from those projected by the Base Case forecast (termed the Baseline forecast in the FEIS) to those projected by the MAC High Forecast would significantly change the extent of the impacts or the mitigation required. Additional analyses were performed for air quality, noise, environmental justice (due to changes in the aircraft noise impacts), and surface water quality. It was concluded that with the exception of air quality at the 2020 traffic levels, no additional mitigation would be required. The MAC High Forecast traffic levels for 2020 represent a 43% increase in enplaned passengers and a 23% increase in aircraft operations over the Baseline Forecast. In contrast, the Baseline Forecast for 2020 represents only a 16% increase in enplaned passengers and a 7% increase in aircraft operations over actual traffic levels in 1996. It appears entirely implausible that the fairly modest increases under the Baseline Forecast would require all the mitigation measures discussed in the FEIS, but the much greater additional increases under the MAC High Forecast would require no further mitigation, other than for air quality issues.

The FEIS states that with a new west terminal in 2020, daily highway traffic would increase about 27.5% due to the MAC High Forecast. It also states that the maximum peak-hour traffic on the most critical segment of the affected regional system would not require additional capacity improvements beyond what is called for in the FEIS. There is no information on how the figure of 27.5% was obtained. As noted above, the MAC High Forecast predicted an increase in enplaned passengers in 2020 of 43% above the Baseline Forecast.

Even if the increase in daily highway traffic of 27.5% is correct, an increase in airport-generated traffic of this magnitude must have a significant effect on highway congestion on critical segments of the highway system adjacent to the airport. Whether this requires additional capacity is beside the point. If no additional capacity is provided, the increase in airport traffic will adversely impact other highway users, impacts that the FEIS should disclose.

According to the 2020 Average Daily Traffic (ADT) projections presented in Appendix F of the FEIS, the West Terminal is expected to generate 74,600 daily trips under the Baseline Forecast. A 27.5% increase would be an additional 20,500 daily trips. A 43% increase would represent an additional 32,000 daily trips. While these volumes may be relatively small compared to overall ADT on the freeway segments adjacent to the airport, their effect on intersection ramp flows, merging and weaving could significantly reduce highway level of service. The magnitude of such effects cannot be known without a more thorough analysis than that presented in the FEIS.

<u>Future Terminal Facilities</u>. The 2010 Long-Term Comprehensive Plan includes expansion of the Red, Gold and Green concourses in the existing terminal buildings, provision of a people mover in the Green concourse on the north side of the terminal complex, and construction of a skyway connector between the Green concourse and the Gold concourse on the south side of the terminal complex. According to the FEIS, the existing Lindbergh and Regional terminals have 68 air carrier and 37 regional aircraft parking positions, with domestic air carrier facilities occupying 1.4 million sq ft in 1993. By the year 2020 the airport was projected to require 83 air carrier and 34 regional aircraft parking positions, and nearly 2.2 million sq ft for domestic air carrier facilities, to meet the projected needs under the Baseline Forecast. Over the same period, regional airline facility requirements are projected to increase from about 31,000 to about 50,000 sq ft, while facility requirements at the Humphrey Terminal, which currently serves international arrivals and nonscheduled operations, are projected to increase from 90,000 to 459,000 sq ft.

The higher traffic growth rates envisaged under the MAC High Forecast have two implications for these proposed requirements. The first is that the level of facilities planned for the year 2020 will now be required much sooner. Under the MAC High Forecast, the traffic level originally envisaged for 2020 is projected to occur by the year 2000. The second implication is that much more extensive terminal facilities will be required by the year 2020 than originally planned.

It is clear that a terminal development of the scale of the proposed West Terminal cannot possibly be constructed by the year 2000, quite apart from the need to obtain legislative approval for such a development. In the near term, the only option open to the Metropolitan Airports Commission is to expand the existing terminal facilities as much as possible.

However, it is highly unlikely that the existing facilities can be expanded sufficiently to adequately handle the level of traffic envisaged for 2010 under the MAC High Forecast, much less that projected for 2020. This suggests that the proposed West Terminal, or some alternative concept, will need to be constructed by 2010, and that the design of these new terminal facilities will need to provide more gates and terminal space than originally envisaged under the 2020 Concept Plan.

Conclusions. Clearly the recent traffic growth, and the implications of this for future levels of traffic in the region, mean that the current plans to expand MSP are at best a short-term fix. Having set out to provide a long term vision of how to meet the air traffic needs of region, that could guide an investment strategy that would result in adequate airport facilities to 2020 and beyond, there is now no agreed vision or even a coherent plan that will get the airport much past the turn of the century. The new runway is unlikely to be open before the year 2000, since the FAA approval of the FEIS was not issued until the end of September 1998.

Given that any new airport could not possibly be opened until some time after 2010, the MAC will need to update the Long-Term Comprehensive Plan to provide for adequate facilities at MSP to handle the expected traffic growth until at least 2010 and perhaps much longer. However, this planning cannot be done in a meaningful way without better information on the likely traffic growth that will occur and whether a new airport will be constructed and when. It is clear that the forecasts used for the dual-track process are no longer credible, thus the very first order of business should be to develop new traffic forecasts that better reflect the traffic trends during the decade since the Airport Adequacy Study.

While opening the new runway will enable traffic levels to continue growing for several more years without excessive airside delays, the terminal building is likely to be experiencing severe congestion by the time the runway opens, much less once it has to handle the additional operations resulting from the new runway. Efforts to resurrect the proposed West Terminal could result in additional terminal facilities by some time between 2005 and 2010, or about the time that MSP runs out of airside capacity again. There appear to be few viable options to increase the airside capacity beyond that point. About the only one would be to construct another parallel runway to the north of the existing runways. However, the opposition to this from the surrounding communities can be expected to be intense, and of course the Minnesota legislature gave those communities a veto over such a runway unless the MAC can figure out a way to keep the 60 Ldn noise contours on the airport property.

9.4.6 Lessons from the dual-track experience

When it embarked on the dual-track planning process in 1989, the Minnesota legislature and the Metropolitan Airports Commission initiated a bold experiment in airport systems planning that squarely addressed one of the most difficult political issues facing airport planning in the world's larger metropolitan regions: namely how long to continue expanding the existing major airport in the face of severe environmental impacts on surrounding communities and the escalating costs of further development on a constrained site, and at what point to make the move to a new airport outside the existing urban area, with all the costs and impacts that this incurs. Recognising that the answer to this question is critically dependent on the future growth in air traffic, the dual-track planning process adopted another innovative technique, the use of a risk-based approach to forecasting demand that attempted to quantify the uncertainties involved in predicting the future, so that decisions could be made in full awareness of the risks of both underestimating and overestimating future demand.

Only time will tell whether the decision to pursue the Long-Term Comprehensive Plan and prohibit the MAC from proceeding with the development of the West Terminal or acquiring land for the new airport will turn out to be a wise resolution of a controversial process or a case of throwing out the baby with the bathwater. At the very least, one must question a decision that was made on the basis of information contained in a Draft Environmental Impact Statement, when the Final EIS released some two and a half years later acknowledges that the forecast assumptions use to generate the information are ludicrously implausible. Of course, such decisions are reversible, and in the light of recent traffic growth, the legislature can always authorise the MAC to proceed with the West Terminal or even to begin acquiring land for a new airport. Unfortunately, since these two options were precluded when the FEIS was prepared, the new terminal was only given superficial consideration and the new airport was not addressed at all. In the light of the discredited forecasts, the lack of attention to any alternatives to the West Terminal as a way of meeting the terminal capacity needs of the airport until a new airport or the West Terminal can be built, and the absence of any analysis of whether the planned runway configuration under the Long-Term Comprehensive Plan can handle the forecast aircraft movements at levels that now appear possible by the year 2020, it would appear essential that a new EIS be prepared. Thus not only have the goals of the dual-track planning process not been achieved, but two and half years have been wasted at a fairly crucial juncture in planning to meet the longer term needs of the Minneapolis/St Paul region.

What went wrong? At the risk of an overly simplistic analysis of a complex issue, several important points can be made:

1. Airport development decisions revolve around political power: the power of affected communities, the power of the business community that needs access to air service, and the power of the dominant airline or airlines at the airport. While it would be naive to imagine that such important decisions could ever be taken in a way that was not subject to the influence of those who have much to gain or lose, the greater the efforts that are made during the planning process to identify and quantify all the uncertainties and risks, and to assess each alternative course

of action in an even handed way, the less likely it is that the final decision will be overly influenced by narrow interests.

2. The need to provide facilities to meet near-term demand can force decision makers into sacrificing long term options in favour of achieving a consensus on immediate actions. The need for additional runway and terminal capacity at MSP is clear, even if a new airport was to be chosen as the long term alternative. Developing a new airport on a greenfield site is a five to ten year undertaking, even after the decision to proceed has been made. Thus the alternative courses of action should not have been posed as expand the existing airport or build a new airport, but rather varying the amount of expansion of MSP in conjunction with building a new airport at different points in time. The later the decision to build a new airport is taken, the more expansion will have to occur at the existing airport.

3. Since the future is inevitably full of surprises, the more options that can reasonably be preserved the better. One of the most important strategies for preserving the option to build a new airport is to identify where this will be located and to acquire the site (or at least acquire the option to purchase the site when needed). If it is later decided that the airport will never be needed, or that another site is preferable, the land can always be released. In the meantime it can remain in its current use. However, acquiring a site after other development has begun in its location will not only be much more expensive, but incompatible land uses will have emerged that will greatly increase the adverse impacts of developing the airport. Naturally, this action does limit other uses for the site and its environs, which imposes costs on those who desire to take advantage of those other uses. The extent to which they should be compensated for these costs, and how best to do this, is a political question that needs to be addressed as part of the planning process. The sooner a clear decision is made as to where a new airport will be located, the less these costs will be and the better able people are to adjust their decisions to this possibility. The worst of all possible worlds is to know that a new airport is needed, but have no idea where it will be built. This will result in development occurring in patterns that will be incompatible with the location of the new airport, no matter where it is built.

4. The current airport planning process does not deal well with uncertainty. Although the dual-track planning process attempted to explicitly identify the uncertainty associated with the future growth in demand, once the forecasting process had been completed, the remainder of the planning process focussed exclusively on a single Baseline forecast. There was no discussion at all in the DEIS about the uncertainty surrounding the forecasts on which its projections of facility requirements and environmental impacts were based, much less the ramifications of this for the magnitude of the possible impacts. Therefore, when traffic growth turned out to be higher than projected, and the FAA itself came up with a Terminal Area Forecast that was dramatically higher than the numbers used in the DEIS, the entire Long-Term Comprehensive Plan could clearly be seen as totally inadequate for the traffic levels now being discussed. Worse, had

the Minnesota legislature been informed of the likely consequences if future traffic demand grows at rates similar to those now being forecast by the FAA, they might well have been much less willing to forego the new airport option.

In conclusion, it is clear that the story of the dual-track airport planning process is far from over. In the short term it appears likely that the airport will get its new north-south runway and Northwest Airlines will get a modest expansion of the existing terminal building, for indeed in the short term that is about all that can be done. For the longer term, the MAC will need to reopen the planning process, prepare updated forecasts, and revisit the options to satisfy these forecasts. Since any plausible forecasts are likely to predict far more traffic in the year 2020 than were considered in the planning for the new runway and West Terminal, it is hard to imagine how any credible planning process could not consider a new airport alternative. Indeed, the 1996 legislation explicitly allowed the MAC to undertake long-range planning to make recommendations to the legislature on the need for new airport facilities.

9.5 THE SEATTLE "FLIGHT PLAN" PROJECT

By the mid to late 1980s, the Puget Sound Region of Washington State was facing a similar set of issues to the twin cities of Minneapolis/St Paul, with the difference that Seattle-Tacoma International Airport (Sea-Tac), the principal air carrier airport serving the region, was not an airline hub in the usual sense of the word. However, the airport had only two closely-spaced runways, and during IFR conditions was effectively a single-runway airport. Steady economic growth in the Puget Sound region, fuelled by such high-technology companies as Boeing and Microsoft, both with headquarters in the region, was generating growth in air traffic that was threatening to overwhelm the capacity of the airport, which is located about 10 miles to the south of downtown Seattle and surrounded by largely residential communities. In 1989 the Puget Sound Council of Governments (the regional metropolitan planning organisation, later renamed the Puget Sound Regional Council) and the Port of Seattle, which operates Sea-Tac Airport, sponsored a joint study to explore future options to meet the air transportation needs of the region. This study was termed the Flight Plan Project, and was guided by a policy committee established by the two agencies and designated the Puget Sound Air Transportation Committee (PSATC).

As expressed in the foreword to the Draft Final Report of the project (PSATC, 1992):

> "The Flight Plan Project is a forward-looking effort that addresses one of many growth-related issues vitally important to our region, as well as the entire State of Washington. To ignore the role that an efficient air transportation system plays in the quality of life we now enjoy and want for the future would be irresponsible. To not recognise or attempt to minimise the social and environmental costs of maintaining such a system would also be irresponsible. At the same time, the solution that is chosen must be cost-effective and technically feasible to assure it is implemented. These are the issues and value-laden trade-offs that made the Flight Plan Project so challenging."

United States Experience 255

There in a nutshell is the crux of the issues facing the airport system planning process in almost every major metropolitan region in the final decades of the 20th century. While the situation at the start of the project was not viewed as having reached crisis proportions, it was recognised that the lead time to decide on long term solutions to provided adequate capacity for the future and then to implement those solutions would require at least a decade. It was also recognised that the decisions that needed to be made would have a profound effect on the shape and location of future development of the region. At the same time, urban development in the Puget Sound region is both constrained by the regional geography and reflects the strong appreciation of the natural and physical environment that forms a key aspect of cultural attitudes in the region. The recommendations of the Flight Plan Project therefore had to provide a balanced solution to a wide range of conflicting objectives.

An analysis of the capacity of Sea-Tac indicated that at current levels of annual traffic, average aircraft delays were of the order of 5 minutes per operation. This was expected to increase to about 9 minutes per operation at the traffic levels forecast for 2000, and escalate rapidly thereafter. It was felt that average delays of 7 minutes per operation represented the most that could be tolerated (the "capacity" of the airport). Even if a new air carrier runway could be provided at Sea-Tac, this level of delays would be reached at traffic levels around 480,000 operations per year, as shown in Figure 9.7, corresponding to the traffic levels forecast for about 2015.

Figure 9.7: Projected average aircraft delay at Seattle-Tacoma International Airport

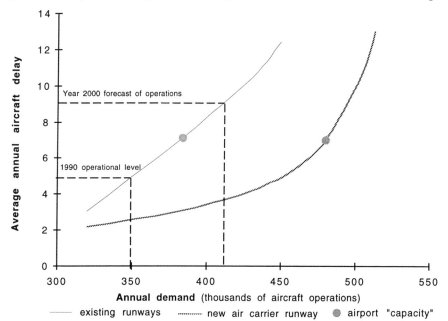

Source: Puget Sound Regional Council/Port of Seattle, The Flight Plan Project - Final Environmental Impact Statement, October 1992.

256 Strategic Airport Planning

Figure 9.8: Flight Plan Project Schedule

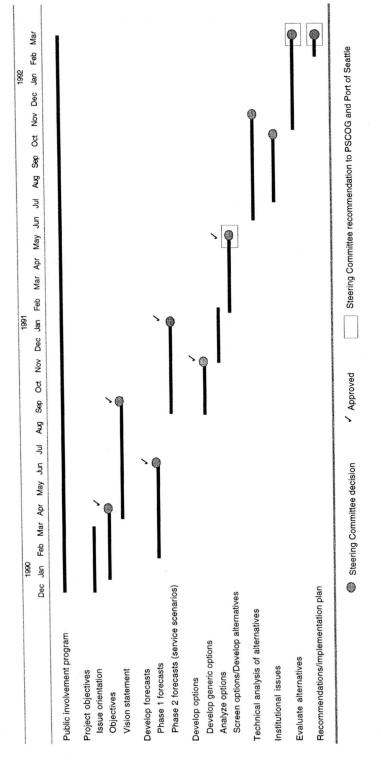

Source: *Puget Sound Air Transportation Committee, Flight Plan Project - Phase II: Development of Alternatives, Final Report, June 1991.*

After an extensive examination and analysis of alternatives, the PSATC came to the conclusion that apart from doing nothing, only three broad alternatives for meeting the region's future airport capacity needs existed:

- replace Sea-Tac with a distant airport
- massively expand Sea-Tac
- implement a regional airport system that divided the traffic between Sea-Tac and other airports.

Doing nothing and the first two airport development alternatives were felt to be infeasible. Therefore the study concluded by recommending a fairly modest expansion of Sea-Tac, adding a dependent runway, followed by the development of a multiple airport system. Initially this would consist of the introduction of scheduled air service at Snohomish County Airport (Paine Field), a general aviation airport located to the north of Seattle, followed by the development of a supplemental airport in the south of the region by about 2010. The specific site for this, and indeed even the county in which it would be located, were left to be resolved by subsequent studies. A key element of the recommendations was the implementation of demand management measures to reduce the need for additional airport capacity.

The work plan for the project was divided into three phases. The first phase lasted from December 1989 to September 1990, and focussed on defining the project objectives and developing forecasts of originating passenger traffic and air cargo demand. These forecasts were revised in the second phase, which lasted until June 1991 and focussed on developing options and defining alternatives. The third phase comprised a technical analysis of the alternatives, an examination of the institutional issues, an evaluation of the alternatives, and preparation of the recommendations and an implementation plan. The work plan elements, milestones, and schedule for the project are shown in Figure 9.8.

9.5.1 Forecast demand

The forecast of air passenger demand prepared during Phase I (KPMG Peat Marwick, 1990) projected that total regional air passengers would increase from 14.5 million in 1988 to 45 million in 2020. The proportion of connecting passengers was projected to increase slightly from 32% in 1988 to 33% in 2020. The forecast also projected the total number of aircraft operations to increase from 316,000 in 1988 to 575,000 in 2020.

The forecast for originating air passengers on major airlines was based on an econometric model relating the number of annual originating passengers to the regional population, per capita income and national average airline yield (average revenue per passenger-mile). The resulting calibration on data for the period 1970 to 1988 gave demand elasticities of 1.04 for population, 0.98 for income and -1.20 for yield, which appear intuitively reasonable and broadly consistent with other studies. The forecasts were based on projections of regional population and income prepared by the Puget Sound Council of Governments, and projections of airline yield from the then current FAA national forecasts. The FAA forecasts only extended to 2001, and yield was assumed to be constant in real terms from 2000. Projected growth in originating passengers on commuter airlines, transborder flights to Canada, and international flights was derived from assumed future growth rates based on recent trends and discussions with airlines. In each case the growth rate was assumed to decline over the forecast period. The percentage of connecting passengers in each category of traffic was assumed to remain constant at 1988 levels.

Projections of air cargo tonnage were based on assumed annual growth rates that reflected the average growth trend over the preceding 20 years. Low and high growth forecasts were prepared using growth rates of 3% and 5% for domestic cargo and 7% and 12% for international cargo. Although the five years prior to the forecast had witnessed a sharp increase in the growth of air cargo carried by all-cargo carriers and package express airlines, there was no attempt to address this in the forecasts.

The projections of the number of aircraft operations were obtained by assuming an increasing number of enplaned passengers per departure, based on the increase for major domestic airline operations projected in the FAA national forecasts, and assuming a continuation of that trend beyond 2000. Separate assumptions were made for the growth in enplaned passengers per departure for commuter airlines and international flights. General aviation operations were assumed to increase in proportion to major air carrier operations at a ratio 1% below that observed in 1988. This appears to have been an entirely arbitrary assumption, that only addressed general aviation operations at Seattle-Tacoma International Airport. In fact, general aviation operations at the airport had been declining steadily from a peak in 1979 to less than half that level by 1988. Military operations were assumed to be constant at 1,000 per year, although they too had declined from about that level in 1975 to less than half that throughout the 1980s. However, general aviation and military combined were only projected to reach 24,000 operations in 2020, or about 4% of total operations. Other air carrier operations, including all-cargo flights, were projected to reach 30,000 in 2020.

No sensitivity analysis of the projected number of aircraft operations was performed at this stage. The forecast of total enplaned passengers was compared to a number of other forecasts and the Flight Plan forecast shown to lie within the envelope of the highest and lowest of these forecasts. However, beyond this there was no attempt to address the level of uncertainty involved in these projections of future demand.

During Phase II a further analysis of trends in average aircraft size was performed and a revised forecast of aircraft operations prepared that assumed the use of larger aircraft in three key markets from Seattle: Vancouver, British Columbia; Portland, Oregon; and Spokane, Washington (PSATC, 1991). This resulted in a reduction in the number of forecast aircraft operations, with the projection for 2020 reducing from 575,000 to 524,000. The Phase II methodology also allowed for projections of fleet mix to be made. No changes were made in the forecasts of passenger or cargo traffic. The resulting forecasts of passengers and operations are shown in Figure 9.9. It was also suggested in Phase II that under the multiple airport system alternatives, the number of aircraft operations would increase to 605,000 in 2020, due to the additional service in some markets from more than one airport reducing the average number of passengers per departure. However, additional analysis of the average number of passengers per departure in other multiple airport markets in the US performed during Phase III did not support this hypothesis, and the total number of air carrier aircraft operations was assumed to be the same for the evaluation of each airport system alternative. The number of general aviation and military aircraft operations did vary with the alternatives, since some alternatives involved introducing commercial service at airports that already had general aviation or military activity.

Figure 9.9: Forecast of air travel demand - Puget Sound Region

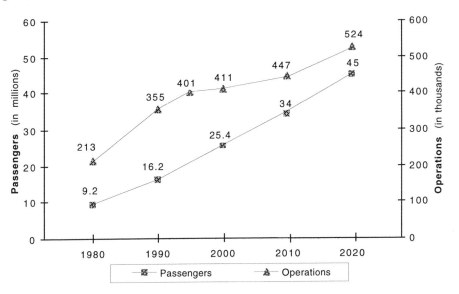

Note: Efficient Sea-Tac operating level is 380,000 operations per year.

Source: *Puget Sound Regional Council/Port of Seattle, The Flight Plan Project - Final Environmental Impact Statement, October 1992.*

There are two aspects of this forecast process that are worthy of comment. The first is that although the critical issue for the Flight Plan project was the runway capacity constraints at Sea-Tac, and thus considerable attention was given to forecasting air carrier aircraft operations, the Phase I forecasts of aircraft operations were based on the total annual passenger demand and trends in average aircraft size at an aggregate level, rather than from an analysis specific markets. Although forecasts of originating passengers were prepared for individual markets, these were not used in the analysis of aircraft size trends. However, air service is provided in specific markets and the flight frequency and thus indirectly the average size of aircraft in each market depends on the traffic volume in those markets. This issue was addressed for three specific markets in the Phase II forecast revisions but a similar approach was not applied to other markets.

The net result of these assumptions gave an increase in the average number of passengers per air carrier operation of almost 100% between 1990 and 2020, resulting partly from increases in average aircraft size and load factor for both the major airlines and the commuter airlines, and partly from an increase in the proportion of the passengers carried by the major airlines. Obviously this has a profound effect on the likely future numbers of aircraft operations and hence the amount of traffic that could be accommodated at Sea-Tac. In view of the importance of this issue to the entire Flight Plan process, it is surprising that a more detailed evaluation of future fleet mix trends in individual markets was not performed.

Such an analysis would have allowed a better assessment of the effect on the total number of aircraft operations of the introduction of commercial air service at secondary airports. As some of the traffic in each market shifts to these other airports under the multiple airport system alternatives, this is likely to affect the average size of aircraft at Sea-Tac. This effect was recognised and addressed in Phase III, but the approach used was to assume average values of passengers per operation at Sea-Tac and the secondary airports based on an analysis of the differences in this ratio between the primary and secondary airports in other multi-airport regions in the US Since such differences are likely to be very market specific, depending on such factors as the total size of the regional market, whether the primary airport is an airline hub, and the number of airports serving the region, this approach could be quite misleading. In fact the analysis of aircraft operations at the secondary airports was based on the assumption that half would be commuter airline operations. However, an analysis of aircraft departures from secondary airports in Southern California and the San Francisco Bay Area performed during Phase II suggested that only about 25% of the aircraft departures were by turboprop aircraft. It is unclear why a much higher share was assumed in the Phase III analysis. Since the largest air travel markets from the Seattle region are at distances that are well beyond typical turboprop range, while air service to closer communities is likely to involve a significant amount of connecting traffic that would favour the use of Sea-Tac, it seems that if anything the share of commuter aircraft operations would be lower than at the California secondary airports. However, this simply reinforces the need for an explicit analysis on a market basis.

The second aspect relates to the methodology adopted to allocate passenger traffic to secondary airports under the multiple airport system alternatives, and hence determine the number of air carrier aircraft operations at each airport in the region. Market areas were defined for each secondary airport and the numbers of annual passengers that would be attracted to each airport at the regional traffic levels forecast for 2000, 2010 and 2020 were estimated, based on assumptions about the share of originating passengers in each market area that would use the secondary airport. Passenger origins in the region were assumed to be distributed in proportion to the population, and different shares were assumed attracted to the secondary airports from short haul, medium haul and long haul air service markets, where short haul markets were less than 700 miles and long haul markets were greater than 1,100 miles. The number of air carrier aircraft operations at those airports was then estimated using the ratios of passengers per operation discussed above. If this exceeded the capacity of the airport, passengers and the associated operations were reallocated to airports that had not reached their capacity.

This approach raises a number of concerns. It is most unlikely that passenger origins in the region are distributed in proportion to population. Apart from variations in household income, which affect air travel propensity, a significant proportion of originating air passenger trips are made by visitors to the region who are staying in hotels, which have a very different distribution from population. The use of defined market areas around each secondary airport ignores the variation in accessibility of each airport from different locations in the region. While there is an average market share that any airport will attract, this is the result of individual choices that balance the relative accessibility of each airport and the air service available there. Without an explicit analysis of the factors affecting these choices, assumptions about airport market share are little better than guesswork. The location of Sea-Tac in the centre of the region would tend to result in it attracting the largest share of any market. This in turn would give it a departure frequency advantage, and possibly result in lower air fares, which would affect the relative attractiveness of the secondary airports. Thus one cannot simply assume the share of particular markets that would use the secondary airports, but rather these need to be derived from an explicit analysis of the air service from the alternative airports. This in turn needs to be economically viable. In other words, the resulting traffic in each market should result in a reasonable load factor, given the assumed aircraft size and flight frequency. This is a complex, dynamic relationship, since changing the assumed flight frequency or aircraft size to adjust the load factor also affects the economics and relative attractiveness of the service, and hence the share of the traffic in the market that uses each airport. In summary, estimates of the number of air passengers attracted to a secondary airport need to be based on an explicit analysis of airport choice process by originating passengers, that reflects the air service that can be economically provided at each airport. Models exist to do this, as discussed elsewhere in this book, but evidently were not used in the Flight Plan project.

9.5.2 Development of alternatives

During Phase II of the Flight Plan project a broad range of system alternatives were defined and grouped into nine categories. Two of these represented versions of the "no-action" alternative required for environmental impact analysis:

- **Base Case A:** No major facility improvements at any Puget Sound airports, except those already underway.

- **Base Case B:** Short-term capital project and policies that may be implemented at Sea-Tac before 2000, including the potential addition of a commuter runway.

It was clear that neither of these alternatives could meet the needs of the region, and they were included only for comparative purposes. The remaining seven categories provided very different ways to meet the future demand for air travel:

- **Expand Sea-Tac:** Full development of the existing airport site, including the addition of a third air carrier runway located roughly within the current airport boundaries.

- **Replacement Airport:** Close Sea-Tac and build a single new airport designed to meet the long term aviation needs of the region.

- **Multiple Airport System:** One airport serving as the primary commercial airport for the region with one or more smaller secondary commercial airports.

- **High-Speed Ground Transportation System:** Development of a high-speed ground transportation system (such as steel wheel or magnetic-levitation trains) linking major urban areas to each other and the airport, diverting a number of trips now taken by air and automobile.

- **Remote Airport:** Development of a second airport operated in tandem with Sea-Tac, with direct ground transportation connection to Sea-Tac (either a close-in airport like Boeing Field, or a distant airport like Moses Lake/Grant County).

- **New Technology:** New aircraft, air traffic control procedures, and other technologies which increase airport capacity.

- **Demand Management:** Pricing and/or regulatory techniques which encourage the use of larger aircraft, flights during non-peak hours, and the diversion of passengers to other travel modes.

A two-stage screening process was undertaken to examine each of the alternatives, determine their technical feasibility and ability to meet the requirements of the PSATC vision statement, and identify the alternatives to subject to more detailed analysis. This process examined existing airports in the region and potential sites for new airports, in order to determine the availability of suitable sites for those alternatives involving the development of airport capacity elsewhere in the region. Threshold criteria were developed to determine the minimum site requirements for supplemental airports handling either regional or domestic and international traffic, and for a replacement airport for Sea-Tac. Sites meeting these criteria were then packaged into options under each of the system alternatives.

The system alternatives were then analysed on the basis of the following criteria:

- **Airspace:** The degree to which each option would be compatible with existing uses or create potential airspace conflicts.

- **Capacity:** The number of aircraft operations that could be accommodated (or diverted) by each option.

- **Ground Access:** The accessibility of the sites in each option to residents of the Puget Sound region.

- **Investment Requirements:** The capital investment needed to implement each option.

- **Economic Impact:** The economic implications for the region and its subareas of implementing each option.

- **Implementation Feasibility:** Major roadblocks that would affect the ability to implement each option.

Following this analysis, those alternatives that were found to have serious problems or could not meet the future requirements of the region were eliminated, and the remaining options were combined into packages that were carried forward for more detailed evaluation in Phase III.

After a series of public hearings on the conclusions of the Phase II analysis, the PSATC approved the following system alternatives for further analysis:

- **Multiple Airport System:** Sea-Tac with or without a new dependent air carrier runway, together with one or two supplemental airports.
- **Replacement Airport:** Close Sea-Tac and construct a new airport at one of three potential sites in the south of the region.
- **Demand Management:** Sea-Tac with no additional runways and a maximum feasible package of demand management techniques, combined with new technologies and alternate modes of transportation.
- **Boeing Field:** Use of Boeing Field as a close-in remote airport to Sea-Tac.

Sites for supplemental airports under the multiple airport system alternative were considered to be feasible at the existing Arlington and Paine Field airports in the north of the region and McChord Air Force Base in the south of the region. Sites for a new supplemental or replacement airport were considered to be feasible in the Central Pierce County, Fort Lewis or Olympia/Black Lake areas in the south of the region. These locations relative to Sea-Tac are shown in Figure 9.10.

Further analysis of the Boeing Field alternative in Phase III resulted in it being dropped from further consideration due to airspace conflicts between Boeing Field and Sea-Tac. The various site options were then combined into the 34 alternatives shown in Table 9.1 for detailed evaluation. The Demand Management option was preserved as the "do nothing" alternative to be presented in the environmental impact statement. In fact, as noted in the draft final report of the Flight Plan project, the forecasts already implied some level of demand management in their assumptions of larger average aircraft size and higher load factors. Furthermore, as new aircraft and air traffic control technologies become available, they are likely to be deployed under each of the alternatives. It is noteworthy that the replacement airport alternatives only considered constructing one three-runway airport in the south of the region, rather than a two-runway airport in the south of the region and a secondary airport in the north. While such an alternative might be somewhat more expensive, it would significantly reduce the ground access travel times and costs for air travellers in the north of the region.

9.5.3 The Sea-Tac Airport Capacity Enhancement Plan

In June 1991, as the second phase of the Flight Plan project was being completed, the FAA published the results of an airport capacity enhancement plan for Sea-Tac (FAA, 1991). This study was undertaken by an Airport Capacity Design Team involving representatives of the FAA, the Port of Seattle, the airlines using Sea-Tac, Puget Sound Council of Governments, Washington State Department of Transportation, and other stakeholders. Capacity and delay analyses were performed for a number of airfield improvements at Sea-Tac, including the construction of a new air carrier runway on the west side of the airfield and parallel to the existing runways. Two options for the new runway were examined, one that provided a

264 *Strategic Airport Planning*

separation of 2,500 ft from the eastmost existing runway (runway 16L/34R), which would require dependent operations under IFR conditions, and one that would provide independent IFR arrival streams to both runways. Delay calculations were performed for each proposed improvement at three levels of operation, a baseline of 320,000 annual operations, that reflected 1989 airport traffic levels, and two future levels of 390,000 and 425,000 annual operations. These two future levels were not tied to specific years but roughly correspond to the forecast operations in the years 1994 and 2004 in the Flight Plan Phase 2 forecasts.

Figure 9.10: Locations of alternative airport sites - Puget Sound Region

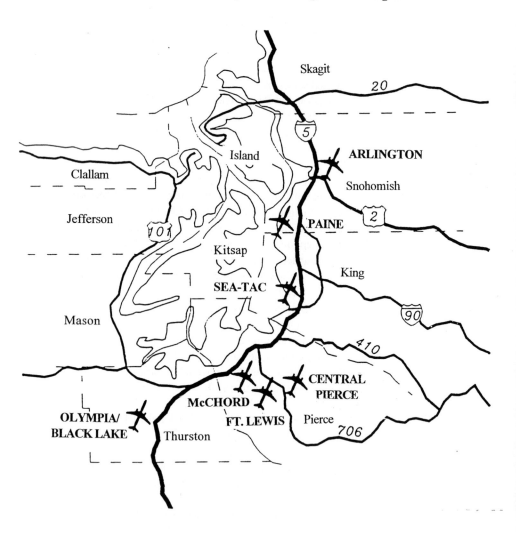

Source: Puget Sound Air Transportation Committee, The Flight Plan Project, Draft Final Report and Technical Appendices, Appendix B - Working Paper #6, January 1992.

Table 9.1: Flight Plan Phase III alternatives

System Alternative	Airport Option
Do nothing	1 Sea-Tac without commuter R/W
	2 Sea-Tac with commuter R/W
Multiple airport system with two airports	3 Alternate 1 & Arlington 1 R/W
	4 Alternate 1 & Paine 1 R/W
	5 Alternate 1 & McChord 1 R/W
	6 Alternate 1 & Central Pierce 1 R/W
	7 Alternate 1 & Olympia/Black Lake 1 R/W
	8 Alternate 1 & Arlington 2 R/W
	9 Alternate 1 & Paine 2 R/W
	10 Alternate 1 & McChord 2 R/W
	11 Alternate 1 & Central Pierce 2 R/W
	12 Alternate 1 & Olympia/Black Lake 2 R/W
	13 Sea-Tac w/ dependent R/W & Arlington 1 R/W
	14 Sea-Tac w/ dependent R/W & Paine 1 R/W
	15 Sea-Tac w/ dependent R/W & McChord 1 R/W
	16 Sea-Tac w/ dependent R/W & Central Pierce 1 R/W
	17 Sea-Tac w/ dependent R/W & Olympia/Black Lake 1 R/W
	18 Sea-Tac w/ dependent R/W & Arlington 2 R/W
	19 Sea-Tac w/ dependent R/W & Paine 2 R/W
	20 Sea-Tac w/ dependent R/W & McChord 2 R/W
	21 Sea-Tac w/ dependent R/W & Central Pierce 2 R/W
	22 Sea-Tac w/ dependent R/W & Olympia/Black Lake 2 R/W
Multiple airport system with three airports	23 Alternate 3 & Central Pierce 1 R/W
	24 Alternate 4 & Central Pierce 1 R/W
	25 Alternate 3 & Olympia/Black Lake 1 R/W
	26 Alternate 4 & Olympia/Black Lake 1 R/W
	27 Alternate 13 & Central Pierce 1 R/W
	28 Alternate 14 & Central Pierce 1 R/W
	29 Alternate 13 & Olympia/Black Lake 1 R/W
	30 Alternate 14 & Olympia/Black Lake 1 R/W
Replacement airport	31 Central Pierce 3 R/W
	32 Olympia/Black Lake 3 R/W
	33 Fort Lewis 3 R/W
Do nothing	34 Alternate 1 & Demand Management

Source: Puget Sound Air Transportation Committee, The Flight Plan Project, Draft Final Report and Technical Appendices, January 1992.

The delay analysis was performed with the use of the FAA Airfield Delay Simulation Model and Runway Delay Simulation Model. The simulations were performed for an average day of the peak month, and the daily traffic profile for 1989 was increased proportionately for the two future traffic levels. This assumed that as delays increased with rising traffic, there would be no spreading of the demand into the off-peak hours. It was found that if no action was taken, delays would rise to about 245,000 hours per year at 425,000 annual operations, or about 35 minutes per operation. The development of a dependent parallel air carrier runway with all weather capability (including a Category II instrument landing system in the predominant direction, a Category I instrument landing system in the reverse direction, and offset approach procedures when weather conditions permit) would reduce this to about 10 minutes per operation. A new independent air carrier runway with a Category II instrument landing system in the predominant direction would further reduce average delays to about 4 minutes per operation.

The study examined the potential benefits of converting an existing taxiway to a commuter runway parallel to the existing runways and 700 ft further west of runway 16R/34L. This was found to produce significant delay benefits in relation to its cost (the estimated delay savings would cover its construction cost in less than two years), but delays would still increase rapidly at future traffic levels.

Various other near term airfield improvements and procedural changes were also analysed and found to result in more modest delay reductions, but still large enough to justify their implementation. It could be expected that these would also increase the delay savings of the new air carrier runway alternatives, although these combined effects were not presented in the Capacity Enhancement Plan. The results of this analysis demonstrate that the ability of Sea-Tac to accommodate future traffic growth is not a fixed capacity, but depends on both the level of delay that the airlines are willing to tolerate and the location and design of any new runway that the airport is able to construct. However, it was also clear from the study that none of the proposed capacity enhancements analysed would allow the airport to handle the full volume of air traffic in the region projected for 2020 by the Flight Plan project.

9.5.4 Evaluation of alternatives

The 34 alternatives defined at the start of Phase III were each evaluated according to 12 criteria grouped into four elements as follows:

Operational/Technical Element

- **Runway Capacity:** Measured in aircraft operations per year.
- **Airspace:** A ranking based on a preliminary review of the amount of interaction or conflict that would occur with aircraft operating to and from other airports or restrictions caused by terrain.
- **Accessibility:** Measured in terms of the percentage of the region's population that can get to a given site in 60 minutes or less and total travel mileage.

Economic/Financial Element

- **Capital Costs:** Estimated land acquisition and construction costs in dollars.
- **Aircraft Delay Costs:** Estimated costs of aircraft delay in dollars per year.
- **Funding:** The ratio of the funds generated over a 20 year period to the capital improvement costs.
- **Economic Impacts:** A ranking based on the level and distribution of economic benefits that would be generated for the region.

Institutional element

- This evaluation adopted a descriptive approach to such factors as the socio-political acceptance of the various alternatives and the role of recent or potential new legislation in implementing the recommended alternative.

Environmental Element

- **Noise Impacts:** Measured in terms of five different criteria, comprising the population exposed to noise levels exceeding 55 and 65 Ldn, the population newly exposed to those noise levels, and the population exposed to a single event Sound Exposure Level (SEL) in excess of 80 dBA.
- **Air Quality:** Measured in tons per year of carbon monoxide and nitrogen oxides emitted from both aircraft and vehicles.
- **Wetlands Impacts:** Measured in acres of wetland affected.
- **Salmon Stream Impacts:** Measured in feet of stream affected.

Several of the foregoing criteria are conceptually fairly simple, although developing the measures in question involved the need to make extensive assumptions or perform analysis of comparable situations. Some measures, such as the annual delay costs and noise impacts were critically dependent on the assumed allocation of passengers to airports and the resulting projections of aircraft operations, which, as noted above, was based on rather questionable analysis. The estimates of airport capacity and aircraft delay were based on the use of annual service volume and annual delay curves developed during the Phase II analysis, adjusted to reflect the simulation results from the FAA Airport Capacity Enhancement Plan for Sea-Tac.

The results of these analyses were presented in summary tables that gave key measures for each criterion for each alternative or group of similar alternatives. Airport-specific data was presented for each site option in terms of the number of runways. These values were then converted in "grades" on a scale from A to D. The basis for assigning these grades was not explained in the draft final report, and several of the grades assigned in the comparison of system alternatives appear to be inconsistent with those in the comparison of site options. Perhaps the more serious concern with this technique is that it tends to cause each criterion to be viewed as equally important, since there is no basis to determine whether an A for one criterion should have the same weight in the decision as an A for another.

None the less, the summary tables do illustrate the principal trade-offs among the alternatives. The replacement airport options involved the most capital cost but had the lowest delay costs and population exposed to noise levels exceeding 65 Ldn. They also had by far the smallest percentage of the regional population within 60 minutes travel time and generated the most air pollution. Multiple airport system alternatives involving a new air carrier runway at Sea-Tac had significantly lower delay costs but exposed a much greater population to noise levels exceeding 65 Ldn than those with no additional runway at Sea-Tac. The somewhat higher capital costs were more than offset by the reduced delay costs. Adding a second supplemental airport reduced delay costs, increased the percentage of the population within 60 minutes travel time, and reduced air pollution, but naturally increased capital costs. However, the reduction in delay costs more than offset the increase in capital costs. There was also a slight increase in the population exposed to noise levels exceeding 65 Ldn, at least in the case of supplemental airports with one air carrier runway.

Of all the supplemental airport sites, Paine Field had the largest percentage of the regional population within 60 minutes travel time, and thus was in the best situation to attract commercial air service in the near term. However, developing the necessary facilities would impact 35 acres of wetland. It was also projected that the potential revenues and other funding sources would not cover the capital costs, although the revenues and other funding for the system as whole, including Sea-Tac, would more than cover the total capital costs.

9.5.5 Recommended system evolution

The preferred alternative identified by the PSATC was a phased multiple airport system involving the introduction of scheduled airline service at Paine Field in Snohomish County and a new dependent runway at Sea-Tac, both actions to be taken before the year 2000. It was also recommended that a site for a two-runway supplemental airport be identified in the southern part of the region for development by 2010, either in Pierce County in collaboration with the military or, failing that, at a suitable location in Thurston County (PSRC/Port of Seattle, 1992). Although the final recommendations did not specify locations for the supplemental airport in the south of the region, the draft final report of the Flight Plan project (PSATC, 1992) identified the preferred location to be either McChord Air Force Base or a new site on Fort Lewis, if coordination with the military could be achieved. If this coordination turns out not to be possible, the draft final report suggested that the third airport should be implemented in either the Loveland area of Pierce County (an area just east of the Fort Lewis Army Base) or the Olympia/Black Lake area of Thurston County.

9.5.6 Implementation

The Flight Plan project was intended to provide a roadmap for the future evolution of the airport system in the Puget Sound region. As such, the recommendations of the PSATC provide input to both the Puget Sound Regional Council and the Port of Seattle as the former updates the Regional Airport System Plan and the latter develops its capital investment plans for Sea-Tac. In fact the Port of Seattle has proceeded with the construction of the new dependent parallel runway. This buys the region some time while the more difficult question of how to proceed to develop the secondary airport capacity is addressed. Apart from the unresolved question of where to provide this capacity in the south of the region, one difficult issue will be the timing of the provision of air carrier service at Paine Field. It is one thing for the PSATC to recommend that this be done by the year 2000 and quite another for this to occur. The airlines will not be in any hurry to provide service at a second airport in the region as long as delays at Sea-Tac are not too severe, and of course they cannot provide service at another airport until the required facilities are available. Thus the update of the Regional Airport System Plan needs to coordinate the timing of the development of Paine Field with the anticipated growth in congestion at Sea-Tac.

Development of the new supplemental airport to the south of the region will be even more problematical. Apart from the issue of coordination with military, it is clear that further studies are required to establish the preferred site and clarify the timing of the need. Once airlines begin providing air service at Paine Field, they will be even less keen on starting new service at a third location, particularly since Sea-Tac is already south of the urban core and Paine Field is closer to downtown Seattle than any of the southern sites.

9.6 CONCLUSIONS

The three-tier structure of airport system planning that has evolved in the United States reflects both the relative roles of the different levels of government, as well as allows system development decisions to be made at the level of government with jurisdiction over relevant aspects of the broader transportation system and its role in economic development. The federal government is primarily concerned with establishing technical standards and allocating airport development funds derived from taxes levied on a national basis. The state governments are responsible for preparing airport system plans that address the air transportation needs of the state and integrate these into the rest of the statewide transportation system. In some cases they also become involved in allocating state aviation tax revenues. Finally, the regional transportation planning agencies are responsible for preparing regional airport system plans that coordinate the development plans for the various airports in the region, place these in the context of the physical and socioeconomic development of the region as a whole, and provide a vehicle to resolve the competing interests of different airports and different communities, as well as ensure that airport development is integrated with the rest of the transportation system. If everything works well, information on development needs flows up from individual airports, through regional and state system plans, to the National Plan of Integrated Airport Systems, where it informs decisions on national funding levels. These funds are then distributed to airports in a way that reflects the priorities and needs identified in the regional and state system plans.

However, in practice there are various departures from this process that reflect local circumstances, varying institutional roles, or the balance of political power. Large airports tend to pursue their development agendas directly with the regional offices of the FAA, and the MPOs and even the states often end up having to incorporate these development plans into the regional and state system plans without much opportunity to influence their outcome. Many smaller or more rural regions do not have transportation planning staff with aviation experience, nor means to fund airport system planning, and thus individual airports tend to work directly with the state. Even some large metropolitan regions lack adequate resources for effective aviation planning within the MPO, leading to a greater state involvement. Where metropolitan regions cross state boundaries, the coordination of regional system plans and state system plans can be problematical. The development of block grant programmes, in which federal funds are distributed by state aviation departments, tends to both increase the importance of the state system planning process, as well as enhance the implementation process.

The increasing number of metropolitan regions served by multiple airports with commercial air carrier service, combined with trends toward privatisation of airport operations or a more revenue oriented approach by public airport operators, suggests a need to strengthen the regional airport system planning process. This will require the resources to support adequate planning staff with aviation expertise within the MPOs, as well as explicit requirements for airport development projects to be consistent with the regional airport system plan before they can receive state or federal funding. This is hardly a radical idea, and would allow airport development decisions to be treated on a similar basis to other transportation modes.

The case studies described above for the California Aviation System Plan and the regional planning studies for Minneapolis/St Paul and Seattle illustrate attempts in three different states to integrate airport, regional and statewide planning in a coordinated way. The approach taken in the California Aviation System Plan represents not only a formal integration of regional and statewide system planning, but a coordinated effort to prioritise the allocation of federal system planning funds to support this strategy. In the case of Minneapolis/St Paul and Seattle, the studies were undertaken jointly by the airport authority and the regional MPO as a way to ensure that the interests of both the airport and the region were given full consideration. In each case, state interest in the outcome led to legislative actions to shape the final decision or require a state agency review of the study findings.

The two regional case studies also illustrate two approaches to addressing the difficult issue of how long to continue expanding the major commercial airport in a region, and when to begin developing capacity at additional or replacement sites. In the case of Minneapolis/St Paul, the dominant role of the hub operations of Northwest Airlines meant that a move to a new site would require virtually all activity to shift to the new airport. Not only would it be very costly to continue operating the existing airport as well as a new airport, but Northwest Airlines would not want its competitors having the advantage of the greater accessibility of the existing airport. In contrast, the composition of the market in the Seattle region allowed for a viable scenario in which new commercial service could be provided from a second airport, while continuing to operate the existing airport. Although the final recommendation of the Flight Plan Project was to leave open the possibility of two additional airports, the economic implications of this were not explored very thoroughly.

This consideration points out the critical importance to airport system planning of careful analysis of both the economics of air service and the effects of the performance of the airport system on the economy of the region. Airlines will not provide service unless it is profitable for them to do so, nor will they move their operations to airports that put them at a competitive disadvantage to carriers operating from existing airports. At the same time, failing to provide adequate airport capacity to meet the air transportation needs of a region has a significant adverse impact on the regional economy and on the costs incurred by residents of the region in their personal travel. Without fully understanding the extent of these costs and foregone opportunities there cannot be a balanced consideration of whether to incur the undeniable adverse environmental impacts of expanding airport activity. Since airport system planning must necessarily base its analysis on projected conditions many years, or even decades, in the future, the value of the entire process hinges on the reliability of the demand forecasts used. The experience of both the Minneapolis/St Paul Dual Track Process and the Seattle Flight Plan Project show that this is an aspect that deserves significantly more attention than it has received in the past.

10

THE UNITED KINGDOM CASE

> "Success is: getting what you want, but
> happiness is: wanting what you get."
> (Masefield, 1986)

10.1 THE LONDON SYSTEM

10.1.1 The airports and their historic roles

The existing airports and others which have been involved in the continuing saga of providing airport capacity for London and the southeast of England are shown in Figure 10.1. At the birth of commercial aviation in the 1920s, London traffic was handled by Hendon in the north and Croydon in the south, though successful demonstrations were held at that time of landings on the river outside the Houses of Parliament.

By the late 1930s, most of the traffic was using Croydon, with Gatwick as a Croydon diversion field. The railway station which serves Gatwick was built for the nearby horse-race course before it became an airport. Northolt also received some civil commercial traffic, though it was primarily a military communications airfield. The site on which Heathrow stands was a small private airfield before the second world war. It was requisitioned rather late in the war, ostensibly as a bomber base, though this quickly changed to Transport Command and there appears to have been a hidden agenda for it to revert to the role of being the main post-war London airport (Dobson-Vida, 1993; Wright, 1996). The site appealed because of the flat and largely unobstructed terrain and the gravel subsoil (D'Albiac, 1957). It was judged to cause less disturbance to residents than any of the other sites being considered at the time. The military excuse allowed the normal planning law to be bypassed, giving a rapid solution which did not need to pay attention to environmental issues. The penalty for this political manoeuvring was that the layout had to be triangular to conform with military practice. In fact, a double triangle in the form of a Star of David was adopted for the main site, with a subsidiary triangle to the north of the main site, on the far side of the A4 trunk road. The plan for these subsidiary runways was abandoned in 1952 on the grounds that the demolition of three villages and the loss of good market garden land could not be justified for the small gain in capacity. The expectation of a very large number of aircraft movements with relatively few passengers led to the central location of the terminals, with access via a tunnel from the A4 trunk road. When the airport opened there were only 6.5 passengers per aircraft movement. The realisation that the cross runways would be largely redundant as aircraft became more tolerant of cross winds apparently came too late to arrange the main access from the east or west.

Figure 10.1: The London area airports

By 1953 the Ministry of Civil Aviation was operating not only these airfields but also Heathrow, Stansted, Blackbushe and Bovingdon. It was becoming impossible to coordinate the airspace for the rapidly increasing movements at all these sites. The Minister proposed that traffic should be concentrated at Heathrow, since it should be able to deal comfortably with the scheduled airline demand. A second airport should be available as a diversion field and to accept summer peak traffic. A third airport should also be available for those movements who might not be able to use the main alternative. Gatwick was chosen as the second airport on the basis of lack of airspace conflict with Heathrow, convenient surface access, proximity to the air routes, absence of fog, and the least disturbance to property, amenity and agricultural land. The third airport was to be Blackbushe. Stansted was to be retained as a reserve and training airport in case demand grew greatly, though only until Gatwick had developed sufficient capacity (White Paper: London's Airports, Cmd. 8902, July 1953). Stansted was rejected as the second airport due to poor surface access and the airspace constraints caused by the many nearby military airfields. A subsequent Public Enquiry established that Gatwick was suitable for the intended purpose, though the Inspector commented that the limited scope of the enquiry did not allow him to judge if it was the best site, and he was concerned about the noise disturbance (Roskill, 1969). Only four years later, the London Airport Development Committee was warning that Heathrow would not be able to cope with anticipated traffic, that developments at Gatwick should be given urgent consideration and that a third airport may be necessary.

In 1960 Blackbushe was closed because of conflicts with Heathrow airspace, while Luton, owned by the local council rather than the Ministry, began substantial commercial operations. The focus moved on to Stansted to combine the roles previously envisaged for itself and Blackbushe. The Estimates Committee found more severe problems at Stansted than the White Paper had identified and also that the capacity problem was worse than expected. It was estimated that, even if Gatwick had two runways as well as Heathrow, a third airport would be necessary by 1972, on the basis that the four runways would only support 104 atm per hour. Yet another committee, consisting of only airline and air traffic control representatives, decided that Stansted should be the third airport since only the northeast quadrant around London could allow a reasonable balance between distance from London and lack of airspace interference between airports, apparently discounting growth at Luton.

Public discussion of development at Stansted was limited to a local Public Enquiry, whose terms of reference were extremely limited. The Inspector found against the site for further development on the grounds of noise, traffic, regional planning, agriculture and house values. He argued that it could only be justified by national necessity, which had not been demonstrated by evidence at the inquiry. He recommended that a much deeper and more extensive review be made, involving experts in all pertinent matters. Despite all these comments, the Government published a White Paper (UK Government, 1967) claiming that a "comprehensive and searching re-examination of the many complex issues" had shown Stansted to have air traffic, surface access and cost advantages over all other sites that had been considered and that noise was acceptable. These advantages were sufficient to designate it as the third London airport despite its admitted shortcomings in local and regional planning. It may be relevant that the British Airports Authority (BAA) was formed by an Act passed in June 1965 to own and manage the main international airports previously owned by the State, including Stansted.

The White Paper was discussed furiously in Parliament, with charges of bad faith, of central government abuse of power and of neglect of the rights of minorities. It was said that insufficient attention had been paid to social costs, ground transport and urbanisation. The President of the Board of Trade acknowledged these concerns and announced in February 1968 that these issues would be investigated by a Commission to be chaired by Lord Justice Roskill. It was to have the dual task of sitting as a Commission of Inquiry to perform an expert, rigorous and systematic study of the many and complex problems and also as an Inquiry where interested parties could be represented by counsel and have the right to cross-examine, both at local site inquiries and before the Commission itself (Sealy, 1976).

The Roskill Commission sat for two years. The Committee consisted of a planner, a consulting engineer, a planning inspector, a professor of aircraft design, the deputy chairman of a very large industrial company and a professor of economics, supported by a research team. It used the technique of Cost/Benefit Analysis (CBA) to make quantitative comparisons between the most promising sites. The results of the CBA are given in Table 10.1, together with the factors considered. There were some factors which it was not possible to value and others over which

Table 10.1: The Roskill Commission cost benefit analysis
Differences from lowest cost site (£ million discounted to 1982)

		Cublington		Foulness		Nuthampstead		Thurleigh	
Row		High Time Values	Low Time Values	High Time Values	Low Time Values	High Time Values	Low Time Values	High Time Values	Low Time Values
1	Airport construction	18	0	32	0	14	0	0	0
2	Extension of Luton	0	0	18	0	0	0	0	0
3	Airport services	23	22	0	0	17	17	7	7
4	Meteorology	5	0	0	0	2	2	1	1
5	Airspace movements	0	0	7	5	35	31	30	26
6	Passenger user costs	0	0	207	167	41	35	39	22
7	Freight user costs	0	0	14	0	5	5	1	1
8	Road capital	0	0	4	0	4	4	5	5
9	Rail capital	3	3	26	0	12	12	0	0
10	Air safety	0	0	2	0	0	0	0	0
11	Defence	29	0	0	0	5	5	0	0
12	Public scientific establishments	1	1	0	0	21	21	61	61
13	Private airfields	7	7	0	0	13	13	27	27
14	Residential conditions (noise, off-site)	13	13	0	0	62	62	15	15
15	Residential conditions (on site)	11	11	0	0	8	8	5	5
16	Luton noise costs	0	0	11	11	0	0	6	6
17	Schools, hospitals & public authority buildings (including noise)	7	7	0	0	11	11	0	0
18	Agriculture	0	0	4	4	9	9	9	9
19	Commerce & industry (including noise)	0	0	2	2	1	1	2	2
20	Recreation (including noise)	13	13	0	0	7	7	7	7
	Aggregate of inter-site differences (costed items only) high and low time values	0	0	197	156	137	128	88	68

Source: Button and Barker, 1975

there was much debate. The Committee were also fully aware of the fundamental weakness of CBA in not being able to deal with the uneven distribution of costs and benefits, but the study set new standards in careful analysis, in clarity and in objectivity.

The Committee found in favour of Cublington, a greenfield site in the northwest quadrant some 60 km from central London. However, the decision was not unanimous. A Note of Dissent was appended from Professor Colin Buchanan, the planner on the Commission. He felt that the decision to be at odds with the fundamental planning concept of the maintenance of a green belt around London, that air travel demand was sufficiently captive for it to be not discouraged by a longer access trip, that the CBA quantification did not properly reflect the way people actually make trade-off decisions, that if a rational national airport plan existed the new airport would only have to consider siting with respect to southeast demand, and that the economic activity associated with a new large airport should be used to re-energise the east side of London. These factors, combined with the conviction that airports are such bad neighbours that they would have to accept being put in the few remaining places on a crowded island where they could be tolerated, led him to favour the coastal site of Foulness, some 90 km from central London. Ultimately, Buchanan's dissent is an example of disagreement over the weighting of the economic and environmental terms in the CBA, together with an attempt to avoid the uneven imposition of disbenefits by minimising them at the expense of not maximising the cost/benefit ratio.

The Government pronounced in 1971 that it accepted the case for further capacity, but took the view that regional planning and environmental issues were of greater importance than had been reflected in the Commission's CBA, to such an extent that it was worth the extra cost of the Foulness solution. In 1973 The Maplin Development Authority was formed to start the site reclamation, but with the oil crisis of that year and the arrival of a Labour Government, the project was put on hold until a full re-appraisal could be carried out. The subsequent review took account of trends for increased passengers per movement, the spreading of charter demand into the shoulder months and the promise of quieter aircraft, together with the possibility of a Channel Tunnel, to paint scenarios of how the airports would develop with and without Maplin. It concluded that the existing runways would suffice until 1990, provided that the airports continued to expand terminal capacity. The Maplin project was therefore cancelled officially in the Airport Policy White Paper of 1978.

Working groups, the Advisory Committee on Airport Policy (ACAP) and the Study Group on South East Airports (SGSEA), were set up as a result of the 1978 White Paper. They concluded that there was going to be a need for more capacity in the southeast by 1990 (DoT, 1979a), and the Government was advised that it should reach a view on which site should be developed and safeguard sufficient land to allow an airport with a capacity of 50 mppa to be developed. SGSEA set out the implications of six sites, similar to those examined by Roskill, including Maplin. On the basis of these studies, the government invited BAA to submit plans for the development of Stansted. BAA accepted the invitation, and in 1980 requested permission for a two-runway airport, having previously been sufficiently far-sighted to buy up the required land.

The application was called in to Public Inquiry. In an unprecedented move, British Airways encouraged Uttlesford District Council, who have planning responsibility for Stansted, to make a counter-application for the necessary capacity to be provided at Heathrow by developing Terminal 5 on the site of the Perry Oaks sewage farm. The government decided to hold a joint enquiry of the two sites, and appointed Graham Eyre, QC as the Inspector. The inquiry convened in September 1981, and ran on for two years. The Inspector's report (Eyre, 1984) took almost a year to write. The first paragraph of the summary of Eyre's overall conclusions endorses the conclusions formed in the critique of 1970s planning:

> "The history and development of airports policy on the part of administration after administration of whatever political colour has been characterised by ad hoc expediency, unacceptable and ill-judged procedures, ineptness, vacillation, uncertainty and ill-advised and precipitate judgements. Hopes of a wide sector of the regional population have been frequently raised and dashed. A strong public cynicism has inexorably grown. Political decisions in this field are no longer trusted. The consequences are grave. There will now never be a consensus. Other important policies which do not countenance substantial expansion of airport capacity or new airports have been allowed to develop and have become deeply entrenched. Somewhat paradoxically, such policies are heavily reliant upon by the thousands of reasonable people who strongly object to airport development. The past performance of Governments guarantees that any decision now will provoke criticism and resentment on a wide scale. I do not level this indictment merely as gratuitous criticism nor in order further to fan the fires of the long history of controversy but to set the context for current decisions which will shape a future that must enjoy an appropriate measure of certainty and immutability".

In attempting to promote a solution which would satisfy the aims stated in the last sentence of the quotation, Eyre made it clear that the issue was not to choose Stansted instead of Heathrow, but that the two sites could form short and long term solutions to the same problem. For the first time, a view was taken of the long term need. The preferred solution was not to develop another major airport at Stansted. The environmental disbenefits were judged to be too strong, as were the airline and user benefits of Heathrow. In Eyre's view, only the timescale required to relocate the Perry Oaks sewage treatment plant justified any development at Stansted. However, he gave no weight to the airlines' arguments that, because they would still have to serve Heathrow, their operations would be fragmented and thus more costly. He could see no evidence from other multi-airport systems to suggest that a similar system in London would not work effectively and successfully.

The government embodied its interpretation of the Inspector's recommendations in its June 1985 White Paper (HMSO, 1985). Permission to develop a fifth terminal at Heathrow was refused but the possibility be kept under review and all possible steps be taken to maintain its leading position in world aviation. Gatwick was expected to become an independent scheduled airline hub, aided by limiting access at Heathrow to airlines previously operating there, but it should continue to have only one runway. It was expected that Stansted would soon develop its own range of scheduled services, as well as capturing some of Gatwick's charters, due to runway capacity limitations at Heathrow and Gatwick. Luton should be encouraged to raise its capacity to 5 mppa to fill in any shortfalls of capacity until Stansted's 8 mppa terminal became available.

Development at Stansted was therefore allowed, but it was limited to a single runway with a cap of 78,000 atm per year and a terminal capacity of 15 mppa. Any further expansion of activity requires an examination by Parliament. It was considered very doubtful whether a second runway would be justified in the foreseeable future. The BAA was required to sell the land it had previously bought for the second runway, in order to give a firm signal to the public that the government was serious in its desire to enforce the environmental capacity limits. It is entirely possible that, if the SGSEA had been able to anticipate the 15 mppa cap, their advice to the Government on site selection might have been quite different. The Government also reaffirmed its policy that a second runway should not be built at Gatwick, though it still declines to be bound by a formal planning agreement between BAA and West Sussex County Council to that effect which does not expire until 2019, on the grounds that it was not a party to the agreement.

Even before the new Stansted terminal had been built, air transport movements (atm) rose much faster than had been expected, actually almost as fast as passenger numbers, giving severe aircraft access problems at Heathrow and in London Area Airspace. As the decade progressed, British Caledonian failed at Gatwick, British Airways was privatised and grew in strength, and American carriers' demand for access to Heathrow became strident. Meanwhile, Terminal 4 opened at Heathrow in 1986, the North Terminal at Gatwick in 1988 and then the Stansted terminal in 1991, each adding approximately 8 million passengers to the system capacity.

London City Airport opened in 1987, with an initial capacity of 1 mppa. The City airport was the prototype for private venture airports, being conceived well before the Government's privatisation Bill was presented. It was thought that an airport would be a catalyst for the redevelopment of the London Docklands, so the Development Corporation and the property developers were in favour, provided that the environmental impact was acceptable and that the airport's obstruction surfaces did not unduly constrain the height to which they could build. In turn, Mowlem, the airport developer, needed to be sure that the safety requirements would not restrict operations to the point where they would be unprofitable for the airlines. The airport was designed around the performance capability of the De Havilland Canada Dash 7 aircraft, which has a capacity of 50 seats and the ability to use an 800 metre runway and a 7.5 degree approach slope as opposed to the more normal 3 degrees (Norris, 1991). The airport was tailored to fit on a disused wharf, the resulting layout seriously limiting runway capacity since there was no room for a parallel taxiway. The airspace considerations also limited activity to approximately 12 movements per hour.

A Public Inquiry was held in 1984, after which permission was given for the development subject to stringent limits on hours of operation, maximum noise levels, numbers of movements per day and per week, and an overall noise budget. The development's alignment with the Government's privatisation policy may have had something to do with the approval of a project that added little capacity to the London system but which certainly offered the opportunity for airlines to compete for traffic generated in the business heart of the city.

Many industry observers felt that it would be difficult for Mowlem to recoup their investment of some £15 million, even if the one mppa capacity were to be reached, but Mowlem presumably had external economic factors to consider as well. Traffic was, in fact, disappointing, only reaching 230,000 passengers in 1990, probably due to poor access and long total travel time with the turboprop aircraft relative to the services available at Heathrow and the other airports. At the same time, the property boom in London came to an abrupt end, and the developments which were to have worked synergistically with the airport failed to materialise. It was, however, already attracting approximately 50% of the UK business passengers and 30% of foreign business passengers travelling to or from the City of London to Brussels, Paris and Rotterdam in 1991. Over 70% of the (single sector) passengers were paying Club fares, compared with an average of only 27% on all Heathrow routes. Some 47% chose London City because it was nearest their home or place of business, compared with 17% at Heathrow (CAA, 1998).

In 1989, the airport brought forward a scheme to allow the BAe 146 four-engine quiet jet aircraft to be accepted. The approach slope was reduced to 5.5 degrees, while the declared runway length for both takeoff and landing was increased to 1,199 metres, starter strips and displaced thresholds being used to ensure obstacle clearance criteria were met with the more stringent safety surfaces. The airport, having less than a 1,200 metre runway, qualifies as Code 2, and therefore effectively meets the required runway strip width without widening the original wharf. The application was helped by surveys showing that the majority of the population near the airport were by then strongly in its favour, by the local planning authority also being in favour, by a demonstration of a BAe 146 operation and by the fact that the differences between the Development Corporation, the property developers and the airport had been resolved before the Public Inquiry in 1990. Approval was granted in 1991, allowing the potential noise contour equivalent to 57 Leq to increase to affect 132 properties compared with none in the earlier permission, raising the maximum noise limit to cope with the extra 10 dB created by the BAe 146 and slightly raising the annual and weekday movement limits to 36,500 and 130 respectively. The main compromise for the airport was to give away the right to build a parallel taxiway over water to the south of the runway. The total cost of the two phases of development was £56 million so reassuring the community that further development would be severely restricted.

Since then traffic has increased to 550,000 passengers in 1995 and 1.16 million in 1997. Foreign operators have taken the opportunity to feed their hubs and others have used it to compete on the dense routes for which they could not get slots at Heathrow. There is a very high level of service in the under-utilised terminal, with 10 minute check-in. Access has improved from the M25 motorway, feeding passengers from the northeast and southeast, and more bus and rail links have been provided, though there is still no rapid transit station at the airport. Most of the property development has now been taken up. Mowlem have sold the airport for £15 million to an Irish businessman and a former chairman of the Irish Airports Authority (Aer Rianta), who has also invested heavily in surrounding property in order to develop airport-related activities. The airport has now been given permission by the planning committee of the local authority (Newham) to double the maximum number of flights to 73,000 per year, provided that the same curfews are kept and no new terminal is constructed (*Airport Business Communiqué, No 96, July 1998, p 23*).

10.1.2 Heathrow developments

In 1997, Heathrow's four terminals handled 57.8 million terminal passengers, some three million more than the BAA quotes as its present stand-limited capacity (BAA, 1995). Approximately 34% were transfer passengers. It also handled 1.15 million tonnes of freight and 431,000 air transport movements (atm). Over 50,000 people work there, 80% of whom travel by car. In contrast, 33% of passengers use public transport despite the low level of service provided by the underground metro system. The demand for additional runway capacity is overwhelming, with up to 30% of the demand for runway slots being denied, and many more airlines not even bothering to apply. With the present runway capacity, the most serious capacity problem is the shortage of stands. The pressure on ground access and terminals is also severe. British Airways (BA) has had to split its maintenance facilities between Heathrow and Cardiff. A new cargo village has been built off-airport.

Some argue that the concept of hard limits to capacity is erroneous; that airports are "elastic-sided", a phrase coined by the BAA Chief Executive, Sir John Egan (Goldman, 1992). It is of interest to recall that, despite historic growth rates of 15% per annum, it was predicted in the late 1960s that passengers would only increase by 4.5%, 3% and 2% annually during the 1970s, 1980s and 1990s respectively; even so, with the maximum hourly capacity predicted to be 64 atm at Heathrow and 40 at Gatwick, a fifth airport was expected to be required by the year 2000 to handle a total of just 44 mppa) with 700,000 atm. Even in 1984, the Eyre report on the Stansted enquiry only anticipated hourly and annual capacities of 66 mppa and 300,000 atm respectively, thus Heathrow could not be assumed to ever exceed 53 mppa. The early estimates were 50% low on passengers while being nearly correct on atm. More importantly, the predictions of annual atm capacity at Heathrow and the central estimates of p/atm were both some 50% low. In evidence to the Terminal 5 Inquiry and to the House of Commons Transport Committee, BAA have estimated that the future sustained runway capacity will be 78 movements per hour, giving an ultimate capacity of 80 mppa with the construction of Terminal 5, limited once more by stands.

An underlying assumption is that there will be no change in the present set of environmental restrictions. The method used to control the impact of aircraft noise on the community relies fundamentally on using the runways in mixed mode, alternating the use between takeoff and landing at 1500 hours each day in any given week when operations are westerly. This has been in use since 1972, and gives the most affected communities relief from noise for half of every day. The easterly movements are also effectively segregated by the 'Cranford Agreement' which protects that close-in community from noise by effectively preventing takeoffs on the northern runway by monitoring noise limits with an adjacent monitor. There are also restrictions between 2300 and 0600 hours on the number and noise quota count (determined by the noise characteristics of the aircraft) of movements, and limits on the allowable noise at monitors under all the takeoff paths of 110 PNdB by day and 102 PNdB at night have been in force since the early 1960s. Aircraft must reach 1,000 ft above the airfield elevation before overflying the monitors, and must continue climbing at least at 500 ft per minute after the monitoring point while making progressively less noise on the ground under the flight path. The path must follow established minimum noise routes until they reach at least 3,000 ft of altitude. These measures were taken partly to compensate for a clause in the 1920 Air Navigation Act removing the right to take legal action over aircraft noise.

British Airways has 50% of night atm and 38% of all atm, incurring fines totalling £56,000 from 51 night and 50 daytime infringements. In contrast, BA paid over £5 million in noise related charges built into the fees at just four other airports in a total of £7 million at 36 airports worldwide, partly due to a policy of continuing to operate Chapter 2 certificated aircraft there (British Airways Annual Environmental Report, 1996). Given that triggering the Heathrow monitors with Boeing 747 aircraft depends critically on the load factor as well as the air temperature and the head wind, it could be argued that this form of noise control penalises the more efficient operator.

These monitors have not been well located and the tracks not too well monitored, so that only some 20% of probable offenders were fined and the limits were only set to control the noisiest aircraft. They are now being relocated to coincide with the flyover point used for noise certification of aircraft, i.e. 6.5 km from the start of roll, and the limit is being reduced by 3 dB by day and 2 dB at night. This seems to be a move by Government to increase the severity of noise restrictions at the airports for which they are responsible, in line with the general trend at airports throughout the UK and Europe. However, the most affected communities are not at the monitoring points, and, now that most aircraft are turbofans certificated to Stage 3 standards, much of the noise comes from arrivals. It is therefore doubtful if the new limits will bring much benefit other than extra revenue from the operators of long haul large aircraft for BAA to spend on noise mitigation measures. These operators will be best able to afford the fines and, in attempting to minimise the fine, will modify their departure technique to make less noise at the monitoring point but increase it elsewhere. The Government is constrained from making the regulations more severe by the need to maintain compatibility with international obligations which imply the ability to allow economic operations by long haul Chapter 2 aircraft until 2001 (*Department of Transport Press Notice 272, 28 August 1996*). The Government is beginning an investigation into imposing noise limits for landings.

The Cranford agreement is a crucial determinant of the runway operating policy, yet it does not appear to exist on paper. Apparently a verbal assurance was given by a senior Government official with Ministerial approval to the Cranford Residents and District Amenities Association in 1952 that overflight of the area to the east of the northern runway would be avoided as far as practical, except during peak periods. At some indeterminate time the undertaking was modified to apply only to takeoffs. Surviving records of the agreement are far from complete and no formal written agreement can now be found (*Hansard, 30 October, 1995*). It seems astonishing that it should have so much power when other more formal planning agreements have been rewritten or are in process of being reconsidered.

Runway pricing is one way of dealing with a shortfall in capacity, managing demand so that it just equals capacity. The most efficient version of this policy is peak pricing, with relative discounts in the off-peak periods. The BAA adopted this policy in the years before privatisation, Heathrow then achieving the reputation of being one of the world's most expensive airports, though the average prices did not justify that reputation. The airline industry complained bitterly that the policy broke the international agreements on user charges, but the prices discouraged virtually none of the major airlines from using Heathrow or, indeed, from continuing to operate in the peak periods, though the US Government took the UK Government to arbitration over the matter. The smaller feeder carriers did find it too expensive, and withdrew service.

When BAA was privatised, the CAA was given the task of ensuring that it did not take advantage of the opportunity to indulge in monopoly pricing at Heathrow or cross-subsidise its other airports from Heathrow revenues. The separate airports were each established as Limited Companies in order to keep the individual accounts transparent. The CAA must take advice from the MMC. In the first two quinquennial reviews, they imposed the so-called RPI - X formula i.e. BAA have had to keep the annual increase in their user charges to X% below the change in UK retail prices, in effect reducing the fees by X% per year in real terms. Since X has been between 8 and 1, the charges at Heathrow have now fallen so that it is one of the lowest among major airports and the House of Commons Transport Committee feel it is in danger of distorting the market. The total average cost per passenger is only £5 approximately, and only the flat pricing of the landing fee, regardless of the weight of the aircraft, reflects that its slots are probably the most desirable in Europe.

There is some concern that the aeronautical income is so low that BAA shareholders might object to further airside investment and that the Terminal 5 proposal has taken precedence over further runway capacity partly for this reason. British Airways actually lobbied the CAA in favour of less severe price controls than it was originally going to apply in the latest review because of the possibility that it would discourage necessary investment. At the latest quinquennial review, the CAA has agreed to limit X to 3 for Heathrow and Gatwick in order to facilitate the construction of Terminal 5. The MMC offered them the alternative of X = 8 but with a sharp increase of fees on the opening of Terminal 5, which would have been more reassuring for those opposed to Terminal 5 that the inquiry has not been decided in advance.

Fleet mix is the key to further capacity on the existing runways. Despite the well publicised capacity pressures at Heathrow, not only is the demand still strong but the system appears to be coping with it. Table 10.2 shows that the scheduled passenger atm have been growing almost as quickly as at Birmingham, even though the latter was establishing its Eurohub in this period. In fact, 62% of the traffic growth was expressed through the increase in movements compared with 38% through the increase in average terminal passengers per aircraft (pax/atm). The increase in atm has been due largely to increased route competition. The highest frequency routes at Heathrow are among the fastest growing, due to increasing competition. Each airline appears to consider it necessary to offer at least seven frequencies per day on these routes. It seems that the airlines have also been competing with frequency on the less-liberalised routes, even when there are only two carriers. Manchester's growth was reflected more in atm than aircraft size because, in contrast to Birmingham, most of its routes were already served by jets by 1990 and it was encouraging small feeder services to its interline hub. Most turboprop operations have now ceased at Heathrow as the airlines use their scarce slots for more profitable routes, and there is increased use of the large twin-engine jets on long haul routes as more carriers enter and more destinations are served. There is a strong tendency for scheduled operators to hold down aircraft size and compete on frequency, and they have been able to do this even at Heathrow. In contrast, charter operators have been trading up from 150 seat aircraft to the Boeing 757 and indeed to wide body aircraft rather than increasing the number of flights, though this had already happened at Gatwick by 1990. Terminal passengers per atm at Heathrow increased from 116 in 1990 to 125 in 1994, after being nearly constant over the previous decade, as the shortage of capacity became increasingly severe. It is predicted to rise to 195 by 2015, with the presumption that a significant number of 600 seat aircraft will then be in use.

Table 10.2: Comparative growth factors, 1994/1990

a) Scheduled

Airport	1994 mppa	Pax factor	Atm factor	Pax/atm
Heathrow	51.202	1.203	1.12	1.077
Gatwick	11.084	0.905	0.92	0.980
Stansted	2.281	5.880	3.13	1.877
Luton	0.484	0.506	0.42	1.199
Birmingham	2.587	1.262	1.13	1.122
Manchester	5.892	1.318	1.28	1.026

b) Charter

Airport	1994 mppa	Pax factor	Atm factor	Pax/atm
Gatwick	9.961	1.133	1.05	1.078
Stansted	0.975	1.27	1.40	0.903
Luton	1.320	0.766	0.43	1.770
Birmingham	2.197	1.525	0.93	1.633
Manchester	8.441	1.797	1.05	1.504

Source: derived from CAA statistics (CAP 592, CAP 645)

Terminal 5 would be fully justified to rebalance the runway and terminal capacities if the passengers per atm does rise in this way, since the present two runways would support 80 mppa, especially if they were operated in mixed mode. Terminal 5 now has the support of BAA as well as BA, the latter expecting that it will have prime use of it. Yet at the time of the Stansted enquiry, BAA were openly hostile to it. They then said that it would be trying to 'put a quart into a pint pot' and that Stansted would widen travellers' choice, cause less noise nuisance, give greater opportunity for industry to expand and offer more flexibility to satisfy long term demand. Also, at the time of the enquiry into Terminal 4, the Secretary of State not only imposed a limit of 275,000 atm per year but also seemed to agree with the Inspector that there should be no further expansion at Heathrow.

The proposal is for a main terminal with remote satellites, the ultimate capacity being 30 mppa with the capability to handle New Large Aircraft (NLA) and have sufficient stands to replace those lost in the central area due to the need to taxi the NLA. The development will require the removal of the Perry Oaks sludge treatment works, the resiting of which will give an additional environmental impact to be considered.

The enquiry started on 16 May 1995 and is expected to last nearly three years, with a further year for a decision. The first phase cannot therefore be in service before the year 2002, yet Heathrow handled 54 million passengers in 1995 with only 125 passengers per atm. Even if it only expands modestly until Terminal 5 comes on stream and if some of the central area capacity is lost due to taxiway modifications, the ultimate capacity with Terminal 5 is clearly greater than the 80 mppa assumed for the governmental studies on London Area capacity requirements, the extra traffic probably being accommodated by an increased efficiency in the use of stands. On this basis, it is easy to see why the BAA maintain that it is not necessary to add a third runway in order to balance capacity, or, indeed, even to move to mixed mode operation, which could give 92 atm per hour (NATS/IATA/BAA, 1994). Operating with more than 80 mppa and a runway capacity of only 80 atm per hour would, however, mean an even greater loss of the shorter and lower density feeder routes which are normally regarded as so important in strengthening a hub and a further reduction in competition. It will also be difficult for the largest aircraft to meet the flyover noise limits, which have just been tightened.

Meanwhile, progress is being made on the access problem, with plans for a whole range of rail links in conjunction with the new privatised train operating companies to complement the Heathrow Express link into London which should open in 1998. The new proposals would give further access to London, Gatwick, and the cities on the rail routes to Manchester and the southwest, raising the percentage of passengers using public transport for access to over 50 from 34 in 1996. Further ideas include computerised on-train check-in, airline ticket sales, duty-free advance purchase arrangements and through-ticketing (Maynard, 1995). Other plans include dedicated bus lanes, cycleways and parking restraints, which should help to improve the 20% of workers who currently use public transport.

Expansion of runway capacity at Heathrow is still possible if more competition is deemed to be desirable. Schemes which have been proposed include lifting the night curfew, using the runways in mixed mode rather than the more environmentally acceptable fashion of dedicating one runway to landings and the other to take-offs (Reid, 1990), using the cross-runway for short take-off and landing aircraft (STOL), and canting and splitting one of the runways for STOL (Ambrose, 1990). The use of Northolt as a Heathrow reliever has been suggested many times (e.g. Brewitt, 1990), including realignment of the runway to avoid conflict with Heathrow (e.g. Beliyiannis, et al, 1990). This option requires both military cooperation in releasing control of Northolt and a complicated and lengthy ground transfer link, and its full use for commercial traffic would reduce still further the Business Aviation access to London. With the exception of mixed mode, most of the operational solutions have been deemed by the National Air Traffic Services (NATS, an independent agency within the CAA) to be of little additional capacity value. Since it both plans and operates the system there seems no easy way of challenging these views.

The CAA has also investigated the addition of a third independent parallel runway on land to the north of the existing site between the A4 trunk road and the M4 motorway. It found that a full length runway could be made to work from the airspace point of view (CAA, 1990a): it was, of course, always difficult to envisage the Government risking the environmental consequences of such a solution, and there must be a suspicion that there was a hidden agenda

286 *Strategic Airport Planning*

Figure 10.2: Third parallel runway options for Heathrow

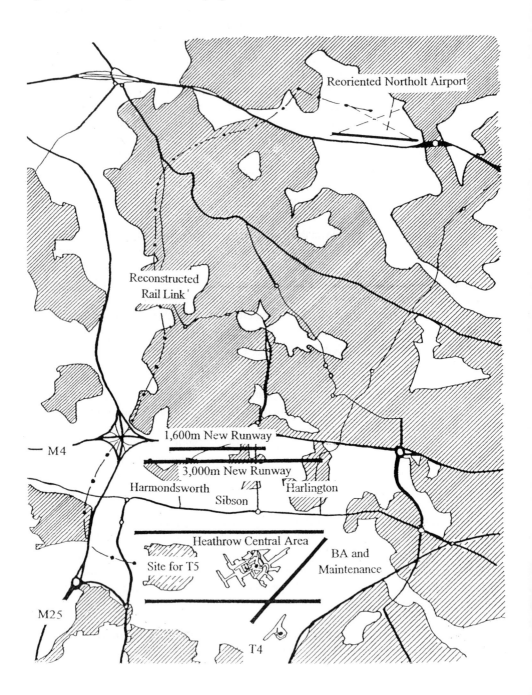

in its being proposed. These studies (CAA, 1990a) rejected an additional STOL runway at Heathrow. However, an independent study (Beliyiannis, et al, 1990) showed that sufficient space exists between the A4 and M4 roads to accommodate a remote terminal serving a remote runway at least 1,600 metres long without substantial demolition of property. There are three villages, several hotels and attractive protected buildings in the area, including old churches and a medieval tithe barn. It was the reserved site for three additional runways in the original 1944 plans, which were explicitly deactivated in 1952 (D'Albiac, 1957). It has however been considered many times since to have potential, not least for the UK-developed V/STOL aircraft proposals of the late 1960s (Quick, 1971). There could be airside and groundside links to the main Heathrow site for passengers, but aircraft would be confined to the remote site. None of the more historically valuable buildings need be affected by this relatively small development. These options are indicated in Figure 10.2.

It can be seen from Figure 10.3 that, in the summer of 1989, 95% of European departures had sector lengths less than 1,000 nautical miles (nm). In fact, all domestic and over 60% of European flights had sector lengths less than 500 nm. There were 830 domestic and 1,955 European weekly departures. Long haul operations accounted for only 15.9% of weekly departures, although they had more pronounced daily peaks. An additional short runway would be able to increase the air transport movement (atm) capacity very significantly without diminishing the short/long haul synergy. The land take and the potential environmental impact of the new runway would be minimised. The release of slots on the long runways would, of

Figure 10.3: Cumulative percentage departures per week at Heathrow

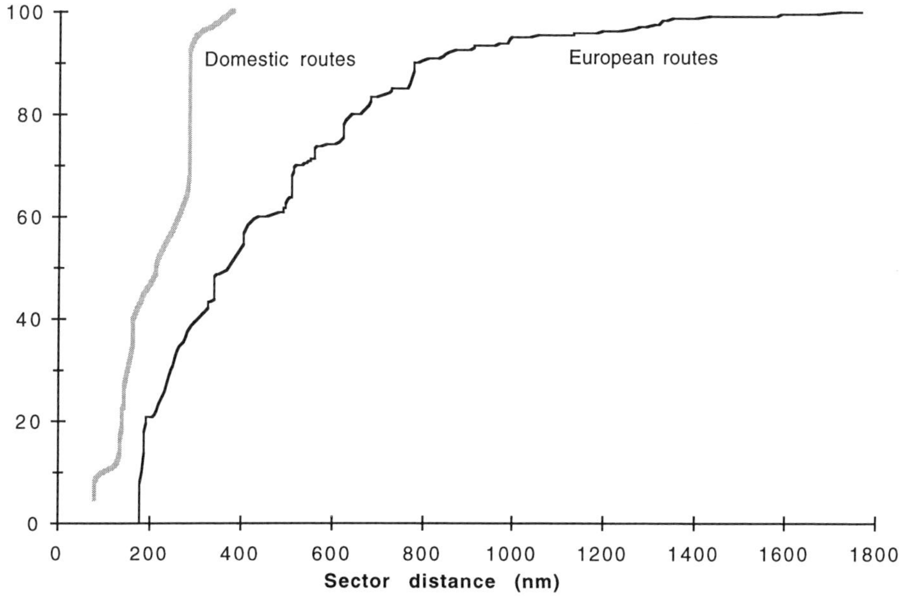

Source: Beliyiannis et al, 1990

course, encourage heavier atm to increase, so increasing both their capacity and their impacts. However, this would probably occur to some extent even if the new capacity were to be provided at another airport: the most desirable slots will always be at Heathrow's long runways, and the longer haul atm will be most able to afford the market price. Any such specialisation of heavy aircraft on the longer runways and lighter aircraft on a short runway could enhance the capacity even further by easing the vortex separation requirements.

The study suggested that even an 1,199 metre (i.e. ICAO Code 2) runway would allow access to some routes which are in danger of being priced out of the market, allowing small airlines to compete and preserving service to UK regions. By taking advantage of the type of operations licensed at London City Airport, almost no demolition need occur, and no existing residences would suffer even 57 Leq from the new runway.

A 1,600 metre runway would require the purchase of approximately 50 houses in Sipson and some 400 houses would suffer between 57 Leq and 63 Leq (45 NNI) from 150,000 Stage III noise certificated atm on the new runway. No houses would be subject to more than 63 Leq, except those already subjected to that level from the existing operations. A 1,600 metre runway would certainly require the M4 Heathrow spur road to be bridged, but the additional expense would undoubtedly be justified by the greater aircraft size and sector length which would then be available, as well as on the grounds of safety. Both the 737-300 and the 757-200 can land in wet conditions on a 1,600 metre runway at almost maximum landing weight, and can reach destinations of approximately 1,400 nm sector lengths with a full load of passengers and bags on an ISA + 10°C day and zero wind on the runway, even with no additional clearway. This is sufficient range to satisfy 100% of domestic and 98% of European departures, only three of the current European destinations not being within reach.

Even a 1,400 metre runway would allow a 757-200 to reach all domestic destinations and satisfy over 70% of European departures, including virtually all of the dense routes. There are, of course, operational questions which require more detailed investigation, such as Category III operation and the need to allow for cargo. It is also likely that growth in traffic will call for the use of the 767 on some routes, which would need a 1,600 metre runway to give the same destination coverage as a 737-300 from a 1,400 metre runway. Needless to say, concentrated operations when full use is being made of the runway length require special attention to safety areas at either end of the runway.

In view of these facts, the adoption by the CAA and the elaboration by the BAA in the RUCATSE studies of a full length third runway as one of the options of Heathrow seems to lead to an unnecessary imbalance between the benefits to the industry and the resulting social costs. There are many more issues to be considered in balancing the costs and benefits, but the runway length holds the key to the operational flexibility. A longer runway might offer greater

freedom of aircraft choice, but also might result in a much lower environmental limit on movement rates and times. Even a 2,000 metre runway, which would give full unrestricted European operations for a 767-300 even in Category III conditions, might significantly worsen the benefit/cost ratio, as well as giving problems with the eastern Public Safety Zone affecting parts of the village of Harlington.

It is possible to fit a 2,500 metre third runway with full terminal facilities for 40 mppa and linking taxiways into the same site, retaining the attractive centre of Harmondsworth, its church and graveyard, together with the hotel and industrial activity along the north side of the A4 (Caves and Brooke, 1993). It would be necessary to demolish some 3,000 houses with a total value (Council Tax Valuation) of £250 million. It is therefore likely that at least £300 million would be required for compensation. The land take would cost a further £120 million. If Harmondsworth residents wished to be bought out, a further cost of some £40 million would be incurred. In addition, some 17,600 additional houses would be brought into the three runway 57 Leq contour compared with a two runway layout with 80 mppa. At £2,000 per house, insulation would cost £35 million, though this is likely to be a low value because no industrial compensation is included. However, the major cost is for clearing the site unless the extreme compensation terms agreed at some US airports is presumed. The resulting discounted cost per extra passenger per year amounts to less than £2 at a real 6% discount rate.

The full third runway scheme investigated by the Department of Transport's working party (RUCATSE) was estimated to cost approximately £1,160 million in acquisition of property and land, out of a total project cost of £3,270 million. An improved alternative layout with overlapping available length of 3,150 metres within an overall length of 4,000 metres was also studied: the total capital cost of this option would be £2,500 million. Either option would impact an additional 56,600 houses with at least 57 Leq. It was estimated that demand would be so strong that the additional capacity would be fully utilised within five years of opening in 2005. The full schemes might then add £5 or £6 per extra passenger per year due to the additional site costs compared with options at Gatwick or Stansted.

A purely market determination would therefore permit even a full runway development at Heathrow, since airline experience has shown that premiums on passenger yields at Heathrow over other airports are much greater than £2, perhaps being as high as £50. However, this only deals with the purely local matter of land use and quality of life. All the national social, economic and environmental issues associated with the desirable quantity and location of air transport activity would still remain to be determined, together with the local non-aviation implications.

The environmental consequences of the local ground traffic associated with a free-standing terminal for the new runway might also be too great to be tolerable, compared with a remote satellite linked underground to the probable Terminal 5. The local disruption due to taxiways linking the old and new sites, and the consequent ground noise and emissions impacts might similarly tip the balance towards rejection at a public inquiry. Accepting the operational restraints of scheme which integrated passenger movement with the existing airport but kept the runway remote, would raise less severe local environmental questions. Using Northolt's runway with a terminal linked by rail into Terminal 5 is a very similar option, which may be more acceptable in that almost no additional noise impact would be caused, particularly if the runway orientation is slightly altered.

Only a lengthy and complex analysis by the industry's stakeholders could clarify the values they would attach to these choices. Unfortunately, there is no equivalently comprehensive and accurate way of valuing the environmental and economic consequences of the choices prior to the acid test of a public inquiry. Any earlier feedback to the industry of the likely reception of the options could only be obtained from an early and ongoing discussion with environmentalists, planners and decision-makers.

Any other solution to the London runway capacity problem is likely to require a full length runway, with the consequent noise generation associated with fully loaded long haul aircraft if the opportunity for long haul to short haul transfers is to be available. It is likely that, in terms of the noise dose to newly affected people per additional passenger, the minimalist Heathrow expansion options would be preferable to expansion at any other existing sites. The short length of the extra runway should offer some inherent protection for those airlines who use it from competitive bidding by the large carriers, thereby allowing more inter-airline competition and preserving access for thinner routes and regional links, particularly if a separate pricing policy were adopted for the extra runway. This may compensate for the disadvantage of not being able to transfer aircraft between the remote and main runways. A final advantage of the new short runway is that it would help to balance the airside and terminal capacities when Terminal 5 is finally constructed, particularly if the encouragement of competition is to be given more than lip-service, and if such competition continues to hold down average aircraft size.

10.1.3 The search for further runway capacity

In parallel with all the changes in airline operation in the London area in the 1980s, a series of exchanges was provoked between the government and its CAA advisors (CAA 1985a, CAA 1985b, CAA 1986b, CAA 1988a, CAA 1988b, CAA 1991c) on the possible ways of coping with the lack of runway capacity at Heathrow, and the value of retaining Traffic Distribution Rules. The trends in passengers per atm and runway capacities were reviewed, as were the effects of resulting capacity shortfalls on regional air services and on the ability of airlines to compete on a route-by-route basis at Heathrow. Various policies to cap movements were examined, i.e. to continue the existing bans on airlines, to ban other classes of service (e.g. thin domestic routes) and changes in pricing policy. After a great deal of vacillation, the Secretary of State for Transport agreed early in 1991 to take the CAA's advice to lift all restrictions on access to Heathrow for passenger airlines, and not to interfere with any of the present practices for determining entry and slot allocation (the CAA had hinted earlier that it might be in favour of a European Commission suggestion to limit frequencies on dense routes to four per day per airline). The previous limit of 270,000 atm per year, set for reasons of environmental protection, was lifted. British Airways immediately declared that a third runway would be necessary at Heathrow.

The CAA's decision to advise the lifting of access restrictions, together with its decision not to interfere with BAA charging policy except at the aggregate level, led it to review its position on regulating runway slots in such a way that regional services are protected. It decided that "it can identify no reasons of aviation policy why airports and airlines should be prevented from putting scarce resources to their most productive use" (CAA, 1993a).

The CAA was also asked to conduct a series of studies of the London area's capacity requirements and the optional ways of providing it (CAA, 1989a; CAA, 1990a). The Authority was again severely constrained by the study objectives as formulated by the Secretary of State for Transport and by its responsibilities as defined by the Civil Aviation Act 1982 and the Airports Act 1986. It was not invited to consider political, environmental, social and other considerations, leaving these for the later consideration of the government. Any ranking of preference for the sites therefore was given solely on the basis of passenger utility, airspace efficiency and the promotion of competition between airlines.

Thus, although the studies firmly established that another runway's worth of capacity would be needed by 2005 and that the airspace would, in principle, be able to handle the resulting traffic, the options examined for fulfilling the need were unrealistic. This was due partly to the constraints on the study, partly to some apparent lack of coordination between the economic and airspace planners within the CAA, partly because no attempt was made to take the options to a preliminary design phase before undertaking the economic and airspace analyses, and partly because no attempt was made to anticipate the behaviour of the airlines as well as the passengers.

The Civil Aviation Authority developed a demand model which not only considered the impact of the geographical distribution of the suggested locations on southeast passengers but also the interaction between regional and London area airports. The first attempt at this modelling was reported in 1989 (CAA, 1989a). Several useful refinements were incorporated in the methodology which led to the predictions of passengers which were used for the analysis of the feasible location options (CAA, 1990a) and the models have been further refined during later studies.

The CAA's model is responding to a particular need to understand the future demand in the London area, and only sets out to address the demand at regional airports to the extent that their growth might influence the London situation. The influence could operate either by improved service offering local competition for London's services or by the nearer regional airports' excess capacity providing a sink for demand spilt from London if the necessary capacity were not provided there. The CAA model's methodology allows both these issues to be addressed, though arguably it performs the latter function better than the former.

Overall total annual demand for the UK is predicted, based on observed relationships between air traffic and economic variables, with some judgement of the likely decline in the rate of growth over time. The model then uses the information on demand generation by zone obtained from the CAA's surveys at airports to distribute the traffic between airports. In some cases, generation is allowed to vary with the amount of air service available and it is also sensitive to the total generalised trip cost. Passengers are categorised as domestic, charter, short haul scheduled and long haul scheduled. Separate models are developed for each category, using airport accessibility, service frequency and cost to explain passengers' preferences. In addition, attraction factors are used to reflect the actual success of an airport in attracting traffic. The models assign traffic to each airport in a given year, estimate the response of the airlines with changes in route starts, route frequency and aircraft size, and then use the new supply characteristics to estimate how the traffic will change in the next year. The CAA separately estimates the airports' capacity, and reflects any capacity constraints by then applying an additional cost to passengers until the number of passengers wanting to use the airport is just equal to the available capacity.

The CAA's demand analysis made it clear that any solution based on developing the airports on the periphery of the region will 'lose' just as many southeast passengers, perhaps 16 mppa by 2010, as by doing nothing. Representative results are shown in Table 10.3. The model assumes that prices at the London airports will still rise and traffic will continue to spill to the more developed regional airports. Fragmentation of the demand among peripheral airports will, of course, put more strain on the limited airspace capacity without contributing very much to the movement of passengers.

The premises to which the CAA had been working were set out in its report (CAA, 1990, p 8). They were based on the view that a greenfield site would be unrealistic and unacceptable. The options remaining to be considered were then:

- additional capacity at an existing major airport in the southeast
- a full runway's capacity at another, or more than one, existing aerodrome
- expand the capacity of 'near' regional airports to take traffic growth from the southeast.

These options were used in the examination of the airspace system but had little influence on the demand and economic analyses. The main conclusion was that, from the point of view of the passengers, the additional capacity should be provided at the most popular airports, substantial numbers of potential passengers being discouraged from travelling if sub-optimum locations were chosen for development.

The Government set up a yet another Working Party, named Runway Capacity in the South East (RUCATSE) to examine the wider issues associated with developing the sites identified in the CAA study, to review the 1985 Government decisions not to allow second runways at either Gatwick or Stansted, to test again the extent to which regional airports could help to meet demand and also to advise on the possible use of redundant military airports or a new airport (Marinair) in the Thames estuary (DTp, 1993b).

The RUCATSE Committee worked in six sub-groups:-

- Development and Environment, concerned with planning issues
- Surface Access
- Noise
- Airport Development, concerned with airport design possibilities
- Regional Airports, considering airports outside the southeast
- Appraisal Methodology

Table 10.3: Traffic projections with and without expansion at Heathrow
a) The base case (mppa)

Year	1995	2000	2005	2010	2015	2020	2025
Heathrow	48.3	55.0	70.0	80.0	80.0	80.0	80.0
Gatwick	22.6	30.0	30.0	35.0	35.0	35.0	35.0
Stansted	3.9	6.5	12.9	19.2	33.2	25.0	35.0
Luton	2.7	4.0	6.0	13.1	20.0	20.0	20.0
Manchester	13.7	18.8	25.6	31.4	38.1	45.0	50.0
Birmingham and East Midlands	5.9	8.2	11.5	13.8	18.6	31.5	45.0
Bristol	1.2	1.4	2.9	4.0	5.8	7.9	10.0
Southampton	0.7	1.3	2.8	3.5	4.9	5.0	5.0
Lydd	0.2	0.9	1.8	2.1	3.2	5.1	5.0
Manston	0.2	0.7	1.4	1.8	2.7	8.7	10.0
Leeds-Bradford	0.7	0.7	0.8	1.0	1.6	3.4	9.3
Newcastle	1.8	2.1	2.4	2.5	3.0	4.8	13.2
Total[1] traffic	117	148	191	234	277	320	362

Note:
1. The columns will not sum to give total traffic as this includes traffic at other UK airports

b) The Heathrow option (mppa)

Year	1995	2000	2005	2010	2015	2020	2025
Heathrow	48.3	55.0	70.0	1092	120.0	120.0	120.0
Gatwick	22.6	30.0	30.0	35.0	35.0	35.0	35.0
Stansted	3.9	6.5	12.9	13.6	23.6	35.0	35.0
Luton	2.7	4.0	6.0	5.1	16.2	20.0	20.0
Manchester	13.7	18.8	25.6	29.0	35.7	43.8	50.0
Birmingham and East Midlands	5.9	8.2	11.5	11.8	15.8	21.9	40.5
Bristol	1.2	1.4	2.9	2.9	4.6	7.0	10.0
Southampton	0.7	1.3	2.8	1.5	3.3	5.0	5.0
Lydd	0.2	0.9	1.8	0.8	1.7	3.0	5.0
Manston	0.2	0.7	1.4	0.7	1.4	2.5	6.6
Leeds-Bradford	0.7	0.7	0.8	0.9	1.1	1.7	3.1
Newcastle	1.8	2.1	2.4	2.5	2.9	3.3	5.3
Total[1] traffic	117	148	191	238	294	337	380

Note:
1. The columns will not sum to give total traffic as this includes traffic at other UK airports

Source: DTp, (1993b)

with wide representation of interested groups. The sub-groups reported to the main committee (Sunderland, 1992). Sensible though the deliberations of the Working Party have been, the amount of resource devoted to the exercise is very small relative to the national economic importance of the issue. The quality of information on which decisions were to be taken was limited; further, the final balance between the various interests was, as before, struck using a methodology known only to the Secretaries of State for Transport and the Environment.

RUCATSE predicted that all the present runways in the southeast would be full by 2020 if no others were built, with 170 mppa by 2015. If a new runway was provided at either Heathrow or Gatwick, it would be needed by 2010, with 195 mppa by 2015 if it were to be at Heathrow. This would offer the greatest benefits to the industry and to passengers but, because it was presumed to be a full length runway, it would have by far the worst noise, land use and demolition implications. A new runway at Gatwick would have the same environmental impact as one at Stansted but would have greater passenger benefits.

The Government response was to direct BAA not to take further the RUCATSE options at Heathrow or Gatwick, and invited it to consider if there were less environmentally damaging options, such as a close parallel runway at Gatwick and also to examine further the gains which might be achieved and the environmental impact involved in making better use of the existing infrastructure at Heathrow. It did not specifically rule out a new runway at Stansted (*Department of Transport Press Notice 032, 2 February 1995*). These suggested investigations of possible new runways at Gatwick and Stansted both fall foul of earlier agreements not to develop further runways, though there is some doubt as to the applicability of that between BAA and the local authority at Gatwick to a short runway.

10.1.4 House of Commons Transport Committee hearings

RUCATSE cast a quite broad net in its attempt to hold a structured and intelligent debate. Though it was a closed forum, there was considerable opportunity for written consultation. The all-party Transport Committee took an even broader spectrum of evidence in its review of UK Airport Capacity (House of Commons, 1996). With respect to the London Area, it concluded that it is important that attempts be made to find ways in which capacity can be increased provided this does not involve disproportionate environmental cost. It recommended that the Department of Transport should consult publicly on the issue of mixed mode at Heathrow because it has advantages, if only in some peak hours, of bringing greater flexibility into the airport's operations and allowing more competition. However, competition would only increase if no airline's hub becomes too dominant, and the greater dominance of Heathrow would tend to reduce competition between airports. The Committee also recommended that charges at Heathrow should be increased, and the revenue used to mitigate environmental impact and improve ground access, particularly if additional runway capacity is found at Heathrow, and not withstanding the CAA's role in controlling BAA's pricing.

The Committee also recommended that the Government should fund National Air Traffic Services (NATS) to study how the air traffic difficulties of using new runways at Northolt and Redhill, and the proposed airport in the Thames Estuary could be overcome. It suggested

Northolt be made a civil airport with reserved slots for the military and VIP traffic, and that the Government, through the CAA rather than an interested party like BAA, should assess the merits of any new runway proposal and whether or not demand should be met. This may not be the best way to ensure that the maximum benefit is achieved: the CAA or at least NATS, is itself an interested party. When the Committee asked the CAA if they had examined Northolt, the reply was: "in a desultory way.......in that it is not Government policy to develop Northolt to a significant degree......the proposition that passengers would interline between Northolt and Heathrow strikes us as not being realistic" (House of Commons, 1995). This seems to reflect a civil service view rather than a private enterprise one. A private developer would be likely to realise that a good link would be required and could be made available. An earlier study (Beliyiannis et al, 1990) found that it would be possible to use existing railway alignments to give an 8 minute link into Terminal 5, so that the passenger would sense very little difference between that and any other terminal interchange.

The Committee clearly felt that there was quite insufficient information available for it to take an intelligent view on the importance of meeting demand, just as it had felt unable to judge the feasibility of the suggested options for the same reason. None of the questions were new, but insufficient resources had been applied to them. Other countries, such as the Netherlands, had seen fit to carry out really detailed studies of the economic and environmental impact of a series of possible development scenarios prior to taking a decision on the extent to which demand should be met, and how. The incentive to carry out this sort of work clearly still is influenced by the ultimate ineffectiveness of the Roskill Committee's efforts, but the direct consequence is a serious loss of information. The one serious study by the CAA into passengers' airport choice modelling indicated clearly the high cost of a wrong decision, or even a delayed one. The costs of further paper studies are minute in comparison.

The Government responded to the Committee's recommendations in a typically conciliatory way. It saw no lack of clarity in policy, which it said was still founded in the 1985 White Paper. It believed the policy to be consistent with its commitments on sustainable development, including making passengers pay the full social costs of travel, and developing a package of proposals for noise mitigation enforcement, including new powers to enable local authorities to take action against designated airports which fail to enforce their noise mitigation schemes. It agreed that noise mitigation costs should be reflected in airport charges, but considered it a matter for each airport what measures it wishes to take: this seems to contradict the Government's recent action to increase the severity of the noise limits at Heathrow and Gatwick. It admitted that there is a case for enhanced guidance on the economic and environmental assessment of airport development proposals, but that this must not prejudice the Secretary of State's quasi-judicial role in decision making following public inquiries. It sidestepped the issue of ring-fencing slots at Heathrow for regional services with the comment that the loss of a service to Heathrow might be compensated by a new link to an international hub: this seems to ignore the plight of the regional passenger wishing to access the Heathrow area of London rather than a transfer flight.

10.1.5 Policies, options and difficulties

Including some wider aspects of Government policy, and presuming that policies have not necessarily been discontinued simply because they have not been restated, it is possible to impute an unweighted listing of objectives to be satisfied by the new runway capacity, against which the Government might have been expected to have weighed the wider issues. These are:

- compatibility with airspace capacity
- suitable site development capability, including access
- promotion of competition among airlines
- promotion of competition among airports
- satisfaction of consumer demand at lowest practical cost
- minimum disruption to airlines
- promotion of regions, their airports and their feeder services
- interlining capability
- minimum environmental impact
- compatibility with local and regional planning.

These objectives should result in full accounting of the development needs arising from airports and their environmental consequences, in making best use of existing facilities, in encouraging regional airports, in supporting the international position of Heathrow and Gatwick, and avoiding subsidy from the taxpayer, as set out in the 1985 White Paper to which the Government still affirms (BAA, 1995). The Government made it clear again in 1995 that it is basically supportive of airport expansion: "The Government is firmly committed to enabling the development of additional airport capacity, where this makes economic, social and environmental sense. The economic benefits to business and industry are clear, as are the advantages of greater opportunities for leisure travel. The quality of airports in the southeast provides the UK and London with a vital competitive advantage, attracts inward investment, and makes the UK a world centre for commerce and tourism. The Government wishes to ensure that capacity can be made available in response to future demands but in such a way that recognises and takes reasonable account of environmental impacts, including the impacts of increased air traffic associated with additional runway capacity."

Apart from this broad indication of policy, the Government sees its role as the final arbiter after a developer has brought forward a proposal and it has been considered at a Public Inquiry. This leaves a private industry developer with the need to judge the timing and content of a proposal, taking a view on the need for capacity, on support by airlines for that view, the likely influence of factors outside the developer's control such as changes in the operating practices of airlines, the positive and negative impacts of the proposed development, the way in which Government might interpret or change the balance of their policies, as well as the financial viability of the investment, including any change in economic regulation of airports. The factors influencing the judgement are now considered.

No matter which strategy be adopted towards the new airport, its success will be more probable if the natural demand for its services can be accurately estimated, and by choosing a location that maximises the utility of the largest number of people. The London area is a classic example of the emergence of a second airport (Gatwick) in the shadow of a successful one (Heathrow). Its success (other than in its original role as a diversion airfield) only came with the onset of capacity problems at Heathrow and the consequent traffic distribution rules which gave only the existing scheduled operators the right to use Heathrow. This long struggle for traffic, shown in Figure 10.4, is entirely natural, given that the first airport in the system (e.g. Heathrow) will either have been located to provide the maximum utility for the greatest number of passengers or the demographics of the city will have changed until it becomes so: indeed, the demographic change may go so far that serious disutilities arise (e.g. Congonhas airport in São Paulo, Munich Riems, Athens, Sydney).

Figure 10.4: Shares of London Airport passengers

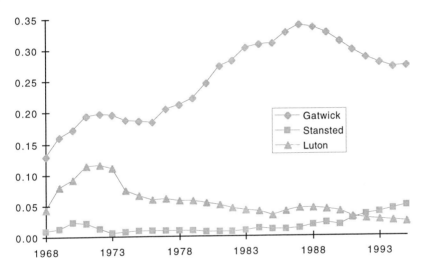

Source: derived by the authors from Civil Aviation Authority annual statistics

If the other airports in the shadow of the main ones could develop services, they might provide a greater utility for their local populations than either main airport, given the ground congestion problems. The experience of the airports north of London, including Stansted, shows how difficult it is to develop such a role in the shadow of Heathrow. Luton has no more traffic now than in the 1970s. The attraction of the major airport causes an out-migration from the local catchment area: this is not balanced by in-migration until the local airport generates a substantial amount of traffic. Analysis of the data from CAA traffic surveys shows that when Stansted (STN) does offer a destination (particularly a scheduled one) it attracts a large percentage of those trips from locations close to it, with even a fairly rudimentary level of service. Despite the limited number of destinations and relatively poor quality of service from Stansted at the time of a survey in 1984, some 15% of all scheduled trips and 30% of all charter trips from Uttlesford, the nearest zone, were made through STN, as shown in Figure 10.5. While the small local market for Stansted resulted in only an approximate 6% share of total London area traffic on the scheduled routes which it served in 1991, Ryanair was able to

298 *Strategic Airport Planning*

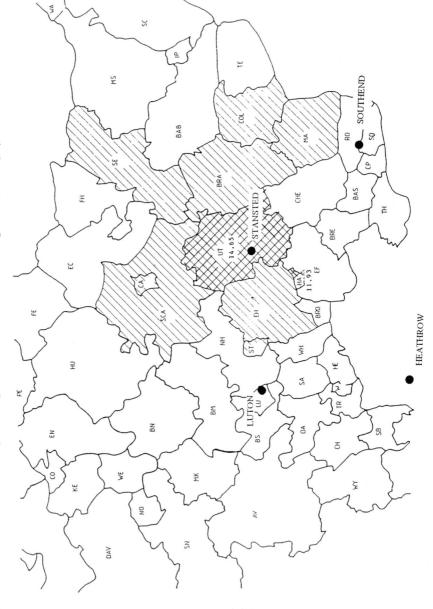

Figure 10.5: Stansted shares of origin zone passengers
a) International scheduled trips through Stansted from surrounding zones, 1984 (%)

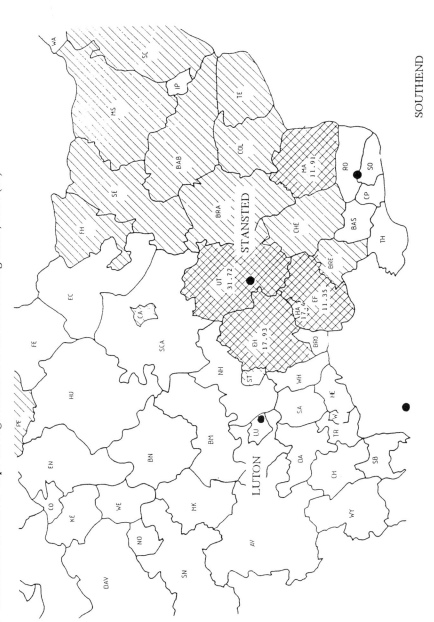

Figure 10.5: Stansted shares of origin zone passengers
b) International charter trips through Stansted from surrounding zones, 1984 (%)

increase share to over 20% to Dublin with a very aggressive low fare policy in a particularly fare-sensitive market. The yields proved insufficient to sustain service through the recession. Only 31% of international scheduled passengers at Luton and 47% at Stansted paid full fares in 1991, compared with 52% at Gatwick, 55% at Heathrow and 79% at London City (CAA 1993b). However, low cost services are now once again boosting London's secondary airports market, with the emergence of Debonair and EasyJet in addition to Ryanair.

The same problem is reflected in the history of services at Gatwick. Even on the densest short haul routes, the scheduled carriers have had a struggle to retain market share, as shown by Figures 10.6 and 10.7. The combination of a low traffic share, withdrawal of regulatory protection, uncontrollable external factors and the need to resort to low fares resulted in the demise of British Caledonian, Air Europe and Dan-Air (Cronshaw and Thompson, 1991; CAA, 1993a). Air Europe's attempt to set up a scheduled network floundered on yields lower than their low costs, even though Gatwick's yields have been better than those at Stansted or Luton. When the Gatwick share of the total London area traffic is related to the share of the frequency offered, in Figure 10.8, the classical relationship is seen to hold; i.e. a low frequency share results in an even lower market share. Where this effect is modified by low fares, the carriers often do not survive the next recession. BAA have tried all sorts of inducements to get carriers to move from Heathrow to Gatwick, but, until 1995 they were not able to tempt even those carriers with no transfer traffic, any more than the Government could get the agreement of the Spanish and Portuguese governments to move their airlines' services there in the days of British Caledonian.

Figure 10.6: Gatwick share of London domestic routes

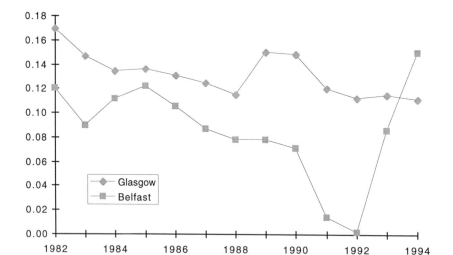

Source: derived by the authors from Civil Aviation Authority annual statistics

Figure 10.7: Gatwick share of London international routes

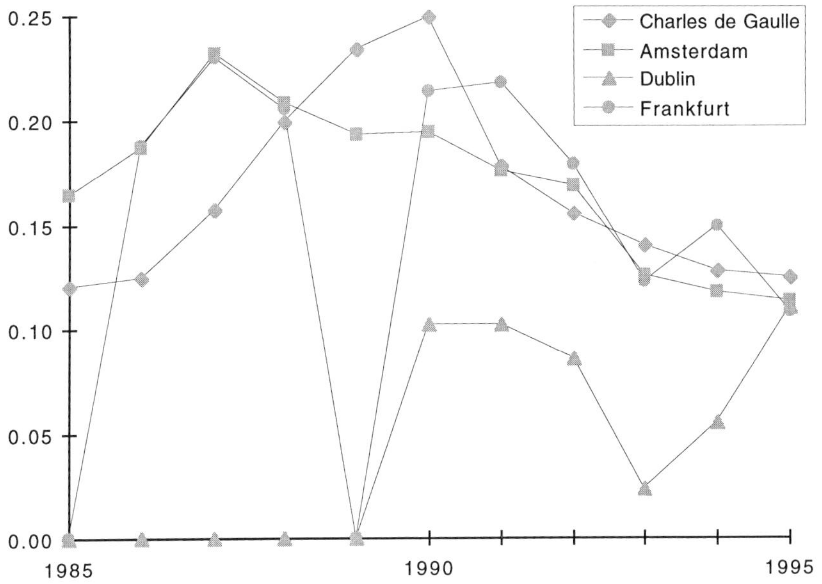

Source: derived by the authors from Civil Aviation Authority annual statistics

Figure 10.8: Effect of frequency on Gatwick share of London short haul market

Source: derived by the authors from Civil Aviation Authority annual statistics and Official Airline Guide.

It is even harder to encourage airlines to use less popular airports if the pricing policies are counterproductive. Heathrow Ltd makes most of its profits from non-aeronautical sources but it does cover its aeronautical costs with aeronautical revenue. It might therefore be expected that those BAA airports with lower levels of traffic and lower levels of utilisation of capacity would have to charge more per passenger on a cost-related basis. In fact, Gatwick and Stansted charge less than Heathrow. The CAA set an overall formula for the three BAA airports together and separate individual formulae for each of the three airports, Stansted's ability to raise charges being limited to (RPI -1) for the next five years. The CAA are also responsible for ensuring that there are no cross-subsidies between the BAA airports. Luton, which finds it has to charge the equivalent of £15 per passenger, has complained to the European Commission that Stansted's charges are below cost, but BAA denies any cross-subsidy. Meanwhile, Luton has succeeded in capturing some low cost airlines, as has Stansted, but it may have had to offer more substantial starting discounts in order to do so.

Analysis by the CAA indicates that price elasticities at Heathrow are low for both airlines and their passengers. Even if BAA had the freedom to charge at market clearing prices rather than cost-related prices, it would take 10 times the Heathrow fee to persuade an average passenger to switch to the next most suitable airport (House of Commons, 1995). The airlines may well choose to serve other high-yielding continental airports rather than other airports in the London Area. British Airways gave the data shown in Table 10.4 as evidence to the Terminal 5 inquiry to illustrate the difficulties facing operators at Gatwick (Rees-Jones, 1995).

Table 10.4: Seat factors and yields at Gatwick relative to Heathrow

Route groups	Seat factor (points)	Revenue yield
UK domestics	- 1	89%
Scandinavia	- 21	78%
France/Belgium	- 11	83%
Spain/Italy/Greece/Israel	-14	89%
Germany/Switz/Austria	- 16	78%

Source: Rees-Jones, (1995).

However, British Airways recently decided to develop a major hub operation at Gatwick, including the relocation of services to the Iberian peninsular and Latin America, ironically reconstructing the British Caledonian routes, but with the benefit of a strong domestic and European route structure supplied by its franchised feeder operators and its takeover of Dan-Air. BA traffic is to increase from under 2 mppa and 700 scheduled flights per week in 1995 to 1,000 flights and some 7 mppa in 1997. This is despite the greater yield it can command from Heathrow. Transfers at Gatwick have risen from 17% in 1994 to 35% in 1996. Until this recent move by BA, no real attempt had been made to use the secondary London airports in a genuine hubbing sense. The geographical distribution of British Caledonian's routes at Gatwick had made this difficult in any case. The traditional use for the secondary airports had been freight and low yield passenger services in common with the generic theories of successful traffic distribution (de Neufville, 1984). To this has been added spoke operations to the hubs of foreign carriers, particularly those who have had difficulty with slots at Heathrow, but also by majors wishing to protect market share.

Site-specific factors constrain the choices for new runway capacity at least as much as the behaviour of airlines and their users. Although Sir Peter Masefield's suggestion for a single large airport on the Isle of Sheppey (Masefield, 1989) may be the best answer for London if the system could be designed with a clean sheet of paper, it will be politically impossible to resist the temptation to make the most of the present infrastructure: indeed, the Secretary of State explicitly instructed the Civil Aviation Authority (CAA) not to bother to investigate green field sites in their search for new runway capacity. In view of the probable continued shortage of runway and airspace capacity, and if there is indeed no case for inhibiting traffic growth, the preferred solution must be one which concentrates the new demand sufficiently to allow the same high ratio of passengers per ATM as that enjoyed by the other major airports, contrary to the situation at London City Airport.

One option is then to develop Luton to a full 40 movements per hour. If a substantial and early investment were made at Luton, it could certainly draw more traffic than predicted in the CAA studies. It has plans to lengthen the runway and for a new terminal for 10 million passengers incorporating a direct interchange with the main Midland railway line into central London. Luton is near central London, airport competition would be provided since it is not owned by BAA, airline competition would be enhanced, its main based charter airlines have a substantial fleet of modern large aircraft, it has the advantage of different weather conditions from the other major airports and the majority of the local community are in favour of expansion. On the other hand further fragmentation of services would be necessary, the single runway configuration would once again limit long term expansion and the possibility of hubbing operations, most of the benefits would fall to Bedfordshire while a substantial proportion of the environmental cost would be borne by Hertfordshire, there must be some doubt on safety at the west end of the runway and the M1 motorway access is often very congested. Meanwhile, Luton's traffic levels continue to be volatile.

The CAA traffic prediction follows from a policy decision based on the airspace incompatibility with full development of Stansted's runway. Time and again, airspace has proved to be an artificial constraint on the provision of runway capacity, yet has been a major factor in the decisions. Yet the CAA is so ambiguous with respect to the implications of the various development options on airspace that one might as well assume all options to be workable if they are given enough warning. To quote (CAA, 1990a, p 39): "... with varying degrees of difficulty, it should be possible to develop airspace and ATC arrangements to accommodate most of the scenarios examined. However, some significant problems would need to be overcome". The report goes on to say that an extra full runway's capacity at any of the major airports should not impose insuperable design problems, though the requirements for the expansion of airspace could be very serious, particularly if both Luton and Stansted were operating at 40 movements per hour.

The remaining options all involve the development of BAA-owned airports to provide at least two runways. The resulting concentration of traffic would support hubbing operations and/or interlining on a large scale and would therefore be more marketable to airlines, provided that the site is not too far removed from the trip origins of passengers. While a two-runway airport would concentrate the environmental and economic impacts, some alleviation of both may be obtained by the adoption of a relatively short second runway: any likely future operation would only have a minority of the long haul take-offs which require extreme runway length. Airline competition on a head-to-head basis, i.e. at the same airport, would be encouraged. Airport competition, on the other hand, would be limited by the BAA's common ownership of the airports in question.

It could be argued that, to give a more balanced system, it would be preferable to develop single runway airports into two runway airports rather than to throw extra weight onto Heathrow. Apart from the environmental arguments, the overriding difficulty is that both Gatwick and Stansted have had that option removed by legal agreements. Gatwick would be the preferable location from the points of view of minimising disruption to the air transport system and maximising the synergy of the present high level of operations. However, the BAA has signed a 40 year planning agreement with the local authority not to build a second runway. This could be overturned by an Act of Parliament, but only at the expense of creating even greater bitterness and suspicion of the value of any future assurances among those opposed to airport development. Even if the political difficulties could be overcome, neither the CAA's choice of location for the second runway, nor any other location which would give a full additional 40 movements per hour, is practical. There remains, at Gatwick, the complicated alternative of providing a short, remote reliever runway at Redhill: this could be linked to Gatwick by a dedicated rail service while legitimately claiming not to be a second runway at Gatwick. It is a possibility which should be given further consideration, despite having been rejected at a recent Public Enquiry on the grounds of environmental impact and mutual interference with Gatwick runway's operations.

A second runway at Stansted may well be preferable to Gatwick on environmental and regional planning grounds, though the synergistic benefits to operations would only come later. The Government would, however, have to make an even greater U-turn if it were to adopt this solution, having required BAA to divest itself of the land it had previously been acquiring in order to safeguard the possibility of further development.

The final option is to further concentrate the industry by developing more runway capacity at Heathrow. Again the Government is in danger of being inconsistent if it favours this solution, in that it encouraged BAA to invest in the early development of Stansted's terminal. Any early significant increase in the capacity of either Gatwick or Heathrow would seriously undermine the financial viability of the Stansted investment. Given free choice, the large majority of airlines and their passengers would use Heathrow, despite its unenviable reputation for congestion. This is shown by the market's behaviour. The passengers' preferences are shown by the CAA estimates that Heathrow's expansion would generate an extra 5 mppa by 2005 compared with the other options. The airlines' preferences are shown by the struggle for peak slots: when British Midland tried to obtain the slots it required to support the new routes it had been granted, it only obtained 30% of them, and only 10% at the requested time (Reid,

1990). British Midland surveyed British business executives in the southeast and found that 45% opted for Heathrow as their preferred airport and 60% believed its capacity must be expanded because of its location and its contribution to the UK's status.

Clearly, there are real and difficult questions of balance between economic, social and environmental issues to be faced if decisions in the best interests of the nation are to be taken. Very little research has been done in some of those areas where the Government rather than the CAA is responsible. It is difficult to see how an intelligent balance can be struck without better knowledge of such issues as: the trade-off between noise impact and benefits to the local economy; the effect of the lack of provision of air service on the national economy; the social and economic consequences of positively encouraging rail for short haul travel. In the absence of this information, actions will always depend on what is politically expedient.

10.2 NATIONAL PLANNING

10.2.1 The airports and their roles

In common with any other case study, the UK can be seen to be a special case in many ways. Internally, the centres of population are close-packed, leading to intense competition from good quality surface transport infrastructure and to competition between adjacent airports for services to common destinations. Many viable domestic routes and all international routes are over water. There is a very strong international dimension to UK air transport. Foreign operators accounted for 30% of the 131 million total passengers (36% of scheduled passengers) handled at all UK airports in 1995 (*CAA, 1995 Airport Statistics*), while just over half the passengers at Heathrow in 1987 were foreign, though only 30% of Gatwick passengers and 10% of Manchester passengers were foreign. Air transport is predominantly used for non-business purposes. Even at Heathrow, with virtually no charter operations, some 60% of both UK and foreign international passengers are non-business, almost half of whom are visiting friends and relatives. On charter flights, virtually all the passengers are non-business. Even on domestic flights, approximately a third of passengers are non-business. The UK has more than twice the international terminal passengers per capita than France or Germany and four times that of Italy. This reflects the historic world role of the UK, the presence of the English Channel and the ready availability of cheap holiday packages. Yet the annual 5% growth in passengers matches that of other developed countries.

Within the UK industry, the distribution of traffic among airlines and among airports is very uneven: BA alone accounts for 66% of the tonne-kilometres used and is one of the world's top five airlines. The group of airports owned by the British Airports Authority (BAA) accounts for 70% of passengers handled at UK airports, Heathrow alone handling 42% of the total. Even on a route-by-route basis the same pattern emerges. Just four airport pairs account for 38% of domestic passengers and 11% of international scheduled passengers, all from Heathrow. In contrast, 10 routes account for 20% of charter passengers, four of the routes being from Manchester.

Economic regulation of UK international air transport is based mainly on separately negotiated bilateral agreements with other governments. While UK negotiators have, since the early 1980s, attempted to develop a liberal approach to the regulation of route access, fares and capacity along the lines indicated in the Civil Aviation Act of 1982 and now being taken for the post 1992 internal services of the European Union (EU), these factors remain subject to close control on many routes. The unregulated charter industry arose partly in response to the requirement to supplement scheduled capacity in unbalanced markets. Domestically, the CAA retains ultimate authority on route access and fares in order to protect thin routes and small operators against predatory action, while maintaining a liberal stance. Government control of the airlines' economic behaviour is now largely in the hands of the Monopolies and Mergers Commission (MMC) and the Office of Fair Trading (OFT), though government dictates the proportion of equity which can be in foreign hands.

The UK appears to be over-equipped with airfields to cater for short haul jet aircraft. This situation is the product of a long history of changing views as to the importance of air transport and the ownership of airfields. It is also a product of standardised airfield building during the 1939-1945 war with runways suitable for bombers (6,000 ft) and for fighters (4,500 ft). Outside London, the only airport with a runway capacity problem is Manchester, but several have either recently added runway length or are considering making applications to do so, in order to make better use of their capacity by tapping into the longer haul European or intercontinental charter destinations. Charter passengers are showing an increasing preference for their local airports as they take more short breaks and as delays increase, but it is by no means certain that the airlines will provide a consistent service: available capacity is a necessary but not sufficient condition for traffic growth. Several of the airports have captured traffic and required terminal expansions, often at the expense of neighbouring airports.

Many of the smaller airports used to operate at a loss, subsidised by their local authority owners. A study in the late 1970s (Doganis, Pearson and Thompson, 1978) found that only Heathrow and Jersey were in profit and suggested that a throughput of up to 3 mppa would be necessary to bring others into profit. It appeared that only a considerable rationalisation of the industry, or continued local subsidy, could resolve the situation. However, the Joint Airport Committee of Local Authorities (JACOLA) managed to introduce a substantial and consistent, phased increase in charges to replace the previous situation in which competitive discounts were being offered to airlines to counteract the oversupply. The formal arrangement was terminated after being ruled as anticompetitive, but, combined with cost control and severe Government constraints on the availability of loans to the public sector, the balance sheets improved. It became clear that airports with throughputs of only 500,000 ppa could be profitable if their facilities were tailored to the demand, but this did nothing to resolve the oversupply of airports able to support commercial air services, or to quieten the calls for an authoritative National Airport Plan which would offer some guidance in the matter.

10.2.2 Attempts at national airport planning

Most of the early UK planning history has been detailed elsewhere, e.g. Caves, (1979). In summary, before 1965, the predominant issues were of ownership. It was expected that the market would determine the size and location of infrastructure. Planning came into favour with the 'Edwards' committee on the state of the British civil aviation in 1969. The recommendations for planning were excellent, dealing with the integration of routes and airports, of aviation and other modes, of aviation and other sectors of economic planning. The report also recommended concentrating air services, putting the ownership of airports in the hands of the most interested community, paying attention to technological changes and creating the Civil Aviation Authority (CAA) to combine the functions of route licensing and airport planning.

The Edwards Committee summarised the objectives of a National Airports Plan as:-

- to achieve a distribution of airports which will meet the need of economically viable or socially-supported air services in every part of the country, without wasting scarce resources in view of the large amounts of capital and areas of land

- to prevent proliferation of airports serving the same traffic areas, and thus to strengthen the economics of the airports themselves, and to provide the basis for a stronger network of air services than would be possible if airlines served two or three airports where one would suffice;

- to ensure greater coordination between airport development and the provision of air services so that investment decisions in neither field are made without the knowledge of decisions in the other;

- to establish forward plans for meeting future requirements, particularly those arising from the rapidly changing operational characteristics of civil aircraft;

- to coordinate airport development with the means of access to airports.

The CAA could not see the need for National Airport Planning, except to tidy up the corners and conflicts left by the separate regional planning studies which they initiated. This was probably due to the fact that UK aviation is dominated by international traffic, with the large majority of this traffic concentrated in the London hub: also the land transport for access to hub airports was apparently highly developed. In fact, the CAA disagreed with all the consultants' recommendations for regional airport development. The government in 1978 took the CAA's advice that maximum use should be made of existing regional infrastructure, even if aviation costs were not minimised. The government tried to structure this advice by developing a four-tier classification of airports, disclaiming responsibility for the two smaller classes.

This could only be seen as a policy rather than a plan, because it failed to take into account the interactions between the regions and the London hub, and the UK does not have a formal regional structure of government. Also, the national government has only a permissory function over local infrastructure planning. The policy did not address some of the primary questions that would normally be considered in good transport planning practice: it did not formulate equal accessibility policies, integrate transport services between modes, resolve conflicts between airports with common catchment areas, or end the mutual suspicion between the national and local authorities.

Perhaps the most serious deficiency of the 1978 White Paper was that it did nothing to quell the fears of those outside the air transport industry that they had been excluded from the debate. It was felt that the community was only allowed to respond rather than to participate (RAeS, 1978). In fact, the degree of consultation was only a small advance on the Edwards Committee evidence, at which time the BAA first proposed the idea of a standing policy advisory council. This would have provided the necessary forum for continuous participation. In fact, the system has developed under internal initiatives and despite conflicting decisions on route licensing between the CAA and the Department of Transport. The latest Airport Bill (1986) effectively admitted the previous deficiencies and explicitly allows individual initiatives to prevail. It was, in any case, more concerned with the privatisation of BAA and setting up public companies to run the major regional airports. Having allowed a certain amount of freedom in route and airport development, the CAA/Department of Transport now have to take a stronger regulatory role to control private monopolies, to ensure adequate standards of safety/environment/security, and to control airport pricing.

Since 1986, central planning of infrastructure provision has once again been undertaken only to relieve congestion around London. The regional airport situation is only considered in terms of possible relief for London, though the airports do, of course, have to conform to normal planning guidelines developed in County Structure Plans and Local District Plans.

As for General Aviation (GA) planning, virtually none was done in the UK or on a European basis in the 1970's. No runway slots are specifically earmarked for GA in the peak at Heathrow, and the situation is just as tight at Gatwick. The lack of space at London's hub airports has caused studies of alternative available 'relievers', but always they are far from the main airports and the centre of London and virtually all of them have environmental limitations. One of three heliports in the river Thames has been closed for environmental reasons.

Figure 10.9 shows regionally how there have been winners and losers. In general, those with the greater market share in 1984 have increased it further. The trends are based in the economies of scale available to both airports and airlines from a concentration of activity, this being sufficient to allow the larger airports to overpower the advantage of local access, given the short distances between competing airports.

Market forces have therefore been shaping the industry despite its rather tight historic regulation. There is no guarantee that the resulting system has produced a distribution of capacity which is appropriate when judged against wider economic, social and environmental goals than those internalised in the air transport system's development. The most important issue is the bias towards the London airports in terms of scheduled services, for which there are strong historic reasons associated with British Airways' base and bilateral agreements, and against which the regional airports have had an uphill struggle. In the charter market, the propensity to fly and to use a regional airport are relatively well balanced across the country. In contrast, the propensity to fly on scheduled services is much lower in the regions than in

London and the southeast, even allowing for the passengers from the regions who use ground access to airports outside their own region (CAA, 1993b).

Figure 10.9: Regional shares of UK international markets
a) Scheduled passengers

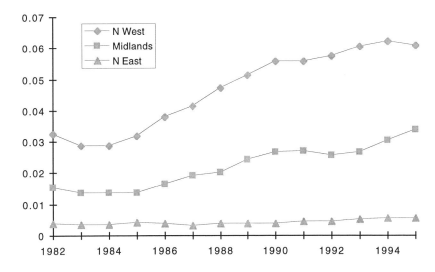

Source: derived by the authors from Civil Aviation Authority annual statistics

b) Charter passengers

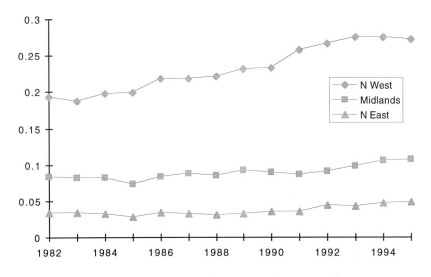

Source: derived by the authors from Civil Aviation Authority annual statistics

10.2.3 The 1986 Airport Privatisation Act

The end of the 1970s had seen the first fruits of deregulation in the USA and the birth of the Thatcher-style free market philosophy in the UK. The early 1980s were characterised by a hiatus in planning, as the laissez-faire policy setting began to gain strength. The 1985 White Paper brought forward plans for privatising BAA, making larger regional airports into Public Limited Companies (even though all the shares would be in the hands of the former local authority owners, and hence the airports would still be subject to public borrowing limits) and raised the issues of how to control private airport monopolies. The subsequent 1986 Airports Act gave effect to these measures, BAA being left intact but being required to form separate Public Limited Companies for each London airport and for the Scottish Group (by then including Aberdeen, Edinburgh and Glasgow as well as Prestwick). Encouragement was given to improve airport ground access generally, for Manchester to develop as a regional hub, and for other airports to develop their own futures.

As the 1986 Act was being written, there were simultaneous policy developments to reduce regulation of route entry, frequency and fares, both on domestic routes and those international routes where liberal Air Service Agreements (ASA) had been signed. It was no surprise when this next major review of airport policy abandoned the national plan in favour of market forces, but with strong governmental influence on the non European Union longer haul role of airports through the route rights negotiation process. Thus airports would now have to hunt for their own roles, both in the airline market place and in government circles. This has been so particularly with the several airport pairs which had led an uneasy co-existence. However, by this time some of the winners were already beginning to appear, and it would be wrong to attribute their success solely to the privatisation Act.

An airport pair which has been much influenced by the change in regulatory regime and the 1985 policies is Birmingham/East Midlands. Birmingham had prepared itself prior to 1985 with a new terminal and had a strong desire to reflect its status as the second largest UK city with a comprehensive range of services, despite being only 150 km from London. The local authority wanted to retain full control of the airport. It used its high quality facilities to start to attract the short haul routes that the 1985 policy gave it the freedom to develop. This, in turn, attracted British Airways to establish a small Eurohub operation in a terminal extension funded by a consortium whose members are the airport, British Airways, Laings the contractors, NCP the car parking company, and Forte the catering company. British Airways was guaranteed a minimum income from the terminal and paid lower passenger charges. This joint funding had never been used before at a UK airport. It overcame the funding constraints of the public sector borrowing limits in exchange for equity shares in the Eurohub terminal rather than in the airport itself. The Eurohub opened in 1991. This and other initiatives have been successful to the point that the airport has received permission to double the terminal and apron capacity to 10 mppa, without going to public inquiry but in exchange for several guarantees, one being not to increase the number of night flights. The site is much more environmentally constrained than East Midlands but an additional 7,000 jobs will be created. In 1996, it put 40% of its shares up for sale for £130 million, and accepted a bid from Aer Rianta, the Irish Airports Authority, partly to ensure funding for the new £400 million expansion. The local authorities retain 49% of the shares (*ACI Europe Communiqué No 80, Dec. 1996/Jan. 1997, pp 1-2*), having bought back British Airway's shares.

East Midlands' local authority owners had always taken a very cautious view on investment, only expanding when overcrowding clearly signalled the need, the airport making a modest contribution to their general income funds. This did not encourage the airlines to make a serious commitment to the airport, though the total potential demand from a wider and lower density catchment equals that of Birmingham. It has lost market share to Birmingham, more gradually in the charter than the scheduled sector. The 1986 Act caused a change in the constitution of the Board, with a greater proportion of the airport's executives beginning to have a stronger voice in decision-making. However, a real change in attitude had to wait for full privatisation in 1994. The National Express coach company bought the airport in competition with a management buyout bid, and retained the full management team. The new company immediately bought up a farm which had been given permission to develop an industrial estate and which posed a threat to the airport's long term expansion possibilities. It also funded a large freight apron expansion, following the arrival of substantial UPS and DHL transatlantic operations, and has undertaken a doubling of passenger terminal capacity, all within two years of the takeover. As traffic develops, it is planned to link with train services run by National Express on the nearby main line to London, in addition to the wide range of inter-city coach services which already hub at the airport, all of which will fit in well with the government's Integrated Transport Policy.

National Express had wanted to gain a controlling interest in Birmingham as well as East Midlands, with a view to coordinating activities between the two airports. East Midlands has permission to extent its runway to give genuine long haul capability and has very little housing development within its noise contours. The intention would have been to bias the long haul, charter and freight towards East Midlands, while encouraging a full range of scheduled European services at Birmingham. This, together with the group's purchase of Bournemouth airport and its involvement in newly privatised airports worldwide, including Stewart airport in New York state, indicate the synergistic initiatives which private enterprise can bring to airports, reaching far beyond the release of funds.

The Bristol and Cardiff airports are sited on either side of the Bristol Channel. The have quite similar catchments which hardly overlap. Their traffic increased almost identically both before and after 1986, neither airport releasing any of their shares to the general public at that time. In the 1990s, despite heavy investment by Cardiff to attract airlines and its success in mounting transatlantic services, Bristol began to grow more quickly. This can be attributed mainly to expert marketing of a constrained site, particularly to the foreign end of scheduled routes and to selling the south west of England to foreign tour companies. Bristol is now expanding its terminal to cope with demand, having seen a proposal by British Aerospace to develop its own existing airfield closer to the centre of the city rejected by the Government after a public inquiry. First Bus, a coach and train operator in the same region of the UK, has recently been bought 51% of Bristol airport, while Cardiff airport has been sold to TBI, a local property company which has since bought Belfast International and Coventry airports. TBI has also bought Sanford international airport in Florida and Scavsta airport at Nyköping, which Ryanair uses to serve Stockholm (*Jane's Airport Review, July/August, 1998, p 4*).

Belfast had been served only by Belfast International airport, some distance outside the city, until Belfast City airport was made available to commercial services by its owners, Short Brothers (later Bombardier) in 1983. Northern Ireland's airports were excluded from the 1986 Privatisation Act, but the Airports Order (Northern Ireland) 1994 passed statutory control passed to the local authority who then became legally entitled to sell Belfast International. The International airport was sold in 1996 to TBI for more than £40 per annual passenger, including compensation to the staff who had already bought the airport from the local authorities a year earlier in association with Mercury Asset Management.

Prestwick had been the nominated airport for Scottish/North American services since the first bilateral agreements due to its long runway and good weather. As a 'gateway' airport, it became part of BAA when it was established. The BAA also acquired Glasgow Abbotsinch airport in 1975. When airlines took advantage of the ruling allowing Abbotsinch to be a gateway in 1990, Prestwick was soon sold by BAA to a local business consortium headed by a Canadian lawyer and capitalist (*Jane's Airport Review, September 1997, p 94*), which has marketed it heavily for transatlantic freight with some success (see chapter 13).

The smaller regional airports in general have faired less well as the larger ones have increased their market share. Any traffic growth that has occurred has been led by the new low cost operators who have appeared in the wake of the recession of the early 1990s and the new route freedoms available under the European Commission's Third Package of liberalisation measures, effective from 1 January 1993. Ryanair has been particularly active, using secondary airports to link London and the UK regions to Ireland and also to compete on the main domestic trunk routes. Stansted, Prestwick and Liverpool have benefited directly from their initiatives, as have other smaller airports as Aer Lingus has responded to Ryanair's low prices. It is too early to say whether these services will stand the test of time and re-energise the smaller airports, but the early signs are encouraging. Other low cost airlines like EasyJet and Debonair have also made a good start in 1996 from Stansted and Luton, not only domestically but also in opening continental destinations to genuine low fare competition.

Another private sector company, Regional Airports, has bought Southend, which has changed hands several times without notably improving its market share, and Biggin Hill, which is the main General Aviation airfield for London with ambitions to enter the scheduled market. Southampton, which has been in private hands for decades, is now owned by BAA, which is likely to meet its first significant competition from a privatised airport in the London area, now that Luton is likely to licence the US Airports Group International (AGI) to run the airport for 50 years. AGI includes Bechtel and Lockheed, and has pledged £170 million for a new terminal which will integrate the airport with a railway station, and put Luton only some 15 minutes from central London. A property company, Tinsley Park, jointly with the local Corporation, has financed a new airport at Sheffield (Woolley, 1997), and KLM (UK) has opened a service to Amsterdam.

10.2.4 London and the regions

The RUCATSE studies had shown yet again that there were only limited opportunities for enhanced regional services to reduce the strain on the London airports. The Government, none-the-less, wishes to maximise the use of the spare capacity in the regions and minimise the provision of new capacity in the overcrowded southeast. This policy is in line with the wishes of the majority of communities, both in the regions and in the southeast, but the airlines and many of the passengers would prefer more capacity in the southeast where the revealed pressure is greatest.

The regional airports certainly wish to play an increased role in the long and short haul scheduled market. The arguments about economic benefit usually presume them to be related to the number of direct links which are available. The passengers' utility may, in fact, be maximised by using high frequency links to hub airports rather than low frequency and perhaps higher cost direct links, but a large majority of them tend to prefer short haul direct links if the frequency is twice daily (Brooke, Caves and Pitfield, 1994). There is still considerable opportunity for the airports to exploit their high quality of service on new scheduled routes, particularly to the continental hubs. Amsterdam has already shown the potential for these links, with service to 21 UK regional airports. This may well be good for the regional airports and their regions' economies, but will result in more traffic for foreign airlines and airports at the expense of the UK economy.

The Government has encouraged these regional ambitions by liberalising the transatlantic route agreements to include all regional airports, together with Stansted and Luton. Manchester and Glasgow, with their longer runways and the good field performance of the long haul twin-engine aircraft, had had some reasonable success in getting scheduled routes started, usually by foreign carriers, even before this ruling in October 1994. These routes provided data on the net benefits of such services on the local, regional and national economy which the CAA (1994) analysed. The positive conclusions, even when the services were provided by a foreign airline, gave useful backing for this more liberal policy. Although some routes have flourished and other new routes have been started, several of the services have been withdrawn due to low yields, the business class passengers apparently preferring to use the higher frequency London services rather than hubbing in the US. Since the ruling, other regional airports have seen the start of transatlantic services, with British Airways showing more interest and Birmingham serving three North American destinations, but it is too soon to say if these and other new routes will be successful.

Many of the regional airports see their main new market opportunity in long haul charters, as the operators increasingly use their long haul twins to exploit the desires of the more seasoned holidaymakers to explore further afield. Orlando is already being served by eight of the regional airports and Manchester is serving nine points in the Caribbean. These charter markets could support service from other airports, but most would have to extend their runways. There is more chance of these charters fulfilling the Government's expectations of direct regional services off-loading the London airports than the long haul scheduled services, but this will not ease the situation at Heathrow. Indeed, it is unlikely that any new route starts from the regions will reduce the demand for runway slots or stands at the congested London airports. It is more likely to hold down the rate of increase of aircraft size.

10.3 THE PLANNING PROCESS

10.3.1 Planning frameworks

The planning process for UK airports is described by the Government as consisting of three stages:

- the Government establishes a policy framework for the market within which airport operators could plan,
- specific proposals are then put forward by airport operators according to their judgement of what the market wants and these are considered at public inquiry;
- the Secretaries of State for Transport and for the Environment then jointly take a view on the acceptability of the proposal in the light of the inspector's report (House of Commons, 1996) and the wider national issues.

These wider issues, when and if they are considered, have three effective modes of expression. Firstly, they may be discussed in private by the Secretaries of State and their civil servants. Policy statements will then be made in Parliament and the grounds for them will be subject to questioning from members. Secondly, the state of any part of the industry may be examined by a governmental Select Committee (usually the Transport Committee), a Standing Commission (e.g. Monopolies and Mergers) or a Special Commission (e.g. Roskill). The Special Committees are convened in a spasmodic way and are often subject to the criticism that they were given the wrong job (Rhodes, 1975). While these bodies can accept evidence from a very wide net, and the results of their deliberations are open to public scrutiny, the deliberations themselves are not necessarily in public. Despite holding public hearings at each short-listed site, and later general hearings, at least one of the Roskill Commissioners felt that the process by no means ensured public acceptance, probably due to the public's lack of trust in quantification by cost/benefit analysis (Keith-Lucas, 1974). Further, although the Select Committees' powers of scrutiny are impressive, their ability to influence government decisions is limited by the overall majority which the government usually enjoys. This is exemplified in the responses of Government to the latest report by the Transport Committee, discussed above, and by the rejection of the Committee's advice in the 1980s that the extra runway should be built at Gatwick.

Thirdly, the public inquiry process, which is almost invariably necessary since any major infrastructure will require a significant change of land use, allows wider issues to be raised publicly. The ensuing debate is always long and comprehensive, though the boundary is usually drawn around purely local issues. To the extent that the combined Stansted/Heathrow Terminal 5 inquiry was concerned with two locales and debated wider implications for the London aviation system, it was an exception to the rule. BAA and others feel that holding discrete inquiries for each proposal does not allow a strategic view to be taken of airport capacity provision. Ultimately the Inspector issues publicly available advice to the relevant Secretary of State, whose eventual decision is taken in the same way and subject to the same sanctions as the first mode of expression.

Some in the industry feel that the Public Inquiry system is the main obstacle to the addition of airport capacity because it is adversarial and the loser loses all. It is also felt that a proper system of compensation for disbenefits would ease and speed the process (Egan, 1992). The delay in approval of plans a makes a nonsense out of some of the design calculations. By the time Heathrow's Terminal 5 opens, and it is hard to believe that permission will not be given, given that BAA has already committed so much time and effort to it and that there is no other obvious alternative which could be brought on stream before the onset of serious congestion, Heathrow will be close to handling the 80 million passengers per year which was supposed to be the design capacity of all five terminals. The cost of the inquiry by day 436 was £70 million, of which local and national government have contributed £17 million. Some of the local authorities have had to stop attending because of the high cost. There had been 21,000 written submissions, and 50,000 questions to 650 people (*Sunday Times, 14 June 1998, p* 16). It is not only in terms of timescale that larger airport issues do not fit comfortably into the Public Inquiry process. It was devised for smaller-scale development issues, where there is a limited number of actors and they do not each have multiple roles. For example, the government was the initiator of the Stansted proposal, the individual government departments had their own responsibilities and the government had its own economic and political policies which could be expressed through its decision, whereas its nominal role in the inquiry process is merely to act as an adjudicator (Toms, 1984). This can easily lead to "a wide-spread cynical resignation to the notion that airport decisions are political, that politics is the power of organised interests" (Sharman, 1984).

There is a fourth mode of expression, namely the Planning Inquiry Commission (PIC), which was created in the Town and Country Planning Act 1968 (Harrison and Williams, 1991). However, despite its retention in the 1990 Act and it being expressly designed to tackle large non-local issues, it has not been invoked. Essentially, the PIC allows a pre-inquiry assessment process, taking into account technical and national issues.

Committees have considered it inappropriate to formulate any long term plan for civil aviation, though there have been many calls for one by those giving evidence. The only remaining opportunity for any such plan to be introduced into the decision-making process is in discussion between the civil servants and the Secretaries of State, where the civil servants would have had to be the sole authors of the plan. There is no evidence of such a plan. Hence it must be concluded that the long term future of aviation infrastructure is being decided by an unstructured interplay of information from four main sources:

1) the CAA offers advice which has been preconditioned to detail the user's interests, expressed as relatively unconstrained long term demand forecasts;

2) the Secretaries of State for Transport and the Environment balance the issues on the evidence from their civil servants;

3) locally-oriented plans are submitted in evidence to public inquiries;

4) all the actors lobby the responsible officials.

The individual airport operators certainly do their own planning, but this is usually based on relatively short term forecasts of the airport's organic growth, and on the art of what is possible in terms of facility provision; evaluation is generally by expected rate of return, tempered by judgements as to the preferred scheme passing the public inquiry hurdle. Any longer term and formalised master planning along the lines of international recommendations (ICAO, 1987a), which themselves derive from the FAA's systemic approach, is often performed confidentially to avoid creating undue alarm about long term expansion, yet one of its main purposes is to ensure compatibility with other local plans. The BAA's laudable attempt to develop and implement a long term master plan for Stansted suffered a serious setback when the government directed it to sell the land which it had bought to safeguard the possibility of constructing a second runway.

As yet, it is rare for an independent airport to go through the full system planning process from the setting of goals through to continuous monitoring of out-turn in relation to the goals, though some do embody elements of the process in their five year corporate planning. It is even rarer for these exercises to identify and quantify a role for the airport relative to other airports or to use the analysis to create a new role, as Manchester has recently done (Bowen, 1991). Indeed, many airport authorities have previously joined with other pressure groups in asking the government for a national airport plan to guide them in the role they should adopt, and there is now the desire in some quarters to see a similar European-wide plan.

It is not clear how the present laissez faire approach to planning can solve the problem of the shortage of capacity in the London area. Substantial commitment is required simply to make a planning application of sufficient magnitude, followed by a far greater commitment to construct facilities sufficiently in advance of the traffic being generated to be able to convince the airlines that they should make full use of the facility when it is available. Meanwhile, the airlines object to financing from profit i.e. to contributing now to investments which will benefit others 10 years later, and are often prepared to accept some congestion costs rather than make access easier for competitors. It would appear that the investment could only be made entrepreneurially when congestion costs become so great that the airlines would benefit from spending on new capacity. This would require a different approach to the price control mechanism imposed on BAA so that prices could rise to the market-clearing level, and some way would also have to be found of feeding the excess rent back into the industry (Pelkmans, 1989).

Solutions of this sort would not benefit competition by opening up access to attractive airports. Nor would they resolve the NIMBY (not in my back yard) syndrome which reflects the frequent incompatibility of local and national, and user and non-user interests, unless the market-clearing price is allowed to be high enough to provide generous compensation. In the public inquiry process, the national need is subsumed into the local traffic predictions and the non-user interests are limited to local environmental effects, the economic impact and alternative land uses. Whilst these are all important and necessary issues, they do not, and cannot represent more than a localised indication of the national consequences of these issues. The national interest requires wider consideration, so, quite properly, the relevant Secretary of State then deliberates on the decision. At this point the UK planning system diverges from the 'best planning theories' appropriate to its context. The national level decision process is subject to uncontrolled lobbying and is ultimately secret unless and until a Select Committee investigates the decision, in contrast to theory which suggests a maximum degree of participation on a wide front and has developed specific evaluation tools for the situation.

Methods such as the Minimum Requirements Approach (Hill and Lomovasky, 1980) have not been given an opportunity to function. Instead, the resulting decisions have perpetuated the disjointed incrementalism in airport development. There is no evidence that the decisions have been informed by other than the local information and central dogma, plus some sort of political judgment of what the consensus would have been if it had been possible to achieve one, coloured by lobbying and the private advice of civil servants. The consequence is that the government's objectives for the system, where they have been stated, have only partially been met. Further, many objectives which it would be normal to include in transport planning studies have not even been stated. Also, the decisions being taken now will only influence the system in the next century, yet little, other than the CAA's demand modelling, has been done to explore what either air transport or the world could be like then.

The power of the Secretaries of State for Transport and the Environment, and the necessity for all plans for major projects to pass through their hands, means that projects based on British soil can only proceed with their consent. However, this filter makes it difficult to retain the 'market force' ethos unless the issues and the methods of judging them are sufficiently transparent: otherwise the entrepreneur is buying a ticket for a very expensive lottery. It is no wonder that the team leader for the planning of the new Denver airport, almost the only new airport in the US to be built in 10 years, was a lawyer with experience in lobbying Congress. In the relatively simple situation of a supermarket chain with one hundred branches applying for planning permission for one more branch, the issues are usually clear and only of local import; the relative commitment is low; the inspector's advice to the SoS is rarely overturned. Where a major airport development is concerned, not only is the relative scale much greater but also the inspector, the SoS for Transport and the SoS for the Environment become effectively three separate filters; further, the arguments balancing national and local interests are much less well rehearsed. Yet, if private initiatives for airport development could be encouraged, in sufficient quantity and quality, it would ensure that useful solutions were not overlooked by early and crude filters of a strategic planning process conducted internally by government (Caves, 1991). They would not, however, avoid the need for compromise which the planning system seems to generate: thus Gatwick's second runway was traded for Terminal 4 at Heathrow, Stansted's second runway was forfeited in order to gain permission for its single runway development and an early opportunity to secure a fully independent second runway at Manchester was lost due to fears of the political consequences of developing into another county. BAA's attitude after various public inquiry experiences is that it is essential to wait until the need for the facility is quite indisputable before attempting to pass the public inquiry clearance point.

10.3.2 Implicit goals for the UK airport system

A major concern about the present planning process is the lack of clarity in policy guidance against which to judge a proposal for major new additions to airport capacity. Many witnesses to the latest Transport Committee investigation called for a clearer statement of policy, including the Aviation Environment Federation (AEF), Manchester Airport and the BAA. The AEF were concerned about how the principles which underpinned planning policy guidance applied to airport development. The BAA thought the Government should be increasingly proactive in determining a clear role for air transport within the national economy, so allowing private industry and local communities to use such policy as inputs to their planning processes, relieving the present high degree of uncertainty. Manchester suggested that policy tests be formulated which any development proposals would need to satisfy. This would require the Government to be much more explicit in its policy formulation, indicating priorities for airport development and which proposals would be preferable. The Transport Committee supported this suggestion (House of Commons, 1996).

The Government believed that its policies were transparent and consistent, that they stemmed essentially from the 1985 Act and conceded only that it may be useful to collect them all together in one statement. Yet there is evidence of some confusion. BAA said in evidence that it weighs the following factors in judging the timing and content of development proposals:

- its own assessment based on its own level of service criteria;
- its airline customers' views on need;
- the likely influence of factors outside BAA's control;
- favourable and unfavourable impacts;
- Government and Regional Planning policy;
- whether the investment could be remunerated.

They mention bilateral negotiations, European Union legislation and global airline alliances among the external factors, but not sustainability, global emissions or UK economic benefits, all factors which would be weighed in the balance by the Secretary of State and on which there is disputed evidence. The British Chamber of Commerce, in its evidence, stated that air should be linked with rail and road, stating that this cannot happen without a clear vision by Government, so implying that they could not infer the Government's policy, at least on that issue. In reply to a question from the Transport Committee on how the CAA's views would change if environmental pressures had to be taken into account in their advice to Government on the preferred location for a new runway, the CAA representative said he could not answer because he found it difficult to forecast the effect of those pressures on the Government and the decisions that they might take. A Department of Transport official admitted in evidence that the policy framework was ".....not complete at present. .... I would not pretend that there is a neat, clearly presented framework already in existence."

The lack of consistency in policy adds to the confusion. Shortly after permission was given for the fourth terminal at Heathrow, there was a comment in the House of Commons, with respect to the 275,000 atm per annum environmental cap, that "It is not playing the game for someone who has received planning permission to come back later and demand that one of the crucial conditions should be broken in order to make more money" (*Hansard, 22 June 1984*). Yet the condition was lifted by Government only a few years later. Equally, inconsistency has caused problems for the developers. A BAA director commented recently that, if BAA had known that the traffic distribution rules (which had limited the categories of user able to operate at Heathrow as well as the total movements) would be scrapped, it would have thought twice about building Stansted (Everitt, 1995).

It is possible to challenge the consistency and clarity of Government policy in the Law Courts, as instanced by the action brought by some of the local authorities affected by noise at Heathrow, claiming that the changes to night noise rules were not clear or consistent with stated policy. The Court of Appeal ruled against the claim, and further ruled that it was sufficient for the Secretary of State to have identified the policy objectives with adequate clarity and to have acted in accordance with them. Further, the court took the view that it was Parliament, not the courts, which should hold him accountable if it wished to query or challenge the policies he had chosen (*Times, 20 August 1996, p 18*).

It was felt by many of the witnesses to the Committee that the inquiry process would not only become less of a lottery, but also be speeded up by testing of a proposed development against policy before taking it to the inquiry stage. British Airways indicated in their evidence that, for a major decision like a new runway, the Government should give a broad view on the acceptability in principle of a proposal. The Airports Policy Consortium (Bendixson, 1991) felt that, for a proposal by a private developer for an airport in the Thames Estuary to be forthcoming, the Government would have to give a clear political lead, as it did with the Channel Tunnel. It would also have to indicate its willingness to give Parliamentary support to the necessary infrastructure links and probably give a view on whether BAA would be an eligible developer. In other words, there should be a two stage clearance process. A process like this exists for railways and some other transport schemes under the Transport and Works Act. A developer makes an application directly to the Secretary of State, having undertaken consultation and an environmental statement. A motion is then moved in both Houses of Parliament by a Government Minister which allows each House to approve the proposals in principle, after which there may be a public inquiry. In essence, it is similar to the PIC process mentioned above.

In parallel with these improvements in the process prior to the public inquiry stage, it was also suggested that much of the factual information could be agreed in advance, so avoiding endless debate on shades of expert opinion. If the Government took the role of assimilating the necessary factual evidence with working parties like RUCATSE, bringing all interested parties together to contribute and to consolidate as common a position as possible, the inquiry could then get on with its prime task of examining the planning issues themselves against the clearly stated policy goals.

The Committee supported these suggestions to make the planning process more balanced, efficient and speedy. It also urged the Government to set policy criteria and to use them to test each alternative scheme for acceptability. The Government saw benefits in maximum consensus prior to an inquiry, but pointed out that this already happens to some degree and is specifically referred to in the Inquiries Procedure Rules. This comment seems to miss the points that, to encourage full representation at the meetings, a neutral forum, e.g. a Government-chaired working party, is necessary, and that funded work is necessary to establish facts rather than to have the developer's work endlessly challenged for bias during the inquiry. The Government does not see evidence that the planning inquiry process is a barrier to operators bringing forward proposals, but does wish to improve the process and will shortly consult on possible changes to the appeals procedures.

10.3.3 The performance of the UK airport system

In retrospect, the development of UK air transport has been a rather classic example of disjointed incrementalism. The piecemeal development of London's runways was made to seem reasonable by the presumptions that airlines would continue to be regulated and that passengers would not have strong preferences between airports. Development of single runway airports was similarly condoned by taking traffic forecasts only 15 years ahead and neglecting the possible need to develop large hubs as the most promising instrument for enhancing the roles of the additional airports. The role of the regional airports was dictated not by the most concerned community as recommended by the Edwards Committee, but by airlines' policies which in turn were shaped by bilateral agreements. BA's international route structure has been protected by these agreements, the emphasis in the negotiations favouring BA and London.

Market liberalisation has been encouraged, despite the distortions to competition produced by lack of capacity in the infrastructure. Runway and terminal air traffic capacity is now well below that required for 'free flow', partly because the recent strong demand for atm was not predicted. Those involved would probably say they had been practising 'the art of the possible'. They might also ask why the methodology should be faulted if it has provided a system which 'satisfices' and which shows every sign of being virile and financially successful. The CAA representative being questioned by the Transport Committee said that there will certainly be problems for the industry and problems for its passengers and losses to the UK economy which could be avoided, but it may well be that in real life the three airports we have got now (in the London area), from a series of stops and starts rather than a grand strategy, is the best which was achievable.

However that may be, it is not clear that the existing planning process has resulted in a system which is broadly satisfactory for the majority of the stakeholders or that it is achieving the Government's apparent goals, even if the airports are making profits. BA is in a poor position relative to its continental rivals with regard to future capacity at its own home base. It is burning 42,000 tonnes of fuel a year by having to hold at Heathrow and Gatwick (British Airways, 1996). Delays in the peak were 11.9 minutes to arrivals, and departures were delayed by 10.8 minutes in 1996, but the true cost of congestion is much greater than this when the extra aircraft required to increase the scheduled flight times to maintain punctuality are taken into account. Smaller airlines attempting to compete in the more lucrative markets find it almost impossible to gain entry. This makes a nonsense of policies which attempt to promote competition in air transport, except for the low cost end of the market which can be expressed through secondary airports.

There had already been advance warning from the US experience for other countries of the need to plan infrastructure with the consequences of deregulation in mind, the US having learned the lesson only in hindsight in the US:

> "We did not initiate an airport plan that aligned well with the new freedom for rates and routes. ... I think this lack of systemwide perspective was a strategic error. Certain requirements must be met for deregulation to yield its full benefits: competition and enforcement of measures by legal action to sustain competition; merger policy that does not foreclose meaningful access to airports; policies to deal with overcrowding of terminals, gates, runways and other facilities, and to provide for new capacity where warranted: and policies to add air traffic controllers and to introduce modern control technologies so that the system functions safely and efficiently. In sum, officials must look at the whole system, as opposed to just looking at the problem that was within the CAB domain." (Bailey, 1989)

It is certainly possible that improved planning policies might have provided a better level of satisfaction for the majority of stakeholders. Better siting of facilities or stronger enforcement of environmental limits or the development of dual runway alternatives might have encouraged more airlines to move from Heathrow. Earlier consideration of relaxing the rules for foreign airlines to serve regional airports could have allowed a more balanced system to develop. Development of short runways at Heathrow and Gatwick could allow regions to feed traffic into UK hubs rather than continental ones, so safeguarding regional access and promoting competition. Better mechanisms for early debate may have diffused the feeling of exclusion experienced by the proponents of sustainability and environmental protection.

The RUCATSE committee, for all its scope, still did not internalise most of the more serious system interactions. Little work has been done on the interactions between supply and demand or between supply and regional economies. Little thought seems to be given to the furthering of airline competition or to the need for peak, rather than annual capacity: The demand distribution model used shadow costs only to control traffic to annual passenger limits rather than discriminating between those services and passengers who could pay the peak fees and the others. The committees were only open on invitation. There was little recognition that the environmental consequences of each site needed to be part of the optimisation process for the site prior to its evaluation against other sites, nor that the environmental debate is global as well as local, even when only local environmental quality is threatened.

Little use has been made of either formal systems analysis or formal planning methodologies in developing the London airport system. Even within the limited and relatively manageable subsystem defined by the air transport industry and its users, there has been neither adequate coordination of effort, or sufficient understanding of the system's costs and dynamics, or any serious attempt to generate preferred solutions based on clear system objectives and their interpretation.

It is probable that the airports could have been developed so as to more nearly fulfil Government's objectives if they had been regarded as an interactive system so that their roles could have been more clearly defined, if the process of evaluating optional schemes had been more inclusive of concerned interests and operational consequences, if the decision-making process had been explicitly recognised as part of the planning process and if the whole planning/decision/implementation process were managed more interactively rather than being seen as discrete and sequential steps. Certainly the BAA feels the present system of airport planning in the UK to be inappropriate. "The planning challenge, I would suggest, is to find some new, more rational, more equitable and, above all, more efficient way of balancing legitimate interests" (King, 1990).

In fact, there has been a move even further away from central planning of the airport systems, in parallel with the move in the western world towards liberalisation and privatisation of air transport operators and airports, at least in the UK. This is associated with the belief that the individual airport authorities can plan adequately without government help and that they would not be able to express their entrepreneurial skills within a preordained framework. National Express, the private sector owner of East Midlands and Bournemouth airports believes that the market will decide and that industry is capable of serving the needs through the existing planning process without Government intervention (House of Commons, 1996).

However, the market could possibly generate either excessive or insufficient capacity. Excessive capacity in the regions could be augmented further if interactions between optional proposed projects result in excessive expectations for each project, with a consequent waste of resources, just as the airlines tend to take an optimistic view of future market shares when placing orders for new aircraft. On the other hand, in the London area, the monopoly enjoyed by BAA may lead to a lack of incentive to expand capacity at a sufficient rate to allow competition between airlines to express itself in the more lucrative markets, especially in view of the difficulty of judging the outcome of any inquiry process. BAA has admitted that its shareholders may not be best pleased if it invested in a short runway at Heathrow for small domestic aircraft whose passengers bought "nothing more than a cup of coffee and the Financial Times...........when 86 per cent of the profits come from shopping" (Wilson, 1995). In addition to these difficulties which face the entrepreneur, it could be objected that a major successful project would itself be making national policy by default.

10.3.4 Future prospects

The tendency for governments to withdraw from central control and planning means that each airport will have to develop or buy in skills in system planning in order to identify and to fulfil its most appropriate role in a European set of airports. Further, the skills will have to include the analysis of the initiatives open to its competitors and the consequent threats. This is not too serious a problem while an airport is considering expansion up to the point where its runways are being fully utilised. The serious difficulty arises when further runway capacity is required. The consequences for land use, for environmental impact, for the interpretation of Government policy and for the limitations which might be put on the project by the several clearance levels of the planning process may then seriously affect its viability. This difficulty is only being faced in the UK in the southeast and at Manchester, although several other regional airports face a smaller version of the problem in their desire to extend runways or reserve land for expansion.

Privatisation clearly has further to run. It may well be that there will be consolidation in the airport industry, for the benefit primarily of their shareholders, which mimics that among the airlines and which will need similar consumer safeguards if local monopolies are formed. Such monopolies, if sensibly policed, might provide the most appropriate solution to the long-running problem of close pairs of airports with surplus capacity, in that they may be able to market the individual airport's strengths in a complementary fashion, as National Express might have done with East Midlands and Birmingham, and safeguard reserve capacity.

As for London, the RUCATSE study concludes that, though it would be desirable that decisions taken now should not be inconsistent with reasonably foreseeable longer term developments, there seems to be no way of reaching longer term certainty that existing airports will be developed, or even of obtaining agreement now of the acceptability of such future development. Breaking up BAA might resolve the issue. So might a viable plan for a new mega-airport in either the Thames or Severn estuary, but it would struggle to draw traffic over the short term. The alternative would be several large airports with two runways each, which could provide competition between airports and so avoid the need for heavy regulation. Either way, the pressure will remain on Heathrow and Gatwick until, and if, charges reflect fully not only the costs of congestion but also the local and regional land use impacts. Policies of intervention on market-clearing or other pricing mechanisms would have to be made clear and definitive before predictions could be made of the timing of the consequent need for new capacity and its preferred location.

These broad perspectives reflect the Government preference for waiting in anticipation of the market providing a solution rather than being proactive. Unfortunately, in a complex and interactive system, there is a tendency for the individual actors to retrench in the face of uncertainty when planning theory would encourage working together towards creative change. However, it is one thing for private enterprise to be creative in its own field or possibly across one other system boundary; it is quite another to draw together the disciplines of such a complicated social and technical system as air transport. The encouragement of any such positive approach requires the creation of a RUCATSE type of forum with a brief to explore solutions which cross boundaries between disciplines, the exploration including the enabling processes necessary for their implementation so that they, as well as the technical characteristics, could be included in the comparative evaluation. It may be that the deliverables of the forum would be the definition of some technological, managerial or planning challenges resulting from the clarification of the planning goals, rather than the presumption that airport planning can only work with the present state of the art of the system's component parts.

The foregoing discussion of the UK's planning process leads to the following general conclusions:

- If the system is to rely on private initiative, there is a need for a modified three stage planning process, the new first phase giving a more concrete policy orientation and allowing the third phase to be less of a lottery. The second phase should be speeded up by the improved first phase and by a directed approach to the technical evidence.

- For this initiative to work properly, the developers would need to include local and national agency representatives in their early planning, which may make the process too transparent for a private developer unless the aim were only to capture an appropriate market share rather than forestall competition.

- Past planning has suffered from a lack of appreciation of how technology, management, regulation would change. Government should have a responsibility to ensure that this dimension is properly addressed.

- There has been a general failure to achieve a transparent balancing of policies in the areas of competition, air transport, regional development and local citizen's rights. This is partly due to a lack of informed research to quantify these issues, and the lack of a specific requirement on the government to be explicit in the first stage policy definition and transparent in the assessment methodology used in the third stage.

10.4 CONCLUSIONS

10.4.1 Heathrow

- Plans have been continuously outdated by poor forecasts of technical progress

- Environmental mitigation agreements have been ossified in out of date operational techniques and technologies, rather than allowing for technical progress and have not encouraged efficient operation

- Peak pricing worked at the margin to ration capacity, but only at the expense of those airlines unable to cross-subsidise the fees

- Monopoly regulation can distort the market and the behaviour of management (e.g. BAA shopping malls)

- Unbridled competition leads to extreme pressure for frequency and hence for runway capacity

- Strategic planning by the airport operator has been made very difficult by continual changes of judgement by government and inconsistency with regard to the upholding of planning agreements

- A Master Plan for Heathrow would have had to be impossibly farsighted to have overcome all its difficulties

- Evaluating options for expansion is made difficult by a lack of clarity in the underlying balance of policies on environment, the economy, the industry and the user.

10.4.2 London

- The limited scope of inquiries to examine non-local implications throws the weight of decision on the Minister, the resulting decision lacking the transparency of the inquiry

- Each successive inquiry has come to a different decision on the location of new capacity, partly due to limited scope but mostly due to changing technologies and views on the weighting of factors, these weightings often not being transparent, and being influenced political judgement to the extent of reinterpreting any cost/benefit analyses.

- The outcome of the planning process can completely alter the economic value of the proposal

- The irony in the planning process is the public cynicism towards it when its main aim is to involve that same public

- The airlines seem as ready to change their minds on the benefits of fragmenting the airport system as do the government, who apparently did not foresee that they would withdraw Traffic Distribution Rules or the consequences of so doing

- The airport roles have been strongly influenced by EU liberalisation and the available capacity

- The secondary airports have struggled in the liberalised settings unless their situation has specific advantages, or they are cross-subsidised, or they can capture and retain cheap carriers

- The Transport Committee hearings provide a suitably broad forum for discussion but have no teeth

- There is a great need for substantive evidence of costs and benefits, which should be sorted out transparently prior to any inquiry. This implies a governmental responsibility.

10.4.3 The UK

The case study of the UK airport system's development shows that the air transport industry continues to have to make do with the infrastructure it could get, rather than getting what it wants, largely because the advice given by the Civil Aviation Authority (CAA) to the Government has been rejected, or only partially accepted. In some cases this has been due to the CAA or its forerunners being given too narrow a remit, the CAA's responsibilities being centred primarily on the industry and its consumers. At other times, the Government, in considering the advice of the Public Inquiry inspectors, have judged longer term aviation benefits to be less important than the effects on the environment; there have also been times when world crises, or changes of the political flavour of the Government, or lobbying, have overturned decisions. The balancing of the wider issues has, in contrast to the CAA's work, not been a transparent process. This is only partly remedied by the mechanism of the Parliamentary Transport Committees. A further source of concern is that, frequently, the planners have misjudged both the industry's behaviour with respect to how it would use the available capacity and the amount of capacity which improvements to the existing system would make available.

All of these effects appear to be evident again in the current investigations into the need for, and the location of, a fifth major runway for London. Though the RUCATSE process generated a welcome improvement in participation, there was little chance for the interests of those airlines wishing to enter the system to be represented, or for an assessment of the amount of additional capacity necessary for competition to be properly expressed. The CAA's terms of reference give it little or no responsibility for wider issues, e.g. multi-modal effects, environmental or other planning matters, nor, indeed for the planning of preferred networks or the airports to support them. Other observations which can be made about the evolution of the UK airport system include:

- there have been conflicts between national and local policy goals
- the law of positive economies of scope appears inexorable in an airport system
- any attempt at system planning has only rubber-stamped existing roles
- privatisation inevitably produces losers as well as winners
- funding and efficiency problems can be solved without fully privatising; corporatisation can work just as well
- route fragmentation does not offload the main airports' runway congestion

10.4.4 UK planning methodology

If the system is to rely on private initiative, there is a need for a three stage planning process, the new first phase giving basic policy orientation

For this to work properly, the developers would need to include local and national agency representatives in their early planning, which may make the process too transparent for a private developer unless the aim were only to capture an appropriate market share rather than forestall competition

Past planning has suffered from a lack of appreciation of how technology, management, regulation would change. There has been a general failure to achieve a transparent balancing of policies in the areas of competition, air transport, regional development, local citizen's rights. This is partly due to a lack of informed research to quantify these issues. It is possible that a national system version of the local participative planning process could also be made to bear fruit, i.e. the format used by RUCATSE could be extended on a voluntary basis.

For this to happen, a permanently appointed body, called, let us say, the Aviation Infrastructure Forum (AIF) would be needed to define the system being managed, to set up the methods for generating goals for the system, to collect information pertinent to the system's planning, to monitor the achievement of the goals and the continuing validity of the assumptions underlying the policies themselves (Masser, 1987), and to give early identification of the need for change.

The previous airport policy review (HMSO, 1985) promised that two standing committees would be set up, one for London airports and the other for the regions, to provide this sort of monitoring. Until the recent round of deliberations in response to the CAA's advice (CAA, 1990b), the published actions of these bodies have been perfunctory. The London committee has suffered a metamorphosis into a task force (RUCATSE) with six sub-committees, to consider those broader issues not included within the CAA brief. Real adherence to the principle of monitoring, with the biennial production of a rolling plan as required of the FAA, would provide an earlier alert to problems, and tighter control of their resolution. The system boundary should include all UK airports and the major European airports, as the Dutch have done in considering Schiphol's future development. The specialisms needed to support the AIF should include, besides familiarity with the UK and European planning and decision processes, worldwide demand forecasting for passengers and freight; the understanding of passenger utility maximisation; worldwide airline strategy; technological forecasting of the vehicle and the atc hardware.

AIF would have to have a sufficiently wide field of influence to include all those factors which have a major impact on the airport system, without overlapping too much with any other similar bodies which may be formed. Its appreciation of the interconnectedness of the system would have to extend at least to a number of concerned bodies and organisations, and it would have to embrace a wide range of specialisms and concepts. Conceptual methodologies should be developed to consider formally the balance between environmental nuisance and user benefit; the balance between national and local economic benefit and the cost of expansion of the system; the acceptable delay and its distribution within the system; the role of scarce resources. It should have the means to commission research in those areas where more information is considered necessary. It must also identify and maintain the necessary links with those other bodies whose decisions influence the air transport system, e.g. road, BR, regional planning.

AIF would be responsible for drawing together goals for the system, for interpreting them as objectives and for developing performance indicators with criteria against which to measure the success of the system. The relevance of the objectives and the criteria themselves should be monitored. These are the most important part of any systems analysis, and where public participation is most vital in a participative democratic society. It would therefore be necessary for the body's deliberations and their outcome to be available to the public in some disciplined manner, probably following UK Select Committee practice, even though there be also formal representation through organisations representing user and non-user interests.

It would also be a responsibility of AIF to ensure that projects offered for entry to an evaluation process be realistic: this implies that some preliminary facility planning be performed. It would then rehearse the evaluation which would eventually be performed in a public inquiry. AIF would formally select the most appropriate evaluation methodology in the light of the its defined responsibility. In doing so, it should bear in mind that evaluation methods which involve public participation call for interactive evaluation reflecting individual's perceptions of their problems, their goal preferences and the alternatives available. AIF should then apply the methodology to all offered projects and to those generated internally by the body. This would not be to pre-empt the subsequent inquiry, but to educate it; to shorten it; to attempt to account for wider national issues prior to the local debate; to ensure adequate quality of information; to provide a forum for 'out of court' settlements. Despite all these inputs which AIF could make, its primary value would rest in its role as a forum for informed debate, in which it would offer the best opportunity for cooperation and consensus. It is to be hoped that the discussions in the AIF would be able to move from conflict, through a compromise stage to a new high ground of 'integrative' solutions: this type of solution, by changing the way in which the situation is regarded and valued, has the characteristics of commanding the assent of all the participants (Vickers, 1965).

11

THE EUROPEAN UNION CASE

11.1 THE AIRPORTS, THEIR UTILISATION AND THEIR SETTINGS

The nature of European air transport operations is very different from the US, even after the Third Package of liberalisation measures which came into force in January 1993 and the Schengen Agreement on the ending of border controls from January 1995. Flights within the EU retain international characteristics, not only in terms of residual border controls in those countries who did not sign the Schengen Agreement (Rouaud, 1991) but also in passenger preferences for home-country airlines and linguistic barriers in marketing and operations. The propensity to travel over a national border is only a quarter as high as within a country. There are many differences between member countries in the regulation of airport and airline operations, of conditions of employment and of salary overheads and other costs. The infrastructure is also fragmented. The airports are owned in a variety of ways, and some have an outright profit motive while others perform a pure service function. Even the declaration of airport capacity varies with the interests of the parties involved.

Given the 15 year lead that the US had in deregulation of air service, it might be possible to anticipate some of the results of the EU liberalisation, particularly in drawing inferences from the US hubbing experience. However, there are other major differences between Europe and the US, which cannot be ignored. The geography (van den Berg, 1990; Villiers, 1989b; Pavaux, 1990), the modal competition (Reynolds-Feighan, 1992) and the nature of the operations and of the deregulation itself (Button and Swann, 1989) are all important enough to warrant caution in interpreting the US experience. In addition, European capacity is currently severely constrained (Chevallier, 1992).

The European geography is almost the inverse of the US despite the general similarities in size, population and wealth. Most of the major economic activity is clustered in the centre, reducing average intra-European distances, the peripheral regions' links being predominantly leisure and pseudo social service routes. The long haul routes are already based in the rather central capital cities, rather than having a coastal site like New York Kennedy as the strongest gateway. The many new Eastern countries will also have weak economies and low density traffic for some time to come.

There is at least one natural fortress hub for each national flag airline in the Union. These hubs have been oriented traditionally to serving long haul markets, particularly those with links dating back to colonial territories. There has not in general been the same migration-based community of interest within Europe, with some exceptions: many Iron Curtain refugees want to explore the new freedoms in Eastern Europe; prisoners from the Second World War have drawn relatives to search for work; armed forces stationed abroad have married; unemployed in poorer peripheral countries have found work in the stronger economies.

Half of all air travel within Europe is satisfied by charter operators, almost all of it attracted by sunshine or snow. The charter and scheduled markets are very specialised in geography and in product. The expected conversion of charter carriers into scheduled carriers (Gialloretto, 1988) has not proved easy, while the mix of high and low yield necessary to support scheduled levels of service is not readily available in the traditional charter markets. It is not inconceivable that charter carriers might develop hubs, following the example of Hapag-Lloyd at Munich, but efficient point-to-point services are normally made viable by skilful massaging of the originating traffic to match the supply. The scheduled industry then has a much smaller traffic base and network of destinations on which to construct a hubbing system than in the US, while having a large number of potential hubs. Even a city like Hanover, with a population of well over half a million, generates more than 40 scheduled passengers per day to only 10 destinations, though there is direct daily service to a further nine destinations. The importance of being able to make connections is indicated by the overall total of 570 scheduled destination markets from Hanover in 1993 (*AEA, 1995*).

The European Civil Aviation Conference (ECAC) is the most inclusive forum for discussion of the European region's air transport problems, representing 35 countries (*Jane's Airport Review, March 1997, pp 30-31*). It recognises that the system is short of capacity in the airspace and at the large airports. Airspace improvements are being driven forward by the European Air Traffic Control Harmonisation and Integration Programme (EATCHIP). The Air Traffic Management system consists of 44 ATC centres in 22 separate systems (Donoghue, 1991), with differing levels of capacity and methods of operation. ECAC has identified serious capacity shortages in the airspace and the runways. The busiest terminal areas (TMA) are London, Paris and Frankfurt, with 2,800, 2,300 and 1,800 flights per day in September 1993; a further 11 TMA averaged 1,000 flights each (ECAC, 1994). Plans to ameliorate the problems have been formulated. In the short term, the emphasis is on better coordination and communication between air traffic sectors, the use of mature techniques to enhance runway capacity and the strengthening of flow control, though airports still only participate voluntarily in this congestion management tool. The long term solution to airspace capacity problems is seen to be the application of advanced technology to allow direct routings by four-dimensional navigation using Global Positioning Systems (GPS) and long term conflict resolution by Air Traffic Control.

The responsible ECAC body to improve the runway and terminal area capacities was, until recently taken into EATCHIP, the Airports/Air Traffic System Interface (APATSI) Project Board. One of its tasks was to analyse the situation at some of the more critical airports in order to assess the benefits of a range of enhancement measures for application in the air and on the ground. The normal procedure in European controlled airspace is to work to Instrument Flight Rules (IFR) and the airspace and runway capacities should be declared appropriately. The capacity available seriously constrains access to the system at many airports. Of the airports over 5 mppa in 1993, 25 will need extra runway capacity before 2005 unless delays are to become worse (AEA, 1995). The 33 largest airports offered two-thirds of the total intra-European scheduled seats, the others being spread among 72 airports with between 1 and 5 mppa and a further 429 airports with less than 1 mppa. The capacity utilisation at the major airports is shown in Table 11.1.

The worst affected airports are the largest hubs now: Heathrow, Gatwick, Frankfurt, Charles de Gaulle, Orly, Amsterdam, Brussels, Berlin, Vienna, Düsseldorf, Manchester, Copenhagen, Zürich, Geneva, Barcelona, Madrid, Milan Linate, both Rome airports and Athens. In 1997, 40% of flights at Athens were delayed with an average delay of 30 minutes. At Madrid, 33% of flights were delayed with an average delay of 21 minutes. 25% of Brussels flights and 22% of Amsterdam flights were also delayed (ERA, 1998). Not all of these delays were caused by airport problems. In the peak months, 50% of the delays at Athens were due to en route atc, 23% were reactionary and a further 19% were due to ground atc, the figures for Frankfurt being 18%, 25% and 13% respectively (*Jane's Airport Review, November 1997, p 7*). Only 2% of flight delays at Athens and 12% at Frankfurt were caused by airport and government problems (*Jane's Airport Review, November 1997, p 7*). Eurocontrol expect the airspace system capacity to have increased by 70% by 2003, so that the main bottleneck in the system will be runways. Paris Charles de Gaulle is opening the first of two new runways in 2000 to bring the capacity up to more than 110 movements per hour. Manchester and Madrid are also in the process of building additional runways on site, as is Amsterdam, though this will be used more to relieve environmental problems than to add capacity. Berlin and Athens are constructing major new facilities at Schonefeld and Spata respectively to overcome their capacity problems, while Milan will make much greater use of the two runways at the second airport, Malpensa. Even after these plans are implemented, there will still be extreme pressure on runway capacity in the system, most particularly at Heathrow and Frankfurt.

The UK experience can be seen as a prototype when considering the planning methodologies which the Commission might adopt as it involves itself in ensuring the provision of appropriate infrastructure in the face of liberalisation and privatisation. Perhaps the most important experience has been that it proved difficult to try to preordain the roles that airports would develop. However, transferring the analysis to the EC setting must allow for the fact that, in the UK, air transport liberalisation has occurred in a setting which has also seen changes in the ownership of the industry as well as of the status of planning in general and in the application of systems analysis in particular. The UK airline industry has, with the privatisation of BA, ceased to have a publicly owned component. Of the non-UK major carriers, only Lufthansa has followed suit, most of the others still being the recipients of government money for restructuring (Feldman, 1997b). Similarly, few of the European airports have followed the UK airports into full privatisation, though some are now fully accountable.

Table 11.1: European airport runway capacity utilisation

	Atm/hour	Mppa	Mppa/atm
Parallel runways			
Heathrow, London	81	54	0.67
Charles de Gaulle, Paris	76	28	0.37
Copenhagen	76	14	0.18
Orly, Paris	70	27	0.39
Munich	70	15	0.21
Frankfurt	70	38	0.54
Brussels	60	11	0.18
Fiumicino, Rome	56	21	0.38
Malpensa, Milan	30	4	0.13
Converging			
Stockholm	66*	13	0.20
Zürich	60	15	0.25
Madrid	30-50*	19	0.38
Vienna	45	8	0.18
Hamburg	42	8	0.19
Dublin	36	7	0.19
Barcelona	[30]	11	0.37
Amsterdam		25	
Single			
Gatwick, London	47	23	0.49
Manchester	42*	15	0.36
Düsseldorf	(30)	15	0.50
Geneva	30	6	0.20
Athens	30*	10	0.33
Linate, Milan	22	10	0.45

Notes:

1. Atm/hour relates to best performance
2. Mppa is per million terminal passengers in 1995, except for airports with less than 13 mppa, where the data are for 1994
3. * denotes that a new runway is expected
4. ( ) denotes environmental cap
5. [ ] denotes an atc/radar capacity constraint

Sources: CAA, 1995a; ACI Annual Statistics

The EC does, like the US FAA but unlike the UK government, have funds to augment airport capacity in accordance with the provisions of the Maastricht Treaty on European Union. The total budget under the Trans European Network (TEN) programme, aimed at developing a transport system capable of moving people and goods efficiently across the whole of Europe, is ECU (European Currency Units) 1,900 million for the period 1994-1999, though the vast bulk is being spent on rail and only some 3% on air infrastructure (Nittinger, 1997). In addition, nearly ECU 600 million has been released for airport projects from the Structural and Cohesion funds between 1990 and 1997 (*Airport Business Communiqué, August/September 1998, pp 17*-19). Airports can compete for these funds with other modes, the member states having the responsibility for bringing forward projects for financial assistance, though they are being viewed by the Commission as a way to kickstart the interest of the private sector (Greenwood, 1998). The Commission selects projects from those brought forward on their value to the Community by cohesion or intermodality or safety, with priority given to enhancing existing capacity, developing new capacity, enhancing environmental capacity and the enhancement and development of airport access (House of Commons, 1996). According to EC officials, if airports do not get a say in the distribution of financial assistance for airport development, then it is the fault of their own national administrations (Frommer-Ringer, 1995).

The EC constructed a Trans-European Airport Network from those airports identified by the member states as of national interest, according to the necessary functions (EC Communiqué V11 - C4/Com. 2/93, 19.11.93). The airports were classified by their primary function:

- Community Connecting Points, of which there are 40, primarily providing links with the rest of the world and having a minimum of 5 mppa or 100,000 atm or 150,000 tonnes of freight per year

- Regional Connecting Points, of which there are 61, linking the regions to each other and to the Community Connecting Points, having at least 1 mppa or 50,000 tonnes

- Accessibility Points, of which there are 136, providing links to the Regional and Community Connecting Points. These must have 250,000 ppa or 10,000 tonnes unless they are on an island or are at least 100 km from the nearest connecting point.

Other airports are eligible if they are remote from the European mainland or are close to having the required traffic levels. Of the 312 airports identified by the member states, 75 have not been included in the network at this stage.

The process of identifying the airports within the network which provide obstacles to the efficient functioning of the network is politically sensitive and subject to approval by the European Parliament. The Plan seeks to limit the impact of policies to support infrastructure developments on the competitive position of other airports. There have been claims that support for Milan's Malpensa airport would attract traffic away from Munich or the proposed new Athens airport at Spata (*Airport Business, Management and Development Section, July/August 1995, p 5*). None-the-less, funds are being released for the projects at Malpensa,

Spata, Berlin, Arlanda and for Lisbon's new airport study, among others. In the case of Spata, 25% of the development is EC funded, while over 50% is to be funded from a package put together by the contractors. In addition, the high priority EC rating helps these projects to qualify for loans: 20% of the cost of the improvements at Malpensa are to be borrowed from the European Investment Bank. It is not at all clear how these particular projects came to be favoured. Though there are goals, no quantitative analyses of comparative Community benefits appear to have been published. Madrid's Barajas airport is to construct a new 4,400 metre runway to increase capacity from 50 to 80 atm per hour which will cost US $400 million with no EU funding. The official UK view is that the UK should get its share of the funding but that it should go to improve ground access rather than to distort competition by directly influencing airport capacity.

In fact, the Plan seems to be little more than an inventory of the present situation. No thought seems to have been given to local over or under supply of airports, to how their roles might change as the airlines' strategies and the bilateral negotiations evolve, or to the uneven implications of environmental impacts, traffic growth and airport ownership.

11.2 POLICY CONSIDERATIONS

11.2.1 Competition and fiscal policies

The relationship between airlines and their governments varies greatly in the application of the regulations on State Aid, on intervention on tariffs and route access and on subsidising air transport for social purposes. In Greece, for example, the law in 1992 still prevented any kind of scheduled service competition with Olympic Airways and required all operators to own at least 51% of each of their aircraft, so preventing leasing (*Flight, 29 July 1992, p12*). The EC does not intend to involve itself in purely domestic issues, but does wish to harmonise those aspects of policy which deny a 'level playing field' for competition between airlines and between modes of transport. It also has a long term objective to harmonise the frameworks governing working conditions, airport charges, environmental impact and the recovery of social costs. The EC also intends to abolish duty-free sales on intra-EU flights and to include air fares within the Value-added tax rules, though initially they will be zero-rated. It has been estimated that these fiscal harmonisation measures could increase ticket prices by some 15% to 25% on scheduled and charter routes respectively and reduce intra-EU air traffic by about 11% (Coopers & Lybrand/AACI, 1990). Meanwhile, some countries are increasing the lack of harmony by imposing unilateral taxes. Sweden has a 25% VAT on domestic air travel, while the UK has imposed a departure tax which must be collected by the airlines and which is greater than the average total cost per passenger at Heathrow (Stockman, 1996).

The EU incorporated the ethos of free and fair competition into its original treaty, but for a long time the airlines were allowed to behave as though they were exempt from the competition rules, just as the US airlines were given immunity from the anti-trust rules in order to set up interline agreements and rationalise tariffs. A series of test cases in the EU courts succeeded in establishing that the competition rules did indeed apply to air transport but only in 1986 (Rapp, 1986), long after the European Commission's Transport Directorate, DG-7, had initiated the construction of a framework for gradual liberalisation of the industry, on the advice of the Competition Directorate, DG-4, explained in their 1979 report: "Air Transport, a Community approach". Consumers were complaining that European scheduled fares were much higher than those of European charter operators or US scheduled operators. Although these differences were explainable to a large extent by differences in input costs, level of service and market size, the pressure from consumers has been inexorable. Their pressure was at least partly justified by EC analyses showing that there was no transparent relationship between cost and fares, that there were excessive profits on some routes and that there was an unacceptable degree of cross-subsidy between routes. However, some of the comparisons are still unreal. The EC believes that airport charges are three times higher in Europe than the US (EC, 1994), apparently forgetting the ticket tax which largely funds US infrastructure, and the fact that US airlines tend to run their own terminal facilities.

The first phase of legislation in 1983 (EEC, 1983b) applied only to a strictly defined regional category of airlines flying international intra-community routes, with limits on aircraft size, sector distance and the airports which could be linked. The major hub airports were specifically excluded, as were airports within 80 km of the hubs. This was effectively mimicking the interstate route and tariff freedoms which had been enjoyed by the commuter airlines during the regulatory era in the US. Not surprisingly, there was not a great rush to serve these routes. Studies by ECAC (1984) had shown the limited potential for viable services on these thin routes at existing fare levels.

The Commission continued to champion liberalisation of the whole industry, convincing the Council of Ministers to adopt an interim Second Package which allowed all carriers some freedom to vary fares and to enter routes (Barrett, 1991). Cargo services were allowed to set their own rates and to operate unlimited third, fourth and fifth freedom services within the Community from February 1991. The full Third Package was adopted in June 1992 (van Hasselt, 1992). This effectively removed the distinction between scheduled and charter airlines, the latter always having been subject to less regulation. National ownership rules were abolished, all carriers with majority ownership in the EU being licensed to fly any fifth freedom routes in the EU. There are now no controls on pricing, but States or the EC may intervene in the event of excessive fares relative to fully-allocated costs or of predatory pricing. Full cabotage rights became available in April 1997, but there are still limits on the extent of intra-EU liberalisation, with non-discriminatory blocking of airport access if there is a shortage of physical or environmental capacity, and with special exemption for social service routes and for carriers in difficulty (van Hasselt, 1994). The EC also recognises the need to protect new thin routes.

Once again, the early experience of the UK in liberalisation could have served to warn the EC of the difficulties which can frustrate attempts to allow the market to express itself. With the necessary caveats, most importantly the UK regulators' desire to give the consumers' interests at least the same priority as the interests of the airlines, the early experience of the UK air transport system has relevant implications for the rest of Europe: in any case, 10 of the 15 most frequently served and densest international routes involve London (CAA, 1993a). The UK had been allowing competition on domestic trunk routes even before its signing of liberal ASA, though only since 1982 has the main carrier been subject to head-to-head competition on these routes from London Heathrow. Although there was an initial increase in the traffic of perhaps 7% in the first year of competition on the routes to Glasgow, Edinburgh and Belfast (LAE, 1984), in the following years the growth was lower (by perhaps 5%) than 'control' routes with little or no competition (Poole, 1986). The stimulus for the initial growth may well have been the greater capacity available and to the greater use of yield management techniques, leading to lower spill rates (i.e. turning away of passengers). Further, once the situation had settled down, there was little evidence that published fares had fallen relative to the 'control' routes: this was true for discounted as well as full economy fares.

The main effect of the attempt to stimulate competition on UK routes appears to have been for the ratio of air transport movements (atm) to passengers to increase much more quickly than on the control routes. Other analysis, using UK/Europe scheduled traffic as the control, showed even less conclusive evidence of either temporary or lasting effects on traffic or fares from competition on domestic trunk routes (CAA, 1987). Similarly, on international routes, a comparison of the history of liberal with less liberal routes between the UK and the European mainland certainly shows that more routes opened as a result of more liberal bilaterals, and nominal discounted fares fell, but, with the exception of the Irish-UK routes (Barrett, 1991), there is little evidence that the supposed extra travel opportunities were converted into long term traffic growth. Table 11.2 compares the traffic growth rates before and after the liberal bilaterals were activated in the period 1984 - 1986 with route groups which remained more regulated.

Indeed, there were serious doubts as to whether the European market could be contestable, given the existing airline and hub dominance and the shortage of capacity (Villiers, 1989). A further study, which transferred the US relationship between route density and number of carriers to the European setting (Pryke, 1991), suggested that the less dense European routes may well have fewer carriers after liberalisation, though real competition might still improve because a second carrier on thin routes at the time would be more likely to be colluding than offering genuine competition.

Attempts by the UK CAA to open up the thinner domestic routes to competition were aborted because of industry protests that open route access would have allowed the major airlines to challenge for traffic on cross-country routes, while the small carriers would have no compensating rights to the thick routes because they could not gain access to Heathrow. Thus competition on the thick routes has, by stimulating extra atm at Heathrow, helped to deny competition on thinner routes without apparently producing major benefits for consumers. It could be argued that there have been significant second-order benefits: the choice of airline per se; increased frequency; improvements in cabin service and airport lounges; punctuality (though this has been affected for all airlines by air traffic control capacity problems). However, unless these benefits can be seen to have stimulated extra use of the services, they must have accrued as consumer surplus, particularly since there was relatively static competition from rail during the period under investigation. There had, therefore, already been advance warning from the experience of the shortage of Heathrow capacity inhibiting competition, of the need to plan European infrastructure with the consequences of deregulation in mind; a lesson only learned in hindsight in the US.

Table 11.2: UK international scheduled traffic
(Annual average growth rates, per cent)

	1980-84	1987-89	(1987-89)-(1980-84)
Most liberal EEC countries			
Netherlands	1.0	7.5	6.5
Ireland	-2.7	22.7	25.4
Belgium	1.5	8.8	7.3
Luxembourg	5.3	4.5	-0.8
Less liberal EEC countries			
Germany	1.4	8.6	7.2
France	1.0	16.7	15.7
Italy	0.1	13.6	13.5
Non-liberal EEC countries			
Greece	1.5	8.5	7.0
Portugal	0.0	14.9	14.9
Spain	2.2	14.9	13.7
Non EEC countries			
Austria	4.4	7.8	3.4
Denmark	2.9	8.8	5.9
Norway	5.4	6.1	0.7

Source: Caves and Higgins, 1993

The major European airlines were already responding to the promise of liberalisation well before January 1993, (Villiers, 1989b). The major airlines were attempting to expand their own national hubs (BA with Terminal 5 at Heathrow, Lufthansa at Frankfurt, Air France at Charles de Gaulle) or find new hubs for expansion (BA and KLM at Schiphol, Air France at Brussels with Sabena, Lufthansa at Munich and Berlin). Manchester Airport was creating a pre-deregulation interline hub with a blend of omni-directional and directional complexes (Bowen, 1991), hoping to mimic US history by transforming into an on-line hub once a major carrier could see the potential and gain the necessary route freedom.

Within the first two months after full liberalisation had been declared officially in the EU in January 1993, only 14 services were started or announced which relied on the new fifth freedoms or of the limited cabotage rights (Avmark, 1993). In view of the UK's experience in the 1980s, it is not surprising that the Third Package took time to stimulate competition or traffic either between countries with existing liberal ASA or elsewhere.

By 1995, however, there were signs that purposeful competition had begun to emerge (CAA, 1995b). Between December 1992 and December 1994, the number of international scheduled routes linking pairs of cities in the 16 countries subject to the EC's air transport rules increased from 500 to 525. The proportion of city pairs served by three or more effective competitors increased from 5% to 7%, the proportion of flights increasing from 19% to 25%. Fifth freedom services increased, 36 new ones being added to the 19 already operated by European airlines before December 1992. BA's Heathrow - Stuttgart - Thessaloniki service appears to be a successful example, though its first attempt via Turin rather than Stuttgart failed to pick up significant fifth freedom traffic. Of the 55 operations, 33 still existed in December 1994, the most successful exploiters of the rights being Alitalia and Finnair, though only seven new cabotage routes were still operating in December 1994, while 12 other new starts had already terminated. This slow uptake of opportunities was despite the legality of change of gauge and starburst connections. Apart from Sabena's daily service between Barcelona and Venice, the only significant seventh freedom services in December 1994 were those operated under the BA code but actually operated as third and fourth freedom services by TAT and Deutsche BA.

Many of the new routes were from secondary airports and, of 183 major city pairs, 32 suffered an increase in concentration of frequency in the hands of the dominant carriers, while only 13 became more competitive. Also, there was an increasing concentration by the major carriers on their own hubs. Meanwhile, non-hubs have been benefiting from increased indirect competition as more airlines extend their own hub's feeders to them. A regional UK airport may not have competitive service to any given hub, but there will usually be indirect competition to each hub from a adjacent regional airport, while the choice of hub to onward destinations must increase the passengers' potential utility.

The EC has recognised that liberalisation requires policing, again without denying the principle of subsidiarity but with the teeth to take action where necessary. It has developed policies on mergers and alliances to avoid local monopolies, on predatory pricing, on the control of slot allocation at busy airports. It has also published a draft directive on ground handling at airports, in response to evidence of very large differences in charges which can be partly linked to lack of competition (*Jane's Airport Review, March 1997, pp 11-12*). These policies attempt to harmonise the treatment of the issues across the Community, just as there are policies to harmonise the other issues in which large variations exist, but the pace of acceptance is itself very variable. There is also uncertainty, as there was in the US, as to the way which the residual regulation is to be effected, i.e. monopoly abuse, anti-competitive practices, market access, non-EC relations. The UK government, despite its long commitment to a 'second-force' airline policy, allowed BA to rescue British Caledonian and later Dan-Air, on the grounds of ensuring BA's global competitiveness, but, in the process, allowing it to further dominate the industry.

The Competition Directorate's control of alliances has so far been subject to variable interpretation of how the rules are to apply. In any case, the EC's responsibility does not apply to alliances whose worldwide turnover is less than ECU 2 billion and Community turnover less than ECU 100 million (*Airline Business, March 1991, p 23*). The rulings imply commitments not only by the two parties but also by governments and even airport authorities, but without guarantees that the commitments would, or even could, be met (de la Rochère, 1994).

In the event of insufficient runway capacity being available to allow competition to express itself, some method of managing capacity has to be adopted. The EC's original proposals on slot allocation were quite draconian, with suggestions that no airline should have more than four slots per route at those airports which were deemed to require full coordination of slots. Their first Regulation in 1993 (95/93) was actually far less drastic, falling back on the well established IATA process which gives precedence to the grandfather rights of the scheduled carriers, but with several new requirements. The individual countries must arrange for independent coordinators to be appointed at the fully coordinated airports, who must establish a pool of slots from those falling vacant and redistribute 50% of them among aspiring new entrants. The priority grandfather rights apply to regular services, whether operated by scheduled or charter carriers. There are 'use it or lose it' clauses to attempt to control abuse of the grandfather rights. Airlines are allowed to transfer slots between routes and between each other, though not officially for payment.

The process has not been very successful in allowing access to new entrants. It has taken Virgin five years to build up sufficient slots at Heathrow to start services to South Africa. The UK CAA has proposed many amendments in order to improve its effectiveness (CAA, 1995). It wants all slots which become available to go into the pool, and for the slots to be prioritised to allow the most promising new carrier on each of the denser routes the opportunity to build up an effective level of route entry. It also sees that slot trading is inevitable, and that it would be better to require that they be registered with the coordinator. Some observers wish to see further controls, not only to tighten definitions of who is a new entrant, of whether an airline is serious rather than simply a piece of paper, and of when slots should be returned to the pool (Harding, 1994), but also to earmark the slots' purpose. If it is accepted that they were originally granted, not so much as a favour to a carrier as in order to serve a route, a case can be made for requiring a slot to be returned when an airline chooses no longer to use it to serve that route. The UK Department of Transport feels that there should a 'super-congested' airport category, at which passenger benefits may be better maximised by prioritising throughput rather airline access. Germany is floating the idea of minimum size/frequency ratios for different routes out of Frankfurt and Düsseldorf, set by local committees of airline and airport representatives (*Airline Business, November 1996, p 10*). As a result of these and other comments, the EC is expected to make many changes to the slot rules in 1997, possibly including a slot market, but stopping short of the US FAA's rules which allow the slots to be owned by other than aircraft operators.

Meanwhile, some countries have used the shortage of slots, or the need to protect weak routes, to keep new entrants out of the busier airports (Balfour, 1994). At the same time, individual airports have taken unilateral decisions on traffic distribution in the face of their capacity shortages. Düsseldorf Airport banned all turboprop aircraft in April 1996, hoping that they would use the 1,200 metre runway at nearby Mönchengladbach, in which they own the majority of the shares. This is being challenged in the courts on the grounds of Mönchengladbach's limitations and the lack of logic in allowing 50 seat jets but banning 66 seat turboprops. If it is not overturned, the decision will set a further precedent for other congested airports to drive out the regional carriers to add to the BAA's use of a high flat-rated peak landing fee.

11.2.2 Sustainability

Until the Single European Act of 1987, there was no explicit Treaty responsibility for environmental protection. The EC's "White Paper on the future development of the Common Transport Policy: a global approach to the construction of a Community framework for sustainable mobility", adopted in December 1992, attempts to reconcile the aims of promoting a sustainable transport system and of promoting competition, arguing that the two aims are compatible if the costs of externalities are fully recovered by imposing the principle of 'the polluter pays'. The EC has made proposals for carbon taxes, with no special treatment for aviation despite the difficulties that would be caused for EU carriers if the EU moved on this issue before a general agreement was reached in ICAO. No action has been taken as yet, partly because the 'Committee des Sages', called to consider how to deal with an air transport industry that was inefficient, financially fragile and in the depths of the early 1990s recession, advised that the situation of most European airlines was so dire that they should be spared any further financial burdens. The EC have also indicated a willingness to move in advance of ICAO in imposing stricter noise limits on aircraft than the present Stage 3 certification levels and also stricter nitrous oxide emission limits than ICAO's 1993 standard (Fergusson, 1995).

The main thrusts of the Commission's approach to sustainability in air transport are to minimise congestion and to maximise the use of the alternative modes which it sees as less polluting. Congestion can be reduced by rationing demand, by attempting to distribute the demand geographically and seasonally, by ensuring best-in-class performance of airports before providing judicious expansion of capacity and by encouraging the use of rail for the shorter distance intercity trips. If the high speed rail services can also be used to access airports, the larger airports should obtain net benefits from improving rail services. The airport industry would prefer to solve the problem by the introduction of much stricter land use controls, but sees some logic in rail substitution. Amsterdam's Schiphol airport, for one, is determined to adopt both approaches. German Bundesbahn is cooperating in the construction of railway stations at Frankfurt, Düsseldorf, Berlin, Cologne and Leipzig connecting directly into the high speed rail network, with agreements to exchange rail and air tickets, to offer reduced fares for the domestic sectors of international trips and even to incorporate taxis into the through ticketing so as to require only one ticket from home to destination airport (*Passenger Terminal World, October 1997, pp 22-26*). Lufthansa is exploring with Air France the possibility of extending the concept across Europe (*Jane's Airport Review, April 1998, p 16*).

Experience with the French TGV high speed train does not fully support the rail substitution theory, nor does the history of cross-channel traffic over the introduction of rail services by Eurostar through the channel tunnel. The air traffic between London and Brussels grew from 1991 to 1996 by almost as much as the traffic to Frankfurt and to Amsterdam, neither of which are linked into the Eurostar services directly. On the other hand, London/Paris air traffic fell by 14% (CAA, 1998). There has been a tendency for airlines actually to increase frequency in an attempt to compete with rail rather than withdrawing service. Certainly this can be justified when a large demand exists in close proximity to a major airport and the high speed train network is not fully linked in with it, as in the case of Heathrow with the Manchester route. Also, there is at least a suspicion that the full costs of the track are not being reflected in the train fares, with the French government reducing its plans for the further extension of the TGV network. It is, in fact, not at all certain that rail fares would be able to match air fares on a perfectly level playing field (Caves, 1976; Veldhuis et al, 1995).

There are grounds for believing that non-aviation interests have already achieved some degree of regulatory capture. The EC appears to have adopted a far from level playing field in assessing air against rail (EC, 1992a; EC, 1992b), which the aviation industry attempted to correct after the fact (AACI, 1992). The EC had only taken an old technology B 727 as an indication of aviation fuel efficiency, with no account of the other modes' circuity, primary energy consumption, energy for track construction, land take and spoil disposal. Nor did it take into consideration the variability in comparative results depending on the terrain, the route length or the route density. Yet the Transport and Tourism Committee of the European Parliament felt the "Green Paper on the Impact of Transport on the Environment: a Community strategy for 'sustainable mobility'", adopted in February 1992, had not gone far enough in proposing general and selective restrictions to air and road infrastructure.

For the system to be sustainable, environmental and economic consequences must be in balance locally as well as globally. It is inevitable that these consequences will be different for each airport in Europe. Airports and their regulators are being forced to take what action they see fit to achieve the appropriate balance for their own local communities. This has resulted in a proliferation of methods for measuring and controlling the consequences, and of judging the correct balance. The EC would like to see at least some harmonisation in the methodology for judging the appropriate balance, but it is difficult to envisage what can be done when environmental sensitivities vary considerably across the Community and when economic overheating can cause as much opposition in some cities as unemployment can cause in others. Yet, if no action is taken, the airport system will be shaped much more by local environmental capping than by central policy. The capping policies are likely to be changeable and uneven, making it difficult for airlines and airports to plan their strategies.

11.2.3 Subsidiarity in third country relations

Outlying regions are very interested in promoting airport growth with new routes, at least until the environmental costs become too great. If the regional airports can show that they have the conditions for local traffic growth and the available airport capacity for local plus transfer traffic, they are in a strong position to negotiate with the carriers. Fruition of the regions' plans will, however, depend on national and EU attitudes to route negotiation, and will require increased participation of the regions and their airports in external air policy (Wassenbergh, 1991). The process could be difficult, because many intrastate conflicts will require resolution, such as the interests of e.g. the west of Ireland, where all Aer Lingus's transatlantic flights to Ireland were required to call at Shannon, but now only half must do so.

The EC eventually wants to be able to conduct all ASA negotiations with third countries to ensure equal opportunities for all EU carriers, to eliminate the nationality clauses which are common in bilateral agreements and to control the harmonisation of fifth freedom rights within Europe (Nuutinen, 1992). This will take a long time to implement, given that each member state has some 70 ASA, often supplemented by confidential Memorandums of Understanding.

Many member states have not respected a 1990 Council Decision which requires them to submit all new ASA for EC ratification and the Commission accepts that practical logistical difficulties require the delegation of negotiating powers to the member states in most cases. The EC believes it should take the lead role if a block position would give better results, if EC principles are at issue or if the negotiations involve a third country block. Thus the Transport Council mandated the Commission to negotiate aviation agreements with ten countries of Central and Eastern Europe in October 1996. Meanwhile, the states are trying to ensure that they have the ASA they want, before there is too much interference from Brussels. In this they are following the example of the Austrian Airlines hub at Vienna which arose from a positive government policy to sign bilateral agreements with all the new eastern European states, a policy which has encouraged an increasing number of multinationals to base their regional headquarters in Vienna (Stadler, 1995).

11.2.4 Transport policy and Community cohesion

The EU's transport policy is designed to facilitate the economic and social union of the member countries by promoting an efficient and comprehensive transport network. Reports from the EC's consultants seem to claim benefits for social cohesion from investing in rail systems throughout the Community as though air did not exist, advising that the required accessibility benefits would accrue from a reduction in journey times if the population centres were linked within a single high speed rail network (Allport and Brown, 1993). They also state that "smaller cities are not so well served by air, with few and infrequent flights, whereas high speed rail would provide a minimum two hour service frequency". In fact, using hubbing, a huge number of small city pairs are already connected by air, and liberalisation will bring many more into the network. Frequency is less important than timing for day-return capability, as the consultants themselves admit. Also, it is one thing to be on a rail network, but quite another to have direct intercity service available, particularly if there is a need to interchange between stations as in London and Paris, and many people do not wish to travel from city centre to city centre. Further, the assumption that it is easier to achieve a longer productive day with a train journey because of the greater ease of working during the trip, does not allow for the shorter travel time and reduced check-in times for air in a single market. Yet the EC has been advised that these 'new' factors in cost-benefit analysis make the case for rail investment incontestable, compared with accounting only savings in travel time and operating cost.

The Organisation of Economic Cooperation and Development (OECD) carried out a large study called Project 33 in 1977 which showed clearly that the most efficient way of improving the accessibility of the peripheral regions of Europe in order to minimise their comparative disadvantage was to invest in air transport (OECD, 1977). The comparative disadvantage has tended to worsen with time (Keeble, Owens and Thompson, 1982), causing the Social Cohesion and Regional Development funds to be used to support the improvement, not only of the roads and railways in the chosen European Transport Networks (TEN), but also of economically marginal airports, for example in western Ireland and in Scotland. It is dubious whether this support for the airports will achieve the cohesion objectives without the sort of subsidy for the small carriers which is available from the Small Community Service Program in the US. Reynolds-Feighan (1994) points out that the EC policy does not have an Essential

Air Services (EAS) component to ensure subsidised low density service. It also does not have the concept of reliever airports, yet provision should be made for private air transport which can pay its way. The Plan favours enhancing regional connecting points to a 'critical mass' so that they become 'hinges' of the airport network, as well as improving existing capacity (particularly to respond to changes due to the Schengen agreement on open borders), making better use of existing capacity including military airports and providing new runways.

There is, indeed, a feeling that the proper role of transport in the Community's economic integration has not been fully understood (Vickerman, 1990)

11.3 AIRLINE STRATEGIES

11.3.1 Airline responses to liberalisation

Airlines have positioned themselves for an increasingly competitive environment by using alliances, both defensively by consolidating home markets (e.g. Air France with Air Inter, British Airways with Dan-Air) and offensively by moving into competitors' markets (e.g. British Airways with TAT and Deutsche BA).

The independent airlines such as Air Liberté, Euralair, Ryanair, Air Europa and Spanair, encouraged by the liberal competition policy, have played an increasing role as new entrants in lowering fares. However, there has been little evidence of the major airlines lowering their full economy fares, probably because of so little new entry on dense routes. British Midland made some impact with its low business fares from Heathrow and Sabena's Skypass allowed unlimited travel for a month between Brussels and Heathrow for four full fares. The airlines became much more cost conscious to the point of confrontation with the unions, but any benefits have gone into justifying 'final' packets of State Aid, often in preparation for privatisation, rather in lower fares. Lufthansa followed British Airways into privatisation, while others were adopting many of the same marketing strategies as the more financially secure airlines as they continued to benefit from state aid. Air France are using their terminal at Charles de Gaulle to offer low minimum connection times to an improved range of destinations. Iberia is competing on price in its domestic market.

One of the strategies for reducing cost has been to form low cost subsidiaries (e.g. Iberia's Viva) and feeder airlines. The latter have sometimes spun off from the flag carrier (e.g. Olympic Aviation), sometimes an equity stake has been taken and sometimes the major airline has franchised its name (e.g. British Airways with Brymon, Cityflyer, Express, Manx and Loganair; Air France with Brit Air).

The subsidiaries have been active in taking up the new third and fourth freedom opportunities, following the example of Air UK and KLM at Schiphol which took advantage of the very liberal ASA between the Netherlands and the UK in the 1980s. The opening of new low density routes has been encouraged by the advent of 50 seat regional jets: Lauda Air is using the Bombardier RJ to expand its Vienna hub with high yield traffic even while route density is low (*Airports International, June 1994, p 22*). Lauda Air also created a second hub at Salzburg and, more interestingly, used the new seventh freedom opportunities to initiate a hub at Malpensa adjacent to its Lauda Air Italia charter subsidiary and linking into Manchester and Vienna (*Airline Business, April 1995, p 14*). This latter experiment has not succeeded, probably due to the poor location of Malpensa in addition to the other problems associated with operating from a foreign base and competing with a subsidised national airline. This may well change as Alitalia builds up services there, following the increase in capacity.

These recent moves suggest that two of the major airlines' preferred ways of dealing with the opportunities and threats of liberalisation are to strengthen home hubs and to locate new ones in competitors' markets, just as the US carriers did after deregulation. It appears that the international nature of these moves has not proved an insuperable marketing problem, particularly if the raiding is by means of an alliance with an indigenous carrier. In the face of the totality of the US deregulation experience and in the light of the European experience to date the questions to be faced are, firstly, is it a good idea to promote hubbing in Europe and, if so, what are the implications for network shape and the consequent infrastructure requirements? The answer depends on which hubbing strategy is adopted.

11.3.2 Hinterland hubs

Strengthening of traditional hubs is certainly possible within the demographic constraints, despite the frequent argument that the main traffic demands will continue to be in the central axis from Manchester to Rome and that hubbing in this very short haul market would be unable to compete with existing direct connections. Many secondary city pairs even in this central zone do not have enough traffic to justify direct service, e.g. Birmingham-Florence. All the major carriers are reforming their hubs to create more frequent and more concentrated complexes. The competition for transfer passengers has accelerated considerably since 1991, the battle being fought in terms of frequency and ease of transfer, one of the most important factors being the Minimum Connecting Times (MCT). The MCTs vary between classes of transfers but, for the main hubbing carriers, they were 60 minutes at Heathrow (90 between terminals 2 and 4) compared with 45 minutes at Frankfurt (even between terminals) and Charles de Gaulle, 40-50 minutes at Amsterdam, 35 minutes at Munich, 40 minutes at Zürich, and 60 minutes at Madrid and Rome (Woolley, 1996). The British Airways hub at Gatwick has a 26 minute MCT. Lufthansa is to have five waves at Frankfurt and three at Munich, KLM is aiming for six waves at Schiphol, Air France has five waves at Charles de Gaulle, Austrian Airlines has four waves at Vienna, SAS has concentrated its hubbing on Copenhagen with five waves as well as feeding its alliance partner's hub at Frankfurt. Swissair is to concentrate its hubbing at Zürich with four waves which will be linked into five waves by Sabena at Brussels (*Air Transport World, July, pp 162-164*). Discounts to the airlines on the normal passenger charges are available for transfer passengers at Amsterdam, Copenhagen and Manchester, while there are no charges at all at Dublin, Madrid, Milan and Paris, helping the airlines to discount the tickets through the hub. The result of these initiatives is to enhance the percentage of passengers who are transferring. In 1995 these percentages were: Heathrow, 34.5%; Frankfurt, 48.0%; Charles de Gaulle, 19.0%; Orly, 13.5%; Amsterdam, 38.7%; Gatwick, 14.0%; Zürich, 30.1%; Vienna, 19.3% (Woolley, 1996). There is no doubt that concentrating demand for service at the major hub airports, and serving it with intensive waves of flights, will put further strains on runway and stand capacity. Continual increases in capacity will be necessary to support this trend, though there is much room for improvement in the average size of short haul aircraft at European hubs, particularly if the fee structure ceases to favour smaller aircraft (Airbus, 1994).

More bypass hubs like Birmingham might also be set up at, say, Hanover or Lyons or Basle (Gialloreto, 1992) to counter this lack of capacity at the main airports and to give the regional airlines the chance to compete against the major carriers. Nice is already being used in this way by Air Littoral, and is realigning its runway to increase apron space and shift the noise footprint (*Jane's Airport Review, April 1998, pp 13-14*). Air Littoral also has a hub at Montpellier, Proteus Airlines has a hub at St Etienne and Regional airlines has one at Clermont-Ferrand (Graham, 1997).

11.3.3 Gateway hubs

The airlines and the peripheral cities may both be interested in attempting to develop gateway hubs at the periphery of Europe as a way of improving the local economy and of extending network coverage. Rome, Athens, or an eastern European capital may be appropriate for southeast Asia traffic, while Madrid or Lisbon are obvious candidates for Latin American traffic, these airports then providing onward services into secondary EU cities (Flint, 1991). However, peripheral gateway hubs are unlikely to divert existing traffic from established and more central gateway hubs unless they offer the passenger greater utility in completing their trip. The experience of American Airlines decision to pull out of many routes to secondary European airports indicates the preference of carriers to concentrate on the main hubs where possible (Katz, 1991). Peripheral gateways are also unlikely to entice passengers away from direct long haul service to the main European cities except by charging very low fares. They therefore have three potential markets:

1) intercontinental line-haul transferring to secondary EU cities, e.g. Atlanta-Shannon-Gothenburg;

2) intercontinental line-haul transferring to those secondary and tertiary EU cities situated between the gateway and the major EU hubs, e.g. Atlanta-Glasgow-Newcastle/Teesside;

3) intercontinental secondary cities transferring to EU hubs and secondary cities, e.g. Baku (Azerbaijan)-Athens-Rome/Naples.

Option 3 is difficult because the third country's secondary cities will tend to link with major European hubs before using emerging gateways, unless they have a special affinity such as the Greek Orthodox religion which is common to many of the ex Soviet Union countries and is encouraging them to use Thessaloniki as their point of contact with the EU. It also implies competition with the hubs in the foreign countries, and is less likely to be successful for North American routes than elsewhere in the short term. Option 1 is likely to be a higher cost solution than carrying the intercontinental line-haul further into Europe unless the bulk of the traffic is concerned with the gateway city. Option 2 has the best chance of success, but will tend to be a relatively small market. The experience of UK regional airports with transatlantic links is ambivalent as noted in the UK case study. Peripheral country airlines who might be gateway operators have been losing traffic to central hubs: indirect traffic to Ireland increased from 40% to 55% of total originations between 1992 and 1993, mostly over London and Manchester, while Olympic finds it hard to compete with fifth freedoms on BA at Heathrow, KLM at Amsterdam, United at Paris and Delta at Frankfurt for traffic between Athens and the US (McMullan, 1993).

There is, in fact, not a great demand between, say, North America and Europe which is likely to be drawn away from the major capital airports. Total North Atlantic passengers in 1987 amounted to 21 million, of whom nearly nine million used New York Kennedy. Very few city pairs with a European capital at one end were not well served, though several other large European cities still lack direct connections. Of the 21 million passengers, only 1.0 million connected airside to Europe at Heathrow, with a further 0.2 million at Gatwick. The figures for 1991 were 1.54 million and 0.27 million respectively, a further 0.9 million connecting with UK services (CAA, 1993b). By 1996, with the increased emphasis on hubbing by British Airways, the figures were 3.3 million, 1.2 million and 1.9 million respectively (CAA, 1998). Few passengers are likely to be tempted away from existing hubs by less intensive operations elsewhere, so new hubs will have to rely largely on new traffic denied access to crowded existing hubs. However, the long haul services are likely to be the last to be affected by congestion at the existing large hubs. Also, for every other European hub with similar connecting passengers, there will be at least another potential gateway attempting to pick up the traffic, so only a very competitive new gateway will capture more than perhaps half a million connecting passengers. In view of the volatility of the soft 'over the hub' traffic, this may well not be enough to justify new long haul gateway facilities, unless the local long haul demand is also very strong or the main hub carrier becomes concerned about yield dilution.

The prospects for gateway hubs in the Eastern European countries are affected by the low density of the local traffic and the problem of airport and airspace congestion, though the airports will certainly have runway capacity to spare for some time. The industry predicts growth in traffic between Western and Eastern Europe from 7 million in 1990 to 27 million in 2010, but this is small compared with the prediction of one billion passengers at European airports by 2010 (IATA, 1992).

11.3.4 Multi-airport metropolitan areas

Multi-airport cities provide the main existing gateways into Europe, and the majority of the major airports in them are already congested (Airbus, 1994). However, the 'hard' local traffic will remain strong at most major hub cities. There will therefore be a strong incentive for competing carriers to raid other airlines' territories by setting up hubs at secondary airports within the same metropolitan areas, where those airports can offer the right conditions. The BA/TAT operation at Orly and the Lauda Air mini-hub at Malpensa are early examples of this trend, but the difficulties should not be underestimated, as shown by the London system case. Also, the balance between economic benefits and environmental disbenefits is likely to be drawn differently in each of the major metropolitan areas.

Experience in the US suggests that not more than two major airlines can hub successfully at a single airport, however large the airport. Since there is relatively little room for expansion at most of the main European airports, a new major hubbing operation could not easily be mounted at an existing major hub. A new hub is likely to need to use secondary airports and, in general, to need a second runway, as shown at Manchester. This airline-based competition is a way of making efficient use of a fragmented metropolitan airport system, if the incumbents at the primary airports do not themselves opt for fragmentation in order to protect themselves against just such a raid on their territory or to overcome congestion at the main airport. Another way is to mount spokes to competing hubs, as has occurred with Air UK's services from Stansted to Amsterdam and Air France's London City route.

Secondary metropolitan airports can therefore be used for either offensive or defensive purposes. A prime example of both is Orly, with Air Inter being Air France's defence and TAT being BA's offence. Despite the aggressive examples from BA, the strategy for most of the European majors appears to be to dominate a whole region rather than raiding the traditionally difficult area of another major airline, even merging for this reason, e.g. Air France/Sabena (Jenks, 1992), though this fell through and Swissair has now taken a 49.5% stake in Sabena (French, 1995). This follows the US carriers' success in developing hubs from the advantageous position of local loyalty and sunk costs (Bailey and Williams, 1988).

11.3.5 Low cost new entrants

These carriers, modelled on Southwest Airlines and other similar airlines in the US, were not expected by conventional wisdom to thrive in Europe. The culture of scheduled passengers still requires quality of service, the costs of international operations and ground handling would dilute price differentials, and congestion and slot controls would negate attempts to maximise aircraft utilisation and minimise turnround time (Eccles and van der Werff, 1995; Humphreys, 1994). It is said that the Ryanair prototype operations, using the same low cost philosophy to link Stansted and Luton with Dublin, together with other Irish points from these airports and other UK regional airports from Dublin, rely uniquely on the Visiting Friends and Relatives (VFR) traffic between the UK and Ireland as well as the over-water setting and the common language, so that it is not repeatable on the continental mainland. That it does not rely solely on these factors is shown by costs and productivity comparable with Southwest and by the fact that 40% of the weekday traffic is travelling on business (Reed, 1995). Business travellers can become very price sensitive in a recession. A study of the change in European market shares between 1982 and 1989 suggests that price differentials appear to have a significant effect on market shares, so that low cost new entrants may be able to penetrate the market effectively (Marin, 1995).

Ryanair already has routes from Stansted to Scotland and plans to open others to the continent, while EasyJet is also serving Scotland from Luton and has announced its intention to expand with the purchase of a 40% stake in TEA-Switzerland. Debonair has a range of continental destinations from Luton which make use of the new cabotage possibilities to link its stations into loops beginning and ending at the Luton base and plans further continental routes to link into its existing stations (*Flight, 1 April, 1998, p 26*). Ryanair is basing its strategy on using cheap secondary airports, avoiding direct airport-to-airport competition with the majors, using Beauvais rather than Paris for its Dublin service, which has prompted Continent Air Paris to use it for Portuguese and Italian routes. Ryanair is even using lower cost alternatives to secondary cities, including Carcassone for Toulouse and Kristianstad for Malmo (*Airport World, Vol. 3, Issue 2, 1998, pp 17-18*). Another innovative airline, EBA Express, opened effectively ticketless routes from Brussels, forcing Sabena to compete on fare (Guild, 1995a). It has now become Virgin Atlantic's vehicle for launching a low cost pan-European airline with 20 Boeing 737-300 aircraft by the end of 1996, initially targetting the routes with the highest fare structures both from the Brussels hub and also exploiting seventh freedom opportunities like Rome-Madrid (Jones, 1996). An interesting example of synergy between low and high cost airlines is the arrangement to carry Sabena passengers on the Brussels to Heathrow route, cutting full economy fares by 50%. It is looking to a new hub at Rome Fiumicino and expansion at Dublin to mitigate the high social costs of employment and shortage of pilots in Belgium (*Flight, 19 August 1998, p 13*). Several other low cost jet carriers have recently entered (and in some cases also exited) the domestic and international EU markets, including Air One, which has secured enough slots at Milan Linate to compete with Alitalia between the main scheduled airports in Rome (Feldman, 1996b). This was only achieved by lobbying to increase the declared capacity at Linate from 22 to 32 movements per hour (Parry, 1996). These new entrants are bringing new equity and skills into the airline industry from their parents' backgrounds in hotels, superstores, shipping, civil engineering (*Avmark Aviation Economist, December 1995, pp 4-7*), car rental and music (*Flight, 22 May 1996, p 15*). In many cases they are using other airlines' operating certificates.

The new equity is important, since Air Europe's attempt to move from charter to a scheduled hubbing operation at Gatwick is considered to have failed due to underestimating the capital required. It had estimated a need for £90 million and only managed to raise £30 million, when in retrospect it needed £250 million plus 30% equity financing in the aircraft (Powell, 1994). The mortality rate for international scheduled routes with more than 3,000 passengers per annum (ppa) at UK non-London airports between 1986 and 1996 averaged 46% with an average life of 2.4 years; for routes above 15,000 ppa it was still 20% (Morris, 1997). This is a very important factor that airports should take into account before enhancing their facilities to satisfy the requirements of new entrants.

The long term success of the low cost carriers depends on the identification of enough niche markets without impinging too seriously on the major carriers' high density markets. Low costs themselves are not a substantial barrier to competition (Sorenson, 1991), as the emergence of low cost subsidiaries of the major carriers has demonstrated. Given the difficulties in accessing the major hubs, the crux of these low cost operations is the availability of low cost and convenient secondary airports. Ryanair even investigated the use of a military airfield in Dublin before deciding to build its own terminal at Aer Rianta's airport. London has Luton as well as Stansted as well as several established airports slightly further from the centre and also Biggin Hill, the latter having upgraded its terminal in anticipation of its first scheduled service. Frankfurt could use Hahn as a reliever airport. Brussels is also served by Charleroi, which is only 50 km away. Rome has Ciampino which is closer to the city than Fiumicino. Athens might still have its old airport when Spata opens. Paris has Villeneuve St Georges. The future use of Berlin's Tegel and Templehof when Schonefeld airport has been expanded is still under discussion. Stockholm has Bromma much closer to the city than Arlanda and is also being served from Scavsta. Similarly, Oslo has Torp as well as Fornebu, the latter being expected to close when the expansion of Gardermoen is complete. Amsterdam has Rotterdam. Many other capacity-constrained airports have less congested partners in close twin cities, e.g. Liverpool for Manchester, Cologne and Mönchengladbach for Düsseldorf, though both of the latter pair have been capped by environmental regulation, as explained above (Odell, 1995).

It may be that other Ryanair-type airlines will also use Southwest's multi-sector, high utilisation strategy to provide direct links between regional cities. Investigations by the European civil aviation conference showed in the 1980s how few single sector routes in Europe were viable, except those with a hub at one end (ECAC, 1984). However, the advent of genuine low fare frequent multi-sector scheduled services using cabotage and fifth freedom rights could make a substantial difference to the analysis. Ryanair believes that its success will depend on releasing demand rather than diverting traffic.

11.4 POWER RELATIONSHIPS

Some of the power relationships which have affected the development of the European air transport network have already been discussed, including lobbies based in modal, 'green' and national interests. Others are mentioned below.

Based carriers can serve their own interests by influencing the declaration of capacity below the potential available from the physical and environmental considerations. The EC's requirement for the independence of the slot allocation process at busy airports attempts to ensure that 'grandfather' rights to slots are not too abused, but the rules make it difficult (at least in theory) for a European government to grant access to a busy airport as part of an ASA without interfering with the slot allocation process. In the US, only four airports have slot allocation: at these airports, foreign carriers are given precedence and some slots are ring-fenced for regional airlines and General Aviation.

As with the US federal authorities, the EU can only make an investment in the system if the State and Local governments wish the project to be constructed. Local environmental opposition at large airports is often strong enough to result in most of the benefits of a new runway being used to mitigate noise impacts rather than to increase capacity. It is usually the communities served by regional airports with spare runway capacity who are most willing to agree to expansion of air traffic, some more than others.

Use of the spare capacity at provincial city airports will depend on the cities' influence in persuading the EU or national negotiators to grant route rights and persuading the airlines to take them up. Those with strong local traffic and the required airport attributes will best persuade the airlines. Unless the national carrier is interested in gateway hub fragmentation, there are only five established scheduled long haul EU airlines other than the flag carriers (Chataway, 1995), though some new low cost entrants are moving into the market (*Airline Business, December 1997, p 9*). A peripheral airport wishing to act as a gateway hub will also need to obtain slots at the European airports it wishes to serve. The main hubs' slots might be more easily available as spokes from their own based carrier's hub than as spokes from the potential gateway, at least until firmer action is taken on slot allocation by the EU.

Lobbying has reached a fine art, not only in attempting to influence the opinion of prominent politicians but also in persuading the state and European Commission civil servants who advise the politicians. Compared to the Councils of Ministers, the European Parliament and the Commissioners have limited powers. Most of the Councils' decisions are resolved informally before they even reach the ministers by the Committee of Permanent Representatives which is made up of the ambassadors of the member states. They, in turn, take instruction from their capitals, having direct access to their premiers (*Financial Times, 11 March 1995, section 3, pp 1-2*).

It was obvious, long before liberalisation occurred in Europe, that it could only be effective if there were residual controls to prevent predatory and monopoly behaviour, to maintain safety, to compensate the consumer in the event of bankruptcy, and, most importantly, to allocate runway slots at the most congested airports so that competing airlines could have equal opportunity of access to them (NCC, 1986). Despite the necessary legislation and definitions of abuse being in place, there is much doubt as to the Commission's ability to offer effective policing (*Airline Business, May 1996, pp 58-61*) or to take sufficiently draconian action, particularly given the need to sustain the European trunk carriers' global competitiveness.

11.5 Towards a Realistic European Air Transport Network

It may be concluded from the above case study that:

- Access to Heathrow and Frankfurt has been, and will be, the key to the success of any EC competition policy to affect full fares

- Air transport is in danger of 'regulatory capture' of the administrators by other modal interests

- Environmental capping is here to stay, hence methods of mitigation become of overwhelming importance, including third party safety

- EU policies of social cohesion may distort the air transport system, while there has been insufficient testing of the role of aviation in social cohesion

- Control of alliances and hub dominance has so far been ineffective

- A pressing regulatory requirement is for a uniform method of slot rationing

- Except for having some funding, the European Airport Network Plan repeats the UK's 1978 Plan mistake of being a passive wish list based in historic patterns of use

- The low density of the scheduled air transport system in Europe makes it more important that airlines should have the infrastructure to hub effectively.

Following the patterns in the US and those already developing in Europe, it is possible to suggest a framework for constructing a base scenario for airline network development. The system will evolve around a pattern of gateway, hinterland and secondary airport hubs (particularly those gaining economies of scope by blending gateway and hour-glass hubbing at provincial points like Birmingham) in competition with new entrant airlines. Any normative planning which does not recognise this, and the likely alliance behaviour of the main European carriers and their global partners (Staniland, 1997), will produce an inefficient system, but political expediency will make interference almost inevitable.

It is likely that strong 'hinterland' hubs, generated under historic international regulation, will grow even stronger in the central areas of Europe to the extent that physical and environmental capacity allow. The existence of 'hard' local traffic at these existing hub cities will encourage alternative hubs to develop at secondary airports in the same metropolitan area, dependent on the communities' attitude to the balance between the economic benefits and environmental disbenefits, but also on the financial health of the secondary carriers and the extent to which liberalisation really allows hub raiding, either by EU or foreign airlines.

Provincial cities in the central area of Europe which have strong economies and 'hard' traffic, should expect to be chosen as by-pass hubbing points, especially if they have the ability to develop second runways. More opportunities of this type might appear as newly revived central regions benefit from overspill investment from the leading economic regions or have successfully restructured their economies (Steinle, 1992). However, some of them may decide that their continued success as cities does not depend on extra accessibility to the air transport system, and hence decide not to expand their airport capacity and not get involved in the bidding for non-EU route rights.

Peripheral regions in the EU, and their airlines, will mostly wish to enhance their share of EU air traffic in order to boost their economic health. Cities in these regions are likely to favour economic growth and the encouragement of hubbing, despite the resulting peaky nature of the airport operations and the increased environmental effects caused by catering for transfer as well as local traffic, since hubbing will produce substantial increases in accessibility. Either 'gateway' hubs or 'hourglass' hubs might be possible, depending on the specific location.

There is only limited potential for economies of scope to attract major airline interest at most existing peripheral gateway airports, unless the airlines can see the possibility of improved control and flexibility in managing their main hubs, perhaps using the gateway as a by-pass. There is even less potential for new gateway airports to enter the system, other than as spokes for third country carriers' hubs.

The remaining hubbing possibility, for cities who wish to boost their economic potential, is to act as an 'hourglass' or by-pass hub, feeding traffic from cities in its regional sector to main EC hubs, the Birmingham Eurohub being a good example. To the extent that the negotiators allow the present and future by-pass hubs to take a gateway role, so the requirement for peripheral gateways will tend to reduce further, and the spiral of comparative advantage towards the centre of the EEC will only be countered if these regional by-pass hubs develop good quality air and ground access links to the most peripheral regions, thus off-setting their exclusion from the major hubs (Graham, 1995).

The consumers' dream of direct service between all city pairs, encouraged by the EU legislation, is unlikely to become reality in the face of the hubbing opportunities outlined above. However, as the hubs develop within the settings defined by the route negotiators, by the local communities, by the airport owners and by the airport/airspace capacity limitations, there is every prospect that many more EU international city pairs will obtain direct services, even if they lose service to their local national hubs through lack of slots or through rail competition.

There are good prospects for low cost direct multi-sector routes to be developed by innovative airlines. Access to the main hubbing points will continue to be provided by the main hubbing carriers or their subsidiaries. System densities will remain quite low relative to the US, so it will be important for the consumer that any available economies of density from hubbing should be captured, rather than being discouraged. There may well be additional airport and airline economies of density to be found in blending gateway hubs with hinterland or hour-glass hubs, leading to a proliferation of these types of operation as ASA between the EU and third countries are loosened.

All these developments may help to keep demand matched to the dispersed runway capacity available. Some presently under-utilised airports with strong local markets can therefore expect to find strong demand. The more they exploit their opportunities, the less likely it is that their smaller neighbours will be able to retain market share.

However successful the regional airports and secondary metropolitan airports become, the preferred networks of the major carriers will require more runway capacity at precisely those major airports which currently suffer the greatest capacity constraints. The expansion of capacity at these airports is necessary to allow the major European carriers the opportunity to play a world role and to offer the hubbing which allows smaller cities to participate in an integrated Europe, while at the same time allowing genuine head-to-head competition at the major airports as well as over-the-hub competition. Ways need to be found to make these airports 'elastic' through the development and application of smarter technology and smarter management, the enhancement of both physical and environmental capacity and the encouragement of multi-modal operations. The technical initiatives will need to go beyond those suggested in the ECAC/APATSI reports of mature and medium term technologies (ECAC/APATSI, 1994; ECAC/APATSI, 1995). Management will have to face the issue of full social cost recovery. The EC's Airport Network Plan requires more power to recognise the full roles of these airports and to encourage environmental mitigation, while leaving the local communities to have the final say on the appropriate level of air activity. The EC could, at least, ensure a level playing field in the education of all parties in the debate with respect to the technical and economic possibilities.

12

NATIONAL AIRPORT PLANNING CASES

12.1 BRAZIL

Brazil is a middle income developing country with a population of 161 million in 26 states at a density of 18 per sq km. It has a GDP of approximately US $500 billion in 1995, giving a GDP per capita of US $3,100. However, 48% of the income is received by just the top 10% of the employed, at an average of some US $25,000 per capita. This compares with an average of some US $8,400 for the next richest 10% and US $600 for the poorest 10% of the population. Since 1993, when the currency was renamed the Real and its value tied to the US dollar, inflation has fallen from an annual rate of 7,000% to single figures. After a decade of little real economic progress, the GDP grew by 30% in 2 years.

There are 46 airlines, with a total turnover of US $5.2 billion. The three largest are the national carriers: Varig, Vasp, Transbrasil, placed 25th, 50th, 66th respectively in the world sales rankings. TAM, the largest regional carrier, is 93rd in the same rankings. Brazilian airlines have a 60% share of their international market, which may reflect the fact that almost twice as many Brazilians fly out as foreigners fly in. There is, however, an almost unlimited opportunity for tourism, from beach holidays to ecotourism, once the infrastructure is in place and the tourist is made to feel secure. One third of the traffic is with other South American countries, most of the rest being with either the US or Europe. The international market is being opened up to competition between the Brazilian national carriers, Transbrasil having several routes into Europe, and even the regional carriers are finding ways of entering the market. TAM has bought control of the Paraguayan carrier LAPSA, renaming it Transportes Aereos do Mercorsul to serve the US and other points in the Southern Cone through Asunción. It has also bought the Paraguayan domestic airline Arpa and gained clearance from the Brazilian authorities to operate its subsidiary (Transportes Aereos Meridionais) as a national rather than a regional carrier (*Flight 30 April 1997, p 12*). The five countries in the Mercosur economic trading group have agreed to liberalise entry to those routes not covered by existing bilaterals (*Airline Business, March 1997, p 14*). VASP, now Vasp Air System, whose chairman owns another 20 companies with a turnover of US $3 billion, has bought controlling interests in Bolivian and Ecuadorian carriers in an attempt to dominate Mercosur markets. All the major carriers are developing links with foreign carriers, notably Varig becoming a member of the STAR alliance.

The domestic market accounted for over 80% of the 51 million passengers handled at Brazilian airports in 1995. There is only a limited competition from other modes for interstate travel, due to the distances and the lack of a quality rail network, though the cheap and efficient coach services provide well for the low value of time market: there is a large demand even for trips lasting up to two days. Within the air mode there is a considerable degree of competition, with four carriers on the densest domestic routes and a thriving air taxi sector. Regional carriers, of which TAM is the largest were, until 1992, very restricted in the size of aircraft and in the location of the sectors they can fly. This did not stop them offering effective competition to the national carriers, because, with the exception of the Ponte Aérea pooled service between Rio and São Paulo, only the regional carriers were allowed to fly into the downtown airports in these cities and in Belo Horizonte. Not only did this allow them to offer an attractive product with Fokker 100 aircraft on these city pairs, but they were also able to provide longer haul services with nominal stops into these same downtown airports, capturing useful market shares even at premium fares. However, in 1998 the trunk airlines are being allowed into the downtown airports while having their monopoly of the Rio - São Paulo air bridge broken (*Airline Business, May 1998, p 11*). All airlines are now able to offer deeper discounted fares, which has resulted in another surge of traffic growth.

There are approximately 5,500 airports in Brazil. The small airports are administered by the relevant municipal or state authority. The 62 larger or more strategically important ones, which include 20 international airports and which process 97% of the country's total traffic, are administered by a federal agency created in 1972 with the title INFRAERO (Empresa Brasiliera de Infra-Estrutura Aeroportuária), reporting to the Department of Civil Aviation, with 20,000 in-house and outsourced employees. This could lead to some conflict, partly because a state tends to have to subsidise an airport during the initiation of commercial service, while having no income from any profitable large airport, because any such airport will be owned and operated by INFRAERO. In fact, though the total revenue was US $750 million in 1995 (*Airports International, November/December 1996, pp 26-27*), only eight of the 62 INFRAERO airports are profitable and some US $500 million a year is being invested in the system. INFRAERO is organised in seven regions to reflect common local characteristics. Income is gained principally from aeronautical charges, from the 29 bonded warehouses and from the commercial opportunities. In addition, central government funds are made available for upgrading and for investment in new facilities.

Traffic is now increasing rapidly. It is expected that passengers will double by the year 2000 and that freight will grow even faster. INFRAERO intends to invest US $2.4 billion by the year 2000. It hopes to fund some 50% itself, the rest coming from local and state governments and from national and foreign private enterprise (Gennari, 1996). The latter source of capital is not considered suitable for aeronautical purposes, since most airports are shared with the military and the revenues are quite small. However, in line with the governments enthusiasm for privatisation which has seen US $27 billion of sales of state-owned companies in the last decade, it is hoped to attract private interest in catering, warehousing, multimodal centres, meteorology, telecommunications and cargo processing, as well as in the commercial development of the airports with facilities such as parking, hotels, duty-free shops, convention centres and shopping centres.

The large nationally-important airports have been planned by specially instituted short-life commissions managed by committees drawn from the various federal, state and municipal bodies who provide the funds. The planning effort at all levels has been supported by a group created in 1977 with the acronym CECIA (Comissão de Estudos a Coordenacão da Infra-estrutura Aeronáutica) within the Directorate of Civil Aviation (DAC) in Rio de Janeiro. This can mean that several different sets of forecasts are generated for the same project. Since 1981, CECIA has also been jointly responsible with the States for developing State Airport System Plans, though curiously these exclude the state capital airports. Most states do not have the expertise to carry out such studies themselves. Since 1986, CECIA has been incorporated in the Institute of Civil Aviation (IAC), which has overall responsibility for the planning of Brazil's aeronautical infrastructure.

The roles of the airports in the system have not usually been in doubt. Some are clearly necessary to support the process of opening up remote areas or to support the emergency services. Others function to support the social and economic cohesion of the widely dispersed states with jet transportation rather than having to spend up to three days on the road, the latter sometimes being completely disrupted due to flooding and subject to security problems. Many towns require air feeder service into state or regional capitals but cannot justify most other links. Long haul inbound tourists need to access their destinations directly: Historically, most tourists wished to visit Rio, but now more of them are wishing to explore the richness of other areas or to take advantage of the shorter flight time to the beaches of the northeast of the country.

As the traffic grows and the system evolves, the airports' roles are becoming more complex. The downtown airports are being brought into the longer haul domestic and even the international market by indirect service which overcomes the regionals' requirement not to compete directly on the trunk domestic routes (*Flight 8 January 1997, pp 34-35*). Towns in the interior are growing to the point where they can justify interstate links. An 'open skies' agreement on air services within the Mercosul group of countries has resulted in 44 INFRAERO airports being designated for international flights between small towns and cities within the signatory states (*Jane's Airport Review, October 1997, pp 45*-47). The international and domestic airlines are changing their route structures and making alliances to take advantage of new business opportunities. Competition is therefore beginning to develop between airports per se, and between the regions they serve, which is beginning to modify the historic roles. This can be seen clearly in the case of the São Paulo airports described in chapter 13.

12.2 CANADA

Canada has a particular need for a strong air transport network. It has only 25 million people in an inhabitable area measuring some 5500 km by 800 km. There are some 50,000 pilots and 20,000 aircraft. The carriers were protected by strong regulation on domestic and international routes until the mid 1980s. Since then there has been a more liberal policy on route entry and on airline alliances, with a large domestic charter sector. The role of the airports is changing with the creation of the free trade area between Canada, the US and Mexico which has liberalised cross-border routes, which has allowed Canadian carriers to obtain equal shares of the cross-border market as it grew by 37% between 1995 and 1997 (*Air Transport World, May 1998, p 84*).

The federal authorities own or subsidise 150 airports. Until recently, they operated 108 of these, with the provincial or municipal authorities operating the remainder. A further 300 airports are both owned and operated by the provinces or municipalities. Another 600 airports are owned and operated privately. Most states have some form of Air Transport Assistance Programme to support air service to isolated communities and to establish airports and navigation infrastructure.

The airports in the federal plan have been classified primarily by the nature of the air routes operated (Lee and Shaw, 1981):

1. Eight International Airports.
2. National Airports.
3. Regional Airports giving direct access to classes 1 and 2, having collected local traffic.
4. Local Commercial Airports used by airlines but not meeting other criteria.
5. Local Airports, i.e. not served by airlines.

These airports are also subclassified by their importance to the community (as measured by originating passengers), level of aviation activity and the importance of the airport to operators (as measured by the operating and maintenance expenditure). The system was developed through a series of regional plans, whose objectives were to "develop the most appropriate alternative locations and roles ... that meet future operational, environmental, economic and social requirements for a specific area" (Young, 1976).

The system has been funded in the past via Transport Canada through parliamentary appropriations from two funds: one as a Self-supporting Airports and Associated Ground Services Revolving Fund which operated on the top 23 airports and was supposed to have full cost recovery, the other was a general fund, the revenue from the remaining airports contributing to the fund. It was realised in 1994 that much of the Canadian transport system was overbuilt, though a study as long ago as 1985 had drawn attention to the fact that cost recovery in the airport system was only 44% and falling (Auditor General, 1985). It was also realised that, despite the above airport structure, there was no framework which defined a clear operational role for the federal government. This had led to federal policies at times being inconsistent with the needs of the local communities, resulting in national and regional imbalances developing with respect to facilities and funding. Considerable federal resources had been used to support a large network of airports serving only 6% of passengers, while the remaining 26 airports handled the other 94% of passengers and covered their own operating costs. Transport Canada was asked to review the potential for commercialisation of a number of its activities in order to improve efficiency and to assure long term viability.

Under a new National Airports Policy (NAP), the federal government is now encouraging corporate status and privatisation. It is retaining ownership but moving from being an operator to a landlord. The 26 largest airports, forming the newly designated National Airports System (NAS), are being leased for 60 years to Canadian Airport Authorities (CAAs) under enhanced accountability principles. There is no government regulation of the funding and pricing freedoms given to the local authorities. Airport authorities were already incorporated in Montreal, Calgary, Edmonton and Vancouver as an initial test starting in 1992, while Winnipeg, Ottawa, Toronto and Victoria had been transferred by mid 1997 (*Jane's Airport Review, September 1997, pp 21-22*). Other airports will be transferred to regional interests, though airports serving isolated communities which are now subsidised will continue to receive support. The aim is for all the 26 large airports to be individually self-sufficient within five years, though this might be difficult given that the three airports New Brunswick only handle 700,000 passengers between them. Others will have their subsidies removed over five years.

An Airport Capital Assistance Program (ACAP) will be introduced to provide financial assistance for safety-related airside capital projects at regional/local airports, funded in part from the lease revenues of NAS airports. In addition, smaller airports acting as satellites to international airports may be transferred to the appropriate CAA.

The new NAP is meant to increase efficiency by allowing local needs to determine priorities and driving out the less viable assets, while ensuring a minimum level of cover in low density areas. The major airports are also freed from political vagaries such as the strength of their local representatives and the national deficit. However, it is expected that the CAAs will be run by local or regional governments in a non-profit manner, so, although many advantages flow from the commercial discipline, the full advantages and disadvantages of a privatised and market-driven system will not be realised. Instead, it is likely that decisions on subsidy and closure will be put more squarely on the shoulders of the most interested communities, while the federal budget will appear to improve.

12.3 GERMANY

The eight main airports in the former western Germany serve regions with some 7% of both population and GDP (Toepel, 1986). However, Frankfurt has some 37% of the passengers. This relative strength is due partly to the central location of Frankfurt and also to Lufthansa's decision to move there from Hamburg when the first generation jet aircraft demanded greater runway length (Wolf, 1982). The long haul service required a progressively stronger feeder network, some of which is now being provided by rail.

The German international airports have always been either private stock companies or private companies with limited liability (Treibel, 1976), with equity participation by federal, state and city governments. The federal government has now embarked on a programme of privatising its stakes, and some cities, notably Munich, are also selling. At the same time, the federal and state governments have seen the need to set out their long term objectives for airports (*Jane's Airport Review, July/August 1998, p 3*). These include:

- ensure the continued strength of Frankfurt, Munich and Düsseldorf
- simplify authorisation procedures for development
- integrate trains and cars into airport planning
- further privatisation
- encourage traffic to shift from major to regional airports
- explore synergy between airports' roles and management

The federal authorities are ultimately responsible for airport planning and design laws but the states have the responsibility for the actual planning and design. The German Airports Association advises on the preparation and execution of regulations and plans. The Association has also become more active in national transport system planning with coordinated forecasts, uniform capacity calculations and air/rail cooperation. The setting of system goals is left to the states. Bavaria, for example, lays emphasis on access to civil aviation for each of its 18 regions and includes airport development in its State Transportation Plans. The states also assumed responsibility for the construction and supervision functions. The federal authorities were initially major shareholders in the airport operating companies, though with only minority participation. This helped to resolve any state/federal conflicts. Meanwhile, the emphasis on state planning has not proved an adequate solution to the environmental debate, since federal law gives communities the right to self-administration, including land use planning.

Federal aviation law also requires an airport master plan to be established, including noise contours, and approved federally. There is, in fact, a two tier approval process. The first one leads to a decision by a state on the suitability of the proposed site from the points of view of land use planning, urban development, noise impact, safety and security. The state takes account of the views of the communities, the counties and the regional planning associations. At this stage, the federal authorities only have the right to overturn the subsequent decision to implement a master plan on the grounds of federal interest in, for example, air traffic control or military facilities. Construction must not start until detailed designs have been presented to all neighbouring communities and all comments subjected to public debate. An appeals procedure can be followed through three levels of administrative court (Toepel, 1988).

Many of the airports are severely capped environmentally. Frankfurt is able to use its cross-runway only for takeoff and only in one direction. If Düsseldorf commissions a new parallel reliever runway, it will be faced with a movement cap on larger aircraft which is more severe than the existing cap, due to the length of the planning process requiring agreements to be made long before the facility can be operational. Leipzig's new runway will be reoriented to minimise noise nuisance as well as being lengthened, despite the inconvenience of having to construct it on the far side of an autobahn from the new terminal.

The new Munich airport was planned because the old airport at Riem created noise problems for 140,000 people as well as reaching physical capacity and raising safety fears due to an accident on a crowded city road in 1960. None-the-less, it took 30 years to move from the beginning of the search for a site in 1963 before the new airport opened. The site was chosen in 1969 by the Bavarian government 30 km from city and cleared by the Ministry of Economics in 1974. The matter was then opened up to the public planning procedure, during which over 26,000 objections were heard, the government gave their approval and the project went through the appeals procedure. Construction started in 1980, but was stopped after a Bavarian Administrative Court ruled that the area of 2,050 hectares was excessive, only starting again in 1985 after it had been reduced to 1,387 hectares. Finally, in December 1986, the federal government gave their approval for the airport with 2 x 4,000 metre runways separated by 2,300 metres and staggered by 1,500 metres, and the changeover from Riem to Munich II took place over the night of May 17 1992, with traffic levels at 12 mppa and 55,000 tonnes of freight, Riem then closing simultaneously. There was no significant impact on the traffic growth due the change of location, though Salzburg and Innsbruck airports both noticed an increase in their traffic as tour operators decided that the access time from the new Munich airport to Austrian resorts was too long.

Even though the site was chosen so that very few people would be exposed to noise, 6,000 households had to be insulated at a cost of over DM 150 million and the design was seriously compromised by mitigation measures. The 2,300 metre separation of the runways was chosen to minimise noise, and the runways are 4,000 metres long partly to avoid the need for reverse thrust. The operations are also compromised by a total curfew between midnight and 0500 hours, with operations between 2200 and 2400, and between 0500 and 0600 hours there may be only 28 scheduled movements or 38 total movements by aircraft certificated to Chapter 3 noise levels. Although other aircraft may operate during the rest of the day, their landing fees are substantially higher than for the Chapter 3 aircraft, with a 50% extra fee for Chapter 2 and 100% for non-certificated aircraft. In addition to the general regulations based on noise certification, there are specific operational regulations which are imposed through the use of 10 noise monitoring units. There is a special hangar for engine checks. Air quality is monitored at two stations for CO, SO_2, NO_x, HC, O_3 and dust. The lowered water table is rebalanced by 132 wells: in addition, the airport system can be isolated, the existing flows have been re-established, there are 247 table indicators with regular chemical monitoring, there is waterproof lining of runways with bacterial purification in gravel layer, and there are deicing stations with recycled meltwater at the runway thresholds. A 230 hectare green belt has been established to compensate for construction with a 200 hectare boundary buffer. Waste energy is used for district heating.

This has all happened at the greenest site that it is likely to be possible to find near a major city anywhere in Europe. The total cost of environmental measures accounted for some 10% of the total project cost. The design of the terminal is not too well suited for hubbing, having been designed in the days when gate-arrival concepts were expected to lead to a better level of service for passengers. At the present throughput of 18 mppa, the terminal and aprons are already close to capacity, yet the revenues are hardly sufficient to pay off the capital charges, due partly to the long delay between the investment and the beginning of the revenue stream.

12.4 GREECE

Greece has a population of some 10 million people, of whom 40% live on the southern mainland dominated by Athens and 30% live on the northern mainland for whom Thessaloniki is the main city. The journey time between Athens and Thessaloniki is nearly six hours by road or rail, so the air link between the two is vital for business traffic. The remainder are islanders, spread between Ionia to the west, the large islands in the southern Aegean, notably Crete and Rhodes, together with the Aegean Sea island groups of the Sporades, the Kiklades and the Dodecanese. The mainland is generally mountainous, so that the whole of the country can only be integrated by using aviation or by relatively slow roads or ferries. The average declared income is approximately US $4,600 per capita. The population is stagnant and, despite the influx of tourism, the GDP only grew by 15% in the decade to 1994. In the same period, the Greek Drachma lost over 60% of its value against the German Deutschmark.

Until the 1990s, Olympic Airways (OA) was the only carrier allowed to operate in Greece, although it created Olympic Aviation as a subsidiary to provide the lower density turboprop services. It still has a monopoly of all ground handling and pays no landing fees. Its schedules tend to incorporate domestic routes as the first or last sectors of international flights, so they are subject to delay and to difficulties of allocating seats to match local market needs. Transit loads are therefore high but the frequencies are greater than would be the case with a totally dedicated jet service, particularly at Thessaloniki. OA used to capture considerable transfer passengers onto the domestic routes at Athens, despite the inconvenience caused by the foreign airlines' terminal being on the other side of the airport. However, many of the tourist islands now have their own lengthened runways able to accept the charter carriers' jets, so Athens traffic has remained at approximately 10 mppa for several years, and the domestic traffic, even on Athens - Thessaloniki, has also stagnated. Competition was allowed on some domestic routes in 1994, and, despite the difficulties the new carriers encountered in starting and maintaining service, traffic has increased on the main Athens - Thessaloniki route by 30%, probably stimulated by their lower fares.

The 40 or so Greek airports are planned and operated by the Hellenic Civil Aviation Authority, in coordination with the military authorities who share the use of many of the airports. They respond to political pressure from the local authorities (nomos), who have considerable influence with the government in such a dispersed democracy, when they bid for facilities to keep up with competition from other nomos for the valuable tourist trade. This has resulted in the encouragement of many direct foreign charter services to the islands, to the detriment of domestic route densities. The lack of income from OA services has not encouraged the Authority to provide more than the minimum necessary facilities at its airports.

Athens airport has become surrounded by development. Most northern departures from the existing main runway would directly overfly Athens city centre if it were not for a tight noise abatement turn. A similar, but less severe turn is required on southern takeoffs to avoid a hospital. Another airport is being built behind the mountains to the southeast of the city at Spata in order to overcome the physical and environmental capacity constraints of the present site. This has been a stop-start project for 20 years, but is now due for an opening around the millennium, despite the lack of traffic growth at the existing airport. It is felt that, with a greatly enhanced facility, Athens will be in a good position to bid for an EU gateway role for traffic from the middle and far east. Much of the funding has been arranged by the contractors, and there is a special departure tax of some US $15 to help fund the Greek contribution, but Greece is relying on the EU for 20% of the total cost (*Flight, 18 August 1993, p 8*). This will need evidence that is adopting policies which reflect the EU principles of competition. This may be one reason for the opening up of the denser domestic routes in 1994. The Greek island services have been excluded from the EU liberalisation rules for the time being. OA will have to become self-funding and forego its monopoly on ground handling.

The present plan is to sell the present airport site for commercial and residential development when Spata opens. This would avoid the problem of assigning roles for the two airports and would help to fund Spata, but would make the Thessaloniki service in particular less attractive for those whose access time would increase. However, it would be possible to retain a useful presence of scheduled turboprop and Business Aviation aircraft at the scale of London City airport's capacity by confining operations to the present cross-runway and adopting some of London City's procedures to overcome the obstacles on the approach to Runway 27. Only very few houses would be within the 57 Leq noise contour and most of the site would still be available for sale or to provide a green 'lung' for the city (Biris, 1995).

Meanwhile, Thessaloniki international traffic has grown by a factor of three in the decade to 1995 while domestic traffic there has declined slightly. Domestic, international scheduled and international charter traffic each accounted for approximately 850,000 passengers in 1995, the domestic traffic being dominated by the Athens route. The charter growth reflects the exploitation of the coastal resort attractions. The scheduled traffic has been boosted by the new foreign airline services to the former eastern block countries, based on business opportunities, shopping and an affinity for the eastern Orthodox religion. This seems to be a further example of a second city, previously underprovided with air service, beginning to express latent demand. Continued development of its traffic depends on OA's attitude vis-a-vis services at the new Spata site or the emergence of serious competition.

The Aegean islands vary enormously in size, in distance from Athens and in the provision of air service. Some of the islands near the Turkish coast have populations of a quarter of a million and have good jet service competing with ferries which take nearly a day to reach Athens, some of the air traffic being associated with military bases. Others, with less than 10,000 inhabitants, have service to Athens by STOL aircraft in compassion with ferries taking eight hours, while those within four or five hours of Athens by ferry tend to have no air service at all. The bulk of the travel is inbound tourism, with a very heavy summer peak. In the offpeak season, the islands are supplied with essential commodities by subsidised ferries, which operate on higher frequencies than would otherwise be necessary in order to allow the necessary passenger travel to Athens and to the main islands of each group. It is also necessary to provide some facility for medical evacuation. Despite the central planning and control of the airlines, the ferries and the infrastructure, there has been no policy for minimum necessary level of service.

The non-tourist travel demand is very low. In 1988, the average propensity to make off-island trips was only 2.1 per head per year, even when there was a good jet air service. The more wealthy categories of resident made only twice as many trips as the poorer categories, but made over 80% of their trips by air. The propensity reduced to 1.5 and 1.4 respectively for those islands with only turboprop air service or no air service. Ferries, whose second class fares were some 60% of the subsidised air fares, are used to partly compensate for reduced quality of air service, but it appears that the additional utility of a good air service stimulates additional trips among the wealthier categories of the island residents. This travel demand by the indigenous population of the islands is quite constant through the year, though the islands' residents induce much more peaky social travel due to their relatives visiting for holidays in the summer. Some 50% of the residents' trips were for recreation, 15% each for work and education, with the remainder divided between shopping, medical and social purposes. On a sample island with good jet air service, air obtained an average 40% market share in competition with a 14 hour ferry service, rising to 70% for social and 90% for medical purposes (Roupakias, 1988).

A study in the early 1980s examined the possibility of increasing the airport infrastructure on the islands and the options for providing a suitable level of service at a reasonable cost. It was shown that an effective solution would be to create minihubs on the main islands of a group, fed by turboprop aircraft, helicopters of hydrofoils, depending on the availability of land for airfields or helipads on the islands and on the distances involved. This would allow the jet links to Athens to operate at an adequate frequency. It was also suggested that the most appropriate way of serving islands with very low density traffic levels in the winter would be to adopt an air version of a dial-a-ride service, backed by an adequate telecommunications network, which would arrange to divert a multistage feeder flight on the rare occasions that it was required. It might even have been feasible for the ferry operators to have fulfilled their service obligations more efficiently and at lower cost by reducing their operations to freight-only in the winter except for the largest islands, and taking a share in the aviation feeder services, but this would have created too many bureaucratic difficulties. OA have now been operating in the suggested hub and spoke manner for some years, since the CAA embarked on a programme of building STOLports for that purpose (*Airport Forum, March 1989, pp 17-19*).

The case shows clearly the need to plan an aviation system so that it can properly fulfil the social and economic integration role expected of it, both between major cities and in low density situations. It shows how airports in the system are influenced by investment in other airports in the same system and the operating strategies adopted by the airlines. The most important set of influences on the coherent planning of the system is the politics of the power relationships between the actors. In this case, the national airline, the ferry owners and the local authorities all have strong influences on the central government which shape airport planning policy as expressed through the CAA. In addition, the involvement of the military and the reliance on central government funds, supplemented by EU funds, also limit the ability to plan with confidence, despite there being only one responsible authority.

12.5 JAPAN

The Japanese air transport system is, like those in the USA and Brazil, dominated by domestic traffic with 75 mppa in 1996, compared to 5 mppa carried internationally on Japanese airlines. Five year airport system plans are drawn up by the Ministry of Transport. There are also five year plans for tourism, rail and shipping, but not for Air Traffic Control. The Fourth (Comprehensive National Development) Plan, approved in June 1987, aimed for a dispersed multi-polar pattern of development, emphasising the development of the nationwide high-speed transportation and information/communication networks through Integrated Interaction Policy. Although each (earlier) plan reflected the major issues discussed at the time, the basic theme underlying these plans has been the spatially equitable growth of a better and more stable living environment. "Transport investment has always been emphasised as the major policy instrument to achieve the basic goals" (Ohta, 1989). "The theme ... was building a country to act as the place for interactions: interactions between people, regions, and nations; interactions between people and nature; and exchanges of information. .... In such a (high-mobility) society, speed, reliability and comfort will be demanded of transportation. ... The (transport) network will enable a choice among several different routes and transport modes. ... In addition to the conventional vertical links between major metropolises, regional hub cities, and regional core cities, it will be necessary to prepare a network that places importance on the horizontal connections among regional cities. ... A major task facing Japan is to promote social overhead capital, including the transport infrastructure, during this century while the economy still has vitality and before the average age of the population rises. ... The major objective of the plan is the construction of a nationwide transportation network that connects the major cities from anywhere in the nation within three hours (or nationwide day-return). To achieve this objective, the plan proposes the construction of the expressway-type road network of 14,000 km, several Shinkansen lines, and about 50-70 commuter airports." The fifth plan, for the period 1986-1990, stressed improvements in road, rail and air infrastructure to allow nationwide day-return between any major city pair. This implied a move towards jets and the lengthening of runways at commuter airports, encouraged by the communities' desire for jet transport. The government does, however, continue to subsidise commuter airline operations into remote communities and the construction of airports for these operations, despite the lack of enthusiasm from the commuter airlines, who really need better access to the hubs for their services from the remote airports to be profitable (*Aviation Week, May 8, 1989, pp 48-49*). Local authorities are expected to cover any operating deficits at these airports. The capital requirement for the larger airports is largely met from an Airport Implementation Special Account managed by the Ministries of Transport and Finance and funded by passenger charges and fuel tax (*Lloyd's Aviation Economist, April 1985, pp 26-29*).

The airports were classified by runway length but are now designated by function. The four airports at the two main gateways of Tokyo and Osaka are all Class 1. Class 2 contains the 26 major domestic airports, with a further 51 minor airports in Class 3. A Special Account funds all the investment in the large airports except Narita, Haneda and Kansai, 75% of the medium and 50% of the small airports. The sixth plan for 1991-1996 expected to spend 3.22 trillion Yen (Momberger, 1993), 68% more than the fifth plan, a further 3.65 trillion Yen being earmarked by the Ministry of transport for the following five years (Odell, 1996), though this has been extended by two years due to the recent budget problems. The funding provides a total of 93 airports in Japan, 47 of which will have runways adequate for jet operations. The central government makes up any annual deficit on all but the small Class 3 airports where cross-subsidy is hard to justify, and also subsidises the purchase of aircraft for the operators of these remote links, as it has always done with the sea ferries (*Flight, 30 September, 1991, p 10*).

Japan began to open up international tourism for its citizens in the late 1980s in a bid to reduce its balance of payments surplus, and at the same time began to move its production facilities overseas (Wijers, 1991). Until then, Japan Air Lines (JAL) was the sole designated international carrier and there was strict control of entry on domestic routes for the two designated carriers, JAL having rights on only five routes. In these circumstances, it was possible to designate the then new Narita airport as Tokyo's international airport, leaving only Taiwan flights at the Narita airport, and designating it as the domestic airport. It was also government policy to centralise all international flights at either Tokyo or Osaka. These policies have gradually been eroded by the need to retain a national share of the markets while bending to external pressure to liberalise route entry. JAL has been privatised. All Nippon Airways (ANA) and Japan Air System (JAS) have been given international status and even compete with JAL on some city pairs.

Congestion at the main international airports has led the airlines to exploit opportunities to start international service from regional airports, a move encouraged by the provincial cities who wish to play more of a world role. By 1991, 12 airports supported international scheduled flights, mostly to Seoul and Guam. Hokkaido's main airport of Chitose, which already handles more than 14 mppa and 0.2 million tonnes of freight, is planned to be the centre-piece of the Aeropolis Project, giving the region a world role. Lack of capacity, even at Osaka's newly opened Kansai airport, leaves the government little option but to add its encouragement to these initiatives, and actively support JAL and ANA subsidiaries' desire to offer the services, to supplement the charters which have been operating for some time. Nagoya and Fukuoka were chosen by American carriers who were denied further slots at Narita. Nagoya has the more central location and a very intensive industrial base, but is very constrained environmentally (Knibb, 1991). It is only one and two hours respectively by high speed train to Osaka and Tokyo. The city is so keen to increase international service that it already planned a new offshore airport.

Competition has also increased on domestic sectors, but this is seriously constrained by lack of runway slots at the major airports, the latter problem creating even worse difficulties for the commuter airlines. The large airlines resort to the use of very large aircraft to compensate for the slot shortage. The Tokyo/Sapporo route has over 10 million seats per year supplied by Boeing 747s with an average size of 550 seats (Wijers, 1991). Controls on fares have recently been relaxed, so that the airlines can now compete with the high speed train fares. New entrants are being encouraged as part of a general trend to liberalise the economy and revive the stagnant economy (*The Economist, January 11, 1997, pp 21-23*). Newly created slots were to be reserved for them at Haneda, but they only obtained 6 of the 40 on offer (*Flight, 12 March 1997, p 8*), so they only offer token frequency competition. The new activity mimics in some ways the initiatives induced by liberalisation in Europe. The new entrants are owned by newcomers to aviation, namely poultry farmers and a human resources training company, as well as an air ticket discount agency (Mollet, 1997). The major carriers are also creating their own low cost subsidiaries, while taking the opportunity to leave the less profitable routes. The new entrants are trying to keep their fares low by leasing aircraft, outsourcing maintenance, hiring foreign crews and using electronic ticketing.

All the major onshore airports are severely constrained environmentally in terms of caps on the number of daily movements, night curfews and objections to any expansion. Tokyo's downtown Haneda airport supports over 40 mppa on a shoreside site. Two of the three existing runways are being rebuilt on reclaimed land, which will give an additional 40,000 annual movements to the existing 190,000 declared capacity. A new airport, owned by the Narita Airport Company, was opened 85 km away at Narita (70 km and often more than two hours by road from downtown Tokyo) in 1978, after a long battle with local farmers. Both airports have a curfew between 2300 and 0600 hours. Narita still has only one runway and a declared capacity of 28 per hour and 360 atm per day, with no more than 75 slots being used in any three hour period (Woolsey, 1997). The former limit is said to be due to air traffic conflicts between Narita, Haneda and two military airfields, but foreign negotiators are suspicious that Haneda has been able to increase its declared capacity more than Narita. The daily limit at Narita is an environmental one, and the acquisition of land for the next parallel runway is subject to a problem common to all transport schemes in Japan, that the Ministry of Transport insists that there must be a willing sale (Knibb, 1991). The 'farmers' have waged a battle with the airport since its inception and, though there are only 16 hectares still to purchase and only six landowners remain involved, the land is in two parcels in the middle of the planned parallel and cross-wind runways. However, approval for a second runway has been granted and it is expected to be available around the turn of the millennium, which would increase the capacity from the present 120,000 to 220,000 atm per year. The airport is now served by dedicated trains with 50 minute running time and check-in facilities at the downtown terminus.

Osaka's Itami airport was the other main entry point. As the owner, the government has direct responsibility for it, as it does for Haneda. It had an environmental capacity of only 200 jet flights per day and a curfew for scheduled flights between 2100 and 0700 hours, yet it handled 22 mppa by 1991 with some 65,000 atm from a site of only 317 hectares. The search for a new airport started in the late 1960s, but the oil crises and a stagnant local economy delayed the planning of the Kansai offshore site until 1981, being completed in 1987 (Kato, 1992). Given that Japan already had experience in off-shore airport building at Nagasaki, it is not surprising that the chosen site was in the bay of Osaka some 40 km from the city, from which there will eventually be access by two roads and two railways, with high speed boats serving other local cities. The first phase, with one 3,500 metre runway, required the reclamation of over 500 hectares from a 20 metre depth of sea water. The total cost was predicted to be US $6.9 billion, but this increased by more than 50% and the opening was delayed until 1994 due to the continual need to compensate for the sinking of the infill, even though subsidence had been predicted to the extent that 900 computerised jacks were installed to counteract differential sinkage along the length of the building (Bailey, 1993). The completion of Phase 1 has provided a nominal capacity of 160,000 atm per year free from any curfew and the unique opportunity in Japan to combine significant volumes of domestic and international passengers. Itami airport remains open for domestic traffic and, if for no other reason than that Kansai is almost at capacity within three years of opening, will continue to play an important role in the system, even though it does not have Haneda's advantage of the restricted levels of domestic services at Narita. When the second runway is available at Kansai, after even more expensive reclamation at 75% of an estimated total cost of 1.56 trillion Yen (US $16 billion), the freedom from environmental capping may divert some of Tokyo's international traffic through the hub.

The second phase of Kansai development will be undertaken by a new joint venture company between the existing Kansai International Airport Company (KIAC) and the local government, due to KIAC's financial weakness (Odell, 1996). In common with Narita, even high user charges have not been able to recover the cost of developing the capacity. In the case of Narita the main problem was the long delay between spending the US $2.7 billion (in 1973 dollars) and beginning to get a return, whereas the extreme cost of reclamation has been the dominant drag KIAC's balance sheet. Even so, it is hoped that much of the funding for the facilities themselves will come from private sources, as it did for the first phase.

It is likely that, with further liberalisation and economic growth throughout the region, all the presently planned new capacity will have been used by 2020. If Japan does not want to let other emerging hubs in the region with large new airports, like Seoul, erode its role as the major hub for the region, or deny its citizens the opportunity to travel, it will have to face ever more expensive solutions to its capacity problems. One option being mooted is a third airport for Tokyo, which would again have to be reclaimed from Tokyo bay, following the tradition set by the new airport at Nagasaki. Another option would be to use the planned development of 500 km per hour magnetically levitated trains to provide similar access times to those available today from the major central cities to the more remote islands of Hokkaido and Kyusho, where there is more space to expand airports without land reclamation (Momberger, 1993). At the same time, the trains would reduce the demand for domestic air travel, as shown by the 20% fall in air traffic between Tokyo and Osaka when the new Nozomi train was introduced in 1992 with a running time of two and a half hours, equalling the door-to-door time by air. The seventh five year plan's US $34.3 billion budget actually includes Kansai's second runway, a third airport for the Tokyo area and an international airport for the Chubu region.

The capacity and cost recovery problems in Japan are made worse by the high population density and hence the extreme sensitivity to noise and other environmental impacts. Hiroshima's new airport, 40 km from town, supplements a city centre airport which is only allowed to accept turboprops. If Osaka's Itami airport had the same 4,000 hectares as Washington Dulles, all of the insulated houses would be within the airport boundary. As it was, 20,000 residents settled out of court in 1986 for US $15 million in compensation for past noise. The prize in this case was the ability to keep the airport open after the inauguration of Kansai, helped by the introduction of quieter aircraft. Despite the use of noise abatement procedures, the government also had to spend US $4.2 billion insulating 162,000 houses within the 75 Weighted Equivalent Continuous Perceived Noise Level (WECPNL) contour at the 16 noise-designated airports, together with another US $1.2 billion for more noise-sensitive buildings within the 70 WECPNL contour or for relocation of households within the 90 WECPNL contour by 1992. The total cost of all the environmental mitigation measures to the end of 1992 was US $8.9 billion. A substantial part of the cost is recouped by a Special Landing Fee, on the 'polluter pays' basis. The problems are unlikely to get any easier, as the residents around the airports are already complaining of being able to 'see' noise and to raise their levels of complaint whenever there is a crash.

Having a centralised planning and funding framework has helped to achieve a stream of investments which has released capacity which it may have been impossible to fund in any other way. It has also been possible to tie government policies on route rights and competition in with policy on airport development, helping to ensure that logical use is made of the scarce capacity. By cross-subsidy and the targeting of government funds to the provision of essential facilities, room has been left for private investment in those facilities with some prospect of making a positive return. On the other hand, the actual performance of the system in terms of the output per unit of investment is poor compared with, say, the UK. This is due, at least in part, to the serious environmental and topographical constraints.

12.6 NORWAY

Norway has only 4.3 million inhabitants, averaging only 12 people per sq km. Almost half the population lives in the catchment area of the Oslo airports. Norway has a high standard of living and high taxes which support the infrastructure needs of such a dispersed population. The average propensity to fly is four trips per year per person, or four times greater than the UK. The State airline, Scandinavian Airlines System (SAS), is owned by the governments of Norway, Sweden and Denmark and so operates three gateways at Oslo, Stockholm and Copenhagen. The international routes were subject to considerable regulation and prices have been high, but the bilaterals with Sweden and the EU were brought into line with EU competition policy during 1993 and new low cost carriers are beginning to offer service. The Norwegian domestic routes were deregulated in 1994, except for subsidised Wideröe Flyveselskap whose services to the outlying coastal destinations are protected until 1997, but the effect was not very great. The two main airlines, SAS and Braathens SAFE, reorganised their route structure but further airline entry is made difficult by these main airlines having 73% of the slots at Oslo's downtown Fornebu airport (*Air Transport World, February 1995, pp 80-81*).

The 45 airports with commercial traffic are owned and operated by the Norwegian CAA, the Luftfartsverket. The CAA has been a self-financing state enterprise since 1993, but ownership of the 26 smaller airports only passed to it from the local authorities in 1996. The large majority of its revenue comes from aeronautical charges at Oslo's downtown Fornebu airport. The government have been subsidising and improving the other modes of transport, so that there are now some opportunities for rationalising the airport system.

The Fornebu site was chosen in the 1930s as an excellent location for a combined landplane and seaplane base. There were few residential areas nearby, yet it was close to the city. In 1996, Fornebu handled 9 mppa, 85% of whom are local and 5 mppa are domestic. The domestic route to Bergen is responsible for 1 mppa and the Copenhagen route is almost as busy, 70% of the latter traffic going beyond Copenhagen. Similarly, 28% of the 600,000 ppa to London are connecting to intercontinental destinations not served by Fornebu with its relatively short runways. It is only 8 km and 15 minutes from the centre of Oslo and therefore has a curfew between 2300 and 0700 hours. The runway capacity is only 36 movements per hour, and the airport is at capacity physically. Nearly 100,000 people are subjected to a noise impact of at least 55 dBA, measured in units similar to the California Noise Exposure Level. Fornebu is soon to be replaced by a new twin runway airport at Gardermoen. This is examined further in chapter 13.

At the other end of the scale from the capital city's airports is the system of Short TakeOff and Landing (STOL) airfields serving the communities along the north coast of the country. Before the mid 1960s, when Wideröe was licensed to operate a service with Beaver floatplanes, these communities were served solely by subsidised shipping. Between 1968 and 1974 the government set up a chain of STOLports, with a few further additions until 1987, in an attempt to ensure that a maximum percentage of the population would be able to make a day return trip to Oslo. In addition, two offshore islands have helipads. By 1983 half a million passengers per year were using the STOL services. The STOLports had 800 metre asphalt runways and a small apron, a terminal building, some rudimentary radio navigation aids and only an Aerodrome Flight Information Service rather than full Air Traffic Control. It became possible to provide a much more reliable year-round service than with the floatplanes. Some 40% of the STOL services traffic was newly stimulated, but the ships lost up to 50% of their passengers because the total out-of-pocket costs for STOL were almost the same as for the ships, even before accounting for the value of time (Strand, 1983). However, the impressive 97% flight completion rate in the difficult terrain and weather has led to three fatal accidents.

Apart from a short period when a few STOLports in the north were served by the Norving airline, Wideröe have been the sole operator into the 25 STOLports, first with Twin Otters, then expanding to Dash 7s and, now that several STOLports have had their runways extended, with Dash 8s. The airline is effectively a privately owned government contractor, receiving an annual subsidy of some US $50 million (Moorman, 1996). Approximately 10% of the passengers using the STOL services are flying from one STOLport to another, 50% are local to a major Norwegian airport and the remaining 40% are feeding the major carriers.

The economic responsibility for operating the STOLports was initially placed on the municipalities, but by 1983 the government was funding the deficits, as was the case with other airports, after the municipalities made the case that the STOLports benefit everyone within a wide area. Now the CAA is to take the STOLports into the rest of the country's airport system. There is, however, some doubt as to whether the STOLports should be supported by other than the local community, since research shows that approximately 75% of users come from the immediate catchment area town and that they gain inward investment and migration compared with those settlements which remain without service. People living within 30 minutes of a STOLport make 250% more trips per capita than those living between 30 and 60 minutes away. Further, 70% of the STOL trips are not paid for by the passengers themselves since, compared with the ships, a greater percentage is for business purposes.

12.7 SPAIN

The Spanish system consists of 40 airports on the mainland, on the islands of the Balearics and the Canaries, and on the north African coast. The airports are managed by Aeropuertos Espanoles y Navegacion Aérea (Aena). Aena was reformed in 1991 from the previous Airport Authority with a remit to be self-supporting financially. Though still providing a central planning and management facility, regrouping functions which had become dispersed in the civil administration, it is to work towards greater decentralisation in economic accountability and to apply management criteria to its operations (Sarandeses, 1993).

The airport functions are very diverse, as are their levels of activity. Aena classified them into six homogenous groups in order to direct appropriate solutions to their common problems:

- large airports dominated by scheduled traffic
- large airports dominated by international tourist traffic
- airports with medium traffic levels, mostly domestic
- airports with medium traffic levels, mostly international charter
- two classes of smaller airport with poor growth characteristics.

In relation to the Trans European Airport Network, the Spanish system can be considered as seven Community Connecting Points, nine Regional Connecting Points and 16 Accessibility Points, the remaining eight local airports not being considered of interest in a European context (San Nicolas, 1995). The main structure of the system planning was actually laid down as a result of a study in the late 1970s when it was realised that, although there was a central responsibility for planning, the system's performance was falling short of achieving a reasonable minimum level of efficiency (de Andres, 1980). The problems were caused in part by lack of a coordinated transport policy, lack of research, budgetary limitations and unexpectedly large increases in tourist traffic and irrational use of the system. The use of the system was being distorted due to the poor surface infrastructure, subsidised fares and local political pressures. The latter pressures surfaced because there was no nationally coordinated policy for airports. They led to isolated initiatives which caused investment in installations whose usefulness or profitability would have been questioned by a rational filtering process. In the process, they had effects on other nearby airports which would have been considered in a more system-wide analysis.

The comprehensive study, reported in 1980, looked forward to the needs of the year 2000. It established the hierarchy of airports, by importance and function, which would allow a rational allocation of funds and providing the necessary 'filter' for the local pressures. The underlying research was important in improving the potential efficiency. A cost/benefit study was performed on each of three network scenarios:

- the continuation of the existing airport system and its uncontrolled use;
- an option which maximised the number of airport-pair linkages;
- an option which concentrated traffic on a limited number of regional airports, with extensive use of feeder services and improved ground transport links to these regional points.

The costs considered were the capital and operating costs, together with the social costs of the passengers' value of time. The benefits were taken to be the revenue generated by the airports.

Costs and benefits were summed over 22 years of traffic growth, forecasts being prepared separately for each network scenario.

The study's remit was not to examine the usefulness of the airports in the system but rather to determine the best use for them. It concluded that the most efficient system was the one which concentrated the traffic at a few principal regional airports. This was predicted to give 60% more traffic than the 'business as usual' scenario. The investment policy was therefore to tend to favour the additions to capacity which would have been necessary had traffic growth followed that preferred scenario, rather than respond to ad hoc growth caused by air services which did not accord with that pattern. The difficulty was that Aena's predecessor had no control over the airlines or the routes they were allowed to fly. This not only made it difficult to influence the fulfilment of the preferred scenario, but meant that they did not include airline costs and revenues in the cost/benefit analysis.

The greater regional accountability which is being fostered by Aena should help in the filtering of local political pressures to provide capacity in advance of revealed demand. However, a new threat to the fulfilment of the system plans has been the arrival of serious competition to Iberia's domestic services and the introduction of lower air fares. The fares are meant to be sustainable through increases in airline efficiency, and Iberia is restructuring and countering the competition with a tighter integration with its subsidiaries and franchise partners (Jones, 1996). In addition, the EU liberalisation is causing a number of low fare airlines and foreign hubbing carriers to enter and stimulate the international market. There is no consistent gateway policy, other than that implied by the shortage of runway slots at Madrid, so the new scheduled services may appear at any of Aena's airports, just as the charter services have always done. This faces Aena with a new set of local challenges which may conflict with previously determined priorities.

13

COMPETING ROLES FOR AIRPORTS

13.1 INTRODUCTION

Airport systems may be defined by the administrative organisation with responsibility for investments or regulation, or by the interactions which determine the outcome of competitive pressures on traffic. Analyses which focus on the former sets of airports without consideration of the latter may ignore important determinants of demand generation and distribution. Thus national system planners should recognise not only the competition between gateway airports within the nation but also the competition with other nations' gateway airports for through traffic. Similarly, regional or state planners must contend with internal competition between the primary and secondary airports and also with other established national airports (Caves, 1993). Municipal airport planners have to recognise regional and national competition if they are to construct realistic future scenarios.

Planners analysing metropolitan multi-airport systems have to confront the additional problem of competition within that area as well as the other interactions. The outcome of the local competition will itself affect the external interactions as the attractiveness of the total local system to airlines and their passengers changes.

These competitive aspects may be quite complex and obscure in themselves, but they are only one element of the complexity of the systems considered here. Complex systems appear to have properties which are disturbingly untidy, regardless of whether they concern physical, biological or economic systems. In economic systems, this untidiness appears to arise from two real effects which are ignored in classical economic theory. Firstly, there are many examples of increasing returns to scale, or positive feedback. This implies a peculiar dependence on initial conditions in the subsequent distribution of growth in the system, so that, though the resulting patterns may not coincide with the best allocation of resources, they become hard to dislodge.

Any expectation that 'hands-off' policies will allow normal economic behaviour to improve the allocation of resources is likely to be incorrect in situations with positive feedback. Intervention will be necessary, but it is difficult to judge how best to do this when the system's components are continually adapting and readapting to each other. Attempts to analyse and control are usually based on applying economic theory which also ignores the second aspect of reality - that perfect rationality based on perfect knowledge does not exist. Instead, decision makers probably form simplified 'internal models' of the problem, discarding the model when it fails in favour of a new and hopefully better one. The models will depend on personal history, imagined notions and cultural perceptions. These ideas, adapted from W Brian Arthur (1993), ought to apply to airport system developments. Several case studies of regional airport systems are now examined to see if anything can be learnt by relating them to these theories of complexity. Traffic histories are presented for most of the cases, the data being obtained from International Civil Aviation Organisation, Airports Council International or UK Civil Aviation Authority statistics.

13.2 GLASGOW, SCOTLAND

Prestwick was the designated Scottish gateway because it was nearest to America and it has a remarkable weather record. However, it had never managed to attract short haul services to feed the long haul flights because it is on the coast some 60 km southeast of Glasgow and there has always been an airport at Renfrew, only 5 km west of the centre. When the old downtown airport became too constrained, the replacement choice was between nearby Abbotsinch and Prestwick. Abbotsinch was relatively constrained operationally when it opened and, rather than drawing industry to Paisley, its operations caused restrictions on the height of buildings and cranes, and a church had to be demolished. There were also construction difficulties due to underlying peat which were well known in advance of the location decision. All the expense in renovating Prestwick had been incurred before the decision on Abbotsinch was taken. The effects of noise from Abbotsinch could have been foreseen, and the possibility of a rail link to Prestwick was, in fact, examined, with a view to consolidate all services there. There was a strong influence from British European Airlines (BEA) who refused to provide domestic feeder services from Prestwick. It would have had to give up its virtual monopoly at Renfrew to join international airlines at Prestwick. In the event, the Minister decided that long haul operations should stay at Prestwick, while short haul operations could develop at Abbotsinch.

The only way of uncovering the decision process of the Minister, to retain a split operation rather than consolidate at Prestwick, would have been via the Select Committee of Estimates, "the only parliamentary body with any interest in governmental efficiency, merely a committee of MPs without professional help and quite inadequate to the task", (McKechin, 1967). The decision implied that both airports would survive and should be run by a single body to avoid competition, though it was recognised that local government does not have the correct machinery for that type of regional airport administration, since only the local administration bodies had representative or financial power and only the government could grant route rights.

In the event, BAA took over both airports, and the split operation survived despite Prestwick being used mainly for Canadian services linking expatriates to their Scottish homeland. In fact, Prestwick satisfied over 90% of that market, whereas traffic between Scotland and the US almost exclusively used Heathrow. Then, in 1990, a court decision forced the government to allow long haul operations from Abbotsinch. Air 2000, a charter operator, had been forced to respect the historic gateway status of Prestwick by making an unnecessary technical stop there despite having the ability with EROPS twin engine aircraft to fly directly to the US from Abbotsinch. The government then relaxed the gateway rules and, despite strong marketing efforts by the new owners of Prestwick, long haul schedules started consolidating at Abbotsinch to benefit from the synergy with the short haul operations.

Figure 13.1: Traffic at southern Scottish airports

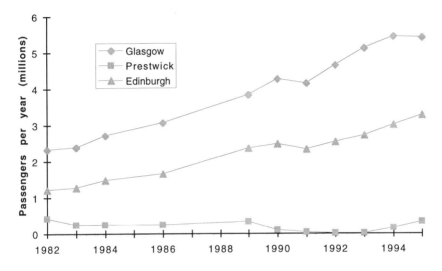

Source: compiled by the authors from UK Civil Aviation Authority statistics

However, in 1995 Glasgow's international traffic dropped below 1994 levels, despite the considerable regeneration of Glasgow as a 'European city of culture' and the emergence of a Scottish version of 'Silicon Glen' in the vicinity. Some of its transatlantic services were deemed unprofitable by the airlines due to the low numbers of high-yielding passengers (*ACI Europe - BAA Souvenir Edition, 1995 pp 72-74*), while Edinburgh, only 60 km away, competed strongly for charters, raising the issue once again of whether a mid-Scottish airport should have been built to consolidate all the demand at one location and improve the route and airport viability (Arbuckle, 1997). Meanwhile, Ryanair has boosted Prestwick's Irish and UK traffic with coordinated train service into the centre of Glasgow and discounted rail fares throughout Scotland. Prestwick also has an increasing long haul freight capability, with investment by Federal Express, its cross wind runway has been restored to full length, and it has had immigration services restored to meet the demand which still exists for diversions from Abbotsinch (*Jane's Airport Review, July/August 1993, p 4*). Meanwhile, the picture has changed again with the arrival of low cost new entrants on the London route at both Abbotsinch and Prestwick.

13.3 BELFAST, NORTHERN IRELAND

Belfast International (Aldergrove) airport, 20 km northwest of the city, took up a civilian role for a second time in 1963 when airline service was transferred from Nutts Corner (15 km west of the city) due to capacity problems. Its traffic consists mainly of domestic scheduled and international charter flights. The main route is to Heathrow, competitive jet service being offered by BA Shuttle and British Midland Diamond services with some 8 round trips per day each. The new owners, TBI plc, have aggressive aviation and non-aviation expansion plans.

Figure 13.2: Traffic at Belfast airports

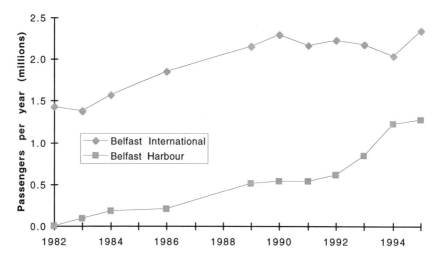

Source: compiled by the authors from UK Civil Aviation Authority statistics

Between the wars, airlines had used an airfield at Newtownards, after having operated for a year (1933) from Aldergrove. However, in 1938 Short Bros and Harland opened a new airport (Sydenham) on their factory 3 km from the city centre and the airlines were immediately attracted there. Shortly afterwards the Short's field was requisitioned for military duties and was not reopened to civil traffic until 1983. Belfast Harbour, as it is now called, is still owned by Short Brothers. It is served by turboprops and the BAe 146 in competition with Belfast International's conventional narrow-body jets, but its traffic has grown in a decade to almost equal its competitor. Despite the handicap of a short runway, some aeronautical obstructions and the very strong jet service from Aldergrove to Heathrow, it now offers competitive service to virtually all UK provincial cities. The provincial routes do not generally justify jets on capacity or speed grounds, so the product is nearly identical into either Belfast Harbour or International (Aldergrove) airports, except for any perceived difference between turboprops and jets. Otherwise, competition is on price, access and regularity, with Belfast Harbour now having gained at least 50% of the regional markets it serves. Due primarily to slot difficulties, it has not managed to provide a consistent service to Heathrow, but it serves London through Gatwick, Stansted and Luton. Bombardier, which owns Shorts, is investing in airside and terminal expansions.

13.4 PORT AUTHORITY OF NEW YORK AND NEW JERSEY

In 1965, when the total New York regional traffic was 25.8 mppa, a report (Port of New York Authority, 1966) was commissioned, describing a re-evaluation of 12 reports since 1957 which had established the need and examined the options for a fourth jet airport. The need for a fourth airport was established to the satisfaction of the New York Port Authority, the FAA, the Aviation Development Council, the tri-state Transportation Committee and the Airline Pilots Association, strongly influenced by two 'Black Fridays' (June 7, 1963; September 24, 1965). However, it was refuted by the Metropolitan Airlines Committee, advised by Dixon Speas, who believed that the use of many reliever airports for 50% of IFR GA traffic would allow the air transport demand to be satisfied.

The passenger and movement forecasts were:

Year	1965	1975	1980
Mppa	26	53.5	65
IFR movements (peak hour)		247	302

The case for need took account of expansion of capacity of existing runways and new runways, and also the expected reduction of capacity due to noise preference runway use, to give the IFR peak hour capacity in the 1970s as:

Kennedy	74
La Guardia	50
Newark	49
Teterboro	12
	185

Despite taking account of alleviations arising from diversion of GA, increased aircraft size, introduction of short haul VTOL and spreading airline schedules, (in fact, they were concentrating), there was expected to be a capacity problem in the mid 1970s.

The New Airport Feasibility Study methodology involved the examination of 23 sites. The airspace preference was for sites in the northwest quadrant. Heavy weighting was given to the effect of accessibility on demand. Splitting service between airports was considered not to be feasible. It was decided that site costs at the airspace-preferred locations would not permit profitable operation even at capacity, due to the heavy debt service requirement. There was State of New Jersey opposition to a new airport in the locations preferable for accessibility and airspace. The conclusions were that there was no site which simultaneously satisfied the criteria in the areas of airspace, economics, accessibility and politics.

It was, however, felt that a federal initiative to create a National Airport System Plan might help.

The traffic levels in 1996, without the site selection being resolved, were as follows:

	Mppa	Atm (000s)
Kennedy	31.2	338
La Guardia	20.7	323
Newark	29.1	431
	81.0	**1,092**

Figure 13.3: Traffic at New York airports

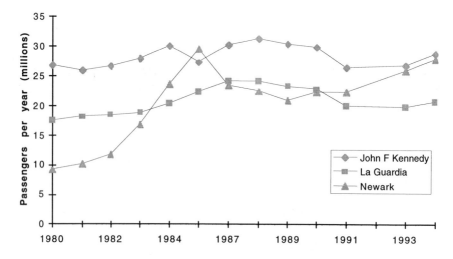

Source: compiled by the authors from Airport Council International statistics

This suggests a very expandable system. The expansion has been obtained by technological and operational improvements, a reduced specialisation in the airports' roles and an increased role for Newark, particularly as an international gateway. This was encouraged by People Express in the first half of the 1980s, and then by the SAS move from Kennedy to cement its alliance with Continental. Many other foreign carriers realised that there was a separate market to be tapped at Newark, with more room for expansion and without diluting their Kennedy traffic (Lefer, 1992). Even so, the three primary airports in the Port Authority system would be under even more strain if two of the airports were not slot-rationed and if the inland hubs and Boston had not begun to take an increasing share of international traffic. Meanwhile, Westchester County airport has taken an increased domestic role despite political and public opposition (*Aviation Week, 31 October, 1988, pp 92-94*) and there are finally indications that Stewart International Airport, some 65 miles north of New York, and already used for freight and charter operations, may be again marketed as the natural overspill location for the metropolitan region's traffic. The UK company, National Express, has won the bid for this first airport in the FAA's voluntary privatisation experiment.

13.5 EDMONTON, CANADA

The Municipal Airport (Muni) opened in 1926 as Blatchford Field, 5 minutes from downtown. Movements peaked at nearly 200,000 in 1979, but had fallen to approximately 130,000 by 1992 on a runway of 5,800 ft, with traffic falling to below 1 mppa. There were four scheduled carriers serving 12 non-stop destinations plus 25 multi-stop destinations. The main services were the Canadian Airlines B737 'Airbus' to Calgary and Air BC's BAe 146 to Calgary and Vancouver. There are obstacles in the form of high rise buildings. The local population is concerned about a repeat of an accident in which a light twin aircraft flew into the operating room on a top floor of a nearby hospital.

Edmonton International opened in 1960 at Leduc, 20 miles to the south, and national and international traffic were transferred to it. By 1992, there were fewer than 1.8 mppa compared with a peak of just over 2 mppa in 1990, the combined traffic at the two airports hardly justifying duplicated facilities. In August 1992, Edmonton International was transferred from Transport Canada on a 60 year lease to the newly created Edmonton Regional Airports Authority (ERAA). The subsequent City Council decision to transfer Muni into ERAA fulfilled the purpose of the Regional Airports Task Force Association formed in 1986 (Edmonton, 1992), but ERAA's plans to rationalise the system by transferring remaining services to Leduc were frustrated by a referendum. The ERAA move was made to stop the erosion of market share relative to Calgary, following public discussion of six options and a consultant's report suggesting consolidation on or after 1997. The principal disadvantage of consolidation at Leduc was seen to be that the Calgary passengers might drive all the way south if they had to drive the first 20 miles to access Leduc. However, a second referendum in 1995 voted 3 to 1 in favour of consolidation at Leduc, which took place in mid 1996. Traffic immediately started to grow, with two low cost carriers and new international services.

13.6 SAÕ PAULO, BRAZIL

Saõ Paulo State generates over 30% of Brazil's wealth with 20% of its population. The main Saõ Paulo airport, Guarulhos, accounts for 70% of the net revenue of the national airport agency (INFRAERO). Nearly all the state is on a plateau between 600 and 800 metres above sea level. It covers nearly 250,000 sq km and 70% of the population lives in the interior of the state, though 90% is urbanised. The state administers 28 airports in addition to the airports in the Saõ Paulo metropolitan area run by INFRAERO and some one hundred other small airfields. The busiest of the state-run airports in 1994 was Ribeirao Preto with 156,000 passengers and 728 tonnes of freight. It had direct service to Saõ Paulo Congonhas airport and to four out-of-state points. By 1996, service had been added to five more out-of-state points as well as three towns inside the state and to Saõ Paulo Guarulhos.

Congonhas was the early Saõ Paulo airport, with a short 1,740 metre runway. This became surrounded by commercial buildings and high cost housing, but by 1980 was handling 5.8 mppa and accepting even Airbus A300 aircraft. In the middle 1970s, a commission was formed by the Federal Government, Saõ Paulo State and Saõ Paulo city to study the future provision of airports for the city. One option was to expand facilities at the previously designated intercontinental airport at Campinas (Viracopos), some 90 km from the central business district (CBD). Some airlines stopped there en route between the northern hemisphere via Rio de Janeiro to Buenos Aires and Santiago, but most passengers preferred to transfer at Rio's Galleão airport for Congonhas. It was decided that a new site was needed much closer to town. There were ecological and cost problems with proposed sites to the south and west and it was decided to expand a previous military base (Cumbica) 20 km from the CBD on the main route to Rio, despite a rather poor weather record. The intention had been that when the new airport, named Guarulhos after the nearest town, opened, Congonhas would be limited to the shuttle flights to the Santos Dumont airport in downtown Rio (*Airport Forum, No 6, pp 18-24, 1981*).

Figure 13.4: Traffic at São Paulo and Rio

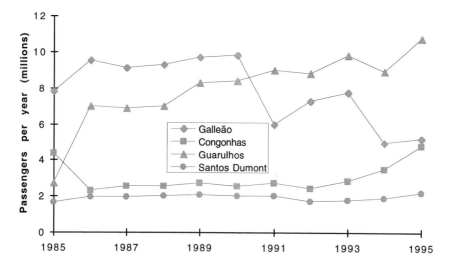

Source: compiled by the authors from Airport Council International statistics

In 1986 Guarulhos was opened with two close parallel runways. It soon became clear that airlines would have to re-equip with Category II avionics, but the new airport prospered, quickly becoming short of cargo space. Congonhas has continued to attract short haul traffic, particularly from the regional carriers, together with the heavier types of general aviation traffic.

The system effects extend to Rio, with intercontinental carriers beginning to overfly Galleão in favour of Guarulhos, as shown in Figure 13.4. However, as INFRAERO has continued to invest in new terminals at both Galleão and Guarulhos, the long term situation is by no means clear. Guarulhos is quite likely to be surrounded by development, as was Congonhas. If the government's economic initiative, the Plano Real, continues to create strong economic conditions, the close parallel runways at Guarulhos will be at capacity by 2005. Congonhas is once again at its environmental capacity. The accident which many had feared for decades finally happened in 1996, with an F100 falling into houses immediately after takeoff. That may well have implications for Guarulhos as well, given how densely populated its surroundings have become. The long term solution may yet rest with the Campinas airport, since, allowing for congestion, the access time is hardly any greater. Further, the three airports of Congonhas, Guarulhos and the busy GA and military airfield of Campo de Marte are very close together and their runway directions conflict. Also, Campinas and the towns further in the interior are the fastest growing in the state, and the airline TAM has already based itself there in order to link Congonhas and Guarulhos with its expanding network.

13.7 PARIS, FRANCE

Paris has gradually moved its longer haul traffic away from built up areas, first from Le Bourget to Orly, then to Charles de Gaulle as capacity problems loomed. The traffic history appears to show a smooth transition, but it has not been without its problems. The government imposed roles on the airports and on the airlines through route licensing which led to split operations (e.g. Swissair feeling it had to continue to serve its customers around Orly as well as obeying instructions to fly from Charles de Gaulle) and to difficult transfer trips. Despite the strong control exerted by the French government, it took 20 years for traffic at Charles de Gaulle to exceed that at Orly, as shown in Figure 13.5.

Figure 13.5: Traffic at Paris airports

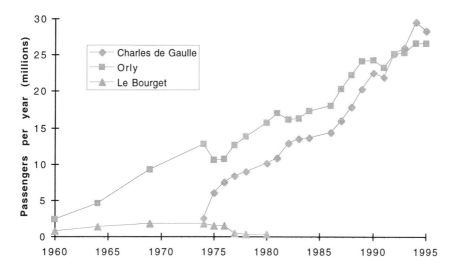

Source: compiled by the authors from Airport Council International statistics

Further rationalisation appears to be possible now that Air France controls Air Inter (as Air France Europe) as well as UTA. Charles de Gaulle has room to expand, has a dedicated intercity rail terminal and Federal Express has chosen to locate its main European hub there (ADP, 1997). Even so, foreign and Orly-based French airlines want to be able to operate more European scheduled services out of Orly in order to serve the densely populated areas in the south of the city. Some feel that the thinner routes can only be served viably from Orly, which would preserve its point-to-point role. However, its other designated role has been to serve only domestic and French overseas territories, except for some grandfather rights. BASE Business Airlines were denied access for a service to Enschede in Holland and initiated service from Toussus-le-Noble, a GA field in the southwest of Paris (Woolley, 1993).

Just as the EU liberalisation was being made effective, the French Transport Minister (Bosson, 1993) announced in October 1993 that Orly was to serve the Iberian peninsula and that the Porto route was to be transferred from Charles de Gaulle. The preferred roles for Orly were to form a pivot for domestic services and to keep its presence in the scheduled international arena, as well as accepting the bulk of charters. The domestic role was to be encouraged by allowing competition on the densest domestic routes prior to the Community 1997 deadline for the liberalisation of cabotage. The international role was to be enhanced with a London service in 1994, but with restrictions on the minimum aircraft size during peak periods, while there was to be an environmental limit of 220,000 atm per year. Although this is allowed under the EC liberalisation regulations, the airlines, and in particular British Airways' TAT subsidiary, challenged it as a device to limit competition to the Air France operations.

There are plans for the improvement of Le Bourget to relieve the two main airports of business aviation (Woolley, 1993), and the first of two close parallel runways at Charles de Gaulle is being built. The preferred role for Charles de Gaulle is a primary Europort to handle European Community and longer haul routes and as a hub base for Air France/Air Inter. The search is already underway for a site for another airport for Paris, which may well have to be over 100 km from Paris (*Flight, 22 May 1996, p 15*).

13.8 MONTREAL, CANADA

Worries about capacity and the environment at the downtown airport of Dorval, 12 km from the city centre, together with the need to cope with the 1976 Olympics and the expected boom in the Montreal economy, caused the decision to develop an expandable and environmentally friendly site some 60 km to the northwest of the city. The environmental benefit was, in fact, arguable, since it took over good quality agricultural land and required the resettlement of 1,000 families. Mirabel airport opened in 1975 with a terminal capacity of 10 mppa, primarily to cater for long haul traffic, all international flights except those to the US being forced to use it together with its 'planemate' apron vehicles for transporting passengers from the remote aprons to the terminal.

The opening coincided with a reduction in the rate of growth of Canadian air traffic, compared to the strong growth of the preceeding decade. Dorval's growth was halted, and the Montreal airports did not mirror the subsequent growth at Toronto. It would be wrong to attribute all the difference in these fortunes to air transport policy. Increasing aircraft range allowed transatlantic flights to overfly Montreal to the larger market of Toronto, leaving Montreal-bound traffic to transfer in Toronto for Dorval, resulting in poor utilisation of Mirabel. The airport has had to rebuild its traffic by intensive marketing concentrated on cargo and charter. Now, with the threat that the Open Skies agreement with the US will tempt carriers to desert Montreal in favour of other cities if they are forced to use Mirabel, from 1997 Aeroports de

Montreal are allowing all airlines back into Dorval if they wish (*Airports International, April 1996, pp 2-3*). The rebuilding at Dorval, which this decision has necessitated, is being funded by an Airport Improvement Fee levied on departing passengers at Dorval for destinations outside the State of Quebec. Mirabel is to be refurbished, including the installation of passenger loading bridges.

Figure 13.6: Traffic at Montreal and Toronto

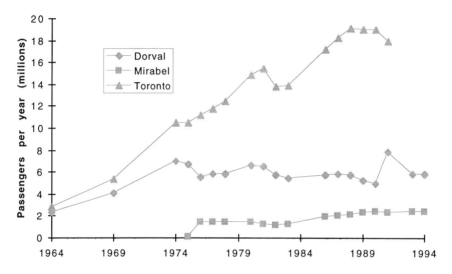

Source: compiled by the authors from Airport Council International statistics

The saga is by no means finished, because Dorval will still have a curfew. Also, the decision to consolidate scheduled services there has been challenged on the grounds that there was insufficient consultation with the community. The objection was ruled as valid in the courts but overturned by the Quebec Court of Appeal because Aeroports de Montreal is not a public body and so does not have to meet the same level of public consultation. This apparently strange ruling may be challenged again in the Supreme Court (*Jane's Airport Review, June 1997, p 4*).

13.9 NORTHWEST ENGLAND

The main airport in the northwest of England is Manchester (MAN). Its traffic has grown through 15 mppa in 1996, having continued to grow even during the recession earlier in the decade, partly due to its 'hub champion' policy. It had no strong based carrier in 1990, but took the lead in encouraging the airlines to coordinate their schedules in order to facilitate interlining. Manchester finally managed to construct an intercontinental dimension to its hub, though only with extreme perseverance and political skill over many years in the face of both US and UK government policies, until the UK Government granted unilateral access to UK regional airports by US carriers (Muirhead, 1995). This has been a factor in encouraging both

British Airways and Lufthansa, with its associates Lauda Air and Business Air, to play their part in allowing Manchester to begin the conversion from its role as its own hub champion to an on-line hub. Manchester airport's ownership continues to be in the hands of the local authorities. It has financed a second terminal and has obtained a positive result from a public inquiry into a scheme for a new parallel staggered runway to support the hubbing operations, as described in chapter 5. With further terminal extensions this will allow a balanced site capacity of over 30 mppa. The local community is generally in favour of the airport and realises its contribution to the economy, but the choice of siting for the second runway was seriously constrained by environmental considerations, so that the hourly capacity will be approximately 65 movements per hour, rather than the 80 that full independent parallel runways would have allowed.

There are two other important airports in the region; Leeds/Bradford (LBA) and Liverpool (LPL). LBA is constrained by topography, by the weather and by its environment. LPL is sited on the Mersey estuary, some 50 miles west of MAN. Figure 13.7 shows the recent history of the LPL and LBA shares of northwest regional traffic, relative to the performance of other secondary airports, i.e. the Gatwick (LGW) share of London traffic and the East Midlands (EMA) share of Midlands traffic. The composition of the traffic at these northwest airports are shown in Figures 13.8 and 13.9.

Figure 13.7: Shares of regional passengers

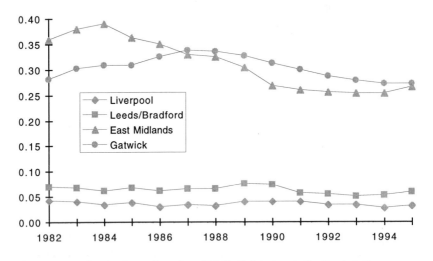

Source: compiled by the authors from UK Civil Aviation Authority statistics

Competing Roles for Airports 383

Figure 13.8: Liverpool shares of northwest passengers

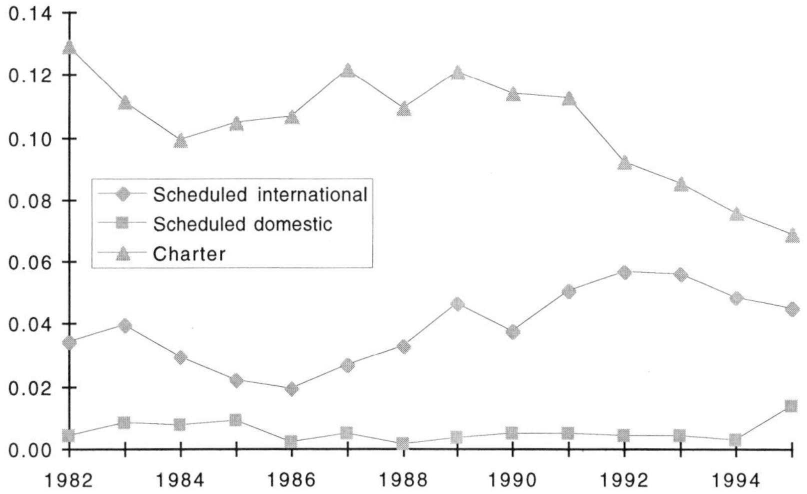

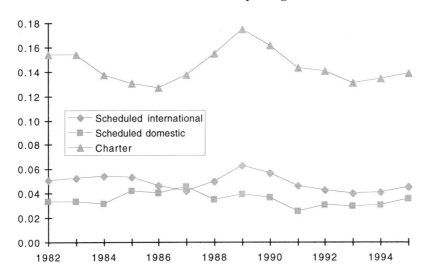

Source: compiled by the authors from UK Civil Aviation Authority statistics

Figure 13.9: Leeds/Bradford shares of northwest passengers

Source: compiled by the authors from UK Civil Aviation Authority statistics

In the 1930s LPL was a thriving airfield with more traffic than MAN, but as the Mersey economy declined, so did its airport's fortunes relative to MAN. The City Council continued strong support for improvements in the face of falling market share, with a new runway and the introduction of 24-hour operation in 1968. The new 2,400 metre runway was located well away from the existing terminal, requiring a 4 km taxi for most takeoffs, and a new closer terminal with a capacity of 1 mppa was not constructed until 1986, though a new apron, control tower and fire station had been completed earlier in the 1980s. Merseyside County Council was by then providing a greater subsidy per boarded passenger than at any other airport in the UK. Neither the improvements nor the subsidy managed to prevent airlines from withdrawing their services, British Eagle ceasing operations in 1968 and British Airways in 1978. The staffing at LPL was reduced by over 50% and the local authority sold the majority of the shares to a consortium headed by British Aerospace in 1990, which reduced the earlier losses and had hopes of turning it into a major European gateway on the strength of its geographic location, the possibility of European development money and its relatively unconstrained site. The marketing plan was not successful, with British Midland Airways dropping the jet service to Heathrow in 1992 to free up Heathrow slots for an international service. Their Manx subsidiary managed to generate only half the previous 100,000 ppa with their turboprops.

LPL has since relied heavily on just four scheduled routes (its niche Irish Sea market and Heathrow) in generating some 400,000 passengers per year. In addition there is a significant freight activity to much the same destinations and it acts as a hubbing point for mail. One or two new thin domestic scheduled routes were started in 1995, and some charters were also introduced, but some of these have been withdrawn. LPL's challenge is to get back some of the traffic which has gravitated to Manchester over four decades. Liverpool is awaiting a decision on a separate inquiry into a major terminal expansion which would offer the combined terminal and runway capacity that Manchester is seeking without the need for a new runway. It is making the case that, since its runway is not well utilised, the region would be better served by it than by a new runway at MAN. There is a significant local traffic base despite the relative economic decline, and the site is less environmentally constrained than MAN. There is the possibility that a fully segregated second runway could be built on land reclaimed from the estuary. Meanwhile, British Aerospace has sold its major interest in the airport.

The Manchester/Liverpool/Leeds-Bradford relationship appears to have been resolved by the market, Manchester making the most of its pole position in terms of traffic, of being the regional commercial centre, and of having local support for the large contribution it makes to the local and regional economy. The other airports serve local roles. If permission had been refused for a second runway at Manchester, they would probably not have benefited a great deal in the short term, rather, Manchester's passengers would continue to grow with larger aircraft and reduced hubbing.

13.10 OSLO, NORWAY

It had been expected for 30 years that the aviation activity would have to be moved from the downtown Fornebu airport to a less environmentally damaging site, but only in 1992 was the decision finally taken to re-engineer the Gardermoen site for opening in 1998, at which date Fornebu should close. A 1968 Commission had concluded that Fornebu should be allowed to grow, while Gardermoen should take the excess non-scheduled traffic as well as its military traffic, and that a new airport should be built at Hobol. The government reserved the land on the site 45 km from Oslo, but slow traffic growth caused the reservation to be lifted in 1981. New forecasts in 1984 suggested that Fornebu would reach capacity in the early 1990s. This led to a new search for a site. Parliament decided in 1988 that it should be at Hurum, 45 km southwest of Oslo, on the basis of shorter flight and ground access distances and that the military facilities would not be affected. However, many felt that Gardermoen would have been a better choice because much of the land was already developed as an airport and it would provide a much needed stimulus to the local economy. In the event, the high cost of developing an airport on the rugged Hurum site and concerns about the visibility and wind characteristics at the site caused the decision to be reconsidered. The decision to develop Gardermoen was taken in 1990 and the master plan was completed in December 1991. After intense debate inside and outside Parliament, permission to construct the airport was given in October 1992 and work started in 1993 for a planned opening in October 1998. Traffic at Fornebu reached 13 mppa and 170, 000 movements in 1996.

The project was helped by the nature of the planning process in Norway (Gaustad, 1996). The government stated a clear set of National Policy Guidelines, which were given the Royal Decree in January 1991. The guidelines covered the objectives for regional planning, settlement patterns, commercial development, land zoning for noise compatibility, environmental policies for ground access, and the control of development at Fornebu after closure. The guidelines focus on the process rather than the criteria, allowing the detail to be decided later. An Environmental Impact Assessment (EIA) was necessary, and the content and extent of it had to be publicly debated and agreed by the Transport and Environment ministries before it could be carried out. The resulting benefit was that, after the scope had been agreed in 1991, no further topics could be brought forward. Once the work had been done there was further public consultation, but this was quickly completed due to the earlier agreement on scope, so that the EIA could be approved in the summer of 1992. Once the project was approved, the Environment ministry put the necessary additional detail on the National Policy Guidelines for both sites.

Gardermoen is 47 km north of Oslo. Prior to 1998 it handled all Oslo's one million charter ppa as well as Air Force traffic. A new private airport company, Oslo Hovedflyplass Ltd (OSL), was set up by the Norwegian CAA in November 1992 to design, build and operate the new airport and to take over the operation of the existing airport and also Fornebu. The CAA owns all the shares. Its first task was to obtain detailed planning consent while the CAA had to acquire the necessary land. The land use plan was approved in June 1993. It required 230 private properties to be vacated, the remaining 75% of the land being acquired from the Department of Defence.

Although the scope of the EIA was in line with the agreed definitions, it has been necessary to set up special projects to deal with the ongoing detail of a cultural heritage plan, the establishment of flight paths, water resource management, a landscape plan, the treatment of waste during construction, compensation for appropriated property, the effect of odours and oil deposits, and the possibilities for the screening of aircraft noise. It has proved difficult to get agreement between the potentially affected communities on prioritising the areas over which there should be least flying. Communities well outside the noise contours have become involved and there is a conflict of interest between residential and recreational areas. This suggests that the benefits of having National Policy Guidelines are partly balanced by the lack of an early voice for the local communities, despite the large amount of local consultation during both phases of the EIA process.

There has been a dispute over the use of normal compensation rules for land acquisition, on the grounds that an unprecedented number of people are being forced to move and that special compensation would be appropriate. The process has been criticised because it has allowed early compensation for those wanting to move, leaving 'ghost' villages for those wanting to stay as long as possible. There has also been conflict between the umbrella Act of Planning and Construction and Acts governing the behaviour of the various state bodies. An example is the use of the Act of Health in the Municipalities to introduce stricter requirements on flight path and surface water studies.

The project is to be self-financing with investment from the CAA and a state investment loan. It is to have a 2,950 metre runway constructed parallel to, but 2.2 km away from, the existing one and new terminals at a cost of US $3 billion, giving an initial capacity of 16 mppa. The existing runway is to be extended to 3,500 metres. The cost includes US $0.5 billion for a new rail link to the centre of Oslo, together with improved roads and US $0.5 billion compensation to the Ministry of Defence for moving a military base off site. The sale of Fornebu is expected to contribute US $300 million. It is hoped that the train, with its 17 minute running time to Oslo's central station and its 10 minute headway, will attract 50% of passengers and 40% of staff (*ACI Europe Communiqué 1996-1997 Gardermoen souvenir edition*).

The advantages of moving to Gardermoen are:

- only 4,000 people will be affected by noise, compared with 45,000 people by the year 2000 if traffic remains split between the two sites,
- the cost of split operations will be avoided,
- it will be possible to serve long haul destinations directly, while allowing convenient transfer to domestic flights,
- the development will double the growth of employment opportunities in the Romerike region, giving a better balance between opportunity and labour availability,

- the alternative uses for Fornebu give almost completely positive contributions in the context of the national policy guidelines,
- there will be no need for curfews.

There are, however, some disadvantages. The airlines will have to construct new facilities, and they believe that some of the southern routes will be vulnerable to rail competition. Most of the population of the Oslo conurbation lives to the south and southwest of the city. Given the resurgence of interest in downtown airports, a case could still be made for selective use of Fornebu with operations with the same character as at London City. Also, there are prehistoric sites at Gardermoen, recreational areas at Jessheim which will be affected by the access roads and railway, and agricultural land will be lost. However, the loss of land amounts to only 2% of the local agricultural area.

The effects of Gardermoen are likely to be felt internationally as well. The possibility of long term expansion and of hubbing effectively between long and short haul, together with the opportunity to establish major maintenance facilities, may attract SAS traffic to the detriment of Stockholm's Arlanda and Copenhagen's Kastrup airports. Arlanda is to have a third runway by the year 2000, but it is extremely sensitive to environmental impact and the long and severe recession in Sweden has weakened its traffic base (*Jane's Airport Review, May 1995, pp 33-35*).

14

TOWARDS IMPROVED STRATEGIC PLANNING

The varied experience in the case studies in chapters 9 to 13 are compared in this chapter to the principles of strategic system planning which were described in chapters 7 and 8. The aim is to test the value of the principles and the practicality of applying them in actual planning situations, and also to identify any areas where the accepted principles may need to be modified.

All the case studies must be read in relation to their political and institutional contexts, particularly the more detailed studies of the US and the UK. The US system accepts planning as an appropriate process to assess the need for infrastructure development, to distribute available public funds and to reach a balance of interests in the local community and between local, regional and national perspectives. The process relies on a culture of active participation in open democratic debate and of public responsibility for infrastructure. The UK has in the last 20 years, largely under a conservative government, moved away from drawing distinctions between public and private roles in the ownership and operation of infrastructure, leaving government with only the same supervisory role as in other industries. In principle, this applies equally to the planning as to the management of airports, though an exception was made in the government's RUCATSE committee which considered the question of additional runway capacity for London. Even then, government really only provided a forum in which the stakeholders generated and debated 'facts'. These and the other case studies do, however, reveal strengths and weaknesses in the way they have addressed the development of their airport systems. Understanding this experience can help to ease the path to appropriate planning for airport development in other countries.

14.1 PLANNING PRINCIPLES AND PROCESSES

14.1.1 The application of planning principles

Each country has a distinct geography, political system and culture. Within their own settings, many countries have developed their own solutions to airport system planning, as is apparent from the cases described in chapters 12 and 13. Planning must depend on its context (Hill, 1986). The Norwegian system has an explicit accessibility objective and has adopted a rather unique technological solution to meeting that objective. The government has accepted a clear responsibility to generate an appropriate set of policies to support this objective. Japan and Spain have both attempted to take an inclusive view of the economic and social consequences of airport planning and to set it within an integrated transport policy. The Netherlands has perhaps taken the quantification of the pros and cons of aviation furthest, in particular with respect to the acceptance of the results of this analysis by the policymakers. Though derived within their own context, each of these national approaches follows from the application of good planning theory. They have this in common with the US, and it can be argued that the same planning principles could be used to advantage in the more *laissez faire* countries, at least in the creation of a firm policy context for a privatised industry to work within.

The most prevalent reason for disappointment with the evolution of airport systems appears to be the lack of commitment to the principles of system planning as described in chapter 7. The examples in the preceding paragraph represent only partial application of the principles which system planning theory calls for. Most countries pay even less attention to the theory, either ignoring it, or only paying lip service to it, or, worst of all, using it as a cover for preordained agendas. The most honest reason for not adopting a rigorous planning process is a concern that the airport system and its relationship to the rest of the transport system, and to the underlying social and economic setting, is too complex to allow useful formal analysis and efficient control.

The inability of rational decision making to cope with complicated social situations was being predicted even during the first applications of systemic analysis to transport planning (Hirschman and Lindblom, 1962). It was pointed out that social conflict often does not allow clarification of objectives; that information is often lacking; that the future can be too uncertain and the problem too complex for human intellect to grasp. In such circumstances, it was deemed preferable to adopt a strategy of 'disjointed incrementalism' towards the improvement of a system, as happened in the UK. This would mean discarding the wider analysis of the system in its environment in favour of incremental, and therefore more easily understood, changes. It would mean pragmatically choosing ends and means simultaneously and in the light of their mutual interactions; it would also mean indefinite iteration of the analysis ⇒ policy loop rather than expecting to achieve a convergent solution in a single pass. The iteration would tend to move away from ills rather than moving towards objectives. The analysis of consequences in any one sector could be allowed to be incomplete, since it would become central for some other sector during an essential but natural process of 'partisan mutual adjustment'. It was maintained that this strategy is not a flight from rationalism or a second-best approach; rather, it avoids inevitable errors which otherwise arise from over-ambitious attempts at comprehensive understanding; it accepts that systems are never complete or in equilibrium, thus not wasting effort on achieving early integration or balance; it makes use of the fact that individuals can often agree on policies when they cannot agree on ends; it relies on the truism that, in the long run, policy choices have as great an influence on objectives as the objectives have on policy choices. In fact, the strategy of 'disjointed incrementalism' is seen by some as more rational than the systemic approach to decision making, "since a rational problem solver wants what he can get and does not try to get what he wants, except after identifying what he wants by examining what he can get" (Hirschman and Lindblom, 1962).

The pragmatic incremental approach has been combined with a systems methodology in Multi Year Program Planning (MYPP) in the US. In MYPP, the difficulties of obtaining any agreement about goals and values in a participatory framework are eased by the discipline of having to spend each year's budget before moving on to later, and progressively less well defined, elements of the longer term plans (Manheim, 1979). The rational (systems) and incremental models have also been incorporated in a 'third approach' by Etzioni in an attempt to give long term direction to cumulative small changes (Starkie, 1987). In this approach, the incremental decisions are nested inside a broader framework provided by high-order, fundamental policy-making processes, which, while exploring the main alternatives seen by each actor, omit the details and specification that the rational model would require if it were to be used on its own. The MYPP and 'third approach' models were early pragmatic attempts to overcome the weaknesses of the programmed versions of the rational paradigm in accounting for the decision-making concerns which caused the poor record of implementation (Flyvberg, 1984). In particular, the earlier evaluations based in monetary terms failed to reflect the one-person-one-vote nature of the democratic decision process.

The need to respond to increasingly turbulent environments was one of the factors which encouraged this new paradigm of adaptive planning. To the extent that it is successful, it is clearly more necessary to adopt this methodology for airport planning as air transport liberalisation and industry privatisation increase the turbulence inherent in the politics of the late twentieth century. It has been realised that allocative planning is more or less superfluous when it is really feasible, whereas it is difficult to apply when it is needed. This has caused planners to try to move towards innovative planning, with an emphasis on the mobilisation of the necessary resources, a concern with institutional change and a basic proactive orientation. If the world is going to continue to move towards liberalisation, privatisation and sustainability, innovative planning will be increasingly necessary, with the participatory style that complements the dispersed power base which is another characteristic of this emerging context.

Also, since the evolution of the system will essentially be driven by large corporations, any attempt to plan for an adequate air transport system in, say, Europe will have to incorporate an understanding of the way in which the corporations do their long term planning. Large organisations are, in fact, integrating programmed Strategic Planning Systems (SPS) methods with Strategic Issues Management Systems (SIMS). The latter is more adaptive to the issues raised by the signals obtained from continuous and comprehensive monitoring of the environment in which their activities are set (Camillus and Datta, 1991). "It is impossible for managers to plan or envision the long term future of an innovative organisation. Instead, they must create and discover an unfolding future, using their ability to learn together in groups and to interact politically in a spontaneous, self-organising manner" (Stacey, 1993). In an environment where changes are rapid and where there is tight interconnectedness between the elements of a system, i.e. the environment is turbulent, organisations cannot properly understand cause-effect chains and cannot predict either the consequences of their own actions or local effects of distant events, the only way to regain control is through a collective and cooperative search for new values and rationales for behaviour (Emery and Trist, 1965).

None of this, however, denies the need for an improved understanding of the airport system and its relationship with the rest of society. System planning may be seen as a continual learning process, providing the context for disciplined thinking, and able to provide inputs to all the larger scale questions given in section 8.1 and repeated here:

- resolving conflicts by providing a forum for discussion
- allocating resources efficiently
- promoting national policy goals, e.g. equal opportunity
- providing the tools to make societal trade-offs between the economy and the need for sustainability, between regions and between local, regional and national priorities
- harmonising analysis techniques, to promote consistency in approach
- establishing best practice, to promote efficiency
- identifying appropriate public/private roles and questioning market place solutions
- addressing global issues to save wasteful and inconsistent considerations of these issues at regional and local levels of planning
- putting necessary enabling processes in place
- improving the planning process.

The underlying principles of 'best practice' system planning are:

- to take a broad definition of the system so as to include all important interactions and stakeholders, e.g. the military and airports beyond the boundaries of the region being planned
- to search for solutions which will allow all stakeholders to achieve a net gain
- to monitor the efficiency and effectiveness of the system, and the continuing relevance of the stated objectives.

14.1.2 Planning processes and frameworks

The case studies make it clear that there have been more failures than successes in the planning of airport systems. Many of these failures have been associated with deviation from the principles of good planning, but many more have been due to the use of inappropriate planning frameworks, so that the process of planning has been subject to distortion, delay and interference. Issues that arise in airport planning, particularly for large projects and for groups of airports, are sufficiently unique that the conventional planning frameworks, based on local planning practice and rational analysis, do not allow the full scope of necessary debate to take place. Equally, with the conventional planning framework, it is difficult to cope with the unequal distribution of costs and benefits across political boundaries.

14.1.3 Participatory planning

Early and inclusive consultation with all affected parties is a valuable tool for reducing conflict. Generally, in advanced democracies, a form of continuous participation by all interested parties is likely to be appropriate (Tietz, 1974). Various examples of these styles have been identified in the case studies. Successful use of them depends on their acceptance in each particular case. In this way, compromises can be made rather than having decisions dominated by technocratic positivism or politics (Torgerson, 1986), and the 'ordinary knowledge' of decision-makers can be taken into account (Breheny, 1986). The participatory style of planning is appropriate where the power in a system is dispersed rather than centralised (Friedmann, 1973), rather than a partnership between planners and citizens with the decision being left to politicians, or delegating all power to the citizens (Larrabee 1970).

There is clearly some requirement for at least initial evaluations to be essentially private, while retaining public legitimacy. Initially, the experts themselves will be in a learning process, and public debate only serves to confuse. It has been suggested that, in the early stages, there is a need for a unit that has the technical capability and the political independence to evaluate the expert discussion and to criticise it (Stevenson, 1972). Meanwhile, it has also been suggested that, whoever takes the decisions, early interaction is necessary in order to generate social information (Wilson and Neff, 1983), preferably through a technique called Social Information Generation (SIG). It is claimed that the public, when involved in SIG, feel that the activity is meaningful and important to them, as well as being acceptable and understandable. Using small group discussion, and given information on the known future implications of changes to their environment, the group members help each other in a peer situation to generate information which it would be difficult to produce in any other way. If necessary, the results can be recycled with further feedforward of information, leading to a gradual change induced by education rather than a perceived revolution. A similar approach is promoted in Strategic Options Development and Analysis (Eden, 1989). Each person's cognitive map is merged to form an aggregated map called a 'strategic map'. The aim is to change the minds of each member of the client group without them feeling they have compromised their position, so producing enough agreement about the nature of the problem that each team member is committed to expending energy on finding a portfolio of actions. The essence is to seek consensus rather than compromise and commitment rather than agreement. Checkland's Soft

System Methodology (Khisty, 1993) complements this approach by setting it in a formal framework of systems thinking, emphasising that changes must be culturally feasible for the people in the problem situation, given the unique history of the specific situation in a particular culture (Checkland, 1989).

This may be an unpalatable conclusion for the stakeholders in a system attempting to adjust to a new environment of untrammelled competition coupled with freedom from government intervention, both of which actually contribute to turbulence (Stubbart, 1985). Yet this search for new values through cooperation may be the only way to achieve a win/win solution without which progress would not be made. The emphasis has to be on joint progress and mediation rather than conflict and confrontation. The theory of negotiating towards a win/win situation is well known (Fisher and Ury, 1986). The principles are to separate the people from the problem, to focus on interests rather than positions, to generate a variety of possibilities and to insist on some objective standard on which to base the results.

Breaking the conventional planning mould and adopting the new paradigm of a common search for understanding, for setting goals and for exploring alternative ways of approaching the future, i.e. emancipated planning (Rosenhead, 1989), is too challenging for many authorities or developers. However, there are some precedents for adopting a more participative and adaptive approach to planning airports despite traditions of adversarial politics. The initiative between the airlines, the airports and the environmental lobby to generate a common methodology to determine a reasonable level of noise nuisance for a given user benefit in Europe (Logan, 1990) is seen as a prototype for this sort of activity. Other prototypes might be the agreement between the developers of London City Airport and the owners of other property developments in Docklands prior to the formal planning inquiry, the mediation procedure which is being used at some US airports to resolve conflicts between noise nuisance and airport capacity (Johnson, 1989), the appointment by Austin City Council of an Airport Advisory Board of 11 citizens and a Citizens' Conversion Task Force representing a cross-section of the city's population to manage airport projects (Austin, 1991). There are also the examples of the airport research foundation coordinated by the city of Nagoya for the development of a new airport (Knibb, 1991) and the mitigation exercise at Vancouver which has allowed necessary expansion with effective environmental safeguards by the use of early community participation which extended even to the generation of the assumptions to feed into the computer modelling of forecasts and the cost-benefit analysis (Matthews, 1993).

These precedents suggest that Vickers, (1965) was correct in predicting that, in the appropriate forum, discussions should be able to move from conflict, through a compromise stage to a new high ground of 'integrative' solutions: this type of solution, by changing the way in which the situation is regarded and valued, has the characteristics of commanding the assent of all the participants.

The concept of some form of Aviation Infrastructure Forum (AIF), described in chapter 10, could actually complement the strengths of private enterprise, rather than supplanting or stifling them. It would do this by making the planning methodology more transparent for decisions of national interest, and by easing the path through that process for worthwhile projects. In particular, it should help to avoid the dilution of the project during its passage through the clearance points of the process, thus helping to ensure that they generate the maximum benefit possible. It would, of course, be possible to bypass AIF and go directly to a normal public enquiry if that were judged to be in a projects' interest. It also would encourage the terms of reference for the inquiry stage to be set by consensus, as in Norway's EIS, rather than by the same authority (the government) which will take the final decision (Farrington, 1984). In contrast, the UK's expert-oriented approach used in the RUCATSE first phase may shorten the subsequent inquiry, but will not reduce the confrontation (Sunderland, 1992).

14.1.4 Power relationships

There are situations where consensus on the issues that need to be considered in societal evaluations is automatic, where the community is totally dependent on the airport and wishes to promote economic growth. More often, and particularly in more developed countries, it has become much more difficult to reach a consensus because

- there are other means of satisfying a maturing travel demand
- people are more concerned about noise and the environment
- the link between increased air transport activity and further economic growth is harder to sustain
- earlier expansion probably occurred in an authoritarian way which will have resulted in a confrontational planning process
- communities are deeply suspicious of the transparency of the decision mechanisms.

In the UK, communities in the north of England have generally been in favour of airport expansion due to their hopes of economic regeneration, so making consensus in planning relatively easy to achieve, e.g. Manchester, Humberside. In the US, careful attention to community relations over many years has allowed airports such as San Francisco International to maintain relatively harmonious relations with surrounding communities. Joint development of scenarios and forecasts by the industry, local business interests and community leaders, formed a key step in the process that led to the decision to expand Minneapolis/St Paul International airport (Hardison et al, 1990), as described in chapter 9. However, it is becoming more usual for confrontational attitudes to be struck based on incompatible world views of the various actors, which have led to substantial restrictions on expansion after a prolonged planning process. Examples of this are the restrictions on movements at Munich and Düsseldorf, the decision against a second runway for Stansted and the long delay in constructing an additional runway at Sydney.

The power of the technical experts should not be underestimated. The interactions among the stakeholders will only be as good as the advice they are given. Several consultants have put a lot of effort into developing multi-media communications packages to improve the information to decision-makers and hence their appreciation and understanding of the issues. The packages are powerful and must be used with integrity. However, it has been found that not all decision-makers are keen to understand much of the complexities, preferring to take the expert's advice, perhaps because the final reporting must still be in the more sterile written medium (Dubbink, Cohney and Frommer, 1995). In fact, there is a general tendency for decisions to be taken on biased and partial models of the system.

It is not easy or appropriate for all societies to adopt all these improvements to the planning processes. Often it is necessary for individuals in positions of power to exert their influence if progress is to be made. Intelligent planners will ensure that they are aware of the influence chains which will dictate the acceptance of preferred plans. Indeed, they may even feel it necessary to build in some features to attract support from powerful quarters, even though they somewhat compromise the project.

It has, indeed, become generally evident that, whether within corporations or governments, planning will only be successful if it recognises the power relationships which shape the broader ongoing debates inherent in the adaptive planning paradigm. Further, it also needs to recognise that the more open the debate, the more likely it is that political antagonisms will be awakened and enabling coalitions will be endangered (Patten and Pollitt, 1983). It is therefore necessary for any useful planning paradigm to be responsive to the implications of democratic theory, including the ways in which individuals and groups take decisions. These involve elements of bias, satisficing, cohesiveness and bolstering decisions. Also, attention needs to be given to the ways in which power is exerted, including the use of coercion, expertise, authority, inducements and constraints.

The understanding of power relationships is particularly important in a free market setting. Both the US and the UK have fully independent airlines, which are to a large extent deregulated except for access to some international routes. Deregulation implies that no control can then be exerted on airlines or on general aviation to use any specific airport, unless the government takes special measures to make an exception. The cases show that the industry was often sufficiently innovative to find ways round regulations even when they were applicable, but now it is even more true that the market place will impose demands on the airports which will distribute benefits to airlines and their users in ways which depend on the relative strength of those stakeholders.

Further, the stakeholders will seek to control the changing environment, rather than simply adapting to it as in the case of responding to deregulation, by dominating the debate and developing direct links to the decision-makers. One possible consequence might be the coalescing of administrators with powerful interest groups against weaker groups, thus maintaining stability despite democratic pressure. Thus human-relation techniques embodied in the adaptive paradigm would be thwarted (Patten and Pollitt, 1983). This is seen clearly in the use by airlines of law firms in Washington DC to communicate with the government officials and politicians, following a formula of $(IA)^2$: 'Influence and Access' on political issues, and 'Information and Advice' on substantive issues (Donoghue and Nelms, 1993). A powerful marketing advantage of a large carrier is its ability to engage all the major law firms, thus precluding them on ethical grounds from being able to represent the carrier's competitors. This is particularly important in view of the poor knowledge of the industry among the political representatives. In the UK, the lobbying ability of British Airways is envied by at least one of its competitors who relies on the equally important power relationship with public opinion (Branson, 1993). In general, airlines are not used to participation, yet their role in achieving agreements with the community is vital.

The present rather acrimonious relationship that tends to exist between airports and airlines requires similar principles to be applied if the liberalised future is to be managed efficiently. The 'grandfather' powers have generally given based airlines the dominant role in the relationship, though any attempt at vertical integration of airports into airlines has so far been resisted. Equally, a powerful airport, particularly if privatised, might take advantage of monopoly powers to allow more benefits to accrue to its owners than to the other stakeholders, as with the use of passenger terminal space for shopping rather than the facilitation of passenger flow. A fine line needs to be drawn between airline equity participation in terminals and any discrimination in access. Some take the view that the correct relationship is for the airport to be the landlord and for the airlines to be short term tenants, the wise manager realising that dedicated facilities are wasteful and that it is essential to diversify the revenue base (*ACI World Report, July/August 1997, p 3*). Others have suggested that legal contracts, tailored on a local basis, would be preferable to legislators taking control (*Jane's Airport Review, January/February 1998, p 3*). Another route to improve the situation is to defuse prejudice by cooperation and an exchange of senior managers, as initiated between Lufthansa and Munich airport (*ACI Europe Communiqué, No 92, March 1998, p 16*).

Particularly interesting relationships exist between the industry and its regulators on the one hand and the political masters on the other. In these relationships, the industry is not so likely to be able to influence outcomes, and may be better advised to put itself in a position to take maximum advantage of political and regulatory trends, as with UK airlines buying into rail and airports investing in rail access.

The type of simulated debate, which the Aviation Infrastructure Forum (AIF) proposed in chapter 10 is designed to encourage, where power strategies are represented in the debate and consensus is attempted, is not entirely original. A 'mock' negotiation between the US and the EC was held in the International Institute of Air and Space Law at Leiden to study bilateral politics (Feldman, 1992). Some time ago an elaborate simulation was performed by the State of Connecticut with teams representing legislators and evaluators (the latter representing society at large); one set of legislators attempted to optimise Connecticut quality of life, the other to maximise state per capita income.

Perhaps the most fruitful of these simulated 'game' approaches (as distinct from analytical 'game' methods of the n-player, non-cooperative, maximum pay-off type used by Hansen (1990)) is the MACTOR method (Godet, 1991). This has been used by both the Paris Airport Authority and the French Electricity Authority to examine future strategies. It focuses on the analysis of the balance of power between the actors, in terms of constraints and means of action. A realistic view of each actor's potential strategies is obtained by synthesising them from each actor's views of the other actors' strengths and weaknesses. In particular, each actor's prioritisation of objectives is identified and used to determine the significance of alliances and the potential for adopting particular tactics. The MACTOR method does, however, stop short of presuming that consensus is possible, except by deferring to dominance.

A combination of MACTOR with other conciliation methods may be more desirable; several methods have been suggested in the literature for taking full account of social information (Wilson and Neff, 1983).

14.1.5 The role of planners

In conventional first generation system planning, planners not only set up the methodology, but also carried out the analyses and advised the decision makers of a preferred solution. As participatory planning has become more accepted, the role of planning professionals has become more that of managing the process, their remit being to ensure that user demands are met fairly and to the extent that society feels to be appropriate, following the reasoning discussed in section 3.1.1. They also still have a strong role in advising on the planning methodology and the techniques to be used, through setting objectives, forecasting and evaluation. They should promote the maximum of transparency and the use of soft techniques of evaluation. This is not to say that quantitative analysis should be abandoned, but rather that it should be integrated into a holistic framework through qualitative methods. Often these rely on ranking of factors and weights (e.g. Paelinck, 1977), which should, in turn, be subject to discussion rather than imposed by the planners' preconceptions of value. The criteria for a successful conclusion in a participatory planning exercise are learning and consensus. The identification of target groups and the implementation of policies are dealt with adaptively during the decentralised and incremental process.

14.2 PLANNING TECHNIQUES AND PRACTICES

14.2.1 Formulating goals and objectives

One of the main causes of the failure to implement airport plans has been the lack of attention to defining clear and feasible objectives that could be accepted by a system's stakeholders. Often the goals have been implicit in the minds of the decision-makers, leaving analysts no firm set of criteria against which to evaluate options. In other cases, objectives have been set dictatorially, perhaps reflecting belief in a simplistic relationship between airport investment and the economy or in the inevitable use by operators of spare capacity wherever it is provided, so that those having to implement and operate the resulting system have had no vested interest in it. Sometimes the project itself has been the objective, almost regardless of the efficiency and effectiveness of the resulting system.

The adaptive planning paradigm emphasises the importance of social interaction. It views the formulation of goals and objectives as part of the social and political dynamics, and subject to change during the social interaction process. The learning is largely inductive, based on the cases under study, and strives for valid subjective knowledge. Little should ever be taken for granted. Not all cities in countries whose economies are based on tourism want airports on their doorstep, even though it might mean a relatively lower standard of living. Indeed, the whole question of whether economic success equates to progress is becoming a subject of serious study (Abell, 1998).

14.2.2 Dealing with the future

It has been realised that flexibility is needed in the face of an uncertain future. The total environment, economically, socially, culturally, technologically and politically, needs to be scanned continually, to identify the obstacles to a safe passage into the future (Trowbridge, 1988). Even the community's view of a project can change dramatically over time, as shown by the case of London City airport, in the face of improved information, changed circumstances and actual experience. The correct interpretation of an adaptive attitude has been promoted as 'to focus on how to enrich and empower future generations by giving them as many resources and as many options as possible'. "A clear appreciation of the plan/decision distinction focuses analytic effort on getting the decision right, with the plans a framework to ensure that the longer term is not sacrificed to short-term advantage .... It is better to attempt to engineer a high level of flexibility .... e.g. liquidity in finance, versatility in military, resilience in ecology, hedging in investment .... A decision package which is in some sense 'multi future robust' is more likely to offer worthwhile options for development than would a single-future robust commitment" (Rosenhead, 1989). This would be operationalised by 'the planning and design of flexible, adaptive facilities and systems'. The theory has recently been practised in the planning for a second Sydney airport (de Neufville, 1990). "Doing dynamic strategic planning is comparable to playing chess well: the planner considers many moves ahead but commits to only one move at a time - moreover, the decision-maker chooses this move to provide flexible response to future challenges, either to protect against threats or to exploit opportunities" (de Neufville, 1995).

However, if it is not known to what circumstances the system might have to adapt, the increased investment to provide the flexibility may be more wasteful of resources than would following an incremental adaptive approach. It may be preferable to adopt the approach suggested in section 7.4.5, of identifying a core strategy which would best cope with the widest range of feasible future scenarios, but delaying implementation of each phase until the risk of being wrong can be minimised. Frequently, and particularly in an entrepreneurial setting, this approach may have to be compromised if it is deemed necessary to commit to an investment in order to consolidate an agreement with an airline, or to ensure that future expansion can occur despite local competition for land use.

The technical options described in section 14.3 are often denied because it is thought legitimate to limit planning horizons to 20 years. "At present virtually all governments are proceeding with characteristic myopia: playing industrial era games; hoping, fearing, denying, blocking and generally not getting the drift of things clearly in view" (Slaughter, 1996). Even though the future cannot be fully predicted, many trends are clearly in view and should not be ignored.

This suggests that the most valuable role of strategic planning is to articulate likely future states of the system in as informed a way as possible. This should flow from a thorough analysis of the factors that shape the evolution of the system and an open-minded assessment of the ways those factors could change and interact. This is no small undertaking and will require appropriate resources to do well.

14.2.3 Multisector influences and evaluation techniques

Many times, the best laid plans have been thwarted by lack of attention to the perspectives (or world views) of those members of society who are not naturally pro-aviation. Historically, with some exceptions, the industry has tended to ignore their views, though tools for taking account of them, such as technology assessment, have been available for a long time. An impressive attempt at a multi-dimensional evaluation was made in 1975 when Boeing performed a comparative analysis of inter-city transport modes which included fuel efficiency, circuity, load factor, safety, emissions and modal split as a function of cost and value of time, as well as drawing attention to a large number of other planning facets which should be taken into account (Schott and Leisher, 1975). A less detailed but more broadly based study was completed in 1990 in the US (TRB, 1990), when, following an earlier panel recommendation (TRB, 1988), the FAA and Department of Transportation requested the Transportation Research Board (TRB) convene a committee to study the strategic choices for US air transport to the year 2040.

The TRB panel evaluated eight strategic options against each of ten factors, rating each option on a scale between -3 and +3. The factors were capacity benefits, capital cost, operating cost, safety, passenger effects, industry effects, environment, local and regional effects, funding and financing, and implementation. They concluded that the 'best' solution must conform to social values, be based on sound economics and find political support locally and nationally. They rejected the strategies of retaining the present system of management, of simplistically planning to build new airports and of reconfiguring the system around transfer-only airports. Clearly many criticisms could be raised against the methodology, not least on the matter of subjective judgements, but the study goes some way to setting air transport acceptably into society, though probably not taking sufficient account of the need to "find political support locally and nationally". In the climate of the next two or three decades, this will only be done by demonstrating that the planned system will be:

- more efficient
- more 'green'
- more appropriate.

A very experienced aviation professional has called for a 'civil aviation benefit impact statement' for all new aeronautical projects, since "the global market test of a transport vehicle must be how safe, economical, environmentally benign it is, not just how technically efficient or fast it is" (Halaby, 1990). Assurances along these lines can only flow from a full transmodal Environmental Impact Assessment (EIA) rather than an industry-based technology assessment. Certainly, as pointed out elsewhere (Yosef, 1991), a far more rigorous analysis than that suggested by the FAA is necessary if the US Executive Order 12291 of 1981, that directs agencies to consider the full costs and benefits to society of their actions, is to be followed correctly. These studies could form the basis of a SWOT (Strengths, Weaknesses, Opportunities, Threats) analysis in future scenarios. This analysis, in turn, would indicate the areas where air could best compete with, and where it could best complement, other modes.

Any societal evaluation of air transport will therefore have to concern itself with at least the following factors:

- resources: energy, materials, land, capital, labour, time
- safety: passengers and third parties
- social: cohesion, accessibility, equity, opportunity, severance
- political: carbon tax, competition
- future technology: aircraft, atc, other modes
- system efficiencies: aircraft, atc, other modes
- planning: land use, spatial distribution of population

in addition to the more traditional concentration on

- environment: noise, air, water, ecology
- demand:, propensity to travel, elasticities, saturation, other modes
- economics: cost, contribution to the economy, stimulation of the economy

if a wider debate is to be initiated. Only such an approach can result in synergistic policy development involving those with the capability to improve the majority of the evaluation factors, thus opening the possibility of a win/win solution.

An evaluation area which is commonly missed is that of the secondary effects of transport. In the area of CO_2 emissions, for example, the emissions come not only from the transport energy but also from the energy used in refining petroleum, in making and maintaining the vehicles, in building the infrastructure and in making the materials from which the vehicles and infrastructure are made (DeLuchi, 1993; Hillsman and Southworth, 1990). The German government purports to adopt a common evaluation procedure for investment across all modes according to criteria relating to the federal economy, regional policy, ecology, etc. The economic items are assessed using cost-benefit analysis, where environmental benefits from such factors as accident, emissions and energy reductions are included (Gand, 1991).

It is necessary to have specialist advice to assess the ranking of alternative plans across all these factors, some of which are difficult to quantify. Any weighting of the factors should reflect the strength of feeling of the various affected parties, and the best way of ensuring this is to have their participation. In adopting participative planning, it should be borne in mind that this calls for evaluation methods which reflect individuals' perceptions of their problems, their goal preferences and the alternatives available. The methods should not, therefore, simply be borrowed from those developed for non-participatory planning (Hill, 1986). In other words, it should allow the adoption of any one of half a dozen other feasible methodologies (Rosenhead, 1989), thus orchestrating a process of learning, recognising the differing world views of the actors and that all the possible accounts of the human activity system are valid: there is no single 'testable' account. It is important that the political context be taken into account, which means that an apparently value-free analysis will not be so productive as a judgmental scale of comparative advantage between alternatives (Friend, 1989).

Other modes are already being scrutinised in this way, either on their own initiative or by policy makers, as cultural and ethical concerns are beginning to critically affect the development and use of technology (Nikolajew, 1991). The Swiss Board of Education commissioned the MANTO study of the implications of advanced telecommunications on society (Rotach, 1988), costing Fr 2.3 million over 3.5 years and involving 40 experts concentrating on those indicators appropriate to Swiss interests - jobs, fuel, office space, and disposal of equipment. The Community of European Railways (CER) has estimated that the social rate of return from the European High Speed Rail Network, due to safety, congestion, pollution and energy savings as traffic switches from road to rail, will add a further 10% to the 9.4% economic rate of return on an investment of between 50 and 90 billion Ecus (AAE, 1989).

14.3 AIRPORT SYSTEM DESIGN

14.3.1 Expansion of existing airports

Expansion of existing airports is often regarded as the last resort, when other solutions have been thwarted by cost, environmental opposition or access difficulties. Too little attention has been given to the possibilities of managing the demand and the environmental impact at existing airports to take advantage of the land use patterns which have grown up with the airport while offering further mitigation of the negative impacts on the community. In fact, one of the general conclusions of the TRB study of strategic choices for US air transport (TRB, 1990) was that, like friction, congestion and delay are inherent in any efficient transport system, implying that there must be the opportunity for an economic rent if a system's efficiency is to be measurable. Many of the studies show the logic of retaining the concentration of activity and allowing the natural law of success breeding success to express itself. The capacity of the primary airport can be maximised by:

- using the economies of density and scale to fund the required mitigation and compensation programmes to make the local environment acceptable,

- recognising the medium term opportunities to employ technological solutions (e.g. reduced or short takeoff and landing aircraft, large aircraft, differential GPS, noise cancelling devices, steeper approaches), and investigating other novel solutions to improving capacity physically, environmentally and in terms of third party safety.

Application of these principles could have favourable implications for many busy downtown airports which are currently constrained (e.g. São Paulo Congonhas, Chicago Midway, Dallas Love Field, Seoul Kimpo, Berlin Templehof, Oslo Fornebu, Montreal Dorval, Washington National, Sydney Kingsford Smith, Seattle, Athens, and the many airports in Africa, Asia and South America which are expected to have strong traffic growth in the next 20 years and for which replacement plans are already being laid. Abandoning these sites just when airports like London City are beginning to demonstrate the potential benefits of adopting innovative airport/aircraft compatibility solutions seems rather short-sighted.

New investment in airports is now much greater than would otherwise be necessary in order to overcome environmental concerns. While much can be done by management of demand and the introduction of new technology to reduce the requirement for new capacity, a choice eventually has to be made as to whether to adopt a very expensive solution like an offshore airport or to accept the high costs of reaching agreement with the local communities about the price of expansion or to settle for no increase in air travel. It may well be that the most disruptive but most efficient solution would be to expand the existing site.

14.3.2 Development of new airports

From the above cases, and particularly the London case presented in chapter 10, it can be seen that there is a variety of reasons for having two or more airports in a metropolitan area. Multi-airport systems tend to work if there is full control of routes, if the original airport closes, if there is a natural second market, or if the original airport cannot provide critical characteristics, e.g. runway length, capacity, curfews. It is hard to get a supplementary airport to work if the original airport has unused capacity, if the access to markets from the alternative sites is inferior, or if transfer traffic is important. All of these factors normally apply.

Subsidiary airports then must be satisfied with:

- spill from capacity-limited major airports: this will be from the weakest routes and airlines in the peak periods
- niche markets, e.g. new entrant carriers, charter operations
- short-haul scheduled traffic for which access is more important than frequency or airline
- operations denied from the main airport by curfews, noise budgets, etc.

There have been many examples of secondary airports succeeding with these traffic characteristics, but their growth has often been sporadic and reliant on the control of route licenses. The only feasible proactive strategy for generating a strong and secure income stream appears to be attracting a carrier with the financial power and stability to set up a hubbing operation to compete in frequency and transfer volume with the main airport, as has now happened with British Airways at Gatwick. Otherwise, there would have to be a very potent regional planning case and/or high social costs to outweigh the air transport system inefficiencies which stem from fragmenting the system and from the quite extreme political interference which would be necessary to dictate to the large carriers, given that market-based pricing would tend to favour the existing airport.

It is much more feasible to establish a replacement airport, unless the existing airport is particularly subject to competition from other airports with qualities that the replacement airport would lack, e.g. access, meteorology. The new airport's access will almost certainly be inferior in the short term, but future demography and multimodal transport provision could overcome this in the longer term. The main difficulty of a replacement airport is likely to be financial, both because of the high and indivisible first cost of the airport and all the support infrastructure and also the need to write off the costs sunk into the existing airport. Most recent replacement airports are finding it difficult to cover their capital repayments from reasonable levels of aeronautical charges.

14.3.3 Role of airports in regional development

The evidence from the case studies ranges from Norway for which the STOLport system is an essential domestic lifeline, though the policies of Japan and the EC of providing the infrastructure to support interregional connections in the belief that this is good for regional development even in a developed setting, and the US policy of essential air service access on the grounds of equity more than economics, to the UK situation where regional links to national hubs are subject to the discipline of the runway slot market place. The latter result is due more to government indifference than the analysis of the CAA, the latter merely concluding that there is no good air transport case for interfering with the hub slot market. Indeed, in other studies, the CAA has found that new regional international links do have positive economic consequences for both the region and the country as a whole.

This is clearly an aspect of planning which depends crucially on context, though, in general, it is true that provision of airports and air services can only be an enabling policy that will, in itself, not produce a magical boost to the economy. Even the impact of airport direct employment will depend on whether these costs are covered by revenues, which will only occur if there is a genuine benefit for users in choosing that airport rather than any other. They may well be better satisfied by road or rail links to a hub, or by air links to a more distant hub, than by the introduction of weak direct service, as indicated by some UK regional experiences and by the passenger choice models presented in chapter 7. The case for using air transport for purposes of social cohesion is more straightforward, though ground transport is usually considered less socially divisive and offers the additional benefit of serving complete corridors rather than isolated settlements.

14.3.4 New technological options

The 1990 TRB study mentioned in 14.2.4 concluded that three strategies were worth further consideration:

- further construction of airports combined with the centralised management of demand allocation;
- letting the market decide about building new airports and introducing new technology, relying on local management to allocate traffic by economic measures;
- revolutionise intercity transport technology, with new aircraft, air traffic control and high speed rail.

Their recommendations with respect to new aircraft were the development of large aircraft and the development of 150-seat RTOL and VTOL aircraft to replace conventional jets, so using airport space more efficiently.

An example of an initiative which might well generate such synergistic environmental, physical capacity and safety benefits is the use of steeper approaches (Caves, Jenkinson and Rhodes, 1998). Similar benefits might also come from further developing synthetic vision systems to reduce the need for aids external to the cockpit in low visibility operations (*European Regional Airlines Report, March 1998, p 17*). A full social debate would also be able to identify and assess intermodal opportunities for the carriage of both passengers and freight like the rail/air freight facility being developed at Frankfurt.

14.3.5 Airport classification and appropriate roles

Airports within a system defined by administrative responsibility are often classified to reflect the type of aircraft they are expected to handle and the role they are expected to play. Unfortunately, the roles have often been dictated by politically motivated international agreements or regional ambitions which have little to do with the actual traffic base and the desire of airlines to serve them. Now that air transport systems are being increasingly liberalised and airports privatised, preferred roles are being identified by the market place. If planners are to assist in creating an effective distribution of capacity across a system, they must understand how the stakeholders are taking their decisions and assess the value of the resulting distribution in the light of overall society needs.

The planners should certainly expect the preferred roles of the airports to change substantially as the new freedoms are exercised, particularly as the globalisation of the carriers and the airports gathers pace. This will have as important an effect on airport roles as aircraft technology has done in allowing airlines to overfly airports which once were busy intermediate stops, forcing them to redirect their marketing to very different activities.

15

CONCLUSIONS

The aim of the book has been to identify the strengths and weaknesses of past strategic planning of airport systems, and to attempt to provide guidance on how the concept of strategic system planning can be used to advantage in the future. The challenge is posed by the changing nature of the air transport industry and the contexts within which it must work. The industry is increasingly being liberalised, privatised and globalised. The intended competition is sometimes rampant and sometimes seriously constrained by lack of physical and environmental capacity or by the behaviour of the operators. The contexts are becoming more sensitive to sustainability issues and to calls for integrated transport solutions to increasing congestion. This chapter offers a summary critique of past planning and some suggestions on how a strategic approach to planning can make an effective contribution to the provision of appropriate airport systems to meet future needs.

15.1 LESSONS FROM HINDSIGHT

15.1.1 Neglect of system planning principles

The case studies show that airport systems have evolved in a broad spectrum of contexts, from completely *laissez-faire* policies to rigid central planning. Some countries have adopted very narrow planning and others have tried to integrate airports into the rest of the transport and local planning sectors. Some countries adopt static planning while others adopt rolling plans and monitor them with feedback loops.

Plans have all too often failed to generate appropriate airport systems, for a variety of reasons. There have often been only weak policies with conflicting goal statements, and inertia in the planning system. Shortage of investment capital and the temporal instability of the social, economic and political setting have often forced fundamental changes in plans. The decision process has sometimes been in danger of information overload, and at other times been unable to cope with soft data and the lack of understanding of the relationships between the factors governing the decision.

Airport system plans have often been made in isolation from the airline route structure. This has led to problems with feeder service access to hub airports, partly because of a tendency to plan national hub airports in isolation from the other airports. Where the air transport system has been planned in an integrated fashion, the airport role has often been wrongly identified because of insufficient understanding of how passengers make choices, how airlines manage their business and how, as a consequence of these decisions, hubs develop and change.

Integration of airports within the total transport system can only be efficient if the advantages and disadvantages of air transport are correctly understood. Historically, very few airport plans have been embedded in plans for transport or for other sectors of the economy. Yet airports have been developed with the express role of avoiding constraints on economic development or because of pressure from local and powerful politicians. The level of understanding of the relationships between airports and these other factors has been inadequate to allow more formal integration.

Most airport systems seem to have evolved with no firm objectives other than to provide capacity to match revealed demand. Some have embodied simple concepts of access to a network (e.g. the US national system). The only plans which explicitly use the concept of time-space accessibility appear to be those of Norway and Japan. Objectives other than accessibility are seldom treated on a basis of quantified criteria. Factors like safety and reliability are usually only treated in a reactive way.

There is often an undue emphasis on forecasting future demand from historic revealed demand data, with little attention to the consequences of changes in technology, operating practices or macro changes in the national and international socio-economic environment. These deficiencies frequently result in the plans being unsuitable for implementation, if they have not already been disrupted by environmental concerns or financial problems. Monitoring of all these external as well as internal factors is essential for successful plan modification.

Monitoring of the implementation and achievement of airport planning has been almost nonexistent, despite the exhortations for this in the guidelines for best planning practice. Certainly a check is kept on the overall system delays, but little is done to monitor the total factor productivity of the complete air transport system or to assess the difference that a change in infrastructure investment would make to that productivity. The planning criteria themselves are seldom monitored to confirm their continuing applicability, the US Essential Air Services Program being something of an exception. The progressive commercialisation of the industry is causing airport accounts to be more free standing, but there is still a great deal of opaque accounting and a lot of the detail is no longer made public. Government agencies should at least be able to monitor indicators of efficiency, effectiveness and economy, where efficiency may be defined as output divided by resources consumed, effectiveness is the extent to which performance criteria are met, and economy is the comparison of actual input costs with planned or expected costs (UK Treasury, 1987).

There has been almost no attempt to assess the full societal balance sheet for air transport which would allow firmer conclusions to be drawn about whether and where society should support further expansion of the system. It is therefore not easy to judge the performance of airport strategic planning, even against the goals implicit in the planning. The extremely political nature of the process even makes it difficult to apply best planning practice to airport system planning, so it would not be surprising to find the outcome to be nonoptimal. There are, in any case, few agreed indicators of system performance, although this is begining to receive increased attention (Bolczak et al, 1997; Gosling, 1998).

Some of the system deficiences identified in the case studies undoubtedly stem from having to take decisions based on a low level of information, leading to a poor quality of debate. More extensive systemic planning, and the analysis required for it, would allow a progressive understanding of how the system works and the needs it satisfies. A hidden danger in the move towards a softer, more overtly political process of decision making, is that the understanding, and the information which supports it, will be lost. All too often the result is that the process degenerates into decisions based on ideology and powerplay made by an inner cabal, with a disillusioned set of stakeholders who feel they have been excluded from a real debate. The planning process itself is then simply used as a justification for policies that are in fact not based on the results of the process.

There are certainly occasions when there is an inevitable lack of information to perform a conventional systems analysis, particularly in low density and emerging economy cases. Deficiencies arise even in an understanding of what policy options are feasible for the provision and operation of the airports, as well as more frequently in areas like predictions of social and economic data, the mix of traffic and its needs, appropriate elasticities, fleet plans and the rate at which operating efficiencies might improve. This could be seen as an opportunity to educate the decision-makers in the risks of being wrong, and the value of taking a pragmatic attitude by retaining flexibility in the face of pressures to take firm decisions.

Commonly used methods of evaluation and decision-taking are shown to lead to inefficiencies in the implementation of appropriate capacity. In particular, the incrementalism and "partisan mutual adjustment" framework, which the case study shows to have been prevalent at least in UK airport planning, may neglect weak political groups, may limit consideration of alternatives, leads to partial solutions, and tends to minimise changes of policy. It is seen that turbulent free-market settings require very different planning frameworks from those which would be appropriate to more settled settings. The early participation of all relevant actors is important. There are methods available for resolving differences and creating win/win outcomes which have been used with some success. The decision and implementation phases can only be predicted if the relative power of the actors and their prospects for forming alliances are given due attention. Techniques for reflecting these factors have been identified in chapter 14.

Major airport development requires societal, as well as local, approval. At the local level, the debate hinges on the economic advantages which might accrue from the project to counter the environmental disbenefits. Monetary compensation and mitigation measures may provide a partial solution to the purely local noise problem, particularly where there is a strong desire by airlines for development. There are, however, many other disbenefits to society to be weighed against the economic benefits. The causal nature of the relationship between supply and demand requires further examination in order to illuminate this crucial debate. The wider issues of national environmental and mobility policy imply an additional need for a wider-ranging and multicriteria assessment of air transport. This is reinforced by the observation that acceptable political decisions are usually those which have synergistic benefits across a range of goals.

15.1.2 Implications of 'market place' planning

Examination of passenger behaviour revealed by air passenger surveys and travel data shows that passengers appear to have strong preferences for airports close to their ultimate trip ends, and for direct flights. Trade-offs can be made between access, direct flights and frequency, depending on trip purpose. Very high frequencies are necessary on indirect flights if a significant share of the market is to be captured. Lower fares available through hubs, which may not be purely cost-related, are shown in the US to greatly modify this behaviour, but it is not currently possible to include fares accurately in analyses of European air passenger behaviour. This behaviour is also distorted in the case of a small airport in the shadow of a major airport: a much higher frequency and lower fare appears to be necessary at the smaller airport if it is to draw more than purely local traffic. Traffic predictions are crucially dependent on the supply decisions of the airlines. Some indications are given in the European case study of likely developments of hubbing and competing services which can be used to assess the likelihood of an airline successfully developing networks at particular airports.

Airlines are seen to have cost structures which favour increased concentration and density of service. Hubbing appears to achieve these characteristics but, more importantly, allows domination of local markets and other marketing advantages which deny any significant contestability for its traffic. Freedoms which come from liberalisation are likely to reinforce the fortress hubs which already exist, resulting in powerful local monopolies and oligopoly on a European scale. This may not be compatible with passengers' preferences. Airports in the shadow of major hubs will find it difficult to convince an airline based at the major airport to fragment service by serving both airports, unless the airline needs to defend its market share.

The debate over the balance between environmental impact and the benefits of air service which flow to users and communities has become characterised by fairly well defined yet unconvincing arguments. Many of the commonly quoted economic benefits are seen to be of dubious veracity and unlikely to impress an independent decision-maker, though further research may show genuine causative effects of supply on demand and economic activity. Similarly, much more detailed research is required to establish the advantages of air transport accessibility for communities. Meanwhile, there are other wider societal benefits which are overlooked, perhaps due to the difficulty of including them convincingly in evaluation exercises. Many of the disbenefits are of real concern. The impact on the local community may lead to capacity limits or financial penalties, in order to control noise and emissions. Society at large is likely to levy taxes to control acid rain and greenhouse emissions, as well as to harmonise standards for acceptable noise impact through legislation. There are strong implications in these policies for the size and distribution of demand, as well as for competition between airports, where the results will depend on how successfully the debate is diffused and converted into a win/win situation.

Airports exhibit very strong economies of density and significant economies of scope. Thus, although it is possible to be profitable with less than a million passengers per year, there is a strong incentive to increase the throughput of passengers and freight. This tendency is reinforced by the requirement under privatisation to maximise the returns to the shareholders. However, it is important, especially under a privatised regime, to ensure that management retain control of the expansion, even where economic rents can be forced up by keeping capacity scarce. This is difficult if reliance is placed on forecasting traffic growth and then expanding infrastructure to satisfy the predicted demand. In a liberalised setting, it has been shown that revealed demand results from a subtle blend of consumer desires and airline strategies. It is necessary to take a realistic view of an airport's appropriate role in the total transport system, bearing community costs and benefits in mind as well as the airport's balance sheet and the needs of its users.

As the airport industry becomes more competitive, it no longer is advisable for an airport to develop reactively in response to any arbitrarily designated role. To do so will almost certainly mean having to accept those roles which other, more proactive, airports have discarded. This will tend to produce traffic with the least desirable characteristics: peaky, low yield, noisy, financially fragile, and volatile. Thus, even those cities prepared to continue with some form of subsidy in return for broader social and economic benefits, may lose patience as the subsidy rises while the traffic fluctuates and the benefits fail to materialise.

Airports will therefore increasingly wish to influence their role in the national and international airport system, bearing in mind that the role needs to be sustainable if it is not to result in wasteful investment. Schiphol sees itself as able to create and to sustain its 'mainport' role in Europe despite the environmental cap at approximately 44 mppa in 2015, by progressively developing a multimode hub to transfer short haul traffic from air to rail (Amsterdam Airport, 1995). Aeroporti di Roma also intends to change its marketing, not only to be more client orientated (aided by Alitalia selling its stake in 1995) but by launching the Leonardo da Vinci airport as 'the hub of the Mediterranean', with terminal expansions which will give very high levels of service at 38 sq metres per peak hour passenger (*Airport Business, April/May 1996, pp 19-23*). Manchester decided in 1990 to change the role as defined in the Development Strategy document of 1985, to become a hub even without a strong based airline. This required major revisions of its Master and Business Plans, calling for a second runway at a throughput of 15 mppa rather than being able to expect that its single runway would support over 20 mppa. Denver's new airport, on the other hand, assumed its role to be a straight replacement of Stapleton's role, but with no operational or capacity constraints. In the short term, its success has been somewhat muted due to the withdrawal of Continental Airlines and the imposition of considerably higher user fees, though delays appear to have been reduced very substantially and there has been no noticeable reduction in traffic (*Flight, 26 April 1995, pp 34-35*).

This discussion of airports' trends and possibilities clearly illustrates the challenge facing airport management teams in an era of airline liberalisation and an ethos of free competition. To the extent that the airports succeed in becoming masters of their own destiny, there is the possibility that they will generate a shape to the system which, while being in the best interest of the airports' owners and probably their own customers, may well not cater sufficiently to the principles of social equity and sustainability. Thus governments are likely to feel it necessary to impose checks on the development of airport systems. If these checks are to be administered with a light hand, airports will increasingly have to be seen to be sensitive to all those wider planning issues which governments would otherwise feel obliged to influence more rigidly.

It has been suggested that, in their search for competitive advantage, airport owners should cease to conform to conventional master planning, which generates a single solution for a long term future. Instead, planning should consciously put the owners in favourable positions relative to new needs and opportunities as they arise, minimising risks by continually adjusting the plan to the real situation. This is clearly desirable for them, but needs to be supported by:

- a knowledge of the real potential in the marketplace, and the probability of realising it;

- a realistic view of the potential role of the airport within the overall air transport system;

- assessment of the likelihood of successful planning approval;

- continuous scanning for risks and opportunities;

- tight management control of costs;

- change of management culture;

- development of further performance indicators to measure progress against goals.

Despite the implementation of these initiatives among the most competitive major airports, even they have often found themselves unable to dictate their role within the overall air transport market. The smaller airports are bound to find it much harder to create the conditions which will allow them to develop their preferred role, even if they have the expertise to use the most effective marketing strategies. It will be crucial to their success that they should be able to identify sustainable roles. They will need to understand the natural roles of other airports, recognising the natural tendency for the strong airports to become even stronger, and being able to judge when it would be better to complement them rather than to compete. For instance, it may be better for an airport and for its community to act as a spoke to a few major hubs, or to remain a predominantly charter or freight airport, than to attempt to develop a range of thin direct scheduled services with fragile airlines. Within the airport's regulatory and environmental constraints, success will depend primarily on the strength of local demand, how it prefers to express itself, and on how attractive the airport makes itself to airlines and to the communities that it serves.

This view is strengthened by the knowledge that the turbulence and uncertainty can only increase as liberalisation develops. The risks will come from the enhanced role of market leaders, doubts about the long term demand trends, government policy for intervention and their implications for costs, the reshaping of the airline industry, opportunities created by new aircraft types, and from the unpredictability of the planning process. A real difficulty is that smaller airports do not have the expertise (or the funds) to perform the sort of risk analysis which the market leaders perform (Volgers, 1992). When this can be done, the potential risks can be ranked and controlled, both those external risks mentioned above and internal risks, e.g. human resources and accidents. The small airports can also do little proactively to influence the level of risk, e.g. by forging an active role in bilateral route negotiations, though it would be wise to press for an enhanced role in these proceedings (Katz, 1991).

15.2 Adapting Creatively for the Future

Most recent thinking in the western world seems to imply that the only realistic future socio-political-economic scenario is a continuation of the current trend toward moderate growth in a free market setting with full social cost recovery. There seems to be a predisposition to accept the myth of predictability, or to incorporate it in justifying the need for flexibility in planning to cope with the expected uncertainty. An adaptive attitude to this type of future is seen to be appropriate.

This attitude of the industry leads to a sub-optimised and surprise-free expectation of the future, accepting maturity and making the best of it. It calls for only incremental change to the established system. It implies that any branching through structural change to a new high ground will happen with much greater resistance than if a creative, optimistic attitude to the future were being taken. Yet it could well be argued that the trends to liberalisation, privatisation and sustainability give rise, not to a surprise-free future, but to an increasingly turbulent and less predictable setting with uncertain understanding of relationships and preferences. The appropriate management response should then be to turn to inspiration as a decision tool (Emery and Trist, 1965; Stacey, 1993; Trowbridge, 1988), aided by a collective and cooperative search for new values and a continuous holistic scanning of the total environment to identify obstacles. This creative attitude requires not only a holistic understanding of all the interactions which might create alternative futures, but also faith that the proposed initiative will work to improve any of the preferred futures. This would mean using some resources now to keep alive a whole range of options for the unknown future, together with the methodologies for taking them into account in planning.

Consequently, it seems that the air transport industry is faced with a dilemma. It can take an adaptive attitude to the future which would lead to a relatively surprise-free future where a maturing industry will find its own level in a sustainable environment. Alternatively, it can take the view that the future will be turbulent and unexpected to an extent that will make it difficult to apply the adaptive approach successfully, and turn it to advantage by encouraging the stimulation of new ideas capable of inducing another major structural change in supply. This could result in shifting the demand curve and the associated present estimates of ultimate saturation to a new level. Even a partial acceptance and response to the latter view might still produce a minor shift to a new 'S' curve within the basically surprise-free future by lifting constraints and increasing overall welfare for a given expenditure.

It is tempting to regard the adaptive attitude as correct for the short term and the creative attitude as more appropriate for the long term. This may be because of the time lag required to overcome the prejudice in favour of the surprise-free future, to certificate new concepts and to modify institutional and infrastructural frameworks. Yet 'tomorrow never comes'. A continual application of the adaptive attitude will never allow an opportunity for large scale creative change. The later the decision to switch to a creative path, the stronger the case for greater utilisation of the sunk investments at the expense of change. There will be some undefined point in time which would be optimum for switching from conventional incremental change to a new path, just as there was between lighter than air to heavier than air, from flying boats to land planes, from piston engines to jets.

Either way, aviation holds a legitimate place in today's socio-economic environment. The balance is now struck uneasily between the users and those suffering from the environmental costs. Aviation has done its fair share of reducing its pollution, but still retains the image of being reluctant to meet its commitments to global 'good housekeeping' policies. Inevitably, it will be asked to do more to tip the scales away from the user and towards the environment.

Although the balance will, necessarily, be struck differently in different settings, the issues which will determine the appropriate capacity and the ability to provide it are:

- the value that society places on further aviation activity locally and nationally
- the extent to which the further activity is judged, to be politically sustainable
- the availability of suitable land
- the extent to which the environmental consequences of expanded capacity can be mitigated
- the power relationships among the stakeholders
- the availability of funding
- the judgement by those voting the funds that the rewards will exceed the risk
- the nature of the framework for granting permission for development
- predictions of new traffic and their robustness
- the rigour of the supporting analyses.

The aviation industry can help to balance the scales in a culture of full social cost recovery by:

- adopting non-confrontational tactics in agreeing policies for ameliorating environmental disbenefits

- carrying out in-depth evaluation studies over the full range of social and economic issues to compare its capabilities against a range of planning policy factors

- developing novel managerial and technological solutions to the problems which emerge from the evaluations

- educating policy makers and the public to the positive results of the evaluations and the effort going into beneficial technological change

- targeting those areas where it has strong natural advantages

- being prepared to withdraw from those areas which give rise to the highest social costs for the least system benefit.

An adaptive evolutionary approach involves 'doing more with less' in order to improve efficiency and capacity in the ways mentioned by Gillen (1993), but it does not stop there. The downstream implications of smart management, full social cost pricing and new information technology must be considered. Also, there must be initiatives to reduce the social costs and improve the social image of aviation. This does not mean simply minimising the score on various accepted indicators of impact. These are often only surrogates for other deeply seated social attitudes (Oster, 1993). It means taking a realistic approach to assessing aviation benefits (Gillen, 1993; Caves, 1994b) and taking a proactive and balanced stance in national level debates on the appropriate role for aviation in society. In general, it means identifying aviation's strengths and maintaining them, controlling the weaknesses, maximising opportunities and averting threats (Caves, 1994c).

The strong preference would be for a limited set of policies which jointly satisfy a majority of these objectives, rather than separate policies dedicated to each area which may be mutually contradictory. However, aviation can only progress if sufficient infrastructure capacity is made available, so it may be preferable to first address that area and then consider the implications of resulting policies in other areas.

Attempts are being made by the industry to improve the capacity situation on all fronts by a series of individual initiatives, mostly without too much attention to the broader aspects of capacity or the combined effects of all the initiatives, and treating non-capacity objectives only as constraints. Aircraft technology has been able to ease some of the environmental worries, both on communities around airports and in the wider debate about pollution and energy, partly through the inherently better performance of Stage 3 aircraft. Also, increasing aircraft size has been able to allow traffic to grow despite limitations on runway capacity, but this could be negated by the frequency competition from which the benefits of competition are supposed to flow. A new large transport aircraft of 600 to 800 seats may not give equivalent system growth that has been obtained from the gradual increase in average size within the existing fleet. While clearly increasing volume per unit, the concept is likely to have adverse implications in many other areas, for example, reducing the benefits which are expected to flow from closer approach spacing because of the wake vortex and noise implications. It will also have financial, environmental and safety implications for infrastructure operators. It is likely that, as in the past, a radical change in technology will be required to cope with the competition-driven need for frequency in an era of scarce physical and environmental capacity.

The overall impression of the industry's present attempts to adapt to the future is that it is more 'clutching at straws' than undertaking the recommended 'collective and cooperative search for new values'. It is being classically adaptive rather than creative, with the potential consequences outlined above.

A more systematic evaluation of options might improve the efficiency of the process of adaptation. If smart management with full social cost recovery (Gillen, 1993) could force a greater collective discipline on the industry, so that investment decisions could reflect all the cross impacts and full social costs, then it is possible, for example, that the overall cost could be minimised by reoptimising the design and operation of aircraft. It is certainly possible to construct a range of feasible shadow prices to reflect likely future charges and to include them in the evaluation of adaptive initiatives. This may indicate the advantages of investing in new technology rather than relying on the smarter management of existing technology. In particular, it may indicate a reoptimisation of the balance between the operating cost of the aircraft and other costs of operation. Instead of the traditional design approach of minimising direct operating cost (doc) subject to certain external constraints, the externalities may be internalised. This approach is illustrated by Morrison's synthesis of aerodynamics and production theory to trade off fuel price, capital cost and operating cost in aircraft design (Morrison, 1984). Benito (1986) indicates how the noise externalities, rather than simple certification constraints, may be incorporated in aircraft design. Due to the uncertainties involved, it would be essential to incorporate risk analysis, as described by Batson and Love (1988), in any minimisation of a direct social cost function.

Intuition suggests that, bearing in mind the massive investment in conventional technology, a minimum modification of that technology in the direction of RTOL could offer an attractive benefit-cost ratio. It has the advantages of catching an existing trend to shorter field performance and of operational demonstration. It has been shown in the UK case study that some 80% of all takeoffs to European destinations at Heathrow could require less than a 1,600 metre runway even with the 1992 fleet mix (Caves, 1994c), if it were in the best interests of airlines to avoid using reduced thrust takeoffs. Enhanced RTOL could bring other synergistic benefits in capacity, safety and noise through the exploitation of steeper approaches. The ability to improve several criteria with a given policy seems to be the key to the adoption of policies and hence to their implementation (May, 1991). This concept has been explored in the case of aviation by Caves (1995). It is entirely possible that a combination of improved RTOL with new approach aids could allow steeper approach and climb-out, so offering the type of noise improvements which are being proposed for 'Stage Three and a Half' regulations. Some of the increase in approach slope is possible even without any change in aircraft technology (Caves, Jenkinson and Rhodes, 1998). Perhaps more importantly, the noise of the larger aircraft could be reduced in this manner, so allowing them to meet single event noise limits. Only detailed research can verify these suggestions.

The development of fully justified policies, rather than the notional ones suggested here, and their successful adoption requires a completely new planning and decision framework in a turbulent, free-market setting. There should be fewer clearance points. The more difficult accommodations should be made early in the process by much more open involvement of all interested parties. There needs to be greater understanding of the decision rules, so that the risk is confined as much as possible to the normal commercial arena.

15.3 RECOMMENDATIONS

Existing models of passenger behaviour provide necessary tools for airport planners, but need much more development in the choice of variables and background information on the decision mechanisms of passengers and how they might change.

More effort is needed to develop models of profit-motivated airline decision processes and airline response to capacity constraints.

If it is going to be possible to answer the question: 'how much air transport is appropriate?', the debate over the costs and benefits of air transport needs to be raised to a level which would be more relevant for political decision-makers, concentrating on alleviation of environmental costs and the quantification of any identifiably unique economic benefits of investment in aviation.

Aircraft technology has the ability to respond innovatively to reduce disbenefits, but this ability will only be tapped in response to legislation or to airlines' specifications which will have to be driven by financial considerations. The perceived need for new technology will turn on the comparison between the costs of change and the costs of environmental protection. Models for the minimisation of system costs are needed, which include the infrastructure and social costs facing the industry in the future. These might well be developed by combining economic analysis with aircraft design optimisation methods.

Maximum advantage should be taken of the present impetus towards integrated transport systems to relieve runway and ground access congestion, but management of the released capacity must ensure that it is used to the benefit of all stakeholders rather than only the incumbents.

The industry must increasingly play its part in evolving attitudes of conciliation and policy solutions which are perceived by society as tending to be more 'green' and appropriate. The policies will be more acceptable if they have synergistic benefits across a range of policy goals.

The necessary understanding of the goals and the ability of air transport's policies to contribute to their attainment can only be obtained by the introduction of an improved method for creating and assessing policy alternatives, with adventurous exploration of alternative futures.

If airport planning is to result in successful implementation, the uncertainties and delays of the traditional planning process must be avoided. This might be achieved by participation in a 'shadow' planning forum which would allow early participation by all actors, mutually agreed goals, explicit recognition of power structures, bargaining within an information-rich framework and the evolution of win/win solutions which it would be difficult for political decision-makers to overturn. The output would be a rolling plan which would be continually reviewed in the forum.

The wide-ranging review of strategic planning in chapter 14 has demonstrated that some form of voluntary collusion together with concerted efforts to further understand the actors and their interrelations could complement the 'relaxation' achieved by liberalisation and privatisation to produce an appropriate path to a sustainable future aviation infrastructure. An open process of this sort would, in effect, internalise the evaluation. The ultimate value of such realistic and open evaluations of air transport would be that the mode's development would become effectively self-regulating, proceeding in a way which the political and planning processes would confirm. Striking the balance would, as it were, be settled out of court, including the difficult area of compensation for those whose benefits do not fully offset their disbenefits. The introduction of such an approach, supported by a rich information base, would, in most countries, require changes in the framework for planning as well as in the methodology, in order to encompass both local and national concerns.

The richness of the information to support this analysis depends on the continued development of the understanding of the characteristics of the actors and the linkages between them. The linkages exist in terms of substantive economic or physical effects and also in terms of power relationships. The latter should receive equal research attention if the iterative learning benefits of planning are to precede events.

The systemic approach to planning, with the need to agree on goals and objectives, establish criteria, and monitor achievement, provides an essential discipline. Factors which need examining in this systemic methodology are:

- the accessibility needs of more peripheral regions, and the ways in which the accessibility should be provided
- the atc capacity needed to encourage regional links and to allow competition: airspace redesign can require a long lead time if new equipment or facilities are required
- the effects of any internationally agreed carbon tax and modal competition on demand
- airline strategic responses to competition and to infrastructure limitations
- the value of aviation to regional, national and supranational economies
- the impact of aviation on the troposphere
- the establishment of societal goals for the environment, for accessibility, and for resource consumption, and the testing of air transport options against the goals.

Strenuous efforts should be made to use the best available methods of predicting, not only traffic and airline behaviour, but also the technological, cultural and political settings, and the links between the settings and the airport users' behaviour. Otherwise the planned system will always be inappropriate. Even the changes of transport policy are amenable to analysis via the understanding of power relationships between the force of circumstance and the establishment's status quo (Starkie, 1987). Given the uncertainties in all these factors, there is no substitute for the painting of rich scenarios and developing robust core strategies which will be able to cope with whichever scenario evolves.

Finally, planning is cheap compared not only to the ultimate project cost but also to the cost of wrong decisions or lengthy delays. The ongoing project/planning cost ratio could easily be of the order of 200:1, in which case the resources devoted to planning are a very good investment if the planning process improves the resulting system benefits by only 1%. An even greater cost might be avoided if participative strategic planning discovers that, at some future date, the marginal returns from the development of the present system reduces to zero, and is able to identify an alternative method of proceeding through the application of systems analysis to the longer term state. The greatest danger of the continual application of disjointed incrementalism is that it is necessarily short sighted and could easily direct the user into a blind alley.

This book has discussed the steps involved in adopting a strategic approach to airport system planning. In summary, they can be grouped into two broad aspects: **process** and **analysis**. The process establishes the institutional framework within which the planning takes place and provides the means by which the concerns of the many stakeholders are identified and addressed. The success of the process can be measured by getting to a decision and being able to implement it. Analysis informs the process, and serves to identify alternative ways in which the system could be developed and the likely consequences of selecting one over another. The success of the analysis takes longer to discover, and is measured by whether the actions finally chosen turn out to be viewed as good decisions or not.

REFERENCES

AACI (1992a). *Airports - Partners in Vital Economies*. Airports Association Council International, European Region, Brussels, November.

AACI (1992b). *Draft Economic Impact Study Kit*. Manchester Airport Plc for Airports Association Council International, European Region, 24 November.

AAE (1989). ICAO Calculates the Cost. *Avmark Aviation Economist*, September, 8-10.

AAE (1993). European 'third package' routes started/announced to date. *Avmark Aviation Economist*, March, 2.

Abbey, D. (1997). The new jet set. *Airline Business*, May, 40-45.

Abell, D. (1998). Leading business beyond the bottom line, in 'Mastering Global Business', Week 8 - Good Citizenship, March, *Financial Times*, London.

ACI (1997). *First ACI Airports Data Survey*, Airports Council International, Geneva.

ACI (1998). *ACI World Report No. 4*, Airports Council International, Geneva, April.

ACI (1998). *Airport economics survey and analysis -1997*. Airports Council International, Geneva, July.

ACI Europe (1993). *The economic impact study kit*. Airports Council International, European Region, Brussels, March.

ACI Europe (1995). *Environmental Handbook*. Airports Council International, European Region, Brussels, January.

ACI/IATA (1996). *Airport capacity/demand management*, 3rd ed. Airports Council International/International Air Transport Association, Geneva.

Acton, R. (1995). *Fleet Planning*. Proof of Evidence for British Airways Plc to Terminal 5 Heathrow Public Inquiry, April.

ADP (1997). *Aeroport de Paris souvenir edition of Airport Business Communiqué*. Airports Council International, Europe.

AEA (1987). *Capacity of aviation systems in Europe; scenarios on airport congestion*. Association of European Airlines, Brussels, November.

AEA (1995). *European Airports - getting to the hub of the problem OR the problem of getting to the hub*. Association of European Airlines, Brussels, April.

Airbus (1994). *The Future for European Airports*. Airbus Industrie, May.

Airport Forum (1981). *São Paulo's hat trick*. No. 6, 18-24.

Alamdari, F. E. and I. G. Black (1992) Passengers' choice of airline under competition: the use of the logit model. *Transport Reviews*, 12(2), 153-170.

Alexandre, A. (1995). The need to reconcile transport and the environment. *Global Transport*, Spring, 15-20.

Algers, S. (1993). An integrated structure of long distance travel behaviour models in Sweden. Paper 930654, *72nd Annual Meeting of the Transportation Research Board*, Washington DC, January.

Allport, R. J. and M. B. Brown (1993). The economic benefits of the European high speed rail network. Paper 930282, *72nd Annual Meeting of the Transportation Research Board*, Washington DC, January.

Alperovich, G. and Y. Machnes (1994). The role of wealth in the demand for international air travel. *Journal of Transport Economics and Policy*, May, 163-173.

Ambrose, M. (1990). The integration of spoke to spoke traffic in the overall traffic system. In: *Third European Aerospace Conference: Civil Aviation Operations - Problems, Solutions and Actions*. Royal Aeronautical Society.

Amin, A., D. R. Charles and J. Howells (1992). Corporate restructuring and cohesion in the new Europe. *Regional Studies*, 26(4), 319-331.

Amsterdam Airport (1995). *Annual Report*. Schiphol Airport, Amsterdam.

Amsterdam Airport (1997). *Environmental Report*. Schiphol Airport, Amsterdam.

Anderson, A. E. and D. F. Batten (1989). Creative nodes, logistical networks and the future of the metropolis. *Transportation*, 14, 281-293.

Anderson, I. and J. Rideout (1992). Making the Most of Change. *IATA Review*, No. 3, 15-18.

Angyal, A. (1941). *A logic of systems*. Harvard University Press. Quoted in: *Systems Thinking* (F. E. Emery, ed.), 1969, pp 17-29. Penguin Books Ltd.

Apostolakis, G. and D. C. Bell (1995). *Modelling the public as a stakeholder in environmental risk decision making*. Paper 0-7803-2559-1/95, Institution of Electrical and Electronic Engineers, London.

Arbuckle, G. (1997). *The viability of one large airport in central Scotland*. MSc thesis, Cranfield University.

Archer, L. J. (1993). *Aircraft emissions and the environment*. Oxford Institute for Energy Studies, Oxford.

Arthur, W. B. (1993). Pandora's Marketplace. In: *New Scientist* supplement: Complexity, 6 February, pp 6-8.

Ashford, N. (1997). Site selection. Lecture notes for short course *Airport Planning Procedures*, Dept. of Aeronautical and Automotive Engineering and Transport Studies, Loughborough University, February.

Ashford, N. and C. A. Moore (1992). *Airport Finance*. Van Nostrand Reinhold, New York.

Ashford, N. and P. Wright (1992). *Airport Engineering*, 3rd ed., Wiley, New York.

Ashley, D., P. Hanson and J. Veldhuis (1994). The competition model for Schiphol airport. In: *The 22nd European Transport Forum, Proceedings of Seminar B, Airport Planning Issues*, pp 37-47.

ASRC (1990). *America's airport capacity needs in the 21st century - a futuristic approach to traffic forecasts*. Aviation Systems Research Corp., Golden, Colorado, August.

ASRC (1991) *Meeting the airport capacity challenge: the continuous hub concept*. Aviation Systems Research Corp., Golden, Colorado, November.

ATAG (1993). *The economic benefits of air transport*. Air Transport Action Group, International Air Transport Association, Geneva.

Auditor General (1985). *Annual Report of the Auditor General to the (Canadian) House of Commons*, Ottawa.

Austin (1991). *An ordinance creating an Airport Advisory Board of the City of Austin, Texas; establishing the duties of the Airport Advisory Board; establishing procedures for the Airport Advisory Board; providing a severability clause; and providing an effective date*. Ordinance No. 841213-0, City of Austin, Texas.

BAA (1995). *UK airport capacity*. Minutes of evidence by BAA Plc to the House of Commons Transport Committee, 6 December, HMSO, London.

BAA (1998). *Annual Report 1997/98*. British Airports Authority, London.

Bailey, E. E. (1989). Comments and discussion of Morrison and Winston (1989), pp 113-115.

Bailey, E. E. and J. R. Williams (1988). Sources of economic rent in the deregulated airline industry. *Journal of Law and Economics*, **31**, 173-202.

Bailey, J. (1993). Land of the rising sums. *Flight*, 10 November, 34-35.

Balfour, J. (1994). The Battle of Orly: the legal dimension. *Journal of Air Transport Management*, **1**(3), 161-164

Ballantyne, T. (1997). Sydney syndrome. *Airline Business*, October, 58-61.

Banister, D., B. Anderssen, J. Berechman and S. Barrett (1993). Access to facilities in a competitive transport market. *Transportation Planning and Technology*, **17**, 341-348.

Barrett, M. (1991). *Pollution control strategies for aircraft*. World Wildlife Fund for Nature, Gland, Switzerland.

Barrett, S. D. (1991). Discussion in: *Transport in a free market economy* (D. Banister and K. Button, eds.) Macmillan, London.

Batson, R. G. and R. M. Love (1988). Risk analysis approach to transport aircraft technology assessment. *Journal of Aircraft*, February, **25**(2), 99-105.

Beliyiannis, A. et al (1990). *The potential for an extra runway at Heathrow - a preliminary feasibility study*. Report TT 9007, Dept. of Transport Technology, Loughborough University of Technology, April.

Belobaba, P. P. and J. van Acker (1994). Airline market concentration: an analysis of US origin - destination markets. *Journal of Air Transport Management*, **1**(1), 5-14.

Ben-Akiva, M., D. Boldne and M. Bradley (1993). Estimation of travel choice models with randomly distributed values of time. Paper No. 930966, *72nd Annual Meeting of the Transportation Research Board*, Washington DC, January.

Bendixson, T. (1991). *Europe One: meeting London's long term airport needs*. Submitted to UK Department of Transport by the Airports Policy Consortium, London.

Benito, A. (1986). Acoustical Design Economic Tradeoff for Transport Aircraft. *Journal of. Aircraft*, April, **23**(4), 313-320.

Bennett, E. (1989). Aeroplane noise - a constraint on airport capacity. *Airport Forum*, No. 4, 10-12.

Betts, P. (1993). Luton accuses BAA of predatory airport pricing. *Financial Times*, 5 June, 26.

Beyer, M. S. (1989). Fleet replacement and additions to meet capacity demands. Paper presented to conference: *How should we deal with the capacity crisis?*, Royal Aeronautical Society, London, February.

Biris, K. (1995). *Athens City Airport options*. MSc dissertation, Dept. of Aeronautical and Automotive Engineering and Transport Studies, Loughborough University, September.

Blow, C. (1996). *Airport terminals*, 2nd ed., Architectural Press, Oxford.

Boeing (1993). *Statistical summary of commercial jet aircraft accidents, Worldwide Operations, 1959-1992*. Boeing Commercial Airplane Group, Seattle, April.

Boeing (1998). *Current Market Outlook*. Boeing Commercial Airplane Group, Seattle.

Boer, E. de (1997). Environmental constraints at a major airport, the case of Amsterdam Schiphol airport. Paper presented to conference: *Airports and the Environment: how to communicate the facts*, Airports Council International, Europe, Brussels, 7-9 April.

Bonnafous, A. (1991). The regional impact of the TGV. *Transportation,* **14**, 127-137.

Borenstein, S. (1989). Hubs and high fares: dominancy and market power in the US airline industry. *RAND Journal of Economics*, **20**(3), Autumn, 344-365.

Borgo, A. and T. Bull-Larsen (1998). Losses: what losses? *Airline Business*, August, 54-59.

Bosson (1993). *French Ministry of Transport, Communiqué on Paris traffic distribution rules*, Paris, 14 October.

Bouw, P. (1995). KLM strategy. Paper presented to *International Airlines Conference*, UBS Global Research, London, 6 March.

Bowen, P. (1991). *The creation of an interline hub at Manchester airport*. MSc thesis, Dept. of Transport Technology, Loughborough University.

Bowers, C. J. (1979). Aeronautical charging policy at UK regional airports. Lecture notes for short course *Airport Finance and Economics*, Polytechnic of Central London, February.

Brander, J. R. G. and B. A. Cook (1986). Air transport deregulation and airport congestion: the search for efficient solutions. *Transportation Research Record 1094*, 18-23.

Branson, R. (1993). *Air transport, the people business*. GPA lecture at Royal Aeronautical Society, London, 2 March.

Breheny, M. J. (1974). Towards measures of spatial opportunity. *Progress in Planning*, **2**.

Breheny, M. J. (1986). The context of methods. Paper presented to Institute of British Geographers Quantitative Methods Study Group Workshop on *Quantitative Methods in Policy Analysis*, Bodington Hall, University of Leeds, 18-19 September.

Breheny, M. J. (1987). The urbanisation impacts of airport development. Lecture notes for short course in *Airport Planning Procedures*, Dept. of Transport Technology, Loughborough University, November.

Brenner, M. A. (1990). The myth of the 'fortress-hub'. *Avmark Aviation Economist*, October, 6-9.

Brewitt, N. (1990). Satellite facilities at Heathrow. In *New runway capacity in the southeast - solutions and impacts*. Royal Aeronautical Society, London, November, pp 4.1-4.3.

British Airways (1996). *Annual Environmental Report 1996, report of additional environmental data*. BA Report Number 9/96, London.

Britton, S. G. and C. C. Kissling (1984). Aviation and development constraints in South Pacific microstates. In: *Transport and Communication for Pacific Microstates* (C. C. Kissling, ed.). The Institute for Pacific Studies, University of the South Pacific, Suva.

Brogan, R. (1993). Public involvement strategies in airport planning. Paper 931139, *72nd Annual Meeting of the Transportation Research Board*, Washington DC, January.

Bromhead, P. (1973). *The great white elephant of Maplin Sands*. Paul Elek, London.

Brooke, A. S., R. E. Caves and D. E. Pitfield (1994). Methodology for predicting European short-haul air transport demand from regional airports. *Journal of Air Transport Management*, **1**(1), 37-46.

Brooks, H. (1971). Remarks on technological forecasting in the next decade. In: *Technical Forecasting for 1980* (E. Weber, G. K. Teal and A. G. Schillinger, eds.), Van Nostrand Reinhold, New York.

Burton, J. and P. Hanlon (1994). Airline alliances: cooperating to compete? *Journal of Air Transport Management*, **1**(4), 209-227.

Butler, S. E. and L. J. Kiernan (1986). *Measuring the regional economic significance of airports*. US Department of Transportation, Federal Aviation Administration, Washington DC, October.

Butler, S. E. and L. J. Kiernan (1988). Transport Benefits from Regional Airports. *Airports Technology International*, pp 38-40.

Button, K. (1993). *Transport Economics*. Edward Elgar Publishing, Aldershot, UK.

Button, K. and D. Swann (1989). European Community airlines - deregulation and its problems. *Journal of Common Market Studies*, **XXVII**(4), 259-282.

Button, K. J. (1988). Transport demands of Scotland's high-technology industries. *Transportation Research Record 1197*, 35-43.

Button, K. J. and P. J. Barker (1975). *Case studies in Cost-Benefit Analysis*. Heinemann Educational Books, London.

CAA (1985a). *London area runway capacity and passenger demand*. CAP 502, UK Civil Aviation Authority, London, January.

CAA (1985b). *Air traffic distribution in the London area - a consultation document*. CAP 510, UK Civil Aviation Authority, London, October.

CAA (1986). *Air traffic distribution in the London area - advice to the Secretary of State*. CAP 522, UK Civil Aviation Authority, London, May.

CAA (1987). *Competition on the main domestic trunk routes*. CAA Paper 87005. UK Civil Aviation Authority, London, March.

CAA (1988a). *Statement of policies on air transport licensing*. CAP 539, UK Civil Aviation Authority, London, June.

CAA (1988b). *Strategies for making good use of airspace 1989-1995*. CAP 546, UK Civil Aviation Authority, London, December.

CAA (1989). *Traffic distribution policy for the London area and the strategic options for the long term: a consultation document*. CAP 548, UK Civil Aviation Authority, London, January.

CAA (1990). *Traffic distribution policy and airport and airspace capacity: the next 15 years*. CAP 570, UK Civil Aviation Authority, London, July.

CAA (1991a). *The need for traffic distribution rules*. CAP 578, UK Civil Aviation Authority, London, January.

CAA (1991b). *Economic regulation of BAA south east airports*. CAP 599, UK Civil Aviation Authority, London, November.

CAA (1993a). *Airline competition in the single European market*. CAP 623, UK Civil Aviation Authority, London, November.

CAA (1993b). *Passengers at London airports in 1991*. CAP 610, UK Civil Aviation Authority, London, January.

CAA (1994). *The economic impact of new air services*. CAP 638, UK Civil Aviation Authority, London, November.

CAA (1995). *Slot allocation: a proposal for Europe's airports*. CAP 644, UK Civil Aviation Authority, London.

CAA (1995b). *The single European aviation market: progress so far*. CAP 654, UK Civil Aviation Authority, September, London

CAA (1998). *Passengers at Birmingham, Gatwick, Heathrow, London City, Luton, Manchester and Stansted Airports in 1996*. UK Civil Aviation Authority, London.

Cadoux, R. E. and J. B. Ollerhead (1996). *Review of the departure noise limits at Heathrow, Gatwick and Stansted airports: additional study of Boeing 747 departures*. CS Report 9539 Supplement, Research and Development Directorate, National Air Traffic Services Ltd., London, October.

CAEP (1989). On the need for increased stringency in the noise certification requirements for subsonic jet and propeller powered heavy aircraft. Working Paper from the International Coordinating Council of Aerospace Industries Associations given to the *4th meeting of Working Group 1 of the Committee on Aviation Environmental Protection*, International Civil Aviation Organisation, Montreal, 25 October.

Caltrans (1998a). *California Aviation System Plan: Background and introduction*. California Department of Transportation, Aeronautics Program, Sacramento, California, Discussion Draft, January.

Caltrans (1998b). *California Aviation System Plan: Capital Improvement Program*. California Department of Transportation, Aeronautics Program, Sacramento, California, January.

Cameron, D. (1997). Out of tune. *Airline Business*, June, 84-87.

Camillus, J. C. and D. K. Datta (1991). Managing strategic issues in a turbulent environment. *Long Range Planning*, **24**(2), 67-74.

Carré, I. (1988). Connection rates in Europe. *ITA Magazine*, No. 48, Special Section, 25-26.

Caves, R. E. (1980). Airport system planning - a UK study. *Transportation Planning and Technology*, **6**, 117-130.

Caves, R. E. (1986). Regional air travel demands - a review. In Royal Aeronautical Society Conference : *Developing European regional air transport - the next ten years*, London, April.

Caves, R. E. (1991). *Airport planning in a liberal environment*. Report TT 9106, Dept. of Transport Technology, Loughborough University, July.

Caves, R. E. (1993). *Airport planning in a liberal setting - methodologies for appropriate airport provision.* PhD thesis, Loughborough University.

Caves, R. E. (1994a). A search for more airport apron capacity. *Journal of Air Transport Management*, 1(2), 109-120.

Caves, R. E. (1994b). Aviation and Society - Redrawing the Balance Part I. *Transportation Planning and Technology*, 18(1), 1-19.

Caves, R. E. (1994c). Aviation and Society - Redrawing the Balance Part II. *Transportation Planning and Technology*, 18(1), 21-36.

Caves, R. E. (1995). Paths to an air transport future: myths and omens. *Futures*, 27(8), 857-868.

Caves, R. E. (1996). Control of risk near airports. *Built Environment*, 22(3), 223-233.

Caves, R. E. (1997). European airline networks and their implications for airport planning. *Transport Reviews*, 17(2), 121-144.

Caves, R. E. and A. Brooke (1993). *The implication of a long third runway at Heathrow.* Report TT 9302, Dept. of Transport Technology, Loughborough University, May.

Caves, R. E. and C. Higgins (1993). The consequences of the liberalised UK - Europe bilateral air service agreements. *International Journal of Transport Economics*, XX(1), February, 3-25.

Caves, R. E., D. Gillingwater and D. E. Pitfield (1985). *The potential for further air cargo development at East Midlands Airport.* Loughborough Consultants Ltd., Report TT/LU/78, Loughborough University, August.

Caves, R. E., L. R. Jenkinson and A. S. Brooke (1996). Determinants of emission from ground operations of aircraft. Paper presented to *20th Congress, International Council of the Aeronautical Sciences*, Sorrento, Italy, September.

Caves, R. E., L. R. Jenkinson and D. P. Rhodes (1998). Development of an integrated conceptual aircraft design and aircraft noise model for civil transport aircraft. Paper presented to *21st Congress, International Council of the Aeronautical Sciences*, Melbourne, Australia, September.

Chang, C. and P. Schonfeld (1995). Flight sequencing in airport hub operations. Paper 950665, *74th Annual Meeting of the Transportation Research Board*, Washington DC, January.

Chatterway, C. (1995). Longhaul freedoms. *Airline Business*, February, 48-51.

Checkland (1989). Soft systems methodology. In: *Rational analysis for a problematic world*, (J. Rosenhead, ed.), pp 71-100. Wiley, Chichester, England.

Cheshire County Council (1995). Statement of agreements and conditions. Document CH35 to the *Manchester Airport Second Runway Public Inquiry*, Manchester, England.

Chesterton, G. K. (1946). *The Napoleon of Notting Hill.* Penguin, London.

Chevallier, J.-M. (1992). An overview of European airport capacity. In: *Airports Association Council International European Region, Airport Capacity Conference*, Manchester, 15-17 June.

Chisholm, M. (1976) Regional policies in an era of slow population growth and higher unemployment. *Regional Studies*, 10, 201-215.

Chou, Y.-H. (1990). The hierarchical-hub model for airline networks. *Transportation Planning and Technology*, 14, 243-258.

Chou, Y.-H. (1992) A cluster analysis of disaggregate decision-making processes in travel mode choice behaviour. *Transportation Planning and Technology*, 16, 155-166.

Christaller, W. (1966). *Central places in southern Germany,* translated by C. W. Baskin, Prentice-Hall, Englewood Cliffs, New Jersey.

Chuter, A. (1995). Tunnel vision. *Flight*, 7 June, 94-96.

CIPFA (1995). *The UK airports industry airport statistics 1994/95.* Centre for the Study of Regulated Industries, Chartered Institute of Public Finance and Accountancy, London.

Clayton, E. (1997). A new approach to airport user charges. *Journal of Air Transport Management*, 3(2), 95-98.

Collins, A. and A. Evans (1994). Aircraft noise and residential property values. *Journal of Transport Economics and Policy*, May, 175-196.

Condom, P. (1993). Airline industry performance: past, present and future. In: *International Air Transport: the challenges ahead*, pp 21-44. Conference of Organisation for Economic Cooperation and Development (OECD), Paris.

Coogan, M. (1994). Technical, institutional and funding considerations. In: *Ground Access to Airports, Proceedings of Two Workshops Sponsored by the Federal Aviation Administration* (G. D. Gosling, ed.). Proceedings UCB-ITS-P-94-1, Institute of Transportation Studies, University of California, Berkeley, December.

Coogan, M. (1995). Comparing airport ground access: a transatlantic look at an intermodal issue. *TR News 174*, November-December, 2-10.

Cooke, A. K. (1995). *Investigating the economic viability of future high speed civil transports by means of technology-based costing and modal choice modelling.* PhD thesis, Cranfield University, October.

Cooper, R. (1990). Airports and economic development: an overview. *Transportation Research Record 1274*, 125-133.

Coopers & Lybrand (1995). Assessment of wider economic benefits: Summary report and appendices. British Airways report BA 15 to *Heathrow Airport Terminal 5 Inquiry*, London, April.

Coopers & Lybrand (1990). *Economic impact of the potential loss of intra-EC duty and tax free sales on the air transport industry.* Report to Airports Association Council International, Paris, April.

Craig, V. (1998). Aviation system planning. Lecture notes for short course *Airport Planning Procedures*, Dept. of Aeronautical and Automotive Engineering and Transport Studies, Loughborough University, February.

Crandall, R. L. (1988). Common ground: unlocking the potential of world aviation. *Aerospace*, August, 16-20.

Creedy, K. B. (1991). The cost of realistic bilaterals. *Interavia*, March, 19-21.

Cronshaw, M. and D. Thompson (1991). Competitive advantage in European aviation - or whatever happened to B Cal? *Fiscal Studies*, **12**, 44-66.

Crowley, R. W. (1973) A case study of the effects of an airport on land values. *Journal of Transport Economics and Policy*, **19**, 144-152.

D'Albiac, J. (1957). London Airport. *Journal of the Royal Aeronautical Society*, **61**, April, 225-237.

de Andres, J. A. (1980). Planning the Spanish airport system. Lecture notes for short course *Airport Planning Procedures*, Dept. of Transport Technology, Loughborough University of Technology, June.

de la Rochère, J. D. (1994). European Community policies on airline concentration. *Journal of Air Transport Management*, **1** (2), 103-108.

de Neufville, R. (1976). *Airport systems planning.* MIT Press, Cambridge, Massachusetts.

de Neufville, R. (1984). Planning for multiple airports in a metropolitan region. *Built Environment*, **10**(3), 159-167.

de Neufville, R. (1990). Successful siting of airports - Sydney example. *Journal of Transportation Engineering*, January, **116**(1), 37-49.

de Neufville, R. (1995). Management of multi-airport systems. *Journal of Air Transport Management*, **2**(2), 99-110.

de Neufville, R. and J. Barber (1991). Deregulation induced volatility of airport traffic. *Transportation Planning and Technology*, **16**, 117-128.

de Wit, J. G. (1995). An urge to merge? *Journal of Air Transport Management*, **2**(3/4), 173-180.

DeLuchi, M. A. (1993). Greenhouse gas emissions from the use of new fuels for transportation and electricity. *Transportation Research*, **27A**(3), 187-191.

Dempsey, P. S. (1990). Airline deregulation and laissez-faire mythology: economic theory in turbulence. *Journal of Air Law and Commerce*, Winter, 305-413.

Dennis, N. (1994). Airline hub operations in Europe. *Journal of Transport Geography*, **2**(4), 219-233.

Dennis, N. (1995). *Changes in airline operations and their impact on airports.* Transport Studies Group, University of Westminster, May.

Dennis, N. (1996). *Competition for connecting traffic between hub airports in Europe.* PTRC Summer Meeting, London, September.

DETR (1997). *Third party risk near airports and public safety zone policy.* (A report to the Department by consultants) UK Department of the Environment, Transport and the Regions, London, October.

Dillingham, G. L. (1996). *Issues related to the sale or lease of US commercial airports.* Report GAO/T-RCED-96-82, US General Accounting Office, Washington DC.

Dobbie, L. (1995). ICAO certification standards for aircraft engine emissions. Paper presented to conference: *Environmental Aspects of Air Transport*, Royal Aeronautical Society, London, September.

Dobson-Vida, M. (1993). A smelly problem at Heathrow. *Airliners*, Winter, 15-17.

DoE (1994). *Planning Policy Guidance 24*, UK Department of the Environment, HMSO, London, September.

Doganis, R. (1991). *Flying off course*, 2nd ed. Harper Collins Academic, London.

Doganis, R. (1992). *The Airport Business*. Routledge, London.

Doganis, R. and H. Nuutinen (1983). *Economics of European Airports*. Research Report No. 9, Transport Studies Group, Polytechnic of Central London, December.

Doganis, R., R. Pearson and G. Thompson (1978). *Airport economics in the seventies*. Research Report No. 5, Transport Studies Group, Polytechnic of Central London, September.

Donald, C. (1996). *European Airports*. Conference organised by UBS Global Research, London, February.

Donoghue, J. A. (1988). The big squeeze - fortress hubs and other barriers. *Air Transport World*, December, 58-65.

Donoghue, J. A. (1991). Sweet harmony. *Air Transport World*, September, 40-43.

Donoghue, J. A. (1997). Airbus: alone at the top. *Air Transport World*, April, 38-41.

Donoghue, J. A. and Douglas W. Nelms (1993). Power and persuasion in Washington. *Air Transport World*, May, 37-48.

DoT (1979). *Report of the Advisory Committee on Airports Policy*. UK Department of Trade, HMSO, London.

Douglas Aircraft Company (1996). *Outlook for commercial aircraft 1995-2014*. Long Beach, California, August.

Dresner, M. and M. W. Tretheway (1992). Modelling and testing the effect of market structure on price: the case of international air transport. *Journal of Transport Economics and Policy*, **XXVI**, 171-184.

Drew, D. R. (1990). Transportation and economic development. Conference Summary. *Transportation Research Record 1274*, 285-290.

DTp (1993a). Consultation paper: *Night flights at Heathrow, Gatwick and Stansted airports: proposals for revised restrictions from 24 October 1993*. UK Department of Transport, London, January.

DTp (1993b). *Runway Capacity to serve the South East; a report of the Working Group*. UK Departments of Transport and the Environment, London, July.

DTp (1997). *Air traffic forecasts for the UK*. UK Department of Transport, HMSO, London.

Dubbink, D., D. Cohney and G. Frommer (1995). Information about noise management: the Interactive Sound Information System (ISIS). In: *Seventh World Conference on Transportation Research*, Sydney, 16-24 July.

Duffy, P. (1996). A Steppe-up in costs. *Air Transport World*, March, 73-75.

Dunbar, J. K. P. (1990). Economic Impacts of Aviation on North Central Texas. *Transportation Research Record 1274*, 223-231.

Dunsire, A. (1980). Implementation theory. OU Paper 15, in *Social Science, Block 3*, Open University Press, Milton Keynes, UK.

EC (1983). Council Directive 83/46/EEC: OJ L237, European Commission, Brussels, August, p 19.

EC (1985). *Official Journal of the European Communities*. No. L 175/41, 5 July.

EC (1996). *Impact of the Third Package of air transport liberalisation measures*. European Commission, Brussels.

EC (1992a). *The Impact of Transport on the Environment - A Community Strategy for "Sustainable Mobility"*. European Commission Green Paper COM (92) 46 Final, Brussels, 20 February.

EC (1992b). Draft Report on the Green Paper on *The Impact of Transport on the Environment*. European Parliament, Committee on Transport and Tourism, DOC EN/PR/206687, Brussels, 12 May.

EC (1993). *Outline Plan of the Trans-European Airport Network*. European Commission Draft Communication VII-C4/Com 2/93, Brussels, 19 November.

EC (1994). *The way forward for civil aviation in Europe*. European Commission Communication, Brussels, 9 June.

EC (1996). *Impact of the Third Package of air transport liberalisation measures*. European Community, Brussels.

ECAC (1984). *Report on the economics of intra-European regional services with small aircraft*. European Civil Aviation Conference. ECAC/CEAC Doc. No. 26, Paris.

ECAC (1994). *ECAC Strategy for the 1990s: Airports Strategy*. Report to Meeting of European Civil Aviation Conference Transport Ministers, Copenhagen, 10 June.

ECAC/APATSI (1994). *Manual on mature air traffic control procedures*. European Civil Aviation Conference Airports Bureau, Paris, May.

ECAC/APATSI (1995). *Document on medium term air traffic control procedures and techniques.* European Civil Aviation Conference Airports Bureau, Paris, August.

ECAC/APATSI (1996). *Guidelines on monitoring and analysis of delays at airports.* European Civil Aviation Conference, Paris.

Eccles, G. and A. van der Werff (1995). American aviation strategies: implications of imposition within a European environment. The case of Southwest Airlines, USA. In: *Annual Meeting of UK University Teachers in Transport Studies*, Cranfield, 5 January, Vol. II, pp 1-12.

Eden, C. (1989). Using cognitive mapping for strategic options development and analysis. In: *Rational analysis for a problematic world*, (J. Rosenhead, ed.), pp 21-42. Wiley, Chichester, England.

Edgell, D. L. Sr. (1990). *International Tourism Policy.* Van Nostrand, New York.

Edmonton (1992). *Edmonton Regional Airports Authority: Annual Report*, Edmonton, Alberta.

Egan, J. L. (1992). *Planning Airports for the Future.* The Handley Page Lecture, Royal Aeronautical Society, London, 23 April.

Emery, F. E. and E. L. Trist (1965). The causal texture of organisational environments. *Human Relations*, **18**, 21-32.

Enders, G. (1996). First airports chosen for private sector. *Jane's Airport Review,* December, 11-16.

Enders, G. (1998). Less noisy neighbours. *Jane's Airport Review*, March, 16-17.

ERA (1998). *e.r.a. (Association of European Regional Airlines) regional report.* May, p 15.

Ernico, S. and T. Walsh (1997). Airport Finance. Lecture notes for short course *Airport System Planning and Design*, University of California, Berkeley, May.

Evans, R. (1989). The West grows old. *Geographical Magazine*, April, 10-14.

Everitt, R. (1995). Presentation to conference: *UK airports policy - developments since Rucatse.* Reported in *Aerospace*, January, 11-13.

Evers, G. H. M., P. H. van der Meer, J. Oosterhaven and J. B. Polak (1991). Regional impacts of new transport infrastructure: a multi-sectoral potential approach. *Transportation* **14**, 113-126.

Eyre, G. (1984). *The Airports Inquiries, 1981-1983.* UK Department of the Environment and Department of Transport, London.

FAA (1968). *Planning the state airport system.* Prepared by a Joint Committee of the Federal Aviation Administration and National Association of State Aviation Officials. US Department of Transportation, Federal Aviation Administration, Washington DC, December.

FAA (1970). *Planning the metropolitan airport system.* Advisory Circular 150/5070-5. US Department of Transportation, Federal Aviation Administration, Washington DC, 22 May.

FAA (1976). FAA airport programs: *Developing the national airport system*, 1 July 1975 – 30 September 1976. US Department of Transportation, Federal Aviation Administration, Washington DC.

FAA (1978). *Aircraft and the Environment.* Report GA-300-104, US Department of Transportation, Federal Aviation Administration, Northwest Region, Boeing Field, Seattle.

FAA (1985). *Airport Master Planning.* Advisory Circular 150/5070-6A, US Department of Transportation, Federal Aviation Administration, Washington DC.

FAA (1987). *National Plan of Integrated Airport Systems (NPIAS) 1986-1995.* US Department of Transportation, Federal Aviation Administration, Washington DC, November.

FAA (1989). *Planning the State Aviation System.* Advisory Circular 150/5050-3B, US Department of Transportation, Federal Aviation Administration, Washington DC.

FAA (1991). *Seattle-Tacoma International Airport - Airport Capacity Enhancement Plan.* US Department of Transportation, Federal Aviation Administration, Office of System Capacity, Washington DC, June.

FAA (1992). *Noise standards: Aircraft type and airworthiness certification*, Federal Aviation Regulations, Part 36. US Department of Transportation, Federal Aviation Administration, Washington DC, September.

FAA (1993). *Aviation System Capacity - Annual Report.* US Department of Transportation, Federal Aviation Administration, Washington DC, October.

FAA (1995). *National Plan of Integrated Airport Systems (NPIAS) 1993-1997.* US Department of Transportation, Federal Aviation Administration, Report to Congress, Washington DC, April.

FAA (1996). *Planning the metropolitan airport system.* Draft update, Advisory Circular 150/5070-5, US Department of Transportation, Federal Aviation Administration, National Planning Division, Washington DC, April.

FAA (1997a). *FAA airport benefit-cost analysis guidance*. Draft, US Department of Transportation, Federal Aviation Administration, Washington DC, 2 June

FAA (1997b). *Airport capital improvement planning: Stewardship for airport development*. US Department of Transportation, Federal Aviation Administration, Office of Airport Planning and Programming, Washington DC, September.

FAA (1997c). *Terminal area forecasts summary – Fiscal years 1997-2010*. Report No. FAA-APO-97-7, US Department of Transportation, Federal Aviation Administration, Office of Aviation Policy and Plans, Washington DC, October.

FAA (1997d). *1997 Airport Capacity Enhancement Plan*. US Department of Transportation, Federal Aviation Administration, Office of System Capacity, Washington DC, December.

FAA (1998a). *FAA aviation forecasts - Fiscal years 1998-2009*. Report No. FAA-APO-98-1, US Department of Transportation, Federal Aviation Administration, Office of Aviation Policy and Plans, Washington DC, March.

FAA (1998b). *Forecasts of IFR aircraft handled by FAA Air Route Traffic Control Centers - Fiscal years 1998-2009*. Report FAA-APO-98-6, US Department of Transportation, Federal Aviation Administration, Washington DC, June.

Farrington, J. H. (1984). A third London airport: options and decision making. *Built Environment*, **10**(3), 168-180.

Feitelson, E. I., R. E. Hurd and R. R. Mudge, (1996). The impact of airport noise on willingness to pay for residences. *Transportation Research*, **1D**(1), 1-14.

Feldman, D. (1996). Commercial magnetism. *Airline Business*, December, 36-39.

Feldman, J. M. (1992). No guts, no glory. *Air Transport World*, January, 64-67.

Feldman, J. (1995). Tasting reform in a mass market. *Air Transport World*, July, 24-32.

Feldman, J. (1996a). Some call it oligopoly. *Air Transport World*, May, 45-47.

Feldman, J. (1996b). Tweaking the establishment. *Air Transport World*, May, 65-68.

Feldman, J. (1997a). Airports for sale. *Air Transport World*, May, 69-72.

Feldman, J. (1997b). For 1997, the big fizz. *Air Transport World*, April, 26-32.

Feldman, J. (1998). From under the boot. *Air Transport World*, April, 63-65.

Fergusson, M. (1995). *Environment and Air Transport in Europe*. In conference: *Environmental Aspects of Air Transport*, Royal Aeronautical Society, London, 19 September.

Field, M. (1998). The role of the Civil Aviation Authority in the 21st century. *Global transport*, Summer, 6-10.

Fischer, M. M., P. Nijkamp and Y. Y. Papageorgiou, eds. (1990). *Spatial Choices and Processes*, Elsevier Science Publishers BV.

Fisher, R. and W. Ury (1986). *Getting to Yes: negotiating without giving in.* Business Books, London.

Flanagan, A. and M. Marcus (1993). The secrets of a successful liaison. *Avmark Aviation Economist*, January/February, 20-23.

Flight Transportation Associates, (1988). *Boston regional airport system study*. Prepared for the Massachusetts Port Authority, May (revised August 1989).

Flint, P. (1991). Roman Holiday. *Air Transport World*, September, 22-26.

Flint, P. (1996). If you can't beat 'em. *Air Transport World*, May, 36-42.

Flyvbjerg, B. (1984). Implementation and the choice of evaluation methods. *Transport Policy Decision Making*, **2**, 291-314.

Fotos, C. (1986). Airport automation to rise with passenger traffic. *Aviation Week*, 3 November, 140, 145 and 147.

Fowkes, T. and J. Preston (1991). Novel approaches to forecasting the demand for new local rail services. *Transportation Research* **25A**, 209-218.

French, T. (1995) A breath of fresh air. *Airline Business*, October, 34-37.

Friedmann, J. (1973). *Retracking America.* Anchor Press/Doubleday, New York.

Friend, J. (1989). The strategic choice approach. In: *Rational analysis for a problematic world* (J. Rosenhead, ed.), Wiley, Chichester, England.

Froggart, J. (1998). Planning implications of privatisation. Lecture notes for short course *Airport Planning Procedures*, Dept. of Aeronautical and Automotive Engineering and Transport Studies, Loughborough University, February.

Frommer-Ringer, R. (1995). Quoted in *Summary of the 5th Regional Airport Conference*, Airports Council International, European Region, Geneva, September.

Fujii, E., E. Im and J. Mak (1992). The economies of direct flights. *Journal of Transport Economics and Policy*, **XXVI**, Part 2, 185-195.

Fuller, R. B. (1983). *Critical Path*. Hutchinson, London.

Gand, H. (1991). Comparative environmental impacts of different modes. *Transportation*, **14**, 139-145.

GAO (1991). *Fares and concentrations at small-city airports*. Report GAO/RCED-91-51, US General Accounting Office, Washington DC, 18 January.

GAO (1996). *AIP Funding for the nation's largest airports*. Report GAO/RCED-96-219R, US General Accounting Office, Washington DC, 31 July.

GAO, (1998). *Airport financing: funding sources for airport development*. GAO/RCED-98-71, US General Accounting Office, Washington DC, March.

Garcia, J. M. de F. (1996). *A methodology for the comparative analysis of airport passenger terminal configuration*. PhD thesis, Dept. of Aeronautical and Automotive Engineering and Transport Studies, Loughborough University, June.

Garrison, W. L., D. Gillen and C. R. Williges (1997). *Impacts of changes to transportation infrastructure and services: How improvements in air transportation infrastructure and services enabled innovations in recreational activities, 1955-1995*. Research Report UCB-ITS-RR-97-4, Institute of Transportation Studies, University of California, Berkeley, July.

Gaustad, A. (1996). Environmental Impact Assessment - the Norwegian experience. Lecture notes for short course *Airport Planning Procedures*, Dept. of Aeronautical and Automotive Engineering and Transport Studies, Loughborough University, February.

Gellman, (1990). *Analysis of airport cost allocation and pricing options*. Gellman Research Associates, Jenkintown, Pennsylvania, 11 April.

Gennari, H. S. (1989). *A General System Planning Methodology (GSPM) applied to National Airport System Planning (NASP) in Middle Income and Economically Active Countries (MIEAC)*. PhD thesis, Dept. of Transport Technology, Loughborough University of Technology, May.

Gennari, H. S. (1996). The Brazilian Airport System. Lecture notes for short course *Airport Planning Procedures*, Dept. of Aeronautical and Automotive Engineering and Transport Studies, Loughborough University, February.

Gethin, S. (1998). Winning cargo business. *Jane's Airport Review*, March, 19-23.

Gialloreto, L. (1988). *Strategic Airline Management - the global war begins*. Pitman Publishing, London.

Gialloreto, L. (1992). Who will survive Euro-deregulation? *Avmark Aviation Economist*, January, 18-22.

Gifford, J. L. (1993). Aviation infrastructure and demand as complex interactive systems. In: *8th International Workshop on Future Aviation Activities*, Transportation Research Board, Washington DC, 13-15 September.

Gifford, J. L., T. A. Horan and L. G. White (1994). Dynamics of policy change: reflections on the 1991 federal transportation legislation. Paper 940814, *73rd Annual Meeting of the Transportation Research Board*, Washington DC, January.

Giles, W. D. (1991). Making strategy work. *Long Range Planning*, **24**(5), 75-91.

Gill, T. (1998). Stampede to market. *Airline Business*, April, 50-53.

Gill, T. (1998). Euro challenger. *Airline Business*, August, 48-51.

Gillen, D. W. and T. J. Levesque (1994). A socio-economic assessment of complaints about aircraft noise. *Transportation Planning and Technology*, **18**(1), 45-55.

Gillen, D. (1993). The Market and Aviation Infrastructure: pricing, productivity and privatisation. In: *8th International Workshop on Future Aviation Activities*, Appendix 3, Transportation Research Board, Washington DC, 13-15 September.

Gillen, D. (1997). Measuring efficiency, productivity and performance of airports: modern business practices for the modern airport. Paper presented to the *76th Annual Meeting of the Transportation Research Board*, Washington DC, January.

Gillen, D. and A. Lall (1997). Developing measures of airport productivity and performance: an application of data envelopment analysis. *Transportation Research*, **33E**(4), 261-273.

Gillen, D. W., T. H Oum and M. W. Tretheway (1985). *Canadian airline deregulation and privatisation: assessing effects and prospects*. Centre for Transportation Studies, University of British Columbia, Vancouver.

References

Gillen, D. W., T. H. Oum and M. W. Tretheway (1990). Airline cost structure and policy implications. *Journal of Transport Economics and Policy*, **XXIV**, 9-34.

Gillingwater, D. (1996). Airports and the environment - the context. Lecture notes for short course *Airports and the Environment*, Dept. of Aeronautical and Automotive Engineering and Transport Studies, Loughborough University, May.

Goddard, J. B. and A. E. Gillespie (1986). Advanced telecommunications and regional economic development. *Geographical Journal*, **152**(3), 383-397.

Godet, M. (1991). Actors' moves and strategies: the MACTOR method. *Futures*, July/August, 605-622.

Goldman, A. J. (1992). The ECAC view. In: *Airport Capacity Conference*, Airports Association Council International, European Region, Manchester, 15-17 June.

Goldstein, B. D. et al, (1992). Risk to groundlings of death due to airplane accidents: a risk communication tool. *Risk Analysis*, **12**(3), 339-341.

Gomez-Ibáñez, J. A. and D. H. Pickrell (1987). Towards an equilibrium in intercity travel choices. In: *Deregulation and the Future of Intercity Passenger Travel* (J. R. Meyer and C. V. Oster Jr., eds.), pp 183-203, MIT Press, Boston, Massachusetts.

Gosling, G. D. (1994). *Development of a recommended forecasting methodology for aviation system planning in California*. Research Report UCB-ITS-RR-94-7, Institute of Transportation Studies, University of California, Berkeley, July.

Gosling, G. D. (1997). Airport ground access and intermodal interface. *Transportation Research Record 1600*, Washington DC.

Gould, R. (1992). New de-icers meet safety criteria. *Jane's Airport Review*, April, 50-52.

Graham, A. (1992). Airport Economics. Lecture notes for short course *Airport Finance*, Dept. of Transport Technology, Loughborough University, November.

Graham, A. (1997). Airport performance indicators. Lecture notes for short course *Airport Finance*, Dept. of Aeronautical and Automotive Engineering and Transport Studies, Loughborough University, December.

Graham, B. (1995). *Geography and Air Transport*. Wiley, Chichester, England.

Graham, B. (1997). Regional airline services in the liberalised European Union single aviation market. *Journal of Air Transport Management*, **3**(4), 227-238.

Greenwood, D. (1998). The European airport network plan. Lecture notes for short course *Airport Planning Procedures*, Dept. of Aeronautical and Automotive Engineering and Transport Studies, Loughborough University, February.

Greff, E. (1996). Aerodynamic design and technology concepts for a new ultra-high capacity aircraft. Paper 96-4.6.3 to the *20th Congress, International Council of the Aeronautical Sciences*, Sorrento, Italy, September, pp 1321-1337.

Grigson, W. S. (1978). The forecasts - fact or fantasy. In seminar: *Airports: Time for decision*. Regional Studies Association, London, 9 January.

Group Transport 2000 Plus (1990). *Transport in a fast changing Europe*. London, December.

Guild, S. (1995). Not so easy. *Airline Business*, June, 68-73.

Gwilliam, K. M. and E. J. Judge (1974). Transport and regional development: some preliminary results on the M62 project. In: *Regional Studies Association, Conference on Transport and the Regions*, London.

Hägerstrand, T. (1987). Human interaction and spatial mobility: retrospect and prospect. In: *Transportation Planning in a Changing World* (P. Nijkamp and S. Reichman, eds.), Gower, Aldershot, England.

Haitovsky Y., I. Salomon and L. A. Silman (1987). The economic impact of charter flights on tourism to Israel. *Journal of Transport Economics and Policy*, May, 111-135.

Halaby, N. E. (1990). How was man meant to fly? *Exxon Air World*, No. 3, 10-13.

Hamsawi, S. G. (1992). Lack of airport capacity: exploration of alternative solutions. *Transportation Research*, **26A**(1), 47-58.

Hannegan, T. F. and F. P. Mulvey (1995). An analysis of codesharing's impact on airlines and consumers. *Journal of Air Transport Management*, **2**(2), 131-137.

Hansen, M. (1990). Airline competition in a hub-dominated environment: an application of noncooperative game theory. *Transportation Research*, **24B**(1), 27-43.

Hansen, M. and A. Kanafani (1989). Airline hubbing and airport economics in the Pacific market. Paper F11-6-2, *5th World Conference on Transport Research*, Yokohama, Japan, 10-14 July.

Hansen, M. and T. Weidner (1995). Multiple airport systems in the United States: current status and future prospects. Paper 950452, *74th Annual Meeting of the Transportation Research Board*, Washington DC, January.

Hardaway, R. M. (1986). The FAA 'buy-sell' slot rule: airline deregulation at the crossroads. *Journal of Air Law and Commerce*, **52**, 1-75.

Harding, D. (1994). *Runway slot allocation: imperfect systems in an imperfect world.* BSc dissertation, Dept. of Aeronautical and Automotive Engineering and Transport Studies, Loughborough University, April.

Hardison, M. F., R. R. Mudge and D. Lewis (1990). Using risk assessment for aviation demand and economic impact forecasting in the Minneapolis - St. Paul region. *Transportation Research Record 1274*, 97-103.

Harrison, T. and R. Williams (1991). Airport planning in a regional and national policy vacuum. In: *Town and Country Planning Association North-West Forum*, Manchester, 30 October.

Hartsfield (1987). *Economic Impact Report*. Hartsfield Atlanta International Airport, Georgia.

Heidner, S. J. (1992). Trends at United States international gateway airports to Europe. *Transportation Research Record 1332*, 30-39.

Helm, D. and D. Thompson (1991). Privatised transport infrastructure and incentives to invest. *Journal of Transport Economics and Policy*, **XXV**, 231-246.

Hensher, D. A. and L. W. Johnson (1981). *Applied Discrete-Choice Modelling*. Croom Helm, London.

Herbert, C. P. and B. R. Custance (1994). Parallel 16L/34R into Botany Bay for Kingsford Smith. *Airports Technology International*, pp 47-51, Sterling Publications, London.

Heymann, H. (1962). The role of transportation in economic development. *American Economic Review*, **52**(2), May.

Hickling, J. F. (1987). *Lester B. Pearson International Airport Economic Impact Study.* James F. Hickling Management Consultants for Transport Canada.

Hill, M. (1968). A goals-achievement matrix for evaluating alternative plans. *Journal of the American Institute of Planners*, **34**(1), 19-28.

Hill, M. (1986). Decision-making contexts and strategies for evaluation. In: *Evaluation of Complex Policy Problems* (A. Faludi and H. Voogd, eds.). Delfsche Uitgevers Maatschappij, Delft.

Hill, M. and C. Lomovasky (1980). The minimum requirements approach to plan evaluation in participatory planning. Paper presented to *First World Regional Science Congress*, Cambridge, Massachusetts, June.

Hillestad, R. et al (1993). *Airport growth and safety - a study of the external risks of Schiphol Airport and possible safety-enhancement measures*. EAC Rand, Santa Monica, California.

Hillsman, E. L. and F. Southworth (1990). Factors that may influence responses of the US transportation sector to policies for reducing greenhouse gas emissions. *Transportation Research Record 1267*, 1-11.

Hirschmann, A. O. and C. E. Lindblom (1962). Economic development, research and development, policy making: some converging views. *Behavioural Science*, **7**, 211-22.

HMSO (1953). *London's Airports*. White Paper, Cmd. 8902, Her Majesty's Stationery Office, London, July.

HMSO (1967). *The Third London Airport*. Cmnd. 3259, Her Majesty's Stationery Office, London, May.

HMSO (1985). *Airports Policy*. White Paper, Cmnd. 9542, Her Majesty's Stationery Office, London, June.

Hogan, P. J. (1990). Why the airports want a 2000 ban. *Avmark Aviation Economist*, February/March, 11-14.

Hong, S. and P. T. Harker (1992). Air traffic network equilibrium: toward frequency, price and slot priority analysis. *Transportation Research*, **26B**(4), 307-323.

Hooper, P. G., and D. A. Hensher (1997). Measuring total factor productivity of airports - an index number approach. *Transportation Research*, **33E**(4), 249-259.

Hoover, J. H. and A. A. Altshuler (1977). *Involving citizens in metropolitan region transportation planning.* Report FHWA/SES-77/11, US Department of Transportation, Federal Highway Administration, Washington DC, June.

Horonjeff, R. and F. X. McKelvey (1994). *Planning and Design of Airports*, 4th ed. McGraw-Hill, New York.

Horowitz, J. L. (1991). Reconsidering the multinomial probit model. *Transportation Research*, **25B**(6), 433-438.

Humphreys, B. K. (1994). Viewpoint: Airline competition in the single European market. *Journal of Air Transport Management*, **1**(2), 121-122.

IATA (1992). *The economic benefits of air transport.* Produced for the Air Transport Action Group by International Air Transport Association Centre, Geneva.

ICAO (1977). *Studies to determine the contribution that civil aviation can make to the development of the national economies of African states.* UNDP/ICAO Project RAF/74/021, International Civil Aviation Organisation, Montreal, March.

ICAO (1985a). *Airport Planning Manual, Part 2, Land Use and Environmental Control*, 2nd ed., International Civil Aviation Organisation, Montreal.

ICAO (1985b). *Manual on air traffic forecasting*, 2nd ed. Doc. 8991 - AT/722/2, International Civil Aviation Organisation, Montreal.

ICAO (1987). *Airport Planning Manual, Part 1, Master Planning*, 2nd ed. International Civil Aviation Organisation, Montreal.

ICAO (1988). Environmental Protection. Annex 16 to the *Convention on International Civil Aviation*, Vol. 1, Aircraft Noise, 2nd ed. International Civil Aviation Organisation, Montreal.

ICAO (1991). *Airport Economics Manual*, 1st ed. Doc 9562, International Civil Aviation Organisation, Montreal.

ICAO (1992). *Investment requirements for aircraft fleets and for airport and route facility infrastructure to the year 2010.* Circular 236-AT/95, International Civil Aviation Organisation, Montreal.

ICAO (1993). Aircraft engine emissions. Annex 16 to the *Convention on International Civil Aviation*, Vol. 2, 2nd ed. International Civil Aviation Organisation, Montreal, July.

ICF Kaiser (1996). *The Central California Aviation System Plan: Interim forecasts.* ICF Kaiser Engineers Group, Prepared for the California Department of Transportation, Aeronautics Program, Sacramento, California, October.

IWG (1990). *CTOL transport aircraft characteristic trends and growth projections.* Industry Working Group, International Air Transport Association, Brussels, January.

Jenks, C. (1992). The correct US analogy. *Avmark Aviation Economist*, February/March, 6-7.

Johnson, K. J. (1989). Towards mediation: an examination of consensus - building techniques applied to the aircraft noise and airport access dilemma. *Transportation Research Record 1218*, 20-30.

Johnson, T. (1993). A study of environmental standards and controls at airports and airfields worldwide. Companion paper to conference: *Aviation, Environmental Regulation and the Future: A Worldwide Perspective*, Airfield Environment Federation, London, 3 June.

Jones, L. (1996). Off to a head start. *Airline Business*, September, 60-66.

Jones, L. (1998). Cheap thrills with no frills. *Airline Business*, February, 28-31.

Jones, S. R. (1981) *Accessibility Measures - a Literature Review.* Report 967. UK Transport and Road Research Laboratory, Crowthorne, England.

Jud, E. H. (1994). Switzerland: rail access to airports and the fly-luggage system. In: *Aviation Crossroads: Challenges in a Changing World* (W. J. Sproule, ed.), Proceedings of the 23rd International Air Transportation Conference, Arlington, Virginia, 22-24 June, American Society of Civil Engineers, New York.

Kaemmerle, K. C. (1991). Estimating the demand for small community air service. *Transportation Research*, **25A**(2/3), 101-112.

Kageson, P. (1994). Effects of internalisation on transport demand and modal split. In: *Internalising the social costs of transport*, Conference of European Ministers of Transport, Office for Economic Cooperation and Development, Paris, pp 77-94.

Kahneman, D. and A. Tversky (1979). Intuitive prediction. In: *Studies in Management Science*, Vol. 12, Forecasting, (S. Makridakis and S. C. Wheelwright, eds.), pp 313-327, North Holland.

Kanafani, A. and A. A. Ghobrial (1985). Airline hubbing - some implications for airport economics. *Transportation Research*, **19A**(1), 15-27.

Kanafani, A. and M. S. Abbas (1987). Local air service and economic impact of small airports. *Journal of Transportation Engineering*, **113**(1), January, 42-55.

Kärrman, L. (1991). Göteborg leading the way in environmental management. *Airport Forum*, No. 3, 46.

Karyd, A. and H. Brobeck (1992). The delusion of social benefits. *Avmark Aviation Economist*, January, 16-17.

Kato, H., (1992). Osaka Kansai International Airport. In: *Transportation Research Circular*, No. 393, April, 30-32.

Katz, R. (1991). The new activists. *Airline Business*, December, 42-44.

Keeble, D., P. L. Owens and C. Thompson (1982). Regional accessibility and economic potential in the European Community. *Regional Studies*, **16**(6), 419-432.

Kefer, H. (1992). New York airports update. *Air Transport World*, May, 80-83.

Keith-Lucas, D. (1974). Airport site selection - Third London Airport - The process of decision. Lecture notes for short course *Airport Planning and Design*, Centre for Transport Studies, Cranfield Institute of Technology, June.

Keller, W. (1998). Presentation to *4th National Aviation System Planning Symposium*, Transportation Research Board, Houston, Texas, March.

Khan, A. M. (1987). Socio-technical factors in air travel - telecommunications interactions: some new insights. Paper presented to the *Annual Meeting of the Transportation Research Board*, January, Washington DC.

Khisty, C. J. (1993). Citizen participation using a soft systems perspective. Paper 930097, *72nd Annual Meeting of the Transportation Research Board*, Washington DC, January.

King, M. (1990). The airport in its surroundings: the planning challenge. Presentation to conference: *Civil Aviation Operations - problems, solutions and actions*, Royal Aeronautical Society, London, 22-24 May.

Kitamura, R. (1989). Formulation of trip generation models using panel data. *Transportation Research Record 1203*, 60-68.

Knibb, D. (1991). Japan's new gateways. *Airline Business*, June, 50-53.

Koppert, B. J. (1992). Keeping the eco-balance right. 1st working session, *2nd Airports Association Council International -Europe Regional Airports Conference*, Faro, 28-30 September.

KPMG Peat Marwick (1990). *Flight Plan Study, Puget Sound region - Phase 1 forecasts*. Final Report, Prepared for the Port of Seattle and Puget Sound Council of Governments, Seattle, Washington, July.

Kroes, E. P., M. A. Bradley and J. Veldhuis (1994). Integrated analysis of air travel demand in Europe. In: *22nd European Transport Forum, Airport Planning Issues, Proceedings of Seminar B*, organised by PTRC, pp 109-119.

Kuijt, P. (1997). Concept for integrating environment in the business strategy. Lecture notes for short course *Airports and the Environment*, Dept. of Aeronautical and Automotive Engineering and Transport Studies, Loughborough University, May.

Lacombe, A. (1994). Ground access to airports: funding and implementation issues. In: *Ground Access to Airports, Proceedings of Two Workshops Sponsored by the Federal Aviation Administration* (G. D. Gosling, ed.). Proceedings UCB-ITS-P-94-1, Institute of Transportation Studies, University of California, Berkeley, December.

LAE (1984). Competition boosts traffic. *Lloyds Aviation Economist*, October, 13.

Larrabee, K. (1970). *Highway project planning with local citizens - a partnership concept*. US Department of Transportation, Bureau of Public Roads, Environmental Development Division, Washington DC.

Le, C. D. and M. L. Shaw (1981). The Canadian Airport System. *Airport Forum*, No. 1, 57-65.

Lefer, H. (1992). New York airports update. *Air Transport World*, May, 80-83.

Lenon, B. (1992). The demographic transition model. *Geographic Magazine*, May, 51.

Levesque, T. J. (1994). A socio-economic assessment of complaints about airport noise. *Transportation Planning and Technology*, **18**, 45-55.

Levine, M. (1990). How can the nonliberalised survive in a liberalizing air transport world? *ITA Magazine*, No. 60, March/April, 3-8.

Levinson, D. M. D., D. Gillen and A. Kanafani (1998). The social costs of intercity transportation: a review and comparison of air and highway. *Transport Reviews*, **18**(3), 215-240.

Lewin, R. (1993). Order for free. Supplement on Complexity, *New Scientist*, 13 February, 10-11.

Litchfield, N., P. Kettle, and M.Whitbread (1975). *Evaluation in the planning process*. Pergamon, Oxford, England.

Lobbenberg, A. and A. Graham, (1995). The strong performers. *Airline Business*, September, 102-107.

Logan, M. (1990). Environmental protection and airport capacity. Paper presented to the *International Air Transport Association's Infrastructure Action Group*, London, 17 October.

Longhurst, J. and D. Raper (1990). The impact of aircraft and vehicle emissions on airport air quality. *Airports Association Council International Europe Seminar*, Brussels, 5-7 February.

Lopez, R. and J. R. Wilson (1991). Tokyo is suddenly just one stop away. *Jane's Airport Review*, May, 20-23.

Lovett, T. (1995). Keynote address to the conference: *Environmental aspects of air transport*, Royal Aeronautical Society, London, October.

MAC/FAA (1995). *Dual Track Airport Planning Process - Draft Environmental Impact Statement*. Metropolitan Airports Commission/Federal Aviation Administration, Minneapolis, Minnesota, December.

MAC/FAA (1998). *Dual Track Airport Planning Process - Final Environmental Impact Statement and Section 4(f) Evaluation.* Metropolitan Airports Commission/Federal Aviation Administration, Minneapolis, Minnesota, May.

MAC/Metropolitan Council (1992). *The 1990 Dual-track forecast process: long-term aviation activity forecasts for the Twin Cities Region.* Metropolitan Council Publication No. 559-92-107, Metropolitan Airports Commission, St Paul, Minnesota, October.

Mackenzie-Williams, P. (1997). Time to measure up. *Airline Business*, September, 98-100.

Madsen, A. (1981). *Private powers.* Abacus.

Mahmassani, H. S. and G. S. Toft (1985). Transportation requirements for high technology industrial development. *Journal of Transportation Engineering*, **111**(5), 473-484.

Makridakis, S. (1995). The forthcoming information revolution. *Futures*, **27**(8), 799-821.

Makridakis, S. and S. C. Wheelwright (1979). Forecasting the future and the future of forecasting. *TIMS Studies in the Management Sciences*, **12**, 320-352.

Manheim, M. (1979). Toward more programmatic planning. In: *Public Transportation: Planning, Operations, and Management* (G. E. Gray and L. A. Hoel, eds.), Prentice-Hall, New Jersey.

Mansfield, Y. (1992). Tourism: towards a behavioural approach. *Progress in Planning*, **38**, Part 1.

Marchi, R. (1998). System Planning. Paper presented to *4th National Airport System Planning Symposium*, Transportation Research Board, Houston, Texas, March.

Marin, P. L. (1995). Competition in European aviation: pricing policy and market structure. *Journal of Industrial Economics*, **XLIII**(2), 141-159.

Masefield, P. (1989). Airports - the 21st century problem. *Airport Forum*, No. 5, 27.

Masefield, P. (1986). Remarks as chairman summing up conference: *European Regional Air Transport - the next ten years.* Royal Aeronautical Society, London, April.

Masser, I. (1987). Policy research for transport planning: a synthesis. In: *Transportation Planning in a Changing World* (P. Nijkamp and S. Reichman, eds.), Gower, Aldershot, England.

Matthews, M. A. (1993). The Vancouver parallel runway project. In: *Aviation, Environmental Regulation and the Future: a worldwide perspective.* Airfield Environmental Federation, London, 3 June.

Mawhinney, B. (1995). Paralysis by analysis? *Global Transport*, Spring, 36-39.

May, A. D. (1991). Integrated transport strategies: a new approach to urban transport policy formulations in the UK. *Transport Reviews*, **11**(3), 223-247.

Maynard, R. (1995). Airline industry and airport policy context (including no Terminal 5). Proof of Evidence for British Airways Plc to *Terminal 5 Heathrow Public Inquiry*, London, April.

McKechin, W. J. (1967). *A tale of two airports.* Paisley Fabian Society, Glasgow, October.

McMullan, K. (1993). Financial distress and strategic disarray. *Avmark Aviation Economist*, June/July, 13-14.

McNally, M. G. and Z.-P. Lo (1993). Prediction of discrete choice via neural networks. Paper 931066, *72nd Annual Meeting of the Transportation Research Board*, Washington DC, January.

McShan, S. and R. Windle (1989). The implications of hub and spoke routing for airline costs and competitiveness. *Logistics and Transport Review*, **25**(3), 209-230.

Meredith, J. (1991). The cost of airport congestion. In: *9th World Airports Conference: Airports and Automation*, Institute of Civil Engineers, London, 10-12 September.

Merton, T. (1987). *The ascent to truth.* Burns and Oates, Tunbridge Wells, p 18.

Metropolitan Council (1988a). *Is the airport adequate? - Report of the Minneapolis/St Paul International Airport Adequacy Study Advisory Task Force to the Metropolitan Council*, Part I: Finding and recommendations, Publication No. 559-88-101A, Metropolitan Council of the Twin Cities Area St Paul, Minnesota, October.

Metropolitan Council (1988b). *Is the airport adequate? - Report of the Minneapolis/St Paul International Airport Adequacy Study Advisory Task Force to the Metropolitan Council*, Part II: Study issues and analysis, Publication No. 559-88-101B, Metropolitan Council of the Twin Cities Area St Paul, Minnesota, October.

Metropolitan Council (1988c). *Twin cities air travel: A strategy for growth.* A report to the community, Publication No. 559-88-102, St Paul, Minnesota, October.

Metropolitan Council (1988d). *Twin cities air travel: A strategy for growth.* A report to the Minnesota Legislature, Publication No. 559-88-125, St Paul, Minnesota, December.

Michie, D. A. (1986). Family travel behaviour and its implications for tourism management. *Tourism Management*, March, 8-20.

Miles, I., S. Cole and J. Gershuny, (1978). Images of the future. In: *World Futures - the Great Debate* (C. Freeman and M. Jahoda, eds.), pp 279-342, Martin Robertson, Oxford, England.

Mollett, A. (1997). Domestic bliss? *Flight*, 5 November, 26-27.

Momberger, M. (1993). Japan: the region's dominant air travel market. *Airport Forum*, February, 30-34.

Moonen, W. and F. Schaper (1994). Quoted in 'Which S-curve are you on?', *Avmark Aviation Economist*, January/February, 7-8.

Moorman, R. W. (1993). The agonies of Air Wisconsin. *Air Transport World*, March, 56-59.

Moorman, R. W. (1996). Fjording ahead. *Air Transport World*, February, 93-95.

Morris, J. M., P. L. Dumble, and M. R. Wigan (1978). Accessibility indicators for transport planning. *Transportation Research*, **13A**, 91-109.

Morris, O. J. (1997). *Regional airport marketing and route development*. MSc thesis, Cranfield University.

Morrison, S. A. (1984). An economic analysis of aircraft design. *Journal of Transport Economics and Policy*, **18**, 123-143.

Morrison, S. A. and C. Winston (1989). Enhancing the performance of the deregulated air transportation system. *Brookings Papers on Economic Activity: Microeconomics*, Washington DC, pp 61-123.

Muirhead, G. (1995). Equal rights. *Global Transport*, Summer, 48-50.

NCC (1986). *Air transport and the consumer: a need for change?* National Consumer Council, HMSO, London.

NASPS (1992). *National Aviation System Planning Symposium*, Minnesota Department of Transportation, Minneapolis, 11-13 July 1991.

NATS/IATA/BAA (1994). *Report of the Heathrow Airport runway capacity enhancement study*. National Air Traffic Services, International Air Transport Association and BAA Plc, London, August.

Ndoh, N. N. and R. E. Caves, (1995). Investigating the impact of air service supply on local demand - a causal analysis. *Environment and Planning A*, **27**, 489-503.

Ndoh, N. N., D. E. Pitfield and R. E. Caves, (1990). Air transportation passenger route choice: a nested multinomial logit analysis. In: *Spatial Choices and Processes* (M. M. Fischer, P. Nijkamp and Y. Y. Papageorgiou, eds.), pp 349-365, Elsevier Science Publishers BV, Amsterdam.

Nelms, D. W. (1997). Hartsfield's Olympian effort. *Air Transport World*, April, 49-56.

Nelson, J. (1980). Airports and property values - a survey of recent evidence. *Journal of Transport Economics and Policy*, **14**, 37-52.

Newcastle Airport (1997). *Economic impact report*. Newcastle International Airport, Newcastle-on-Tyne, England.

Nijkamp, P. and I. Salomon (1989). Future spatial impacts of telecommunications. *Transportation Planning and Technology*, **13**(4), 275-287.

Nijkamp, P. and S. Reichman (1987). Transportation planning in a social context. In: *Transportation Planning in a Changing World*, Gower, Aldershot, England.

Nikolajew, V. (1991). The new technological paradigm of intelligence-based production. *Futures*, October, 828-848.

Nittinger, K. (1997). What Europe's airlines need to compete. *Interavia*, January/February, 34-36.

Norris, P. (1991). Implementing the London City STOLport. Lecture notes for short course *Airport Planning*, Dept. of Transport Technology, Loughborough University, November.

Norris, B. B. and R. Golaszewski (1990). Economic development impact of airports: a cross-sectional analysis of consumer surplus. *Transportation Research Record 1274*, 82-88.

Nuutinen, H. (1992). Negotiating with one voice? *Avmark Aviation Economist*, October, 6-7.

Nyaga, R. S. (1989). The economic impact of air transport to the community. *ICAO Bulletin*, January, 23-25.

O'Toole, K. (1997). European lead. *Flight*, 30 April, 34.

Odell, M. (1995). None the wiser. *Airline Business*, October, 28-33.

Odell, M. (1996). Mirror images. *Airline Business*, June, 57-60.

OECD (1977). *The future of European passenger transport*. Organisation for Economic Cooperation and Development, Paris.

Ohmae, K. (1995). *The end of the Nation State*. Harper/Collins, London.

Ohta, K. (1989). The development of Japanese transportation policies in the context of regional development. *Transportation Research,* **23A**(1), 91-101.

Oki, T., K. Kawate and A. Toyama, (1997). Asia's economic development and the role of transport, telecommunications and tourism. *Transport Reviews*, **17**(4), 311-353.

Oldham, C. C. (1998). The California Aviation Capital Improvement Program Process. Presentation to: *4th National Aviation System Planning Symposium*, 25-27 March, Houston, Texas.

Ollerhead, J. B. (1995). Assessing the impact of aircraft noise upon the community. Paper presented to conference: *Environmental Aspects of Air Transport*, Royal Aeronautical Society, London, September.

Open University (1983). *Course T 241, System Behaviour Module 1*. The Open University Press, Milton Keynes, England.

Orlick, S. C. (1978). Airport/community environmental planning. *Journal of Transportation Engineering*, **104**, No. TE2, March, 187-199.

Oster, C. V. Jr. (1993). Airline industry outlook and infrastructure needs. In: *8th International Workshop on Future Aviation Activities*, Transportation Research Board, Washington DC, 13-15 September, pp 76-81.

Oum, T. H., A. Zhang and Y. Zhang (1996). A note on optimal airport pricing in a hub and spoke system. *Transportation Research*, **30B**(1), 11-18.

Oum, T. H., J.-H. Park and A. Zhang (1995). *The effects of airline codesharing agreements on international air fares*. Faculty of Commerce, University of British Columbia, February.

Oum, T.H. and Y. Zhang (1990). Airport pricing - congestion tolls, lumpy investment, and cost recovery. *Journal of Public Investment*, **43**, 353-374.

Owen, F. M. W. (1991). Competing for the global tele-communications market. *Long Range Planning*, **24**(1), 52-56.

Ozires, da Silva (1976). Suplemento especial - Embraer. *Jornal do Brasil*, 3 December.

Paelinck, J. (1977). Qualitative multicriteria analysis: an application to airport location. *Environment and Planning A*, **9**, 883-895.

Park, K.-C., Y.-K. Pang and K.-S. Lee (1994). *An analysis of partner selection criteria in air carrier alliances*. pp 1-16. Korean Air Transport Research Institute, Seoul.

Parry, H. (1996). Europe's cost crisis. *Airline Business*, May, 62-64.

Patten, J. and C. Pollitt (1983). Power and rationality: theories of policy formulation. Paper 8, in *Course D336, Policies, People and Administration*, Open University Press, Milton Keynes, England.

Pavaux, J. (1990). Air transport in Europe. *ITA Magazine*, No. 61, May/June, 3-9.

Paylor, A. (1991). The EEC looks beyond chapter 3 controls. *Jane's Airport Review*, July/August, 16-17.

Pearson, A. (1997). North Atlantic: shifting sands. *Airline Business*, August, 46-51.

Pearsons, K. S., D. S. Barber, B. G. Tabachnick and S. Fidell (1995). Predicting noise-induced sleep disturbance. *Journal of the Acoustical Society of America*, **97**(1), January, 331-338.

Pedoe, N. T., D. W. Raper and J. M. W. Holden, eds. (1996). *Environmental management at airports - liabilities and social responsibilities*. Thomas Telford, London.

Pelkmans, J. (1989). Discussions to the internal EC market for air transport: issues after 1992. In: *Transport in a free market economy* (D. Banister and K. Button, eds.), Macmillan, London.

Pennington, G., N. Topham and R. Ward (1990). Aircraft noise and residential property values. *Journal of Transport Economics and Policy*, **24**(1), 45-59.

Pereira, A. J., J. P. Braaksma and J. J. Phelan, (1995). Interpreting airport noise contours. Paper 950212, *74th Annual Meeting of the Transportation Research Board*, Washington DC, January.

Phillips, L. T. (1987). Air carrier activity at major hub airports and changing interline practices in the United States' airline industry. *Transportation Research*, **21A**(3), 215-221.

Piers, M. A. et al (1993). *The development of a method for the analysis of societal and individual risk due to aircraft accidents in the vicinity of airports*. Report CR 93372L, National Aerospace Laboratory (NLR), The Netherlands.

Pilling, M. (1991). Airlines face heavy bill for going green. *Interavia*, May, 10-14.

Poole, M. A. (1986). Liberalization of air services: the UK experience. *ITA Magazine*, No. 36, June-July, 9-13.

Poole, M. A. (1995). *Supplementary proof of evidence for Hillingdon Borough Council to Heathrow Airport Terminal 5 Inquiry*. Report HIL 32, London, December.

Poole, R. W. Jr. (1992). Privatisation: the record to date. Ch. 5 of *Airport Finance*, N. Ashford and C. A. Moore, van Nostrand Reinhold, New York.

Port of New York Authority (1966). *Airport requirements and sites to serve the New Jersey-New York Metropolitan Region*, New York, December.

Port of Seattle (1990). *Final package of mediated noise abatement actions for Seattle-Tacoma International Airport agreed to by the Mediation Committee on March 31*, Port of Seattle and Mestre Greve Associates, Seattle, Washington.

Post Office (1976). *Final report of the business communication studies on demand for person-to-person interactive communications systems.* Vol. 1: Aims and Methods. DOC T/ELT (76) 41. Produced on behalf of the Long Term Studies Working Group, by Long Range Studies Division, Post Office Telecommunications, Cambridge, England.

Powell, C. (1994). Financial barriers to market entry. *Journal of Air Transport Management*, 1(3), 187.

Pred, A. R. (1976). The inter-urban transmission of growth in advanced economies: empirical findings versus regional planning assumptions. *Regional Studies*, 10, 151-171.

Prins, V. and P. Lombard (1995). Regulation of commercialised state-owned enterprises: case study of South African airports and air traffic and navigation services. *Journal of Air Transport Management*, 2(3/4), 163-171.

Pryke, R. (1991). American deregulation and European liberalisation. In: *Transport in a free market economy* (D. Banister and K. Button, eds.), Macmillan, London.

PSATC (1991). *Flight Plan Project - Phase II: Development of alternatives*, Final Report, Prepared for the Puget Sound Council of Governments and Port of Seattle, Puget Sound Air Transportation Committee, Seattle Washington, June.

PSATC (1992). *The Flight Plan Project*, Draft Final Report and Technical Appendices, Prepared for the Puget Sound Regional Council and Port of Seattle, Puget Sound Air Transportation Committee, Seattle, Washington, January.

PSRC/Port of Seattle (1992). *The Flight Plan Project - Final Environmental Impact Statement*, Puget Sound Regional Council/Port of Seattle, Seattle, Washington, October.

Purdy, G. (1994). *Rebuttal proof of evidence on third party risks*, Manchester Airport Proposed Second Runway Public Enquiry, Documents MA 1045 and MA 1046, Manchester, England, August.

Quick, J. H. (1971). A BAC view of V/STOL. Presentation to conference: *Aviation's Place in Transport*, Royal Aeronautical Society, London, 12-13 May.

RAeS (1978). *Report on future airport policy, the summary of a joint symposium of Royal Aeronautical Society and the Chartered Institute of Transport.* Royal Aeronautical Society, London.

Raper, D. (1996). Airports and air quality. Lecture notes for short course *Airports and the Environment*, Dept. of Aeronautical and Automotive Engineering and Transport Studies, Loughborough University, May.

Rapp, L. (1986). Procedures for setting air fares, and the rules of competition of the EEC Treaty. *ITA Magazine*, No. 37, September, 9-11.

Rawls (1971). *The theory of justice*. Harvard University Press, Boston.

Reed, A. (1990). HST: Threat or alternative. *Air Transport World*, July, 42-44.

Reed, A. (1995). Southwest style in Europe. *Air Transport World*, September, 63-64.

Rees-Jones, G. (1995). The need for Terminal 5. Proof of Evidence for British Airways Plc to *Terminal 5 Heathrow Public Inquiry*, London, April.

Reid, A. (1990). The case for additional runway capacity at Heathrow. In conference proceedings: *New runway capacity in the south east - solutions and impacts.* pp 9.1-9.7, Royal Aeronautical Society, London.

Reiss, P. C. and P. T. Spiller, (1989). Competition and entry in small airline markets. *Journal of Law and Economics,* XXXII, October, S179-S202.

Renz, E. (1996). Munich Airport. Lecture notes for short course *Airports and the Environment*, Dept. of Aeronautical and Automotive Engineering and Transport Studies, Loughborough University, May.

Reynolds-Feighan, A. J. (1992). EC and US air traffic patterns: a comparative spatial analysis. Paper presented to *32nd European Congress of the Regional Science Association*, Brussels, August.

Reynolds-Feighan, A. J. (1994). The role of regional airports in regional economic development in Ireland and the UK. Paper presented to *Regional Science Association British-Irish Section Annual Conference*, Dublin, September.

Rheingold, L. (1991). Big splash, small ripples. *Air Transport World*, February, 104-112.

Rhodes, G. (1975). *Committees of Inquiry*. George Allen & Unwin, London.

Robertson, J. A. W. (1995). Airports and economic regeneration. *Journal of Air Transport Management*, 2(2), 81-88.

Robins, K. and M. Hepworth (1988). New technologies and the future of cities. *Futures*, April, 155-176.

Roenqvist, R. and H. Jonforsen (1997). Addition of a runway at a major airport can take several years to clear all the hurdles. *ICAO Bulletin*, September, 5-6, 25.

Rosenberg, A. L. (1997). ETA: 1999. *Civil Engineering*, May, 56-58.

Rosenhead, J. (1989). Introduction: Old and new paradigms of analysis. In: *Rational analysis for a problematic world* (J. Rosenhead, ed.), pp 1-20, Wiley, Chichester, England

Roskill, Lord (Chair) (1969). Report: *Commission on the Third London Airport*. HMSO, London.

Rotach, M. C. (1988). MANTO - a research project of telecommunication applications for the future information society. *Transportation*, **14**, 377-393.

Rouaud, H. (1991). The end of community controls in air transport. *ITA Magazine*, No. 67, July/August, 9-15.

Roupakias, D. (1988). *A transport demand model for the Greek Islands*. MSc dissertation. Dept. of Transport Technology, Loughborough University, September.

Rowan, G. (1991). Sound of airport's growth is too loud for neighbours. *Toronto Globe and Mail*: Report on Business. Section B, 2 December, 1-02.

Roy, R. (1976). Myths about technological change. *New Scientist*, 6 May, 281-282.

Ruddock, L. (1992). *Economics for construction and property*. Arnold, London.

Russon, M. G. and C. A. Hollingshead (1989). Aircraft size as a quality of service variable in a short-haul market. *International Journal of Transport Economics*, **XVI**(3), October, 297-311.

Ryan, D. (1998). Dallas Fort Worth. Presentation to *4th National Airport System Planning Symposium*, Transportation Research Board, Houston, March.

Rydin, Y. (1993). *The British planning system*. Macmillan, London.

Salomon, I. (1986). Telecommunications and travel relationships: a review. *Transportation Research*, **20A**(3), 223-236.

Salomon, I., H. N. Schneider and J. Schofer (1991). Is telecommuting cheaper than travel? *Transportation*, **18**, 291-318.

San Nicolas, J. (1995). The Spanish airport system. Lecture notes for short course *Airport Planning Procedures*, Dept. of Aeronautical and Automotive Engineering and Transport Studies, Loughborough University, November.

Sarandeses, A. M. (1993). Aeropuertos Espanoles - a challenge to the future. *Airports Technology International*, London.

Schipper, Y. (1997). *Airline deregulation and external costs: a welfare analysis*. Presentation to 37th Congress of the European Regional Studies Association, Rome, 26-29 August.

Schmidt, K. (1996). The zero option. *New Scientist*, 1 June, 33-37.

Schott, G. J. and L. L. Leisher (1975). Common starting point for intercity passenger transportation planning. *Astronautics and Aeronautics*, July/August, 38-55.

Sealy, K. R. (1976). *Airport strategy and planning*. Oxford University Press, Oxford, England.

Shade, N. T. (1990). Sound insulation and thermal performance modifications: case study for three dwellings near BWI airport. *Transportation Research Record 1255*, 12-18.

Sharman, F. A. (1986). Airport propositions. *Built Environment*, **10**(3), 187-195.

Shaw, R. (1979). Forecasting air traffic - are there limits to growth? *Futures*, June, 185-194.

Shaw, S. (1990). *Airline marketing and management*, 3rd ed., Pitman, London.

Simon, H. A. (1955). A behavioural model of rational choice. *Quarterly Journal of Economics*, **69**, 99-118.

Sinha, A. N. and E. S. Rehrig (1980). *Use of separate short runways for commuter and general aviation traffic at major airports*. Report MP-80W 21, Mitre Corporation, May.

Slater, D. (1974). Underdevelopment and spatial inequality. *Progress in Planning*, **4**.

Slater, K. (1993). *A method of estimating the risk posed to UK sites by civil aircraft accidents*. Report CS9345, Civil Aviation Authority, London, October.

Slaughter, R. A. (1996). Long term thinking and the politics of reconceptualisation. *Futures*, **28**(1), 75-86.

Slovic, P. (1987). Perception of risk. *Science*, **236**, 280-285.

Smith, E. J. (1991). A technique for analysing the risks of aircraft crashes in the vicinity of an airport. In: Proceedings of Spring Conference *Air Transport Safety*, (D. E. Corbyn and N. P. De-Bray, eds.), pp 27-44, Safety and Reliability Society, Bristol.

Smith, M. J. T. (1989a). *Aircraft noise*. Cambridge University Press, Cambridge, England.

Smith, M. J. T. (1989b). On the need for increased stringency in the noise certification requirements for subsonic jet and propeller-powered heavy aircraft. Paper No. WG1/4. To the *Fourth Meeting of Working Group 1 of the ICAO Committee on Aviation Environmental Protection, CAEP 2*, Montreal.

Smith, P. (1998). Management of planning. Lecture notes for short course *Airport Design*, Dept. of Aeronautical and Automotive Engineering and Transport Studies, Loughborough University, February.

Snape, D. M. (1990). Aircraft exhaust emissions - a future constraint on airport capacity? An engine manufacturer's perspective. Paper presented to *International Civil Airport Association Europe Seminar*, Brussels, 5-7 February.

Somerville, H. (1992). Airlines, aviation and the environment. Paper presented to conference: *Air transport growth - How will airports manage?* Royal Aeronautical Society, London, 21 October.

Somerville, H. and G. Meades (1995). Airline audits and environmental management. Paper presented to conference: *Environmental Aspects of Air Transport*, Royal Aeronautical Society, London, September.

Sorenson, N. (1991). The impact of geographic scale and traffic density on airline production costs: the decline of the no-frills airlines. *Economic Geography*, 67, 333-345.

SRI (1990) *A European planning strategy for air traffic to the year 2010*. SRI International, for International Air Transport Association, Geneva.

Stacey, R. (1993). Strategy as order emerging from chaos. *Long Range Planning*, 26(1), 10-17.

Stadler, G. (1995). Fly via VIE. *Airports Technology International 1994/95*, pp 35-39, London.

Staniland, M. (1997). Surviving the single market: corporate dilemmas and strategies of European airlines. *Journal of Air Transport Management*, 3(4), 197-210.

Starkie, D. (1987). Configuring change: reflections on transport policy processes. In: *Transportation Planning in a Changing World* (P. Nijkamp and S. Reichman, eds.), Gower, Aldershot, England.

Steinle, W. J. (1992). Regional competitiveness and the single market. *Regional Studies*, 26(4), 307-318.

Stevenson, G. M. (1972). *The politics of airport noise*. Duxbury Press, Belmont, California.

Stockman, I. (1996). Ramping up the price. *Airline Business*, November, 53-58.

Strand, S. (1983). *Impact studies and transportation as policy instruments: the case of the Norwegian STOL system*. ITE Research Note 0-876/E-728, Institute of Transport Economics, Oslo, 7 July.

Street, J. (1991). Draft EIS for expansion of DFW airport includes USD 150 million in mitigation. *Airport Forum*, No. 3, 40-41.

Stubbart, C. (1985). Why we need a revolution in strategic planning. *Long Range Planning*, 18(6), 68-76.

Sunderland, R. (1992). Finding solutions - RUCATSE. Presentation to conference *Airport Capacity*, Airports Association Council International, European Region, Manchester, 15-17 June.

Taneja, N. K. (1976). *The Commercial Airline Industry*. Lexington Books, Lexington, Massachusetts.

Taylor, M. and C. Kissling (1983). Resource dependence, power networks and the airline system of the South Pacific. *Regional Studies*, 17(4), 237-250.

Teitz, M. B. (1974). Towards a responsive planning methodology. In: *Planning in America: learning from turbulence* (D. R. Godshalk, ed.). American Institute of Planners, Washington DC.

Thomas, C. (1996). Noise related to airport operations - community impacts. In: *Environmental management of airports* (N. T. Pedoe, D. W. Raper and J. M. W. Holden, eds.), pp 8-34, Thomas Telford, London.

Thomas, V. F. (panel leader) (1993). Aircraft and engine manufacturers. *8th International Workshop on Future Aviation Activities*, Transportation Research Board, Washington DC, 13-15 September, pp 50-54.

Thorning, S. P. (1990). Trends in airport's corporate strategy and structure. *ITA Magazine*, No. 62, July/August, 8-11.

Tinch, R. (1995). *The valuation of environmental externalities*. Summary of report to the UK Department of Transport, London, April.

Toepel, W. O. (1986). The institutional scene in European airports: the role of airports in the aeronautical system. *ITA Magazine*, No. 39, November.

Toepel, W. O. (1988). German airport system planning. Lecture notes for short course *Airport Planning Procedures*, Dept. of Transport Technology, Loughborough University, November.

Toms, M. (1984). Airport development and the planning process: the Stansted case. *Built Environment*. 10(3), 181-186.

Toms, M. (1993). Airport charges: the future. *Airports Technology International*, pp 66-67, Sterling Publications, London.

Torgerson, D. (1986). Between knowledge and politics: three faces of policy analysis. *Policy Sciences*, 19, 33-59.

Tornquist, G. E. (1973). Contact requirements and travel facilities. *Lund Studies in Geography*, Section B, No. 38, Lund, Sweden.

TRB (1988). *Future development of the US airport network.* Transportation Research Board, Washington DC.

TRB (1990). *Airport system capacity - strategic choices.* Special Report 226, Transportation Research Board, Washington DC.

TRB (1991). *Winds of change.* Special Report 230, Transportation Research Board, Washington DC.

TRB (1992). *National Aviation System Planning Symposium,* held on 11-13 July 1991in Minneapolis, Transportation Research Board, Washington DC.

TRB (1993). Aircraft and engine manufacturers panel. In: *8th International Workshop on Future Aviation Activities,* Transportation Research Board, Washington DC, September.

TRB (1994). *Public investment in the aviation transportation infrastructure.* Report 55, Transportation Research Board, Washington DC, 27 October.

TRB (1996). *Comments on draft AC 150/5070-5 dated April 1996* by the Aviation System Planning Task Force. Transportation Research Board, Washington DC, 25 September.

Treibel, W. (1976). Germany. In: *Airports: the challenging future,* pp 17-29, Institute of Civil Engineers, London.

Treitel, D and E. Smick (1996). All change. *Airline Business,* July, 34-36.

Trethaway, M. (1996). Giving airports an incentive. *Airport World,* Issue 3, pp 36-40.

Trowbridge, A. V. (1988). 'Titanic planning' in an uncertain environment. *Long Range Planning,* 21(2), 96-90.

TSUG (1995). Remarks by M. Brown in *Transport Statistics Users Group Newsletter 34,* March, p 7.

Turbeville, W. C. (1996). Investment opportunities: balancing risk and return in the airport environment. Paper presented to *Airports and Privatisation,* Washington DC, June, pp 2-14, First Conferences, London.

Twiss, B. (1976). Economic perspectives of technological progress. *Futures,* February, 52-63.

UK House of Commons (1995). *UK Airport Capacity.* The Civil Aviation Authority minutes of evidence to the Transport Committee, 23 October, p 13, HMSO, London.

UK House of Commons (1996). *UK Airport Capacity.* 2nd report of the Transport Committee, Volume 1, 8 May, HMSO, London.

UK Treasury (1987). *Economic Progress Report,* No. 188, January/February, 6-7.

United Nations (1974). *Elements of tourism policy in developing countries.* Report TD B/C 3.89/Rev 1.

Uyeno, D., S. W. Hamilton and A. J. G. Biggs (1993). Density of residential land use and the impact of airport noise. *Journal of Transport Economics and Policy,* 27, 3-18.

Vaagen, E. P. (1993). Environmental charges - making the polluter pay. *Airports Technology International,* pp 58-59, Sterling Publications, London.

van den Berg, D. (1990). The regional airline development in Europe - a comparison with the USA. *ITA Magazine,* No. 62, July/August, 13-19.

van Hasselt, L. (1994). Prospects for changes in the regulation of international civil aviation. *Journal of Air Transport Management,* 1(2), 83-88.

van Leeuwen, B. (1992). Airport investment challenges. Presentation to conference *Airport Capacity,* Airports Association Council International, European Region, Manchester, 15-17 June.

Vandyk, A. (1993). IATA warns of higher charges for airlines. *Air Transport World,* February, 96-97.

Veldhuis, J. (1992). *Impact of liberalisation on European airports.* Transportation Research Circular, No. 393, Transportation Research Board, Washington DC, April.

Veldhuis, J. (1996). The Schiphol planning process. Lecture notes for short course *Airport Planning Procedures,* Dept. of Aeronautical and Automotive Engineering and Transport Studies, Loughborough University, February.

Veldhuis, J. (1997). The competitive position of airline networks. *Journal of Air Transport Management,* 3(4), 181-188.

Veldhuis, J., A. Schmitt, H. Brobeck and A. Karyd (1995). *Fast trains: competition and competitive power.* Background information paper to International Civil Aviation Organisation Committee for Aviation Environmental Protection WG 4, August.

Vickerman, R. W. (1990). Transport infrastructure in the European Community: new challenges, regional implications and evaluation. *21st Annual Conference, Regional Science Association,* British Section, Liverpool, September .

Vickers, G. (1965). *The Art of Judgment,* p 208, Chapman and Hall, London.

Villiers, J. (1989). For a European air transport policy, Part two, Air transport in Europe. *ITA Magazine,* No.57, September/October, 3-20.

Volgers, A. L. (1992). Risk management becomes a strategic issue. Lecture notes for short course *Airport Finance,* Dept. of Transport Technology, Loughborough University, November.

von Wrede (1995). Air traffic and environment. Presentation to conference: *Environmental Aspects of Air Transport,* Royal Aeronautical Society, London, September.

Walker, K. (1997). Bespoke fortunes. *Airline Business,* January, 32-35.

Wassenbergh, H. A. (1991). The globalisation of international air transport. *ITA Magazine,* No. 67, July/August, 3-8.

Watson, A. H. (1965). Potential traffic demand in the London area. *Journal of the Royal Aeronautical Society,* **69**(652), 211-215.

Watts, R. (1965). Scheduling of commercial air movements: the airline point of view. *Journal of the Royal Aeronautical Society,* **69**(652), 216-218.

WCED (1987). *Our common future.* World Commission on Environment and Development, published in 1993 by Oxford University Press, Oxford.

Webb, R. (1998). Competition and regulation in air transport. *Proceedings (of the Chartered Institute of Transport),* 7(2), 16-25.

Weisbrod, G. (1990). Economic impacts of improving general aviation airports. *Transportation Research Record 1274,* 134-141.

Wellman, G. (1998). The Los Angeles airports. Paper presented to *4th National Airport System Planning Symposium,* Transportation Research Board, Houston, Texas, 25-27 March.

Wesler, J. E. (1983). Airports and environmental capacity. *7th World Airports Conference,* Thomas Telford, London.

Wesler, J. E. (1988). Noise abatement and airport capacity. *Airports Technology International,* pp 26-28.

Wheatcroft, S. (1994). *Aviation and tourism policies: balancing the benefits.* World Tourism Organisation/Routledge, London.

Wheatcroft, S. F. (1990). Financing - the problem. Paper presented to *Third European Aerospace Conference: Civil Aviation Operations - Problems, Solutions and Actions,* Royal Aeronautical Society, London, 22-24 May.

Wheeler, C. F. (1989). Strategies for maximising the profitability of airline hub-and-spoke networks. *Transportation Research Record 1214,* 109.

Whitaker, R. (1995). Going to market. *Airline Business,* December, 46-47.

Wijers, P. J. (1991). Coping with dramatic air transport growth. *ICAO Journal,* January, 10-15.

Williams, M. L. (1988). Emissions and air pollution. Lecture notes for short course *Airports and the Environment,* Dept. of Transport Technology, Loughborough University, April.

Williams, F. and D. V. Gibson, eds. (1990). *Technology Transfer.* Sage Publishers, London.

Wills, G. et al (1972). *Technological forecasting.* Penguin, Harmondsworth, England.

Wilson, D. (1995). Presentation to conference: *UK airports policy, a new agenda.* Reported in *Aerospace,* July, 22-25.

Wilson, G. W. (1967). Towards a theory of transport and development. In: *Transport and Development* (B. S. Hoyle, ed.), Macmillan, London.

Wilson, R. M. (1988). Taxiway separation standards for future aircraft. *Airports Technology International,* pp 102-104.

Wilson, T. and C. Neff (1983). *The social dimension in transportation assessment.* Gower, Aldershot, England.

Windle, R. J. (1991). The world's airlines. *Journal of Transport Economics and Policy,* **XXV,** 31-69.

Wolf, P. (1982). Airport planning in the Federal Republic of Germany: a critical review. *Airport Forum,* No. 3, 57-64.

Wolfe, H. P. and D. A. NewMyer (1985). *Aviation industry regulation.* University Press, Southern Illinois.

Woolley, D. (1993). Profitable Paris airports plan five years of heavy investment. *Airport Forum,* No. 3, 18-20.

Woolley, D. (1996). European airports vie for hub status. *Jane's Airport Review,* October, 11-13.

Woolley, D. (1997). Private sector investors take over UK airports. *Jane's Airport Review,* March, 15-21.

Woolsey, J. P. (1997). The lone runway. *Air Transport World,* April, 58-61.

Wright, A. J. (1996). Heathrow 50. *Aircraft Illustrated*, June, 3962.

Yates, C. (1992). Planning action needed for growth. *Jane's Airport Review*, July/August, 17.

Yosef, E. B. (1991). Flawed economic analysis undermines US noise regulations. *Air Finance Journal*, No. 130, September, 42-44.

Young, H. (1976). Development concept for a Canadian national airports plan. In: *Airports: the Challenging Future*, pp 31-55, Institute of Civil Engineers, London.

Zeverijn, A. P. (1997). Airport privatisation in red, white and blue: Dutch perspectives on changes in aviation landscapes. Lecture notes for short course *Airport System Planning and Design*, University of California, Berkeley, May.

INDEX

Airports Council International (ACI) 3, 4, 10, 25, 30
access   18, 31, 33, 36, 37, 51, 56, 59, 62, 87, 98, 125, 138, 150, 154, 161, 169, 185, 195, 199, 224, 228, 233, 244, 274, 280, 290, 307, 333, 339, 350, 374, 379, 396, 408, 410
    facilities 37, 87, 161, 165, 228
    highway (& *see* roads) 35, 63, 161, 180, 224, 250
    modes 36, 161, 180
    rail 36, 58, 73, 120, 132, 161, 225, 285-303, 340, 365, 372, 386, 397
accessibility 34, 51, 84, 104, 180, 187, 199, 227, 260, 270, 291, 307, 342, 351, 375, 390, 401, 408, 418
accidents 25, 63, 89, 227, 359, 368, 377
air quality 110, 120, 174, 224, 249
air traffic control 11, 32, 48, 91, 157, 303, 330
Air Transport Association 4, 103
aircraft size (& *see* frequency)   43, 50, 62, 71-74, 82, 95, 98, 134, 159, 182, 241, 258, 283, 291, 314, 335, 380
airline alliances (& *see* globalisation) 79, 318, 343, 356
airline behaviour   9-13, 19, 30, 32, 33, 36, 48-51, 56, 62, 73-83, 88, 94-100, 133, 143, 152, 165, 171, 175, 183, 209, 225, 252, 257, 266, 269, 278, 279, 281, 296-309, 313, 316, 329, 334, 343-349, 353, 356, 360, 363, 371, 372, 374, 378, 380, 381, 387, 394, 396, 403, 408, 410, 412, 416
Airport and Airway Development Act (US) 7
airport costs 133, 137
airport finance
    charges   49, 77, 87, 94, 97, 111, 127, 133, 137-140, 143, 203, 282, 294, 302, 306, 323, 335, 338, 344, 360, 366, 404
    cost recovery 11, 35, 87, 138, 152, 201, 352, 356, 414
    funding options 201
    subsidy 19, 28, 57, 84, 206, 306, 342, 357, 384
airport finance (& *see* privatisation) 10, 67, 136
Airport Improvement Program (US) 141, 149, 179, 195, 201, 222, 225, 231
airport ownership (& *see* globalisation, privatisation) 17, 94, 334
airport revenues 18, 137-140, 145, 201, 228
airport roles (& *see* gateways, reliever airports) 18, 40, 94-99, 132, 273, 325, 405
airport systems and operators
    Aer Rianta 49, 98, 132, 280, 310, 348
    Australia 31, 33, 140
    Berlin 130, 166, 331, 334, 337, 340, 348, 402
    British Airports Authority 32, 174, 275, 305
    Buenos Aires 169, 378
    Glasgow 12, 102, 166, 310, 336, 372-373
    London area 16, 32, 55, 273-296, 300, 312, 316, 320, 322
    Montreal 12, 22, 52, 166, 357, 380, 402
    National Express 130, 132, 311, 322, 376
    New Zealand 33, 48
    Port Authority of New York and New Jersey 164, 174, 222, 375
    São Paulo 31, 109, 166, 297, 354, 377, 402
    TBI 311, 374
    Tokyo 109, 127, 169, 363-366
    Washington 1, 76, 166, 216, 222, 254, 258, 263, 366, 396, 402
airports
    Aeroporti di Roma 57, 411
    Amsterdam/Schiphol   25, 27, 56-58, 65, 77, 78, 80, 98-99, 102, 107, 112, 121, 126-127, 131-132, 138, 144, 146, 153, 162, 169, 170, 184, 188, 193, 312, 313, 327, 331, 337, 340, 343, 344, 346, 347, 349, 411
    Athens 153, 166, 202, 297, 331, 345, 360, 402
    Atlantic City 33
    Austin 12, 106, 167, 394
    Belfast International 311, 374
    Billund 79
    Birmingham 33, 49, 58, 130, 132, 136, 153, 203, 283, 293, 310, 323, 345, 351
    Bogota 120

Bournemouth 311, 322
Bristol 293, 311
Brussels 58, 132, 153, 167, 331, 332, 337, 344
Burbank 33
Cardiff 281, 311
Charles de Gaulle 36, 56, 58, 102, 153, 331, 332, 337, 343, 344, 379
Chep Lap Kok 133, 169
Cincinnati 84
Copenhagen 32, 78, 102, 120, 130, 144, 153, 331, 332, 344, 367, 387
Dallas Fort Worth 3
Dallas Love Field 102, 402
Denver 12, 56, 57, 96, 98, 133, 166, 169, 226, 317, 411
Dubai 138
Düsseldorf 12, 63, 132, 331, 332, 339, 340, 349, 358, 395
East Midlands 35, 102, 107, 130, 132, 153, 293, 310, 311, 322, 323, 382
Frankfurt 12, 36, 58, 73, 78, 83, 132, 144, 157, 161, 330, 337, 339, 344, 346, 348, 350, 357
Gatwick   58, 98, 128, 132, 139, 140, 153, 162, 165, 167, 273-305, 314, 317, 321, 331, 344, 346, 348, 382, 403
Heathrow   22, 49, 62, 74, 78, 82, 96, 105, 106, 111, 128, 139, 140, 141, 146, 153, 157, 165, 192, 273-305,
    308, 314-324, 332, 334, 336-339, 340, 343, 344, 350, 384, 416
Kansai 133, 169, 363
Kennedy 33, 132, 154, 167, 330, 346, 375
La Guardia 169, 375
Las Vegas 138
Lisbon 153, 165, 203, 334
Liverpool 107, 130, 153, 312, 349, 382
London City 33, 142, 157, 279, 280, 288, 300, 303, 347, 361, 387, 394, 399, 402
Los Angeles 1, 3, 12, 64, 157, 166, 230, 236
Luton 98, 136, 140, 165, 273-305, 312, 347, 374
Macão 120, 169, 178
Madrid 12, 78, 132, 153, 203, 331, 334, 344, 345, 370
Manchester   25, 57, 76, 94, 117, 122-126, 130, 131, 132, 140, 147, 161, 184, 203, 283, 293, 305-313, 314-323,
    331, 337, 347, 349, 381, 395, 411
Manila 109, 165
Miami 77, 132
Milan Linate 331, 348
Milan Malpensa 3
Minneapolis/St Paul 144, 237-254, 270, 271, 395
Moscow 27, 33
Munich 12, 25, 64, 96, 102, 111, 127, 132, 157, 166, 169, 297, 330, 332, 333, 337, 344, 358, 395, 397
Naples 131, 345
Newark 84, 110, 167, 375, 376
Nice 58, 79, 169, 345
Orlando 147, 313
Orly 58, 98, 331, 344, 346, 379
Osaka 120, 363-366
Oslo 11, 161, 165, 203, 349, 367, 385, 402
Philadelphia 98
Prestwick 12, 59, 310, 372
San Diego 76, 165, 169, 226, 230
San Francisco 1, 121, 185, 221, 230, 236, 260, 395
Seattle 91, 119, 216, 254-271, 402
Shannon 98, 107, 143, 341
Southend 130, 312
Stansted 54, 109, 132, 140, 142, 165, 173, 189, 273-305, 314-324, 347, 374, 395
Sydney 121, 129, 144, 156, 165, 297, 395, 399, 402
Toronto 33, 103, 112, 117, 153, 357, 380
Vancouver 104, 122, 170, 258, 357, 377, 394
Victoria 357
Washington Dulles 366
Washington National 166, 222, 402

Zürich 10, 78, 102, 121, 143, 161, 331, 344
Airports Council International 3, 143, 372
benchmarking 53, 146
benefit-cost analysis 172, 230
benefits
    direct 10, 21, 28, 48, 101, 113, 146, 197, 202, 262, 294, 303, 313, 330, 342, 344, 349, 396, 404, 410
    distribution of 13, 27, 37, 42, 46, 51, 64, 67, 116, 149, 176, 202, 228, 233, 267, 277
    indirect 29, 102, 103, 410
    net 28, 103, 105, 114, 117, 122, 146, 175, 313, 314, 377, 392
buses (see access, modes) 125, 134, 162, 228
California 35, 63, 107, 149, 186, 195, 204, 216, 230, 231, 232, 233, 234, 260, 270, 367
capacity
    airport   5, 18, 39, 52, 61, 74, 77, 79, 82, 85, 88, 90, 94-100, 152-162, 163-167, 204, 227, 255, 291, 296, 331, 346, 402-405
    airspace 7, 10, 48, 61, 87, 91, 165, 176, 204, 262, 275, 285, 296, 330
    enhancement 9, 171, 226, 229, 240, 263, 331, 333, 352
    environmental   10, 22-26, 29, 49, 57, 62-66, 99, 109-112, 119-128, 177, 204, 340, 358, 365, 378, 380, 382, 385, 402, 403, 414
    operational 53, 57, 61, 240, 359, 416
    runway   12, 61, 73-75, 83, 87, 93, 94, 153-157, 173, 192, 251, 255, 257, 263-269, 277-294, 303, 330-334, 339, 349, 364, 375, 382, 403, 415-417
    stand 62, 83, 96, 97, 153, 157-160, 174, 345
    ultimate 8, 174
capital   4, 7, 11, 20, 32, 67, 129, 131, 133, 137, 139, 145, 148, 149, 151, 175, 201, 206, 222, 225-230, 236, 267, 354, 360, 363, 408
cargo (& see freight) 7, 25, 30, 31, 37, 71, 82, 102, 120, 134, 167, 176, 230, 241, 257, 281, 288, 355, 378, 380
catchment area 70, 297, 308, 367
charter carriers 28, 31, 75, 81, 202, 330, 339, 360
charter flights 305, 374
choice models 56, 180, 182, 183, 185, 404
Civil Aeronautics Board (see US Civil Aeronautics Board)
Civil Aviation Authority (see UK Civil Aviation Authority)
competition
    airline (& see runway slots) 9, 19, 27, 30, 55, 73, 88, 334, 343-348, 353, 363
    airport 9, 19, 27, 32, 57, 294, 310, 318
    barriers to 13, 31, 40, 59, 75, 94, 118, 122, 153, 329
    modal 30, 37, 58, 83, 152, 197, 203, 228 334, 340, 363, 400, 402
complexity 37, 177, 208, 209, 371, 372
computers, use of 156, 214
congestion (& see delay)   11, 29, 35, 49, 50, 53, 92, 94, 109, 114, 137, 139, 141, 146, 152-162, 175, 187, 204, 222, 224, 269, 297, 304, 308, 320, 330, 340, 346, 347, 364, 402, 407
Congress (see US Congress)
context 3, 5, 8, 9, 17, 35, 36, 39, 45, 64, 97, 114, 163, 195, 196, 199, 317, 390-392, 401, 404
cost-benefit analysis (see benefit-cost analysis) 176, 342, 394, 401
cost-effectiveness 240
costs
    construction 4, 12, 36, 67, 121, 133, 139, 167, 206, 222, 226, 228, 266-268, 363, 386
    marginal 50, 137, 138, 139, 143, 155, 203, 224, 342, 419
    operating 10, 18, 49, 53, 81, 82, 88, 93, 133-137, 145, 151, 228, 343, 348
culture 191, 347, 373, 389, 390, 394, 412, 414
data
    availability   13, 27, 29, 34, 50, 67, 82, 101, 103, 105, 106, 107, 121, 147, 155, 158, 166, 179, 185, 187, 200, 208, 214, 216, 227, 262, 305, 306, 348, 362, 386, 414
    collection 40, 125, 132, 164, 171, 210
    processing 49, 138, 160, 355
decision
    analysis 8, 176, 177, 209, 214, 243, 252, 267, 270, 277, 327, 390-399, 409, 413
    making 8, 9, 15, 32, 41, 54, 61, 67, 72, 84, 112, 114, 126, 173, 179, 210, 212, 240, 295, 322, 395, 409, 413
delay   50, 52, 61, 91, 96, 152-162, 175-177, 204, 213, 227-229, 255, 263, 266, 315, 327, 331, 360, 366, 393, 402

demand 1, 9, 69-72
   elasticity 70, 181, 190
   management 9, 13, 61, 152-155, 188, 257, 352, 403
   saturation 143, 182, 188, 189, 190, 401, 413
demographics (*see* prediction)
density 16, 26, 54, 75, 77, 79, 82, 92, 137, 198, 201, 203, 237, 285, 311, 330, 336, 341, 343, 346, 348, 350, 352, 353, 357, 360, 362, 366, 409, 410
   population 353, 360, 366
   traffic 54, 75, 79, 82, 92, 137, 330, 336, 343, 346, 350, 360, 362
Department of Transportation (*see* US Department of Transportation)
deregulation (*& see* liberalisation) 19, 28, 30, 54, 55, 56, 76, 88, 89, 90, 94, 97, 98, 107, 117, 154, 188, 221, 237, 310, 321, 329, 336, 344, 396
disbenefits 21-27, 99, 101-112, 116, 127, 161, 173, 176, 240, 249, 267, 288, 402
dual-track planning process (*see* airports, Minneapolis/St Paul)
duty-free shops 355
economic analysis 52, 417
economic efficiency 52, 53, 201
economic impact 53, 60, 102, 103, 105, 108, 176, 177, 187, 304, 316
economies of density 49, 75, 352, 402, 411
economies of scale 52, 98, 104, 134, 136, 137, 140, 141, 155, 200, 308
efficiency 27, 52, 141, 179, 326, 392
employees 19, 53, 106, 121, 125, 130, 131, 171, 354
energy 22, 24, 29, 34, 50, 59, 110, 112, 127, 140, 175, 191, 341, 359, 393, 401
environmental
   mitigation 49, 60, 65, 111, 114, 119-128, 167, 170, 249, 282, 295, 324, 350, 352, 359, 366, 394, 402
environmental impact (*see* air quality, energy, noise, regulation, safety, water quality)
   global 11, 21, 29, 35, 41, 49, 83, 89, 131, 193, 197, 204, 318, 321, 338, 340, 350, 392, 400
   statements 213, 241, 314, 407
equity 31, 32, 80, 118, 145, 199, 208, 306, 310, 343, 348, 358, 397, 401, 404, 412
escapable costs 143
Eurocontrol 92, 331
European Commission 24, 28, 30, 34, 58, 95, 144, 290, 302, 312, 334, 350
European Community (*& see* European Union) 31, 62, 333-335, 340-342, 369
European Union 11, 31, 61, 62, 75, 79, 80, 88, 112, 164, 200, 202, 306, 310, 318, 325, 329-352, 361, 362, 367, 370, 380
evaluation 28, 115, 127, 149, 170-177, 209, 222, 232, 257, 259, 260, 262, 263, 267, 316, 321, 323, 327, 398, 400, 409
externalities (*& see* environmental impact, social benefits, social costs) 30, 37, 63, 118, 139, 140, 199, 340, 416
fares (*& see* deregulation) 13, 17, 41, 50, 55, 56, 58, 59, 70, 75, 76, 98, 103, 106, 154, 180, 182, 184, 211, 300, 306, 310, 334-339, 340, 343, 345, 348, 350, 354, 360, 362, 364, 369, 370, 373, 410
Federal Aviation Administration 1, 6, 7, 8, 10, 12, 24, 36, 40, 43, 45, 61, 67, 86, 91, 93, 94, 95, 101, 102, 103, 118, 141, 149, 150, 152, 156, 157, 163, 165, 169, 172, 173, 174, 176, 179, 192, 195, 196, 197, 201, 207, 210, 216, 220-229, 231, 233, 242, 243, 247, 248, 249, 251, 253, 257, 258, 263, 266, 267, 269, 316, 327, 333, 339, 375, 376, 400
financing 30, 132, 201, 222, 316, 348, 400
forecasts (*& see* predictions) 7, 43, 55, 69, 74, 75, 82, 83, 85, 95, 137, 165, 171, 172, 185, 186-194, 211, 213, 216, 223, 228, 232, 241, 247, 251, 257-260, 263, 264, 271, 315, 320, 324, 355, 358, 370, 375, 385, 394, 395
freight (*& see* cargo) 28, 34, 41, 58, 64, 87, 105, 144, 146, 159, 161, 171, 174, 210, 303, 311, 327, 333, 354, 359, 364, 373, 376, 384, 405
frequency 17, 43, 46, 50, 56, 59, 61, 71, 73-79, 80, 82, 88, 94-100, 106, 108, 153, 159, 180-184, 185, 228, 259, 261, 283, 291, 300, 310, 313, 324, 334-340, 342, 344, 362, 364, 403, 410, 415
gateways 8, 18, 57, 77, 79, 80, 89, 102, 184, 202, 230, 345, 346, 351, 363, 367
General Aviation 50, 84, 86, 308, 312, 349
globalisation 10, 13, 30, 79, 98, 132, 189, 353, 405
goals 18, 28, 40, 46, 48-51, 69, 129, 150, 169, 173, 176, 183, 195-201, 206, 223, 229, 230, 234, 243, 252, 308, 316, 318-320, 323, 326, 327, 334, 358, 363, 391, 394, 398, 409, 412, 417
government role 204
grants 7, 20, 108, 118, 144, 152, 179, 201, 222, 225, 228
growth rates 3, 20, 72, 188, 202, 248, 250, 258, 281, 336, 337
helicopters 23, 84, 205, 362

hubbing 31, 49, 50, 55, 56, 76-79, 94-97, 144, 158, 165, 183, 302, 313, 329, 343-347, 348, 351, 359, 370, 382, 384, 387, 403, 410
implementation 4, 15, 42, 46, 67, 124, 126, 169, 177, 196, 205-207, 208, 210, 216, 223, 240, 257, 266, 270, 322, 391, 398, 400, 408, 412, 416
innovation 14, 188
interest rates 67
International Air Transport Association (IATA) 13, 28, 30, 40, 53, 61, 64, 92, 102, 107, 133, 139, 147, 152, 153, 189, 285, 339, 346
International Civil Aviation Organisation (ICAO) 1, 2, 3, 7, 8, 11, 18, 22, 24, 26, 64, 83, 90, 92, 93, 104, 111, 112, 139, 140, 143, 145, 159, 204, 205, 211, 340, 372
inventory 46, 72, 73, 170, 171, 185, 223, 334
investment 10, 20, 67, 133-136, 147, 178, 201, 206, 225, 233, 342, 350
labour...4, 133-137, 191, 384, 386, 401
land use controls 196, 340
  easements 117, 119
landing fees 124, 127, 137, 143, 283, 339, 359, 360
level playing fields..4, 30, 35, 88, 149, 152, 203, 334, 341, 352
liberalisation 4, 10, 16, 28, 31, 32, 56, 62, 75, 77, 88, 98, 148, 184, 186, 193, 202, 205, 312, 320, 322, 325, 329, 331, 334-340, 342, 350, 361, 364, 366, 370, 380, 391, 410, 412
load factor...64, 74, 81, 103, 155, 158, 185, 242, 259, 261, 263, 282, 400
maintenance 11, 24, 61, 88, 93, 104, 143, 145, 147, 148, 150, 157, 206, 277, 281, 356, 365, 387
market share 18, 19, 38, 49, 54, 58, 73, 77, 81, 88, 90, 97, 143, 181, 184, 260, 300, 303, 308, 311, 322, 324, 326, 347, 352, 354, 362, 377, 384, 410
master planning..6, 7, 8, 14, 16, 43, 45, 46, 170, 173, 178, 188, 196, 220, 224, 241, 316, 412
mobility (& see sustainability) 34, 127, 162, 188, 199, 340, 341, 410
models 65, 70, 71, 76, 79, 82, 153, 171, 174, 180-186, 187, 190, 194, 209, 214, 241, 291, 372, 391, 396
monitoring..................6, 22, 24, 52, 53, 91, 112, 119, 122, 124, 153, 160, 173, 179, 205, 281, 282, 316, 327, 359, 391
Monopolies and Mergers Commission 140, 306
monopoly 27, 31, 33, 36, 53, 55, 56, 60, 73, 77, 90, 94, 132, 139, 140, 142, 204, 206, 283, 322, 338, 350, 354, 360, 361, 372, 397
National Plan for Integrated Airport Systems 225
Networks (& see hubbing)..76, 80, 84, 89, 148, 180, 191, 194, 326, 352, 363, 410
new large aircraft 82, 83, 84, 157, 160, 284
noise 10, 12, 22, 29, 50, 52, 62, 82, 85, 87, 93, 109-112, 114, 115-128, 140, 150, 167, 176, 188, 190, 209, 213, 227, 237, 240, 247, 249, 251, 267, 274, 279, 280, 281-290, 294, 305, 311, 319, 340, 345, 349, 358, 360, 366, 372, 375, 385, 394, 401, 402, 403, 410, 415, 416
objectives 7, 16, 20, 34, 40, 42, 43, 47, 50, 64, 69, 129, 133, 141, 146, 149, 154, 170, 173, 175, 179, 196, 200, 223, 229, 235, 255, 257, 291, 296, 307, 317, 319, 321, 322, 327, 342, 356, 358, 385, 390, 392, 397, 398, 408, 415, 418
parking...121, 126, 134, 138, 157, 159, 160, 161, 162, 174, 228, 244, 250, 285, 310, 355
participation 15, 31, 33, 46, 80, 122, 170, 172, 179, 192, 210, 240, 308, 317, 326, 341, 358, 389, 393, 397, 401, 409, 417
passengers
  behaviour 50, 56, 71, 73, 74, 137, 143, 155, 160, 166, 171, 180-182, 186, 237, 241, 247, 257, 279, 291, 302, 381, 397, 400, 411
  business 106, 144, 173, 279, 305, 313, 378
  international 31, 56, 77, 79, 136, 144, 166, 173, 202, 250, 300, 305, 309, 336, 347, 348, 361, 373, 380, 384, 411
  traffic 1, 3, 25, 31, 38, 41, 52, 54, 56-59, 71, 81, 98, 134, 136, 139, 142, 143, 144, 155, 166, 180-184, 186-189, 202, 228, 241, 248, 257-261, 274, 293, 296-304, 306, 313, 330, 336, 338, 344-347, 360, 361, 368, 371, 373, 377, 384, 386, 410
  transfer......30, 49, 55-58, 76-80, 94-97, 144, 161, 186, 202, 281, 285, 290, 295, 300, 344, 360, 377, 380, 386, 411
  transit..31, 121, 127, 166, 225, 228
peaking 74, 95, 134
peaks 52, 57, 63, 74, 85, 94, 97, 139, 152-157, 161, 173-175, 258, 266, 274, 282, 294, 304, 308, 320, 331, 339, 361, 375, 377, 380, 403, 411
performance indicators 46, 130, 137, 327, 412

450   Strategic Airport Planning

planning process (*& see* context, decisions, evaluation, goals, implementation, inventory, monitoring, objectives, participation, perfomance indicators, regulatory capture, stakeholder interactions)   4-9, 14, 21, 37, 39-48, 90, 99, 114, 170, 172, 175-180, 189, 196, 200, 205, 207, 208, 211, 220-225, 229, 230-236, 240, 247, 252, 270, 314-328, 389-400, 409, 413, 417-419
politics 45, 166, 315, 362, 375, 391, 393, 394, 397
pollution 10, 22-25, 30, 34, 59, 62, 110, 116, 123, 126, 140, 161, 165, 203, 204, 268, 402, 414, 415
Port Authority of New York and New Jersey (*see* airport systems and operators)
prediction..52, 69, 183, 186, 189, 192, 194, 211, 303, 346
   demographics 13, 191, 297
   scenarios................. 16, 43, 45, 95, 106, 171, 187, 189-194, 210, 247, 277, 295, 303, 369, 371, 395, 399, 400, 419
   supply 8, 42, 46, 50, 69-71, 77, 84, 107, 153, 182-192, 198, 211, 291, 321, 330, 334, 410, 413
   technology 1, 13, 18, 24, 40, 42, 45, 48, 59, 63, 74, 83, 92, 93, 95, 106, 111, 129, 131, 156, 160, 162, 173, 182, 183, 187-189, 192, 198, 204, 215, 216, 224, 324, 326, 330, 341, 352
   tourism..3, 29, 59, 102, 103, 106, 109, 198, 199, 296, 353, 360, 361, 363, 364, 398
   uncertainty............... 9, 13, 22, 31, 55, 67, 98, 126, 130, 142, 189, 194, 241, 242, 253, 258, 278, 318, 323, 338, 413
preferences 13, 32, 42, 171, 188, 291, 304, 320, 328, 329, 401, 410, 413
pricing 34, 76, 77, 78, 89, 95, 98, 121, 131, 137-140, 144, 148, 152, 154, 165, 175, 184, 188, 282, 290, 294, 302, 308, 323, 335, 338, 357, 403, 415
propensity to fly 106, 187, 191, 308, 367
Public Inquiry...278, 279, 280, 296, 315, 325
public interest...147
public participation (*& see* participation) 210, 223, 327, 328
public transport (*& see* access, mode)...120, 121, 125, 146, 161, 281, 285
rail (*& see* access, rail) 18, 58, 83, 107, 203, 224, 333, 342, 404
rapid transit (*& see* access, rail) 280
regulation
   economic (*& see* rights) 9, 13, 17-21, 30-33, 36, 41, 48, 55-58, 71-83, 88-90, 94-99, 132, 138-147, 152, 154, 173, 176, 180-184, 186, 202, 206, 208, 221, 225, 233, 237, 241, 244, 250, 254, 268, 274, 278, 310, 313, 318, 329, 331-341, 343-348, 353, 356, 361, 367, 374, 375, 379, 397, 399, 403, 408, 410
   environmental 10, 13, 21-27, 33-38, 49, 52, 62-65, 109-112, 115-128, 161, 165, 169, 170-178, 188, 189, 197, 204, 206, 208, 247, 254, 261, 263, 271, 273, 277, 279, 281-291, 294-296, 305, 308, 316, 318, 321, 326, 331-335, 340, 346, 349, 351, 352, 356, 358, 361, 365-367, 378, 380, 382, 385, 387, 394, 400, 402-405
regulatory capture 128, 341, 350
reliever airports 87, 119, 165, 205, 240, 343, 348, 375
residual cost... 28, 141, 143
returns to scale 139, 371
revenues... 11, 53, 94, 137-141, 144-146, 155, 228, 282, 302, 354, 356
rights 18, 31, 48, 57, 79, 81, 90, 123, 132, 138, 147, 152, 155, 202, 275, 310, 324, 326, 335-339, 341, 349, 351, 364, 367, 372, 380
roads (*& see* access, highway) 25, 43, 51, 72, 123, 125, 126, 127, 138, 161, 244, 273, 285, 288, 318, 327, 341, 355, 359, 360, 363, 365, 402, 404
route fragmentation... 77, 156, 166, 184, 303, 326, 347, 349
RTOL..405, 416
RUCATSE 182, 192, 288, 289, 292, 294, 313, 319, 321, 323, 326, 327, 389, 395
runway slots 32, 53, 73, 83, 89, 95, 141, 148, 152-155, 158, 207, 281, 290, 308, 314, 350, 364, 370, 404
safety 10, 17, 19, 23, 25, 28, 29, 34, 42, 50, 53, 64-66, 84, 88-90, 92, 112, 118, 121, 141, 150, 177, 183, 185, 196, 204, 227, 279, 280, 288, 303, 308, 333, 350, 358, 359, 402, 405, 408
scenarios (*& see* prediction)... 106, 114, 171, 190, 191, 192, 194, 213, 270, 351, 370
security 30, 50, 65, 71, 84, 96, 136, 150, 161, 174, 199, 212, 213, 308, 355, 358
SIMMOD 91
social benefits... 28, 29, 49, 197, 199
social costs 4, 9, 15, 29, 63, 116, 152, 197, 275, 288, 295, 334, 348, 369, 403
stakeholders (*& see* goals) 14, 15, 21, 42, 45, 46, 48, 50, 54, 69, 149, 156, 163, 170, 172, 176, 178, 205-209, 212-215, 231, 263, 289, 320, 321, 389, 392-398, 405, 409, 414, 417, 419
standards 6, 8, 11, 24, 28, 42, 53, 65, 82, 84, 94, 114, 134, 141, 143, 146, 160, 164, 178, 203-205, 208, 216, 220, 223, 226, 269, 277, 282, 308, 411
STOL..133, 199, 205, 285, 287, 361, 368
subsidy..11, 16, 19, 28, 29, 32, 56, 57, 84, 107, 152, 206, 296, 306, 342, 357, 368, 384, 411
supersonic transport... 22, 83, 187
supply (*see* prediction)

sustainability 17, 34, 49, 51, 197, 203, 318, 321, 340, 391, 392, 407, 412, 413
system planning 4-9, 14, 15, 39-48, 59, 69, 85, 88, 115, 169-180, 185, 195-200, 205, 207, 211-216, 219-226, 230-236, 255, 269, 316, 322, 326, 358, 369, 389, 392, 398, 407, 409
  intermodal 37, 126, 203, 224, 318
  metropolitan 3, 7, 8, 36, 37, 45, 58, 60, 87, 93, 154, 161-166, 173, 176-179, 191, 195, 196, 200, 207, 220, 224, 233, 236, 237, 254, 255, 270, 273-296, 346, 347, 351, 352, 371-388, 403
  regional (*see* metropolitan)
  scope 5, 6, 39, 40, 42, 52, 77, 80, 90, 109, 115, 116, 122, 162, 168, 175, 182, 184, 190, 209, 220, 221, 274, 321, 325, 326, 351, 385, 386, 393, 411
  top down..171
technology 1, 13, 18, 24, 40, 42, 45, 48, 59, 63, 74, 83, 92, 95, 106, 111, 129, 131, 155-162, 173, 182, 183, 187-189, 192, 198, 204, 215, 224, 324, 326, 330, 341, 352, 400-405, 408
terminals 33, 36, 49, 57, 61, 73, 82, 89, 92, 96, 121, 132, 134, 137, 138, 144, 150, 153, 157-162, 167, 173, 204, 226, 237, 240, 244, 247-254, 277-284, 287, 289-306, 310-312, 318, 320, 330-332, 335, 343, 348, 359, 368, 374, 379, 380, 382, 384, 397
time, value of 70, 71, 106, 180, 354, 368, 369, 400
tourism (& *see* prediction)........................... 3, 29, 59, 102, 103, 106, 109, 198, 199, 296, 353, 360, 361, 363, 364, 398
transfers (*see* passengers, transfer) 20, 30, 49, 56, 57, 58, 62, 78, 79, 94, 95, 97, 105, 126, 131, 144, 157, 161, 198, 202, 240, 281, 285, 290, 295, 300, 339, 341, 344, 351, 360, 377-380, 386, 403
UK Civil Aviation Authority 10, 16, 26, 27, 49, 54, 61, 82, 94, 95, 102, 103, 106, 131, 132, 136, 139, 140, 154, 166, 167, 175, 182, 184, 187, 280, 283-288, 290-295, 297, 300-309, 313-320, 325-327, 332, 335-337, 339, 340, 346, 357, 360, 362, 367, 368, 372-374, 382, 383, 385, 386, 404
uncertainty (*see* prediction) 9, 13, 22, 31, 55, 67, 98, 126, 130, 142, 189, 194, 241, 242, 253, 258, 278, 318, 323, 338
US Civil Aeronautics Board...30, 321
US Congress.. 6, 67, 222, 317
US Department of Transportation 36, 89, 149, 216, 221, 227, 230, 263, 400
US Federal Aviation Administration (*see* Federal Aviation Administration)
US National Plan of Integrated Airport Systems... 220, 222, 225, 269
water quality...23, 29, 64, 110, 120, 122, 249